Graduate Texts in Mathematical Physics

Editor
Elliot Lieb

Volume I	*Essays on Mirror Manifolds*
Volume II	*Quantum Groups*
Volume III	*Perspectives in Mathematical Physics*
Volume IV	*75 Years of Radon Transform*

International Press

International Press Publications

Series in Mathematical Physics

Essays on Mirror Manifolds, Edited by S.T. Yau
Quantum Groups, S. Sternberg
75 Years of Radon Transform, Edited by S. Gindikin and P. Michor
Perspectives in Mathematical Physics, Edited by R. Penner and S.T. Yau

Series in Geometry and Topology

L^2 Moduli Spaces on 4-Manifold with Cylindrical Ends, C. Taubes
The L^2 Moduli Space and a Vanishing Theorem for Donaldson Polynomial Invariants, J. Morgan, T. Mrowka, D. Ruberman
Lectures on Harmonic Maps, R. Schoen and S.T. Yau
Lectures on Differential Geometry, R. Schoen and S.T. Yau
Lectures on Low-Dimensional Topology, Edited by K. Johannson
Chern, A Great Geometer, Edited by S.-T. Yau

Series in Algebraic Geometry

Proceedings of the International Symposium on Algebraic Geometry and Related Topics, Edited by Y. Namikawa, K. Ueno, and J.-H. Yang

Series in Analysis

Integrals of Cauchy Type on the Ball, S. Gong
Proceedings of the Conference on Complex Analysis at the Nankai Institue, Edited by L. Yang

Series in Applied Physics

Proceedings of the Second International Conference on Applied Physics, Edited by D.H. Feng and T.-Y. Zhang
Physics of the Quantum Electron Solid, Edited by S.-T. Chui

Collected and Selected Works

The Collected Works of Freeman Dyson
The Collected Works of C.B. Morrey
The Collected Works of P. Griffiths
Selected Works of V.S. Varadarajan

QUANTUM GROUPS

from coalgebras to Drinfeld algebras

a guided tour

STEVEN SHNIDER and SHLOMO STERNBERG

Bar Ilan University

Harvard University

International Press Incorporated, Boston
P.O. Box 2872
Cambridge, MA 02238-2872

International Press Company, Ltd., Hong Kong
401 Lee Fat Building
30-36 Jardine's Crescent
Causeway Bay, Hong Kong

Library of Congress Catalog Card Number: 93-080921

Shnider, Steven
Sternberg, Shlomo
Quantum Groups from Coalgebras to Drinfeld Algebras

ISBN 1-57146-000-4

Typeset using Latex
Printed on acid free paper, in the United States of America

This book is dedicated to our wives, Marilyn and Aviva, for their enduring love and loving endurance.

PREFACE

The star triangle relation started out at the end of the last century as a device for solving problems in electrical network theory, and, as such, was taught in the standard high school electricity courses. In the hands of Onsager it became a tool for locating certain critical temperatures, and more importantly, as a device for proving commutation properties of certain "transfer matrices" which arise in statistical mechanics on a lattice. In the work of Lieb, Yang and Baxter in the late 1960's and early 1970's, the method of producing such commuting matrices was greatly advanced, leading to breakthroughs in the study two dimensional lattice statistical mechanics and the quantum mechanics of many particle systems on the line. The method hinges on the use of a parametrized family of matrices, $\mathcal{R}(\lambda)$, satisfying certain cubic relations. One of these equations which has become known as the Yang-Baxter equation is reminiscent in form to the celebrated braid relations of Artin and in some cases reduces to the Artin relations as the spectral parameter λ tends to infinity.

In the late 1970's the the central importance of the Yang Baxter equation in the quantum inverse scattering method was emphasized by Fadeev, Sklyanin, Takhtajan and their collaborators in Leningrad (now St.Petersburg). In the early 1980's Kulish, Reshetikhin, and Sklyanin realized that certain special (and limiting) cases of solutions of the Yang Baxter equation could be found via representation theory. In fact, one should really think of these $\mathcal{R}$ matrices as being of the form $\rho(R)$ where ρ is a representation of B, the universal enveloping algebra of a Lie algebra, and where $R \in B \otimes B$. The Yang Baxter equation is then expressed entirely in terms of R: It is a homogeneous cubic equation in $B \otimes B \otimes B$.

In the mid 1980's Drinfeld and Jimbo, independently, discovered the full significance of the Yang Baxter equation in relation to the universal enveloping algebra, which has not only a multiplication $m : B \otimes B \to B$, but also has a comultiplication $\Delta : B \to B \otimes B$. If we think of the elements of B as differential operators (on functions on a group), then this comultiplication is dual to function multiplication and is just the extension of Leibniz's rule expressing the action of a differential operator on the product of two functions in terms of differential operators acting on each factor. It is this comultiplication which allows us to take the tensor product, $V_1 \otimes V_2$ of two representations to get a third. For this standard comultiplication, the usual identification, s, of $V_1 \otimes V_2$ with $V_2 \otimes V_1$ commutes with the action of B. In particular, we get a representation of the symmetric group, S_n, as intertwining operators for the representation of B on the n−th tensor power, $V^{\otimes n}$, of any B−module, V. This is a representation of S_n because the comultiplication on B is cocommutative: it satisfies $s \circ \Delta = \Delta \circ s$. Suppose that we modify the comultiplication so that it no longer is cocommutative, but is only cocommutative up to conjugation by R. Then the expression for an intertwining operator between $V_1 \otimes V_2$ and $V_2 \otimes V_1$ becomes $s \circ R$. We will no longer get a representation of the symmetric group on $V^{\otimes n}$, but the Yang-Baxter equation guarantees that at least we get a representation of the braid group. So the Yang-Baxter equation has to do with modifying the comultiplication.

In a certain sense, the dual of the universal enveloping algebra of a Lie algebra is (approximately) the ring of functions on a Lie group (with the given Lie algebra). Here the multiplication is ordinary pointwise multiplication of functions, hence commutative. There is a comultiplication (at least for representative functions) which carries the group law. So the dual to modifying the comultiplication on the enveloping algebra is to modify the multiplication on functions on the group, making this multiplication non-commutative. This is exactly what happens in quantum mechanics where functions (with their commutative multiplication) thought of as classical observables, are replaced by operators with their non-commutative multiplication. Hence the term "quantum groups".

In his famous address to the international conference at Berkeley, Drinfeld gave a systematic approach to quantum groups in terms of an extension of Gerstenhaber's deformation theory. (He used the deformation parameter h to suggest Planck's constant and the connection to quantum theory.) The corresponding infinitesimal objects associated to a quantum group were called "Poisson groups"', (he had called them "Hamilton-Lie" groups in his first paper on the subject). They are closely related to the solutions to the classical Yang Baxter equation which appear in the theory of completely integrable Hamiltonian systems. As we mentioned earlier, one source of the great interest in the Yang Baxter equation is the important work of Fadeev, Sklyanin and Takhtajan on proving the complete integrability of quantum systems using the quantum inverse scattering method. A solution to the Yang Baxter equation with an auxiliary (deformation) parameter can be considered as a quantization (deformation) of a corresponding solution to the classical Yang Baxter equation. The theory of quantum groups formalizes the passage from classical complete integrability to quantum complete integrability. Drinfeld even suggested that the theory of quantum groups, i.e., the deformation of universal enveloping algebras and their duals, "should be regarded not only as a technical tool for construction of solutions to the (quantum) Yang Baxter equation but also as a natural language for the quantum method of the inverse problem." In the literature, the term "quantum group" is sometimes used to refer only to the "quantization" of the function algebra on groups, and sometimes to include the dual setting of the deformed universal enveloping algebra or more generally to situations where the multiplication or comultiplication need not be commutative or cocommutative. We will adopt this last, broadest, terminology.

Each quantum group gives rise to a series of representations of the braid groups by the algebraic method outlined above. On the other hand, one can find a geometrical origin for braid group representations, the monodromy representation for a flat connection (integrable system of partial differential equations) on a space whose fundamental group is the braid group. The possibility of interpretating $s \circ R$, the R matrix composed with transposition, in this way was discovered by Kohno, who established that these two classes of representations are equivalent for the class of quantum groups arising from a simple Lie group. In an absolutely remarkable paper, Drinfeld gave a new proof of Kohno's theorem by constructing deformations of universal enveloping algebras (of simple Lie algebras) within a broader class of algebraic objects. For these objects, not only the cocommutativity, but also the coassociativity holds only up to conjugation. So in addition to $R \in B \otimes B$ giving the weakened cocommutative law, there is also a $\Phi \in B \otimes B \otimes B$ modifying the coassociative law for comultiplication. The Φ and R are subject to rather complicated looking identities whose meaning only becomes clear in the framework of category theory. The corresponding dual object, which arises more naturally in category theory was introduced by Majid.

To relate to category theory, we must backtrack to the early 1960's and a penetrating series of papers by Stasheff on associativity laws on Hopf spaces. One of Stasheff's results was the convexity of certain polyhedron, K_n, whose vertices can be labelled by the legal ways of inserting (n-2) parenthesis pairs into n letters. The two dimensional faces of this polyhedron are either quadrilaterals or pentagons. MacLane gave a fundamental reinterpretation of this result: Suppose we impose a tensor product on a category, $\mathcal{C}$, axiomatically by simply defining it as a functor from $\mathcal{C} \times \mathcal{C}$ to $\mathcal{C}, (X, Y) \mapsto X \otimes Y$. Then $X \otimes (Y \otimes Z)$ and $(X \otimes Y) \otimes Z$ need not be the same. Indeed, each vertex of the Stasheff polyhedron, K_n, corresponds to a possible way of building up a tensor product on n objects. Each edge of K_n corresponds to a shift of the form $(\bullet(\bullet\bullet)) \leftrightarrow ((\bullet\bullet)\bullet)$ where each $\bullet$ may be a clump of subtensor products, and this shift might be occurring inside more inclusive parentheses. (See for example, Figure 5.3 in the text, for a picture of the three dimensional K_5 with its fourteen vertices, twenty one edges, six pentagonal faces and three quadrilateral faces.) Suppose we now axiomatically impose a natural transformation $a_{X,Y,Z} : (X \otimes (Y \otimes Z)) \mapsto ((X \otimes Y) \otimes Z)$. This associates a transformation to each edge of K_n, and the question becomes: is this consistent? If we compose around a closed path, do we get the identity transformation? It follows immediately from Stasheff's result that it is enough to check this on pentagons, i.e. on K_4. (The quadrilaterals correspond to naturality conditions.) This result is known as (the first) Mac Lane coherence theorem. Drinfeld's equation on Φ is the translation into algebraic terms of the pentagon condition in category theory.

A category with a tensor product, and identity object under tensor product and an associativity satisfying the needed coherence conditions is called a monoidal category. (See chapter 6 for the precise definitions.) The second coherence theorem of MacLane has to do with the imposition (again as an additional axiom) of a natural involution $X \otimes Y \to Y \otimes X$. This time there is a condition related to a hexagon diagram. Joyal and Street extended MacLane's theorem to the case when $X \otimes Y \to Y \otimes X$ is not a involution. Then there are two hexagon conditions. When translated into algebraic terms, they yield Drinfeld's equations involving Φ and R. When combined, they give Drinfeld's generalization of the Yang Baxter equation.

In 1988 it was observed independently in a number of places (including our group here at Tel Aviv under the directorship of Prof. Yuval Ne'eman) that certain moves of Feynman like graphs that had originated in high energy physics in the 1960's and resurfaced as monodromy constraints in rational conformal field theories were identical to the Stasheff-MacLane moves in category theory, yielding still another connection between category theory and physics, and stimulating a large body of research.

It is amazing to see so many different areas of science interact with one another at this level of depth, but this very tangle of interconnections makes a coherent presentation very difficult. We have therefore had to make some choices. Here is a list of topics which are not seriously discussed (or not discussed at all) in our book: Poisson groups or Lie bialgebras appear only once, in a brief section at the end of in Chapter 11. We do not discuss complete integrability, classical or quantum, or the quantum inverse scattering method at all. These matters are treated very clearly in Prof. Takhtajan's beautiful lectures, [1153]. We do not describe the applications of quantum groups to knot invariants or go into much detail about representations of the braid groups. We refer the reader to the books by Kauffman, [634], or Atiyah [61], for this important topic. We do not discuss the representation theory of quantum groups, referring the reader to the papers of Andersen, DeConcini, Kac, Kazhdan, Lusztig, and Rosso listed in the bibliography and to the recent book by Lusztig [811]. Two major contributors to the development of the function algebra approach to quantum groups whose work is not mentioned here at all are Manin and Woronowicz, but one could add very little to their exposition and we direct the reader to the original papers.

Quantum groups is a very active area of research, and any serious treatment of the omitted topics would double the size of an already thick book. In compensation for these omissions, we do provide an extensive bibliography with some 1200 entries.

Here is a description of the path that we do follow. The first three chapters form an introduction to bialgebras and Hopf algebras. Influenced by Majid, we adopt the categorical point of view (including reconstruction) early on. Chapter 3 ends with two gems: Ree's shuffle algebra which has applications to computer science in the abstract theory of languages, and Chen's iterated integrals. We make use of both of these tools in a technical setting later on.

In Chapter 4 we pause from pursuing the main goal in order to give a highly non-trivial application of the theory developed so far - the Liulevicius-Zelevinski approach to the representation theory of the symmetric groups. This approach yields many of the technical combinatorial results in this representation theory as natural outgrowths of the Hopf algebra structure. The whole subject is clearly explained in Zelevinski's book which we follow very closely. But it is not as broadly known as it should be, even among experts. So if we succeed in making some positive propaganda for this approach we will be satisfied.

In Chapter 5 we return to the main theme. We describe the Stasheff polyhedra and prove the coherence theorem by an algorithmic construction of the convex polyhedron by a sequence of truncations. (This truncation construction is of significance from the point of view of toric varieties where the $n-$ simplex corresponds to complex projective n space and truncations correspond to blowing up operations.)

In Chapter 6 we present the MacLane and MacLane-Joyal-Street coherence theorems for monoidal and braided categories, and applications to constraints on certain graphs arising in strong interaction physics as described above.

Chapter 7 is a mini course on statistical mechanics on a two dimensional lattice. It is self contained, but only goes as far as motivating the star triangle relation and the Yang Baxter equation. In the main we follow Baxter's book and the article by Fadeev and Tatkhadjan. We end with a summary of the original paper on the star triangle relation in electrical network theory.

In Chapter 8 we introduce the main object of study for the remainder of the book - Drinfeld algebras and the corresponding dual objects, Majid algebras. The point of view in this chapter is that of Majid, and most of the results are taken directly from his papers. A functor from any category to the category of vector spaces gives rise to a coalgebra, C, as we explain in Chapter 2. If this category is monoidal, and the functor respects tensor product up to a natural equivalence, then C inherits a multiplication $m : C \otimes C \to C$ which is consistent with the comultiplication. But this multiplication need not be associative. It is only associative up to "convolution" conjugation by an element, Φ, in $\mathrm{Lin}(C \otimes C \otimes C, k)$ where k is the ground field. Dually (in finite dimensions), if we set $B = C^*$, this makes B into an algebra which has a compatible comultiplication that is coassociative only up to conjugation by Φ. These are exactly what we have called Drinfeld algebras.

The MacLane pentagon condition translates into a cubic equation on Φ. If we keep the same functor from our monoidal category to the category of vector spaces, but change the natural transformation relating the tensor products, then the multiplication gets modified in the Majid algebra, or, dually, the comultiplication gets modified in the Drinfeld algebra. In the Drinfeld algebra setting, this modification is given by conjugation by an element in $B \otimes B$, i.e.

$$\Delta' = T \Delta T^{-1}, \qquad T \in B \otimes B.$$

This modification by conjugation is what Drinfeld calls "twisting".

The rest of the book is devoted to deformation theories associated to Drinfeld algebras. The classical deformation theory of associative algebras is due to Gerstenhaber in a famous series of papers in the early 1960's. If $m(t)$ is a curve of multiplications on A with $m(0)$ the original associative multiplication on the algebra A, then we can think of the "infinitesimal" deformation $m'(0)$ as an element of $\mathrm{Hom}(A \otimes A, A)$. Gerstenhaber's observes that the associativity of the deformed multiplication implies that $m'(0)$ is a cocycle in the Hochschild cohomology (see Chapter 9 for definitions). Two infinitesimal deformations differ by an infinitesimal "change of coordinates"

Figure 0.1: The Gerstenhaber-Schack (double) complex. The entry at position (p,q) is $\mathrm{Hom}(B^{\otimes q}, B^{\otimes p})$. Horizontal arrows pointing to the right and the vertical arrows pointing upward (not drawn) are explained in Chapter 9.

if they differ by a coboundary. Thus H^2 describes the "infinitesimals" up to equivalence. There is a quadratic expression in $m'(0)$ coming from the associativity condition which is a cocycle. We can extend this putative infinitesimal deformation to one higher order in t if and only if this "obstruction cocycle" is a coboundary. Thus the obstructions to going to one higher term in the power series lie in H^3. Of course, dualizing gives an analogous theory for coalgebras.

In a recent series of papers, Gerstenhaber and Schack have extended the original Gerstenhaber deformation theory to a deformation theory of bialgebras. Let B be a bialgebra. If we deform the multiplication the infinitesimal will lie in $\mathrm{Hom}(B \otimes B, B)$. The infinitesimal deformation of the comultiplication lies in $\mathrm{Hom}(B, B \otimes B)$. So we introduce a "double complex" and follow the convention of Gerstenhaber and Schack by placing $\mathrm{Hom}(B^{\otimes q}, B^{\otimes p})$ at position (p,q), column p, row q. (We do not follow their convention about total degrees. For us the total degree at (p,q) is $p+q$ as usual.) This double complex has arrows moving up one step and to the right one step as usual. See Figure 1.

The infinitesimal deformation of the multiplication is located at position $(1,2)$, and the fact that the deformed multiplication is associative implies that it is sent to zero by the Hochschild differential represented by the vertical arrow. This is just the old Gerstenhaber deformation theory for as-

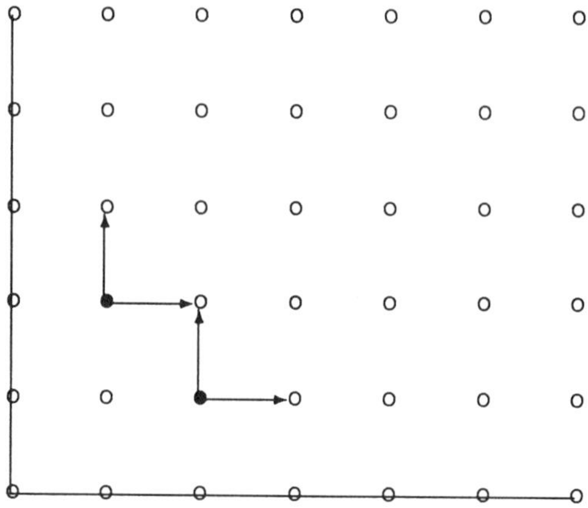

Figure 0.2: The bialgebra deformation complex. The bottom row and left hand column are deleted. The infinitesimal deformation of the multiplication is at position (1,2) and of the comultiplication is at position (2,1).

sociative algebras. Dually, the infinitesimal comultiplication is located at $(2,1)$ and the (co)associativity of the deformed comultiplication says that it is sent to zero by the coalgebra differential, introduced by Cartier in the 50's, which is represented by the horizontal arrow. The delightful discovery of Gerstenhaber and Schack is that the compatibility of multiplication and comultiplication implies that two contributions at $(2,2)$ (from $(1,2)$ horizontally and from $(2,1)$ vertically) mutually cancel. Thus the sum of the infinitesimal multiplication and comultiplication give a cocycle for the"total complex". We refer the reader to Chapter 9 for details, and content ourselves here with a pictorial outline. See Figure 2.

From here the theory "follows the book". In the (associative and coassociative) bialgebra deformation theory we must delete the bottom row and leftmost column in Figure 1. Then an infinitesimal bialgebra deformation, as we have just seen, is a cocycle of total degree three $= p + q$ with two components corresponding to multiplication and comultiplication. Two infinitesimal deformations differ by an infinitesimal "change of coordinates", and hence should be considered equivalent, if they differ by a coboundary in this deleted complex, see Figure 3.

Substituting the putative infinitesimal deformation, i.e. a cocycle of total degree three, in the three structure equations, associativity, coassociativity, and compatibility generates a cocycle of total degree four with three components . See Figure 4. The deformation can be extended to one higher degree

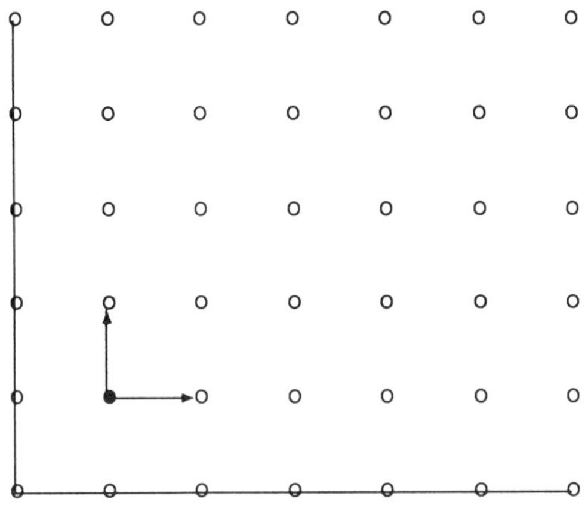

Figure 0.3: Two infinitesimal deformations are equivalent if they differ by a coboundary in the restricted complex

in t if and only if this so called "obstruction cocycle"is a coboundary. This, in outline, is the Gerstenhaber Schack theory of bialgebra deformations: The infinitesimals are cocycles of total degree three and should be regarded as equivalent if they differ by a coboundary. The obstruction to extension is a cohomology class in the total complex of total degree four. All this takes place in the complex of figure 1 with the bottom row and leftmost column deleted.

Now what happens if we start with an associative and coassociative bialgebra, but allow the deformed object to be a Drinfeld algebra? Amazingly, all that has to be done is to include the bottom row! Indeed, the horizontal arrow from position $(2,1)$ may no longer send the infinitesimal comultiplication to zero, since the multiplication need not be associative. The infinitesimal deformation now includes the infinitesimal of Φ which lies in $B^{\otimes 3} = \mathrm{Hom}(k, B^{\otimes 3})$ and so is located at $(3,0)$. The fact that the comultiplication is coassociative up to conjugation by Φ translates, at the infinitesimal level, into the assertion that the contributions as $(3,1)$ coming horizontally from $(2,1)$ and vertically from $(3,0)$ cancel. The pentagon identity on Φ considered as a fourth structure equation introduces a fourth component to the obstruction cocycle. At the infinitesimal level the pentagon identity implies that the image of the infinitesimal of Φ in position $(4,0)$ under the coalgebra differential vanishes. So the situation is as described by Figure 4.

For Drinfeld algebras there is a broader definition of equivalence. One must not only allow for change of coordinates, but also for twisting. The

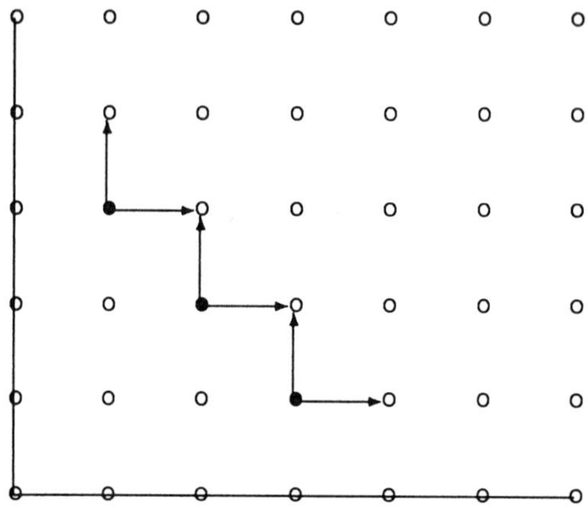

Figure 0.4: The obstruction to extending the infinitesimal one higher power in t is a cocycle of total degree four as drawn. If this cocycle is a coboundary the extension is possible.

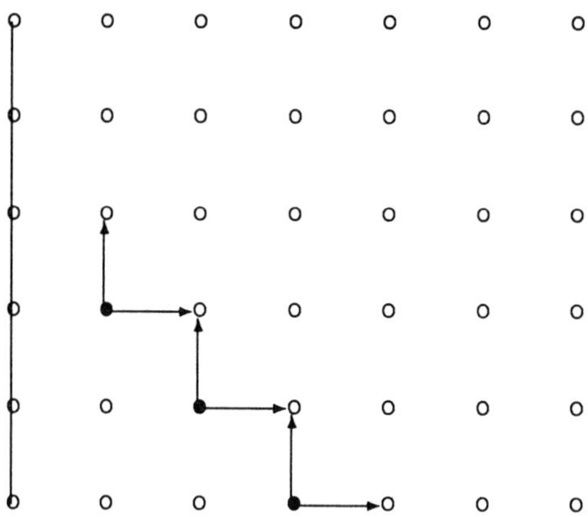

Figure 0.5: Bialgebra to Drinfeld algebra deformation. The bottom row is enabled but not the left column. The deformed comultiplication need not be associative, so the horizontal arrow at (2,1) need not give zero. The infinitesimal of Φ lies at position (3,0).

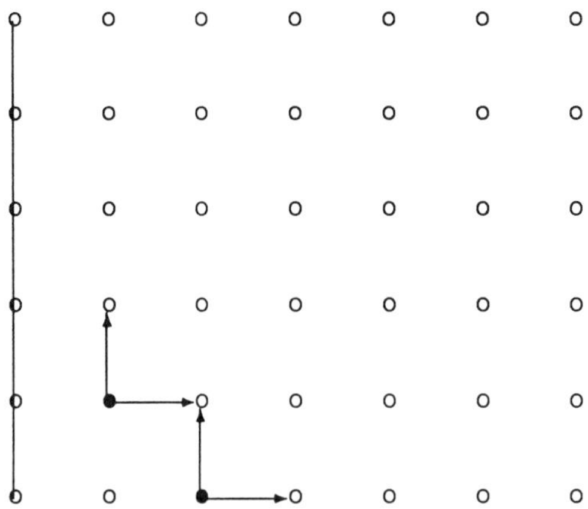

Figure 0.6: Equivalence in bialgebra to Drinfeld algebra deformations. The infinitesimal of the twist lies at position (2,0).

infinitesimal of the twisting is located at position $(2,0)$. So allowing for this modified infinitesimal equivalence, two three cocycles of the total complex give equivalent deformations if they differ by a total coboundary as indicated in Figure 6.

Once again, there is a quadratic expression in a putative infinitesimal deformation which is a cocycle of total degree four. The vanishing of the corresponding total cohomology class determines whether or not the infinitesimal deformation can be extended to a power series of one higher order in t. See Figure 7.

If we were deforming to Majid algebras, we would enable the leftmost column instead of the bottom row. In this book we resist the temptation of considering algebraic objects whose deformation theory from bialgebras enables both the bottom row and the leftmost column.

Details of the computations described in Figures 1-7 are given in Chapter 9. One still must compute the relevant cohomology groups. In Chapter 10 there is a new proof that for the case $B = U(\mathcal{G})$, the universal enveloping algebra of a Lie algebra, the cohomology groups can be identified with certain Chevalley Eilenberg cohomology groups. In chapter 11 we use this fact to prove various existence and uniqueness theorems for the deformation of $U(\mathcal{G})$ when $\mathcal{G}$ is simple, and for the study of the role of the R matrix for deformations of $U(\mathcal{G})$ for general Lie algebras. We conclude with a short discussion of Lie bialgebras, quasi-Lie bialgebras, and Poisson-Lie groups. Most of the theorems in this chapter are due to Drinfeld. Several of the

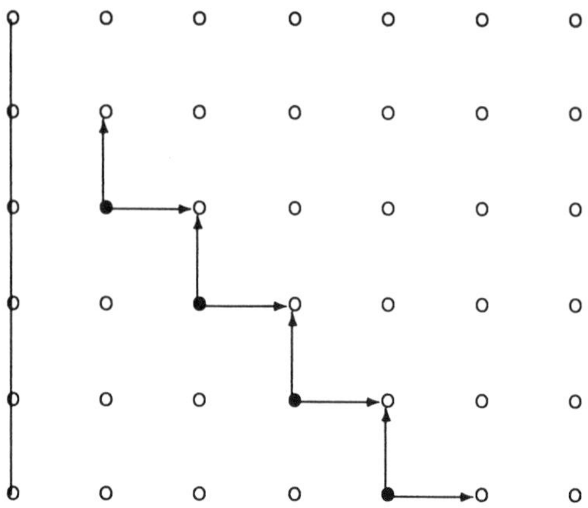

Figure 0.7: The obstruction cocycle to extending an infinitesimal bialgebra to Drinfeld algebra deformation to one higher power in t. If the cocycle is a coboundary, the extension is possible.

proofs and one of the uniqueness theorems are due to Shnider.

In Chapter 12 we present some of the material in Drinfeld's remarkable paper proving Kohno's theorem via a study of the Knizhnik-Zamolodchnikov equations. In particular, this gives an alternative proof of one of the existence theorems in Chapter 11.

In Chapter 13 we go back to general deformation theory. It would seem natural to study Drinfeld algebra deformations when we start with a general Drinfeld algebra as the undeformed object, rather than a bialgebra as in Chapter 9. However this introduces some major complications, requiring modifications of the very notion of a double complex. The ideas in this chapter are due to the authors, to Markl, to Markl and Stasheff and to Markl and Shnider.

This book grew out of a set of lecture notes from the 1992-1993 Ne'eman-Sternberg seminar at the Sackler Institute of Advanced Studies at Tel Aviv University. Our thanks to the participants: Maxim and Sasha Braverman, Dennis Gaitsgory, Dan Levy, Alex Stolin, Sara Westreich, and Paul Zusmanovich for many comments and useful suggestions. In particular, Dennis Gaitsgory's critical comments significantly improved the exposition in Chapter 11 and Maxim Braverman made some very helpful suggestions for improving Chapter 10. We thank Martin Markl and Jim Stasheff for allowing us to include their work, some of it unpublished. The last four sections of Chapter 13 describe joint research of the first author and Martin Markl and we thank him for his help in preparing the presentation. Paul Zusmanovich has our special thanks for taking upon himself the Herculean task of most of the work in preparing the bibliography.

Shlomo Sternberg would like to thank the Gelbart Research Institute of Mathematical Sciences for its hospitality during the period when the manuscript was being completed. Steve Shnider's research was partially supported by a grant from the Israel Science Foundation adminstered by the Israel Academy of Sciences and Humanities.

Contents

Chapter 1

Coalgebras and addition laws

1.1 Addition laws.

The origin of our subject lies in the discovery in the eighteenth and nineteenth centuries of "addition laws" for special functions. These laws were understood to be generalizations of the famous laws for the sine and the cosine of the sum of two angles. We list a few:

Addition formulas:

1. Trigonometric functions.

$$\begin{aligned}\sin(x+y) &= \sin x \cos y + \cos x \sin y \\ \cos(x+y) &= \cos x \cos y - \sin x \sin y\end{aligned}$$

2. Weierstrass elliptic function.

$$\wp(x+y) = \frac{1}{4}\left\{\frac{\wp'(x)-\wp'(y)}{\wp(x)-\wp(y)}\right\}^2 - \wp(x) - \wp(y).$$

3. Jacobi elliptic functions. A typical addition formula (in Glaisher notation for the twelve basic functions) is:

$$cs(u+v) = \frac{cs(u)ns(v)ds(v) - cs(v)ns(u)ds(u)}{ns^2(v) - ns^2(u)}.$$

4. Bessel functions.

$$J_n(x+y) = \sum J_m(x) J_{n-m}(y).$$

5. Legendre polynomials. (Calc. Integ.(1815) pp.262-269 Whittaker and Watson, [1197] p.326.) This is difficult to understand in its original formulation. But here is a translation into semi-modern language: Express a rotation R in terms of its Euler angles

$$R = R(\phi, \theta, \psi).$$

Express the Euler angles of the product of two rotations in terms of their Euler angles,

$$\begin{aligned} R &= R_1 R_2 \\ R(\phi, \theta, \psi) &= R(\phi_1, \theta_1, \psi_1) R(\phi_2, \theta_2, \psi_2). \end{aligned}$$

The expression is messy:

$$\cos\theta = \cos\theta_1 \cos\theta_2 - \sin\theta_1 \sin\theta_2 \cos(\phi_2 + \psi_1) \tag{i}$$

$$\begin{aligned} e^{i\phi} &= (e^{i\phi_1}/\sin\theta)[\sin\theta_1\cos\theta_2 \\ &\quad + \cos\theta_1 \sin\theta_2 \cos(\phi_2+\psi_1) + i \sin\theta_2 \sin(\phi_2+\psi_1)] \end{aligned}$$

and an even more complicated expression for ψ. For each integer u there is a unique irreducible representation of dimension $2u+1$. In terms of a suitable orthonormal basis, we will have a matrix expression

$$T^u_{mn}(R) = T^u_{mn}(R(\phi, \theta, \psi)).$$

This matrix entry is of the form

$$T^u_{mn}(R(\phi, \theta, \psi)) = c(u, m, n) e^{i(m\phi + n\psi)}\, P_u^{-n,m}(\cos\theta) \tag{ii}$$

where c is a constant involving powers of i and quotients of factorials involving u, m and n. The fact that T^u is a representation implies that

$$T^u_{mn}(R_1 R_2) = \sum T^u_{mj}(R_1) T^u_{jn}(R_2). \tag{iii}$$

If we substitute (i) and (ii) into (iii) we get Legendre's addition formula.

But clearly the key unifying idea is (iii). So let T be a finite dimensional representation of a group G and T_{mn} its matrix elements relative to some basis. These matrix entries are functions on G. More generally, a function, f, on G is called a function of type T if it is a linear combination of the T_{mn}. This definition is now independent of the choice of basis. Let $C = C(G,T)$ be the space of functions of type T. Then (iii) reads

$$T_{mn}(ab) = \sum_j T_{mj}(a)T_{jn}(b). \tag{1.1.1}$$

The expression on the right of (1.1.1) is a sum of products of a function of a with a function of b, both functions coming from C. Thus we can think of the sum on the right (before evaluating at specific elements a and b) as being an element of $C \otimes C$. Thus (1.1.1) really provides a map

$$\begin{aligned} \Delta \quad &: C \to C \otimes C \\ \Delta f \quad &= \Sigma f_{(1)} \otimes f_{(2)} = \Sigma f_{i1} \otimes f_{i2} \end{aligned}$$

where the middle expression is shorthand for the rightmost expression and where

$$f(ab) = \Sigma f_{(1)}(a) f_{(2)}(b).$$

(The convenient shorthand was introduced by Sweedler, [1141] and we shall refer to it as *Sweedler notation.*)

Explicitly, in terms of the basis $\{T_{mn}\}$ of C we have

$$\Delta T_{mn} = \Sigma T_{mj} \otimes T_{jn}. \quad \text{(matrix comultiplication)} \tag{1.1.2}$$

Conjugation by a constant matrix shows that this definition is independent of the choice of basis.

A vector space, C, with a linear map

$$\Delta : C \to C \otimes C$$

is called a *coalgebra* and the map Δ is called *comultiplication* because the dual diagram

$$m : A \otimes A \to A$$

is usually called a multiplication. Since the multiplication on G is associative, it follows that the comultiplication on $C(G,T)$ is *coassociative* in the sense that

$$(\Delta \otimes id) \circ \Delta = (id \otimes \Delta) \circ \Delta \tag{1.1.3}$$

as maps from C to $C \otimes C \otimes C$. In other words the diagram in Figure 1.1 commutes.

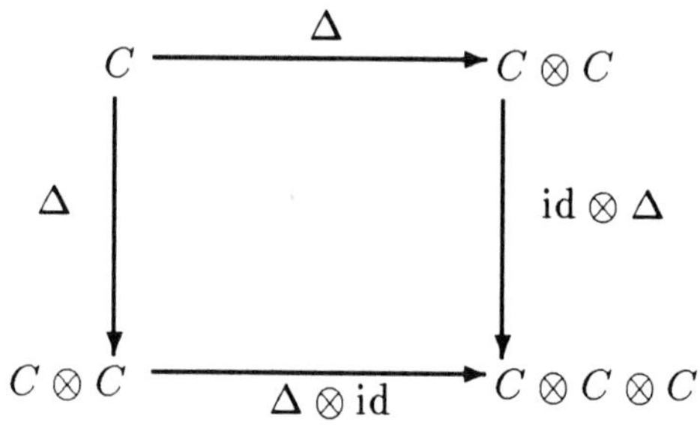

Figure 1.1:

1.2 Matrix comultiplication.

For the proof of (1.1.3) we only used the associative character of the group G. So we could replace G by any monoid with an associative multiplication. In particular we can consider the space of n by n matrices, and consider the matrix entries T_{mn} as functions on the space of matrices. Then the matrix comultiplication formula (1.1.2) applies. For example, in the two by two case where the matrix entries are a, b, c, d with $T_{11} = a, T_{12} = b$ etc., the comultiplication is

$$\begin{aligned} \Delta a &= a \otimes a + b \otimes c \\ \Delta b &= a \otimes b + b \otimes d \\ \Delta c &= c \otimes a + d \otimes c \\ \Delta d &= c \otimes b + d \otimes d. \end{aligned}$$

So we have a four dimensional coalgebra.

(If we think of the x and y in the trigonometric addition formulas as placeholders, with the x indicating placement before the tensor product sign and y after, and we set $a = d = \cos$ and $b = -c, c = \sin$, we recover the trigonometric identities.)

We are to think of the elements a, b, c, d in the above equations or the T_{mn} in (1.1.2) as linear functions on the space, A, of n by n matrices. In other words, $C = A^*$. Now A is an algebra, i.e. we have a map $A \otimes A \to A$. But then taking the transpose of this map gives us our comultiplication $C \to C \otimes C$. So, as long as we are dealing with finite dimensional spaces, the dual of an algebra is a coalgebra and vice versa. (Here, for the moment, we are using the words "algebra" and "coalgebra" in the broadest possible sense, i.e. just the existence of the multiplication or comultiplication maps.) If the algebra A is associative, the coalgebra $C = A^*$ is coassociative and vice versa. Since we are more familiar with algebras than with coalgebras, it is sometimes more convenient to prove theorems in the algebra setting and then dualize.

For example, given any algebra, we can define the "opposite" algebra, A^{op} where $A^{\mathrm{op}} = A$ as a vector space but we simply multiply the elements in reverse order: if $m(a \otimes b) = ab$ then $m^{\mathrm{op}}(a \otimes b) = ba$. Or, more abstractly,

$$m^{\mathrm{op}} = m \circ s$$

where $s : A \otimes A \to A \otimes A$ is the switching operator

$$s(a \otimes b) = b \otimes a. \tag{1.2.1}$$

Clearly

$$m^{\mathrm{op}}(m^{\mathrm{op}}(a, b), c) = c(ba)$$

while

$$m^{\mathrm{op}}(a, m^{\mathrm{op}}(b, c)) = (cb)a.$$

So if A is associative, then so is A^{op}. Dualizing, if we are given a coalgebra C, we can define the "opposite" coalgebra C^{op} where $C = C^{op}$ as a vector space, but where

$$\Delta^{op} = s \circ \Delta.$$

We can now *conclude* that if C is coassociative, then so is C^{op}. For example, we can consider the opposite coalgebra of the algebra of matrix multiplication where

$$\Delta T_{mn} = \Sigma T_{in} \otimes T_{mi} \quad \text{(opmatrix comultiplication)}. \tag{1.2.2}$$

If we write $V = k^d$ for the standard d dimensional vector space over the field k then the space of all matrices can be identified with

$$V \otimes V^* \sim \mathrm{Hom}(V, V).$$

The decomposable elements of the form $x \otimes u, x \in V, u \in V^*$ correspond to the rank one linear transformations:

$$(x \otimes u)(z) = \langle u, z \rangle x.$$

Let $\{e_i\}$ be a basis of V and $\{e^i\}$ be the dual basis of V^*. The linear function T_{mn} on $V \otimes V^*$ is then given as

$$T_{mn} = e^m \otimes e_n$$

and we can write the opmatrix comultiplication as

$$\Delta(e^m \otimes e_n) = \sum_i e^i \otimes e_n \otimes e^m \otimes e_i.$$

We the get the comultiplication

$$\Delta = \Delta_{min} : V^* \otimes V \to (V^* \otimes V) \otimes (V^* \otimes V)$$

given by

$$\Delta(\phi \otimes f) = \Sigma e^i \otimes f \otimes \phi \otimes e_i, f \in V, \phi \in V^*, \tag{1.2.3}$$

This class of coalgebras is minimal in a certain categorical sense which we shall explain in the next chapter. But it is a useful exercise to check the coassociativity of this comultiplication directly.

1.3 Units and counits.

It is usual to assume that an associative algebra has a unit. Let $\mathbf{1}$ denote the unit element of the algebra A, so that

$$\mathbf{1}a = a\mathbf{1} = a$$

for all $a \in A$. Define the map

$$u : k \to A, \quad u(r) = r, \; r \in k.$$

Then the preceding equation can be written as

$$m(u(r) \otimes a) = m(a \otimes u(r)) = ra \tag{1.3.1}$$

for any $a \in A$.

Conversely, suppose that we start with a map $u : k \to A$ which satisfies (1.3.1). Then the element $\mathbf{1} = u(1)$ is a unit element of A, where 1 is the unit in k.

Thus we may *define* a unit in an algebra to be a map $u : k \to A$ such that (1.3.1) holds for any $r \in k$.

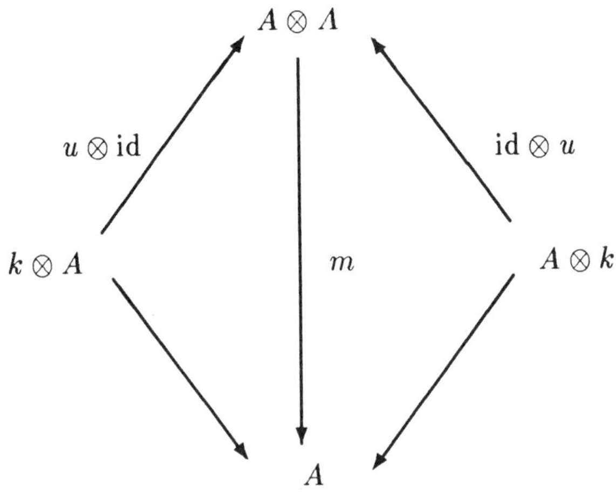

Figure 1.2:

In terms of diagrams, we can express (1.3.1) as saying that the diagram in Figure 1.2 commutes: The two lower diagonal arrows are the natural isomorphisms of $k \otimes A$ and $A \otimes k$ with A, true for any vector space.

Dually, a *counit* for a coalgebra, C, over k, is defined to be a linear map

$$\epsilon : C \to k$$

satisfying

$$(\epsilon \otimes id) \circ \Delta = (id \otimes \epsilon) \circ \Delta = id \tag{1.3.2}$$

under the identification of $k \otimes C$ and $C \otimes k$ with C.

Let us write the comultiplication, Δ, in the Sweedler notation:

$$\Delta(c) = \sum c_{(1)} \otimes c_{(2)}.$$

Then (1.3.2) becomes

$$\sum \epsilon(c_{(1)})c_{(2)} = \sum c_{(1)}\epsilon(c_{(2)}) = c. \tag{1.3.3}$$

In terms of diagrams, equation (1.3.2) says that the diagram in Figure 1.3 commutes.

The duality between the definitions of unit and counit become clear in terms of the diagrams: The diagram in Figure 1.3 is obtained from the diagram in Figure 1.2 by reversing all the arrows and by replacing u by ϵ and m by Δ.

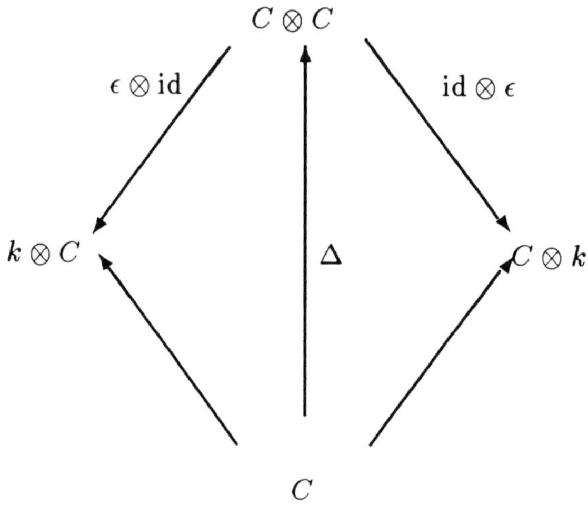

Figure 1.3:

Let us go back to our addition formula examples: If $T : G \to \mathrm{Aut}(V)$ is a finite dimensional representation of a group G, we introduced the coalgebra of representative functions. There is another coalgebra associated to the representation: Namely, we can consider the algebra $\mathrm{End}_G(V)$, the algebra of all linear transformations which commute with all the $T(a), a \in G$. Then $C = \mathrm{End}_G(V)^*$ is a coalgebra. We can think of a group as a special kind of category - a category with only one object. In the next chapter, after reviewing the basic definitions of category theory, we shall see what the analogues of these last coalgebras are, when we replace a group and a representation by a category and a functor.

Chapter 2

Comodules and categories

2.1 Comodules and their morphisms.

In this section we assume that our coalgebras are coassociative with counit. A (left) module over an algebra A with unit is a vector space M over k together with a map $\mu : A \otimes M \to M$ satisfying an obvious associative law and condition relative to the unit. Dually, a (left) *comodule* over a coalgebra C is a space M together with a map $c : M \to C \otimes M$.

The coassociative law is expressed by the commutativity of the diagram in Figure 2.1. (If we reverse all the arrows in Figure 2.2, replace c by μ, C by A, and Δ by m, we would get the diagram expressing the associative law for left modules.)

The consistency of c with the counit, ϵ, is expressed by the commutativity of the diagram in Figure 2.2.

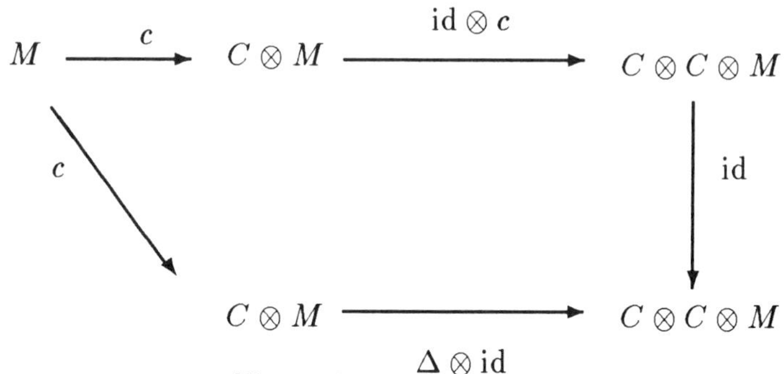

Figure 2.1:

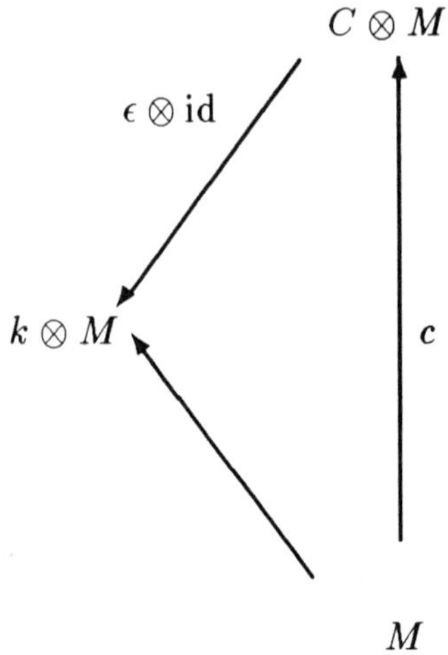

Figure 2.2:

If we have several comodules, M, N etc. we will write the comultiplication as c_M, c_N and so on. A comodule *morphism* is then a map $f : M \to N$ such that

$$c_N \circ f = (id \otimes f) \circ c_M. \tag{2.1.1}$$

The collection of all C comodules and comodule morphisms forms a category which we shall denote by $\mathbf{Com}(C)$.

2.2 Categories.

We briefly recall the basic definitions:

A *category* $\mathbf{C}$ consists of the following data:

(i) A set, $Ob(\mathbf{C})$, whose elements are called the *objects* of $\mathbf{C}$,

(ii) For every pair (X, Y) of $Ob(\mathbf{C})$ a set, $\mathrm{Hom}(X, Y)$, whose elements are called the *morphisms* or *arrows* from X to Y,

(iii) For every triple (X, Y, Z) of $Ob(\mathbf{C})$ a map from $\mathrm{Hom}(X, Y) \times \mathrm{Hom}(Y, Z)$ to $\mathrm{Hom}(X, Z)$ called the *composition map* and denoted $(f, g) \rightsquigarrow g \circ f$.

These data are subject to the following conditions:

(iv) The composition of morphisms is *associative*

(v) For each $X \in Ob(\mathbf{C})$ there is an $id_X \in \mathrm{Hom}(X, X)$ such that

$$f \circ id_X = f, \ \forall f \in \mathrm{Hom}(X, Y)$$

(*for any* Y) and

$$id_X \circ f = f, \ \forall f \in \mathrm{Hom}(Y, X)$$

(*for any* Y).

It follows from the axioms that id_X is unique.

The category **Set** is the category whose objects are ("all") sets and and whose morphisms are ("all") maps between sets. For reasons of logic, the word "all" must be suitably restricted to avoid logical contradiction.

In contrast to the category **Set**, in many of the categories we shall consider, the spaces $\mathrm{Hom}(X, Y)$ will have the structure of a vector space over a ground field, k (or more generally the structure of a module over a ground ring). In these cases the map $(f, g) \rightsquigarrow g \circ f$ will be bilinear. Such categories are called *k-additive.*

2.3 Functors and morphisms.

If $\mathcal{C}$ and $\mathcal{D}$ are categories, a *functor* F from $\mathcal{C}$ to $\mathcal{D}$ consists of the following data:

(vi) a map $F : Ob(\mathcal{C}) \to Ob(\mathcal{D})$

and

(vii) for each pair (X, Y) of $Ob(\mathcal{C})$ a map

$$F : \mathrm{Hom}(X, Y) \to \mathrm{Hom}(F(X), F(Y))$$

subject to the rules

(viii)

$$F(id_X) = id_{F(X)}$$

and

(ix)

$$F(g \circ f) = F(g) \circ F(f).$$

(This is what is usually called a covariant functor. A contravariant functor would have $F : \mathrm{Hom}(X,Y) \to \mathrm{Hom}(F(Y), F(X))$ in (vii) and $F(f) \circ F(g)$ on the right hand side of (ix).)

Here is an important example, valid for any category $\mathcal{C}$. Let us fix an $X \in Ob(\mathcal{C})$. We get a functor

$$F_X : \mathcal{C} \to \mathbf{Set}$$

by the rule which assigns to each $Y \in Ob(\mathcal{C})$ the set $F_X(Y) = \mathrm{Hom}(X,Y)$ and to each $f \in \mathrm{Hom}(Y,Z)$ the map $F_X(f)$ consisting of composition (on the left) by f. In other words, $F_X(f) : \mathrm{Hom}(X,Y) \to \mathrm{Hom}(X,Z)$ is given by

$$g \in \mathrm{Hom}(X,Y) \mapsto f \circ g \in \mathrm{Hom}(X,Z).$$

Let F and G be two functors from $\mathcal{C}$ to $\mathcal{D}$. A *morphism*, σ, from F to G (older name: "natural transformation") consists of the following data:

(x) for each $X \in Ob(\mathcal{C})$ an element $\sigma(X) \in \mathrm{Hom}_{\mathcal{D}}(F(X), G(X))$ subject to the "naturality condition"

(xi) for any $f \in \mathrm{Hom}_{\mathcal{C}}(X,Y)$ the diagram in Figure 2.3 commutes.

The two functors F and G are called *isomorphic* if $\sigma(X)$ is an isomorphism for each X; in other words there exists a morphism τ from G to F such that $\tau(X) \circ \sigma(X) = id_{F(X)}$ and $\sigma(X) \circ \tau(X) = id_{G(X)}$ for each $X \in Ob(\mathcal{C})$.

One last definition: A functor F from $\mathcal{C}$ to **Set** is called *representable* if there is an $X \in Ob(\mathbf{C})$ such that F is isomorphic to $F_X = Hom(X, \cdot)$. Then X is unique up to isomorphism and is called a *representative* or a *representing object* of F.

2.4 The forgetful functor.

We return to the study of coalgebras and associated categories. Let C denote a coalgebra and **Com** (C) its category of comodules, as before. Each object

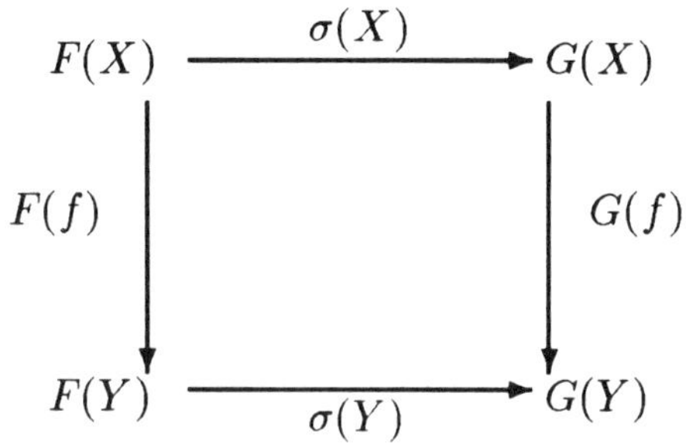

Figure 2.3:

in **Com** (C) is a vector space, so if we simply forget about the additional structure, we get the *forgetful functor* F which is a functor from **Com** (C) to **Vec**, the category of all vector spaces. So $F(M)$ is just M, but regarded merely as a vector space.

Let V be any vector space. We can consider the functor ${}_VF$ from **Com** (C) to **Vec** which assigns to each M the vector space $V \otimes F(M) = V \otimes M$. To each vector space V consider the set $\mathcal{N}\mathbf{at}(F,{}_V F)$, the collection of all morphisms (or natural transformations) between F and ${}_VF$. So an element of $\mathcal{N}\mathbf{at}(F,{}_V F)$ is a collection of linear maps $\sigma = \sigma_V$:

$$\sigma_V(M) : M \to V \otimes M \tag{2.4.1}$$

one for each C comodule, M, which are consistent in the sense that if

$$f : M \to N$$

is a comodule morphism, then

$$\sigma_V(N) \circ f = (id \otimes f) \circ \sigma_V(M) \tag{2.4.2}$$

as vector space maps.

If W is a second vector space, and $S : V \to W$ is a linear map, then $(S \otimes id) \circ \sigma_V$ defines an element of $\mathcal{N}\mathbf{at}(F,\ {}_WF)$. In other words, we have a functor

$$\mathcal{N} : \mathbf{Vec} \to \mathbf{Set} \qquad V \to \mathcal{N}\mathbf{at}(F,\ {}_VF) = \mathcal{N}(V).$$

This functor looks terrible, but it really has a simple description. In fact, we claim that $\mathcal{N}\mathbf{at}(F,\ {}_VF)$ has a one to one correspondence with $\mathbf{Lin}(C, V)$, the space of linear maps from the coalgebra, C, considered as a vector space, to the vector space V.

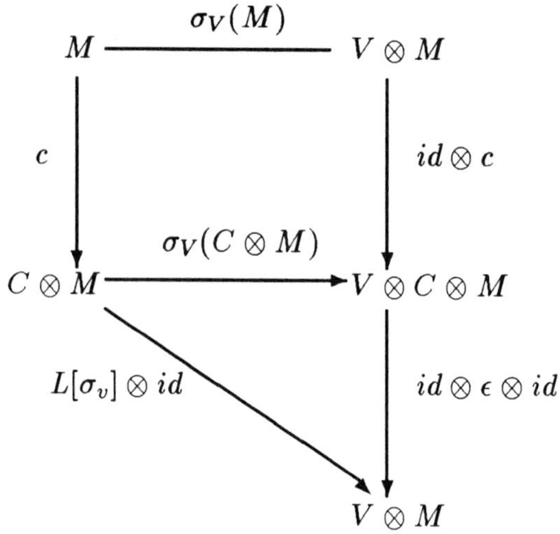

Figure 2.4:

Indeed, given any linear map $L : C \to V$, define the element $\sigma_V[L]$ of $\mathcal{N}\mathbf{at}(F,\ _VF)$ as the composition

$$M \xrightarrow{c} C \otimes M \xrightarrow{L \otimes id} V \otimes M$$

for any comodule, M.

Conversely, given an element σ_V of $\mathcal{N}\mathbf{at}(F,\ _VF)$ define the map $L = L[\sigma_V]$ as the composition

$$C \xrightarrow{\sigma(V)_C} V \otimes C \xrightarrow{id \otimes \epsilon} V\ .$$

A straightforward computation shows that the assignments $L \rightsquigarrow \sigma_V[L]$ and $\sigma_V \rightsquigarrow L[\sigma_V]$ are inverse to one another. Indeed,

Indeed, consider the diagram in Figure 2.4. The top square commutes because the comodule map $M \to C \otimes M$ is a comodule morphism, and hence the top square is just an instance of Figure 2.3 applied to the morphism σ_V. The bottom triangle commutes because all the comodule structure on $C \otimes M$ is, by definition, on the C factor in the tensor product, and hence

$$\sigma_V(C \otimes M) = \sigma_V(C) \otimes id$$

and $L[\sigma_V] = (id \otimes \epsilon) \circ \sigma_V(C)$ by definition. Going from the upper left hand corner to the lower right along the two arrows on the the left is $\sigma_V[L[\sigma_V]]$

by the definition of $\sigma[L]$. Going along right hand path to $V \otimes M$ and then straight down is σ_V since $(\epsilon \otimes id) \circ c = id$ by Figure 2.2. Hence

$$\sigma_V[L[\sigma_V]] = \sigma_V.$$

Conversely, if we start with L, then

$$\sigma_V[L](C) = (L \otimes id) \circ \Delta : C \to V \otimes C$$

and so

$$\begin{aligned} L[\sigma_V[L]] &= (id \otimes \epsilon) \circ (L \otimes id) \circ \Delta \\ &= (L \otimes id) \circ (id \otimes \epsilon) \circ \Delta \\ &= L \end{aligned}$$

since $(id \otimes \epsilon) \circ \Delta$ is the canonical identification of C with $C \otimes k$.

So what we have shown is:

Proposition 2.4.1 *If we start with the category* **Com**(C) *and construct the functor from* **Vec** *to* **Set** *as described above, then this functor is representable and the representing object is C, considered as a vector space.*

2.5 Coalgebras as representatives.

The main point of this section is to prove a sort of converse to the proposition of the preceding section. Namely, suppose we start with any category $\mathcal{C}$ and any functor F from $\mathcal{C}$ to **Vec**. We can then construct the functor

$$\mathbf{Vec} \to \mathbf{Set}, \quad V \to \mathcal{N}\mathbf{at}(F, {}_V F)$$

as above. Suppose that this functor is representable with representing object C. We claim that then C can be given the structure of a coalgebra so that $\mathcal{C}$ maps to **Com**(C) and F to the forgetful functor. In other words, a coalgebra can be defined purely categorically. Here are the details:

For any pair of vector spaces define a composition

$$d_{V,W} : \mathcal{N}\mathbf{at}(F, {}_V F) \times \mathcal{N}\mathbf{at}(F, {}_W F) \to \mathcal{N}\mathbf{at}(F, {}_{V \otimes W} F)$$

by

$$d_{V,W}(\sigma, \tau) = (id \otimes \tau) \circ \sigma$$

so $d_{V,W}(\sigma,\tau)$ is given as the composite

$$F \xrightarrow{\sigma} V \otimes F \xrightarrow{id\otimes\tau} V \otimes W \otimes F .$$

The composition, d, is associative in the sense that if $\alpha \in \mathcal{N}\mathbf{at}(F,\ {}_Z F)$ where Z is a third vector space, then

$$d_{V,W\otimes Z}(\sigma, d_{W,Z}(\tau,\alpha)) = d_{V\otimes W,Z}(d_{V,W}(\sigma,\tau),\alpha) \tag{2.5.1}$$

both sides of (2.5.1)corresponding to the composite morphism

$$F \xrightarrow{\sigma} V \otimes F \xrightarrow{id\otimes\tau} V \otimes W \otimes F \xrightarrow{id\otimes id\otimes\alpha} V \otimes W \otimes Z \otimes F .$$

Also, the standard identification of $k \otimes F(M)$ with $F(M)$ gives a canonical element

$$e : F \longrightarrow k \otimes F, \quad e \in \mathcal{N}\mathbf{at}(F, {}_k F).$$

Now suppose that we are given a representing object C for the functor

$$\mathbf{Vec} \rightarrow \mathbf{Set} \qquad V \rightarrow \mathcal{N}\mathbf{at}(F, {}_V F).$$

This means that for every vector space V we have an isomorphism

$$\rho_V : \mathbf{Lin}(C,V) \xrightarrow{\sim} \mathcal{N}\mathbf{at}(F,\ {}_V F). \tag{2.5.2}$$

Let us take $V = C$ in (2.5.2) and define

$$\iota = \rho_C(id). \tag{2.5.3}$$

Then for any V and any $f \in \mathbf{Lin}(C,V)$ we can write

$$f = f \circ id.$$

The equivalence between the functors $\mathbf{Lin}(C,\cdot)$ and $\mathcal{N}\mathbf{at}(F,\ {}_\bullet F)$ implies that

$$\rho_V(f) = (f \otimes id) \circ \iota. \tag{2.5.4}$$

We claim that C inherits the structure of a coalgebra. We must determine a map $\Delta : C \rightarrow C \otimes C$. Under the equivalence of functors, this should correspond to an element of $\mathcal{N}\mathbf{at}(F,\ {}_{C\otimes C} F)$. We already have a preferred element,

$$\iota \in \mathcal{N}\mathbf{at}(F, {}_C F).$$

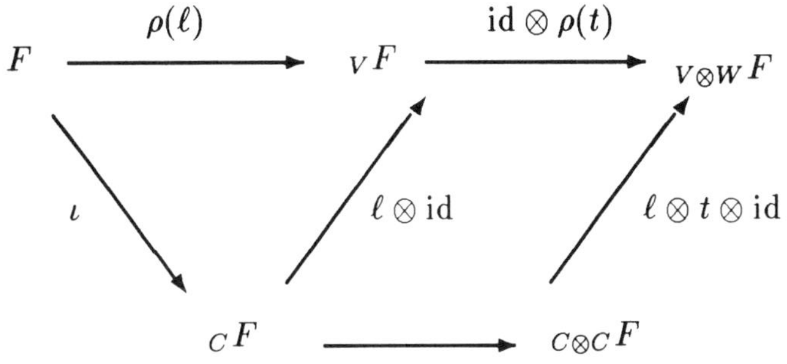

Figure 2.5:

So we may apply the operation $d_{C,C}$ to get an element

$$d_{C,C}(\iota, \iota) \in \mathcal{N}\mathbf{at}(F, \ {}_{C\otimes C}F)$$

and then take Δ to be the corresponding element of $\mathbf{Lin}(C, C \otimes C)$. Concisely, Δ is defined by

$$\rho_{C\otimes C}(\Delta) = d_{C,C}(\iota, \iota). \tag{2.5.5}$$

Similarly, define $\epsilon : C \to k$ by

$$\rho_k(\epsilon) = e. \tag{2.5.6}$$

We claim that Δ makes C into a coassociative coalgebra; that the coassociativity of Δ follows from (2.5.1) Indeed, let $\ell : C \to V$ and $t : C \to W$ be linear maps and $\rho(\ell) = \rho_V(\ell)$, $\rho(t) = \rho_W(t)$ the corresponding natural transformations. We claim that

$$d_{V,W}(\rho(\ell), \rho(t)) = \rho((\ell \otimes t) \circ \Delta) = ((\ell \otimes t) \otimes id) \circ d(\iota, \iota). \tag{2.5.7}$$

Indeed, this follows from the diagram in Figure 2.5 which is clearly commutative.

The composition of the top row is $d_{V,W}(\rho(\ell), \rho(t))$ by definition, the composition of the left two rows on the the bottom is $d_{C,C}(i, i) = \rho(\Delta)$.

Applying (2.5.7) first to to $\ell = \Delta$ and $t = id$ and then to $\ell = id$, $t = \Delta$ we see from the associativity of d that

$$\rho((\Delta \otimes id) \circ \Delta) = d(d(\iota, \iota), \iota) = d(\iota, d(\iota, \iota)) = \rho((id \otimes \Delta) \circ \Delta).$$

Thus $(\Delta \otimes id) \circ \Delta = (id \Delta) \circ \Delta$, in other words, C is coassociative. A similar argument gives the identities for the counit.

Figure 2.6:

Another way of "seeing" the coalgebra structure is as follows: Suppose we take $V = k$. Then the representative property of C says that $\mathbf{Lin}(C, k) = C^*$ is identified with $\mathcal{N}\mathbf{at}(F, F)$, the set of "endomorphisms" of the functor F. But this has a natural composition. So C^* has a natural multiplication, making it into an algebra. Therefore C has the structure of a coalgebra.

2.6 Opmatrix comultiplication as the minimal example.

For example, let us consider the simplest possible category, consisting of a single object, X and a single morphism id_X.

So we can "draw" this category, $\mathcal{C}$, as a single dot standing for the unique object, X and a curved arrow, pointing from X to itself, standing for id_X; cf. Figure 2.6

A functor F from $\mathcal{C}$ to **Vec** amounts to the choice of a single vector space, $F(X)$. We may drop the (X) in the notation for this vector space and call it by the same name, as the functor, namely F. So we can think of our functor as the single vector space, F. For any vector space V, the functor ${}_VF$ is just the choice of the single vector space $V \otimes F$. Then $\mathcal{N}\mathbf{at}(F, {}_VF)$ is just $\mathbf{Lin}(F, V \otimes F)$, the space of all linear maps from F to $V \otimes F$. But $\mathbf{Lin}(F, V \otimes F)$ is isomorphic to $\mathbf{Lin}(F^* \otimes F, V)$, under the isomorphism

$$Lin(F^* \otimes F, V) \ni L \mapsto l \in Lin(F, V \otimes F)$$

where

$$\langle id \otimes \phi, lf \rangle = L(\phi \otimes f), \; f \in F, \phi \in F^*. \tag{2.6.1}$$

Thus we may take our representing element of **Vec** to be

$$C = F^* \otimes F.$$

This suggests that C is the coalgebra of matrix comultiplication. Let us see how the above definitions work in this case to actually give the coalgebra of opmatrix comultiplication. Let e_i be a basis of F and $\{e^i\}$ the dual basis of F^*. The first step in the definition of the coalgebra structure on the representing vector space was to find ι. So we must find the map $\iota = l : F \to V \otimes F$ which corresponds to the $L = id$ in (2.6.1), where $V = F^* \otimes F$. So we want to find

$$\iota \in \mathbf{Lin}(F, F^* \otimes F \otimes F)$$

such that

$$\langle id \otimes \phi, \iota(f)\rangle = \phi \otimes f$$

for all $f \in F$ and $\phi \in F^*$. Clearly

$$\iota(f) = \Sigma e^i \otimes f \otimes e_i \tag{2.6.2}$$

satisfies this requirement, since $\phi = \Sigma\langle \phi, e_i\rangle e^i$. So we have the definition of ι.

We now compute $d(\iota, \iota)$ as the composition $d(\iota, \iota) = (id \otimes \iota) \circ \iota$:

$$d(\iota, \iota): \qquad f \mapsto \sum e^i \otimes f \otimes e_i \rightsquigarrow \sum\sum e^i \otimes f \otimes e^j \otimes e_i \otimes e_j. \tag{2.6.3}$$

Let us consider the isomorphism of

$$\mathbf{Lin}(F, F^* \otimes F \otimes F^* \otimes F \otimes F) \xrightarrow{\sim} \mathbf{Lin}(F^* \otimes F, F^* \otimes F \otimes F^* \otimes F)$$

given by (2.6.1) where

$$V = F^* \otimes F \otimes F^* \otimes F.$$

Substituting the expression (2.6.3) for $d(\iota, \iota)$ gives

$$\begin{aligned} \langle id \otimes \phi, d(\iota, \iota)(f)\rangle &= \langle id \otimes \phi, \sum\sum e^i \otimes f \otimes e^j \otimes e_i \otimes e_j\rangle \\ &= \sum\sum e^i \otimes f \otimes e^j \otimes e_i \langle \phi, e_j\rangle \\ &= \sum e^i \otimes f \otimes \sum \langle \phi, e_j\rangle e^j \otimes e_i \\ &= \sum e^i \otimes f \otimes \phi \otimes e_i. \end{aligned}$$

Thus the map $d(\iota, \iota)$ given by (2.6.3) corresponds to the map

$$\Delta : F^* \otimes F \mapsto F^* \otimes F \otimes F^* \otimes F$$

given by

$$\Delta : \phi \otimes f \mapsto \Sigma e^i \otimes f \otimes \phi \otimes e_i.$$

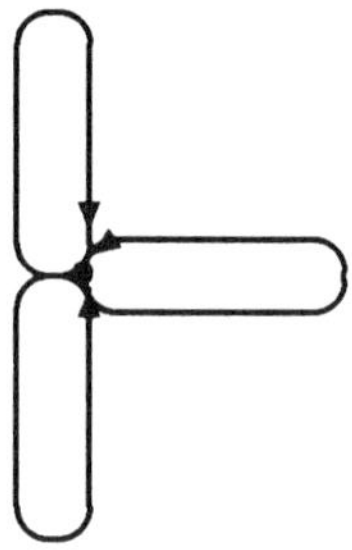

Figure 2.7:

This coincides with our formula for opmatrix comultiplication (1.2.3).

Let us see what the construction gives when our category has one object but several arrows, so can be described by Figure 2.7. In other words, we can think of our category as an associative monoid, for example a group.

A functor F now consists of a vector space (which we shall continue to denote by F) together with a collection of linear transformations, $\{t\}$ of F, one for each arrow, which give a representation, T, of the monoid on F. Now $\mathcal{N}\mathbf{at}(F,\ _V F)$ consists of all $l \in \mathbf{Lin}(F, V \otimes F)$ satisfying

$$(id \otimes t) \circ l - l \circ t = 0\,, \forall t.$$

Thus the corresponding L satisfies

$$\begin{aligned} L\{t^*\phi \otimes f - \phi \otimes tf\} &= \langle id \otimes t^*\phi, lf\rangle - \langle id \otimes \phi, ltf\rangle \\ &= \langle id \otimes \phi, [(id \otimes t) \circ l - l \circ t]f\rangle = 0. \end{aligned}$$

So if Ad^*t denotes the transformation on $F^* \otimes F$ defined by

$$Ad^*t(\phi \otimes f) = t^*\phi \otimes f - \phi \otimes tf,$$

we see that L vanishes on the subspace B of $F^* \otimes F$ spanned by all the $Ad^*t(F^* \otimes F)$, and hence our representative space C is given by

$$C = F^* \otimes F/B.$$

But $B^0 \subset F \otimes F^*$, the annihilator space of B, is the subspace A consisting of those endomorphisms of F which commute with all the t. In other words $A =$ the centralizer of the representation T, which is an algebra, and C is the dual coalgebra, $C = A^*$.

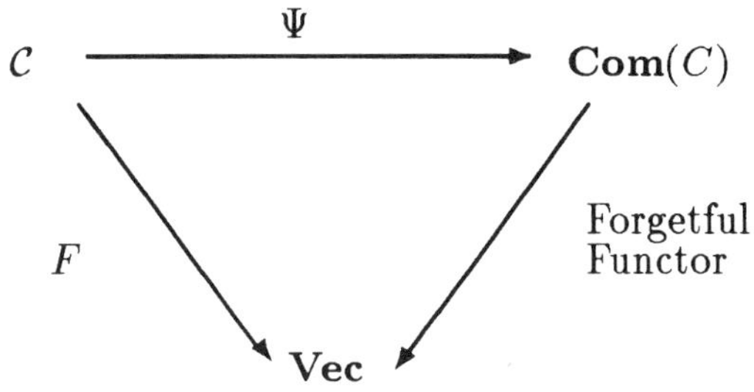

Figure 2.8:

2.7 Reconstruction.

Having worked out these two examples, let us return to the general theory and summarize what we have done. Starting with a category $\mathcal{C}$ and a functor $F : \mathcal{C} \to \mathbf{Vec}$, we have produced a coalgebra C as the representing object for the functor

$$\mathbf{Vec} \to \mathbf{Set},\ V \to \mathcal{N}\mathbf{at}(F,{}_V F).$$

We have done this under the assumption that this functor is representable. But representability can be established for a very broad class of functors, cf. [828].

The coalgebra C is clearly determined up to isomorphism. We have a preferred natural transformation $\iota \in \mathcal{N}\mathbf{at}(F,\ {}_C F)$ corresponding to $id \in \mathbf{Lin}(C, C)$. In particular, for every

$$X \in Ob(\mathcal{C})$$

we have a map

$$\iota(X):\ F(X) \to C \otimes F(X)$$

The associativity of d, together with the definition of Δ shows that this map makes $F(X)$ into a C comodule. So we have the commutative diagram of Figure 2.8.

In this diagram, Ψ is the functor which assigns to each $X \in Ob(\mathcal{C})$ the comodule $F(X)$ with the comodule structure given by $\iota(X)$ as above.

Chapter 3

Bialgebras and Hopf spaces

3.1 Bialgebras.

Suppose we consider the space, $F(G)$, of all functions on a finite group G. It is a finite dimensional vector space with two operations; ordinary multiplication of functions which exists on the space of all functions on any set, and a comultiplication which exists because $F(G)$ is a finite dimensional representation space of G (the "regular" representation). These two operations are consistent in the sense of the following definition:

Let B be a vector space which is both an algebra, with

$$\text{multiplication } m : B \otimes B \to B \text{ and unit } u : k \to B$$

and is also a coalgebra with

$$\text{comultiplication } \Delta : B \to B \otimes B \text{ and counit } \epsilon : B \to k.$$

We can give $B \otimes B$ the structure of an algebra where

$$(a \otimes b) \circ (c \otimes d) = ac \otimes bd.$$

More diagrammatically, we define

$$M : (B \otimes B) \otimes (B \otimes B) \to B \otimes B$$

as the composition

$$(B \otimes B) \otimes (B \otimes B) \stackrel{I \otimes s \otimes I}{\longrightarrow} (B \otimes B) \otimes (B \otimes B) \stackrel{m \otimes m}{\longrightarrow} B \otimes B$$

where s is the twist map

$$s(b \otimes c) = c \otimes b.$$

(Later on in this chapter, when we consider algebra and coalgebras which are graded ("superalgebras" and "supercoalgebras") we will modify the definition of the twist map so it reads

$$s(b \otimes c) = (-1)^{(degb)\cdot(degc)} c \otimes b. \tag{3.1.1}$$

If $\mathbf{1} = u(1)$ is the unit element of B, then $\mathbf{1} \otimes \mathbf{1}$ is the unit of $B \otimes B$. In other words the unit,

$$u_{B \otimes B} : k \to B \otimes B$$

is given by

$$u_{B \otimes B} = u_B \otimes u_B.$$

We can now require that the maps $\Delta : B \to B \otimes B$ and $\epsilon : B \to k$ be algebra maps. So we want

$$\Delta(xy) = \Delta(x)\Delta(y), \quad x, y \in B$$

and

$$\epsilon(xy) = \epsilon(x)\epsilon(y), \ x, y \in k.$$

These two equations say that Δ and ϵ carry multiplication into multiplication. The condition on the units is that

$$\Delta \circ u = (u \otimes u) \circ \delta$$

where $\delta : \ k \to k \otimes k$ is the standard identification and that

$$\epsilon \circ u = id$$

as maps from k to k. In terms of Sweedler notation with

$$\Delta(x) = \sum x_{(1)} \otimes x_{(2)}$$

we can write the first of these conditions as

$$\Delta(xy) = \sum (-1)^{(degx_{(2)})\cdot(degy_{(1)})} \ x_{(1)}y_{(1)} \otimes x_{(2)}y_{(2)}. \tag{3.1.2}$$

In the ungraded case the signs disappear and (3.1.2) simplifies to

$$\Delta(xy) = \sum x_{(1)}y_{(1)} \otimes x_{(2)}y_{(2)}.$$

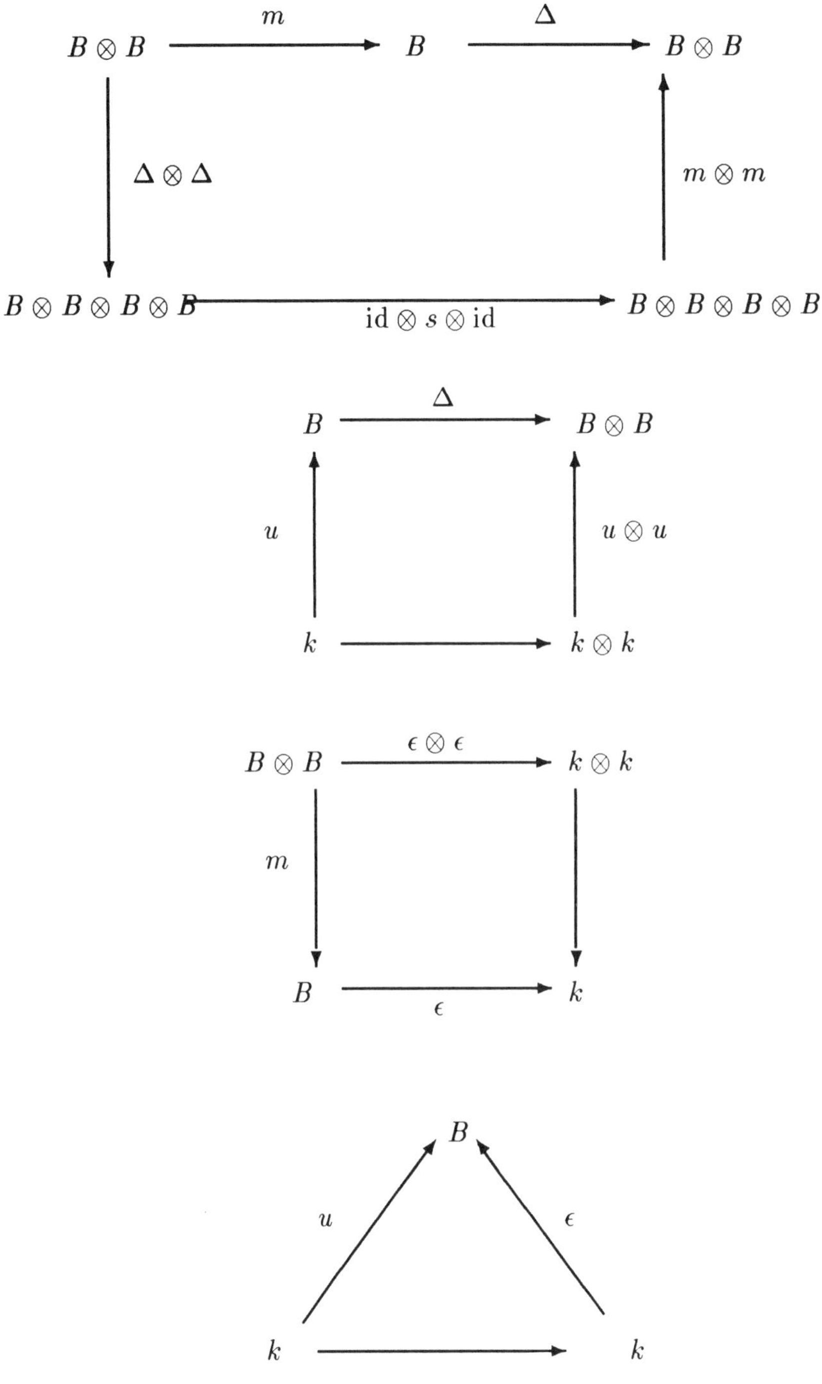

(**1**)

(**2**)

(**3**)

(**4**)

Figure 3.1:

Diagrammatically, the condition that Δ be an algebra map is that the diagrams (**1**) and (**2**) in Figure 3.1 commute, while the condition that ϵ be an algebra map is that diagrams (**3**) and (**4**) commute (unlabeled maps are the natural identifications):

We can read the commutativity of the diagrams in Figure 3.1 another way. We can make $B \otimes B$ into a coalgebra by defining the maps

$$B \otimes B \to B \otimes B \otimes B \otimes B$$

and

$$B \otimes B \to k$$

as the composite maps

$$B \otimes B \xrightarrow{\Delta \otimes \Delta} B \otimes B \otimes B \otimes B \xrightarrow{s_{23}} B \otimes B \otimes B \otimes B$$

and

$$B \otimes B \xrightarrow{\epsilon \otimes \epsilon} k \otimes k \xrightarrow{\sim} k \ .$$

In the first of these formulas the symbol s_{23} means that we apply s to the second and third tensor positions, that is (in our example)

$$s_{23} = id \otimes s \otimes id.$$

Then the commutativity of diagrams (**1**) and (**3**) in Figure 3.1 says that m is a coalgebra morphism while the commutativity of (**2**) and (**4**) says that u is a coalgebra morphism.

Thus we make the following definition: a *bialgebra* is a vector space B which is both an algebra and a coalgebra and such that all four diagrams above commute. This is equivalent to requiring that (Δ, ϵ) be morphisms of the algebra structure or that (m, u) be morphisms of the coalgebra structure. More generally, we may want to replace the ground field, k, by a commutative ring, K and allow B to be a module over K. We will still use the word "bialgebra" in this more general situation. The example that we started with, where $B = F(G)$ is the space of functions on a finite group, clearly satisfies the axioms, and so is a bialgebra. We will make much use of this example to illustrate various constructs.

3.2 The cohomology of Hopf spaces

But the place where the notion of a bialgebra was first isolated was in Hopf's investigations extending Elie Cartan's computation of the cohomology of Lie groups. Cartan's computation showed that the cohomology of a compact Lie group (over the reals) is the same as that of a product of odd dimensional spheres, that is that it is generated multiplicatively by generators in odd dimensions. Thus, for example, SU(2) is the three sphere, S^3, while $H^*(SU(3))$ has a generator in dimensions 3 and 5; more generally

$$H^*(SU(n)) = \bigwedge(x_3, x_5, ..., x_{2_{n-1}})$$

is the exterior algebra with generators in dimensions $3, 5, ..., 2n - 1$. See, for example [181].

The cohomology of any space X is always an algebra (under cup product). If we are given a multiplication,

$$\mu : X \times X \to X$$

then

$$\mu^* : H^*(X) \to H^*(X \times X)$$

gives a comultiplication when we identify $H^*(X \times X)$ with $H^*(X) \otimes H^*(X)$ using Künneth's formula. (We are considering cohomology over the reals, or more generally, over any field k.) So any time we are given a map $\mu : X \times X \to X$ we have a comultiplication $\Delta = \mu^*$ on $B = H^*(X)$ and the diagrams (**1**) and (**2**) of Figure 3.1 commute. (The map $u : k \to H^*(X)$ is the map which sends $1 \in k$ to the fundamental class in $H^0(X)$ whose pullback to any point is the fundamental generator.)

The counit ϵ comes in when we have a preferred base point, $* \in X$. (In the case of a topological group we would take $* = e$, the identity element.) The map $\epsilon : H^*(X) \to k = \mathbf{R}$ is then given as the pullback under the map pt. $\to *$ where we identify k with $H^*(\text{pt.})$. All four diagrams in figure 3.1 commute, as can easily be checked.

The axioms relating Δ to ϵ needed to make $H^*(X)$ into a coalgebra, namely

$$(\epsilon \otimes id) \circ \Delta = id$$

$$(id \otimes \epsilon) \circ \Delta = id$$

translate into the requirement that the composite maps

$$X \xrightarrow{i_L} X \times X \xrightarrow{\mu} \quad i_L(x) = (*, x)$$

and

$$X \xrightarrow{i_R} X \times X \xrightarrow{\mu} \quad i_R(x) = (x, *)$$

both induce the identity on cohomolology. This follows from the very mild requirement that both of these maps be homotopically equivalent to the identity. (For the case of a group these maps *are* the identity when we take $* = e$.) A space with a multiplication and a base point satisfying these conditions:

$$\mu \circ i_L \sim id \sim \mu \circ i_R \tag{3.2.1}$$

is known as a *Hopf space* or an H-space, for short.

3.3 Graded bialgebras over a field.

The bialgebras $B = H^*(X)$ that arise as the cohomology of H-spaces satisfy additional conditions which simplify their study:

(i) They are **Z** graded in the sense that

$$B = \bigoplus_{n \in \mathbf{Z}^+} B^n$$

is a graded vector space and the multiplication and comultiplication preserve the grading:

$$m(B^m, B^n) \subset B^{m+n} \quad \text{and} \quad \Delta B^n \subset \sum_{k+\ell=n} B^k \otimes B^\ell$$

(ii) the multiplication is associative and is supercommutative in the sense that

$$ab = (-1)^{dega \cdot degb} ba.$$

(iii) the maps ϵ and u are graded, i.e. $\epsilon = 0$ on all B^n with $n > 0$ and the image of u is in B^0.

If the space X is connected (as we shall assume for the rest of this discussion) then we have in addition that
(iv)$u : k \to B^0$and $\epsilon : B^0 \to k$ are inverse isomorphisms.

Let us call a bialgebra satisfying these four conditions a *graded-asc* bialgebra. (The "a" stands for associative and the "sc" for supercommutative.)

What Hopf showed was that any finite dimensional graded-asc bialgebra is an exterior algebra over odd generators. We will sketch the key ingredient in his argument in a moment.

First we examine some of the simplifications introduced by conditions(i)-(iv): Suppose we choose some $z \in B^n$ and write

$$\Delta z = \sum x_0 \otimes y_n + \sum x_1 \otimes y_{n-1} + \ldots + \sum x_n \otimes y_0$$

where the subscripts indicate the degrees, and the sums are over terms of a given type. Applying $\epsilon \otimes id$ kills all the terms on the right hand side except the first and the condition $(\epsilon \otimes id) \circ \Delta = id$ implies that the first term is $1 \otimes z$. Similarly the condition $(id \otimes \epsilon) \circ \Delta = id$ implies that the last term is $z \otimes 1$. So

$$\Delta z = 1 \otimes z + \sum x_1 \otimes y_{n-1} + \ldots + \sum x_{n-1} \otimes y_1 + z \otimes 1.$$

An element in an (arbitrary) bialgebra is called *primitive* if

$$\Delta z = 1 \otimes z + z \otimes 1. \tag{3.3.1}$$

Let x and y be homogeneous elements and let us write

$$\omega = (-1)^{(deg x)\cdot(deg y)}.$$

The element

$$xy - \omega yx$$

is called the *supercommutator* of x and y. We claim

Proposition 3.3.1 *If the homogeneous elements x and y are primitive, then so is their supercommutator.*

Proof. We compute:

$$\begin{aligned}
\Delta(xy - \omega yx) &= (x \otimes 1 + 1 \otimes x)(y \otimes 1 + 1 \otimes y) \\
&\quad -\omega(y \otimes 1 + 1 \otimes y)(x \otimes 1 + 1 \otimes x) \\
&= xy \otimes 1 + x \otimes y + \omega y \otimes x + 1 \otimes xy \\
&\quad -\omega yx \otimes 1 - \omega^2 x \otimes y - \omega y \otimes x - \omega 1 \otimes xy \\
&= (xy - \omega yx) \otimes 1 + 1 \otimes (xy - \omega yx).
\end{aligned}$$

This same argument in the ungraded case (with $\omega = 1$) shows that

Proposition 3.3.2 *The space of primitive elements of an ungraded bialgebra is closed under commutator.*

Notice that we did not use associativity or (super)commutativity in the proof of the two preceding propositions.

Let us consider the possible graded associative bialgebras, B, with the property that B is generated *as an algebra* by a single element, z, of positive degree. Suppose that z is of degree n. The element z must be primitive because there are no elements x_k, y_{n-k} of lower positive degree.

If z is of odd degree then $z \cdot z = -z \cdot z$ so $z^2 = 0$ (when the ground field is not of characteristic 2, as we shall assume). Then

$$(\Delta z)^2 = \Delta(z^2) = 0.$$

So we have one graded bialgebra on a single generator for each odd value of n, isomorphic to $H^*(S^n)$.

If z is of even degree, the possibilities for a graded algebra generated by z are either that $z^k = 0$ for some k, or$z^k \neq 0$ for any k, i.e. that the algebra is the polynomial algebra generated by z. (Notice the simplification introduced here by the grading assumption. A general commutative algebra generated by a single element z would by of the form $k[z]/(P)$ where (P) is the ideal generated by the polynomial P. The grading condition implies that P is a homogeneous polynomial. In one variable this means that either $P = 0$ or $P(z) = z^k$ for some k.)

Let us show that the coalgebra conditions do not allow $z^k = 0$. Indeed, suppose that $k > 1$ is chosen to be the smallest integer for which $z^k = 0$. Then, since all the powers of z are of even degree, we can apply the binomial formula to

$$\begin{aligned} 0 &= \Delta(z^k) = (\Delta z)^k = (1 \otimes z + z \otimes 1)^k \\ &= 1 \otimes z^k + kz \otimes z^{k-1} + \ldots + kz^{k-1} \otimes z + z^k \otimes 1. \end{aligned}$$

The first and the last terms on the right vanish. But there can be no cancellation among the remaining terms since $z, z^2, \ldots, z^{k-1}$ are linearly independent. This contradiction shows that $z^k = 0$ is impossible.

Thus the only graded associative bialgebras on one variable are $\bigwedge[x]$ where x is of odd degree and $k[x]$, where x is of even degree. If we assume that the algebra is finite dimensional as a vector space, the case $k[x]$ is excluded. Hence we have proved

Proposition 3.3.3 *The finite dimensional graded Hopf algebras with one generator are exactly $H^*(S^n)$ with n odd.*

This is the first step in the proof of Hopf's theorem. The remainder of the proof is a straightforward induction on the number of generators. But this first step already has a topological consequence, namely: no even dimensional sphere can be an H-space.

The actual facts, one of the triumphs of topology in the 1950's, is the result, due to Adams, that the only connected spheres that are H-spaces are S^1, S^3 and S^7 . We will return to the subject of H-spaces in Chapter 5.

In any graded algebra $B = \bigoplus_{k\geq 0} B^k$ we let I denote the ideal consisting of the elements of positive degree, so that

$$I = \bigoplus_{k>0} B^k.$$

We let P denote the space of primitive elements. The following proposition is useful in allowing us to conclude that certain graded bialgebras are associative and supercommutative.

Proposition 3.3.4 *Let B be a graded bialgebra over a commutative ring, K, with $B^0 = K$. If*

$$P \cap I^2 = \{0\}, \tag{3.3.2}$$

then B is supercommutative and associative.

Proof. Associativity. For any three elements $x, y, z \in B$ let $a(x, y, z)$ denote their associator:

$$a(x, y, z) = x(yz) - (xy)z.$$

We wish to show that this vanishes identically, and we know that it vanishes if any one of the three elements lies in $B^0 = K$. Suppose we have proved that it vanishes if the sum of the degrees of the three elements is less than n. We wish to pass from n to $n + 1$. Now

$$\Delta a(x, y, z) = \Delta x(\Delta y\Delta z) - (\Delta x\Delta y)\Delta z =$$

$$= a(x \otimes 1 + 1 \otimes x + \Delta_+(x), y \otimes 1 + 1 \otimes y + \Delta_+(y), z \otimes 1 + 1 \otimes z + \Delta_+(z))$$

where we are also using a to denote the associator on $B \otimes B$. In expanding this out, the terms involving the Δ_+ for any of the three elements will vanish since they involve elements with total degrees less than n in both tensor factors. So

$$\begin{aligned}\Delta a(x,y,z) &= a(x\otimes 1+1\otimes x, y\otimes 1+1\otimes y, z\otimes 1+1\otimes z)\\ &= a(x,y,z)\otimes 1+1\otimes a(x,y,z).\end{aligned}$$

So $a(x,y,z)$ is primitive. But it clearly belongs to I^2 unless two of the elements belong to B^0 in which case it vanishes trivially. So it belongs to $P \cap I^2$ and so it vanishes.

Commutativity. The same argument, applied to the supercommutator $[x,y] = xy - (-1)^{(degx)(degy)}yx$.

3.4 Antipodes and Hopf algebras.

Let us consider our model example, the bialgebra $B = F(G)$ of all functions on a finite group, G. For any $f \in F(G)$, let us define Tf by

$$(Tf)(a) = f(a^{-1}), \quad a \in G.$$

Then under the identification of $B \otimes B$ with $F(G \times G)$ the comultiplication is given as

$$(\Delta f)(a,b) = f(ab).$$

Thus the element $h = (id \otimes T) \circ \Delta f$ is the function given by

$$h(a,b) = f(ab^{-1}).$$

The multiplication map $m : F(G \times G) \to F(G)$ is given by

$$m(u)(a) = u(a,a), \quad u \in F(G \times G).$$

So

$$(mh)(a) = f(e)$$

if h is as above. In other words, the composite

$$m \circ (id \otimes T) \circ \Delta(f)$$

is just the constant function, $f(e) = \epsilon(f)$. But for any $r \in K$, the constant function $v \equiv r$ is the element $u(r) \in F(G)$. This leads to the following general definition of *antipode* for a bialgebra, B over a ring, K: It is a map $T : B \to B$ such that the composite map

$$B \xrightarrow{\Delta} B \otimes B \xrightarrow{id\otimes T} B \otimes B \xrightarrow{m} B$$

equals the composite map

$$B \xrightarrow{\epsilon} K \xrightarrow{u} B$$

where u is the unit and ϵ the counit.

The abstract idea behind this definition is as follows. Let C, $\Delta : C \to C \otimes C$ be a coalgebra and A, $m : A \otimes A \to A$ be an algebra, both over the same ring, K. Then $\mathrm{Hom}_K(C, A)$ has the structure of an algebra where the product, $f \star g$ of $f, g \in \mathrm{Hom}_k(C, A)$ is defined to be the composite map

$$C \xrightarrow{\Delta} C \otimes C \xrightarrow{f \otimes g} A \otimes A \xrightarrow{m} A.$$

If the comultiplication on C and the multiplication on A are associative, then this multiplication is associative. The element $u \circ \epsilon : C \to A$ (where $\epsilon : C \to K$ is the comultiplication of C and $u : K \to A$ is the unit of A) is a unit for this multiplication. This because a bialgebra axiom says that $(\epsilon \otimes id) \circ \Delta$ is the natural isomorphism of C with $K \otimes C$ and an algebra axiom says that $m \circ (u \otimes id)$ is the natural isomorphism of $K \otimes A$ with A.

The definition given above for the antipode says that T is the inverse in the $\star$ multiplication of the identity map, $id : B \to B$.

A *Hopf algebra* is defined as a bialgebra which is associative and coassociative and which has an antipode.

In the spirit of the preceding section we will now prove that if B is a graded associative and coassociative bialgebra with $B^0 = K$ then an antipode T exists and is unique in the class of graded maps. In fact, suppose that C and A are graded with $C^0 = A^0 = K$ and let $G(C, A) \subset \mathrm{Hom}_K(C, A)$ denote the subset consisting of graded maps f with $f_0 = id$. Then

Proposition 3.4.1 *$G(C, A)$ is a group under $\star$.*

Proof. We need to show the existence of an inverse, g, to any $f \in G(C, A)$. We have $g_0 = f_0 = id$. We want $(f \star g)_n = 0$ for $n > 0$. Let us write, for $u \in C^n$,

$$\Delta u = u \otimes 1 + 1 \otimes u + \Delta_+ u \tag{3.4.1}$$

where $\Delta_+ u$ is defined to be the remaining terms, and so consists of sums of terms of the form $u_k \otimes u_l$ with $k + l = n$, $0 < k < n$, $0 < l < n$. Assuming that we have defined $g = \bigoplus_{k<n} g_k$ on $\bigoplus_{k<n} C_k$ so that $(f \star g)_k = 0$ holds for $0 < k < n$. We define g_n by

$$g_n(u) = -f_n(u) - m \circ (f \otimes g) \circ \Delta_+(u), \quad u \in C^n.$$

The expression on the right makes sense since it involves only the g_l with $l < n$. So we can recursively define the g_n so that $(f \star g)_n = 0$. QED

In particular, the identity map belongs to $G(B, B)$ and hence has an inverse so we conclude

Proposition 3.4.2 *A graded associative and coassociative bialgebra has an antipode which is unique in the class of graded maps.*

3.5 Self adjoint bialgebras

Suppose that B is a graded bialgebra over the commutative ring, K, and is endowed with a non degenerate symmetric bilinear form $\langle , \rangle : B \otimes B \to K$ for which B^k and B^l are orthogonal if $k \neq l$ and which reduces to ordinary multiplication on $B^0 = K$. (In particular we thus are assuming that K is an integral domain.)

We will assume that B is evenly graded , i.e. that $B^{2k+1} = (0)$ for all k, so that there are no odd components. We say that the bialgebra is *self adjoint* if the maps

$$\Delta : B \to B \otimes B, \quad m : B \otimes B \to B$$

are adjoints of one an other as are the maps

$$\epsilon : B \to K, \quad u : K \to B.$$

Suppose that B is a self adjoint bialgebra and let $y \in I \otimes I$ and $x \in I$. Then

$$\langle x, m(y) \rangle = \langle \Delta x, y \rangle = \langle \Delta_+ x, y \rangle,$$

since B^0 is orthogonal to all the $B^k, k > 0$. So x is orthogonal to all the elements $m(y), y \in I \otimes I$ if and only if $\Delta_+ x = 0$, i.e. x is primitive. But $m(I \otimes I) = I^2$. So we have proved that

$$P = (I^2)^\perp. \tag{3.5.1}$$

In particular, if the scalar product is positive definite so that there are no isotropic elements, (3.3.2) holds, and so we have proved:

Proposition 3.5.1 *Every positive definite self adjoint bialgebra is associative and commutative. Since $\Delta = m^*$, it is also coassociative and cocommutative.*

Let x^* denote the adjoint of multiplication by x so

$$\langle x^*y, z\rangle = \langle y, m(x\otimes z)\rangle = \langle \Delta y, x\otimes z\rangle.$$

So if $x \in B^k$ then

$$x^* : B^n \to B^{n-k}.$$

In particular, x^* vanishes on B^n if $n < k$ and it acts on B^k as scalar product by x under the identification of B^0 with K. By the commutativity and associativity of multiplication, all the operators of the form x^* commute with one another.

It will be useful for us to have a formula for the action of x^* on a product yz. For this let us use the Sweedler notation $\Delta x = \sum x_{(1)} \otimes x_{(2)}$. Then we claim that

$$x^*(yz) = \sum x^*_{(1)}(y)x^*_{(2)}(z). \tag{3.5.2}$$

Another way of writing this equation is as follows: let us define $(u\otimes v)^*$ to be $(u\otimes v)^* = u^* \otimes v^*$ as operators on $B\otimes B$. Then we can write (3.5.2) as

$$x^* \circ m = m \circ (\Delta x)^*.$$

But taking the adjoints of both sides of this equation, we get

$$\Delta \circ m(x,\cdot) = m(\Delta x, \cdot)\circ \Delta$$

which is just the consistiency of the multiplication and the comultiplication. So we have proved (3.5.2). If p is a primitive element, so $\Delta p = p\otimes 1 + 1\otimes p$ then there are only two terms on the right of (3.5.2) which becomes

$$p^*(yz) = (p^*y)z + y(p^*z). \tag{3.5.3}$$

In other words,we have

Proposition 3.5.2 *p^* is a derivation if p is primitive.*

3.6 Antipodes of self adjoint bialgebras

For use in the next chapter, we need to collect some general facts about the antipode for graded selfadjoint bialgebras. Recall that every graded associative and coassociative bialgebra has an antipode which is unique in the class of graded maps. So we may speak of "the" antipode of a graded self adjoint bialgebra. As in the preceding section, we will assume that our bialgebras are evenly graded.

Proposition 3.6.1 *Let B be a graded bialgebra over a commutative ring, K with $B^0 = K$ and let T denote its antipode. If the multiplication and the comultiplication of B are commutative, then T is a bialgebra morphism and $T^2 = id$. If, in addition, B is a self adjoint bialgebra then T is an isometry relative to $\langle , \rangle$.*

Proof. If A and B are bialgebras with antipodes T_A and T_B then it follows immediately from the definitions that $T_A \otimes T_B$ is an antipode for $A \otimes B$. To be extra precise, let $s_{23} : A \otimes A \otimes B \otimes B \to A \otimes B \otimes A \otimes B$ be the map which switches the two middle positions:

$$s_{23}(a_1 \otimes a_2 \otimes b_1 \otimes b_2) = a_1 \otimes b_1 \otimes a_2 \otimes b_2.$$

So the comultiplication on $A \otimes B$ is $s_{23} \circ (\Delta_A \otimes \Delta_B)$ while the multiplication is $(m_A \otimes m_B) \circ s_{23}$. Then with this multiplication and comultiplication it follows from the definitions that $T_A \otimes T_B$ is an antipode for $A \otimes B$. Notice that if $A = B$ then

$$[m \circ (m \otimes m) \circ s_{23}](a \otimes b \otimes c \otimes d) = acbd$$

while

$$[m \circ (m \otimes m)](a \otimes b \otimes c \otimes d) = abcd.$$

So if the multiplication is commutative, we have

$$m \circ (m \otimes m) \circ s_{23} = m \circ (m \otimes m).$$

Similarly, if the multiplication is cocommutative then

$$s \circ (\Delta \otimes \Delta) \circ \Delta = (\Delta \otimes \Delta) \circ \Delta.$$

To say that T is an algebra morphism means that

$$m \circ (T \otimes T) = T \circ m,$$

so we wish to establish this equation. The assertion that $T \otimes T$ is the antipode for $B \otimes B$ means that

$$(m \otimes m) \circ s \circ (id \otimes id \otimes T \otimes T) \circ s \circ \Delta \otimes \Delta = (u \otimes u) \circ \epsilon \otimes \epsilon.$$

Applying m to both sides and using the commutativity of the multiplication to drop an s gives

$$m \circ (m \otimes m) \circ (id \otimes id) \otimes (T \otimes T) \circ s \circ (\Delta \otimes \Delta) = u \circ (\epsilon \otimes \epsilon),$$

or

$$m \circ (m \otimes [m \circ (T \otimes T)]) \circ s \circ (\Delta \otimes \Delta) = u \circ (\epsilon \otimes \epsilon).$$

This equation says that $m \circ (T \otimes T)$ is the inverse of m in the group $G(B \otimes B, B)$. So to prove that $T \circ m = m \circ (T \otimes T)$ it is enough to prove that $T \circ m$ is an inverse for m in $G(B \otimes B, B)$, i.e. that the composition

$$B \otimes B \xrightarrow{\Delta\otimes\Delta} B \otimes B \otimes B \otimes B \xrightarrow{s} B \otimes B \otimes B \otimes B \xrightarrow{m\otimes m} B \otimes B \xrightarrow{T\otimes id} B \otimes B \xrightarrow{m} B$$

equals $u \circ (\epsilon \otimes \epsilon)$. But the consistency of multiplication with comultiplication says that $(m \otimes m) \circ s \circ (\Delta \otimes \Delta) = \Delta \circ m$, so we can replace the above diagram by

$$B \otimes B \xrightarrow{m} B \xrightarrow{\Delta} B \otimes B \xrightarrow{T\otimes id} B \otimes B \xrightarrow{m} B.$$

Now the portion of this diagram starting with the second term, B composes to $u \circ \epsilon$ by the defining property of T. Hence the whole composition is $u \circ \epsilon \circ m = u \circ (\epsilon \otimes \epsilon)$ since $\epsilon \circ m = \epsilon \otimes \epsilon$ is one of the bialgebra axioms.

A similar argument shows that T is a coalgebra morphism. So T is a bialgebra morphism. We wish to show that $T^2 = id$. Since id is the inverse of T under the $\star$ multiplication of the group $G(R, R)$, it is enough to show that

$$T^2 \star T = u \circ \epsilon.$$

But

$$\begin{aligned} T^2 \star T &= m \circ (T \otimes T^2) \circ \Delta \\ &= m \circ (T \otimes T) \circ (id \otimes T) \circ \Delta \\ &= T \circ m \circ (id \otimes T) \circ \Delta \\ &= T \circ u \circ \epsilon \\ &= u \circ \epsilon, \end{aligned}$$

where we have used the fact that T is an algebra morphism to pass from the second line to the third, and that T is the identity on B_0 to pass from the fourth line to the fifth. Now suppose that B is self adjoint. If we take the adjoint of the defining equation for T, we see that T^* satisfies the same equation. Since the antipode is unique, this implies that $T = T^*$. But $T = T^{-1}$ so $T^* = T^{-1}$ and so T is an isometry. QED.

3.7 The universal enveloping algebra

Let G be a Lie group with Lie algebra, $\mathcal{G}$. As we are no longer dealing with finite groups, the space $\mathcal{F}(G)$ will be infinite dimensional, and we no longer will have the identification of $\mathcal{F}(G \times G)$ with $\mathcal{F}(G) \otimes \mathcal{F}(G)$ and so $\mathcal{F}(G)$ will not be a coalgebra when G is not finite. Instead, we have the identification

$$\mathcal{F}(G \times G) \sim \mathcal{F}(G) \hat{\otimes} \mathcal{F}(G)$$

where $\hat{\otimes}$ denotes a completed tensor product: finite sums must be replaced by infinite series in the definition of Δ. This is already apparent in the examples 2,3 and 4 of the addition laws cited at the very beginning of Chapter 1. We could, of course, modify the definition of coalgebra to take this into account, and study topological coalgebras with completed tensor products replacing tensor products in the definition of comultiplication. But in the case of a connected Lie group we have an alternative: The "dual object" to the space of analytic functions on G is the universal enveloping algebra, $U(\mathcal{G})$, thought of as the space of distributions on G supported at the identity element, e. We can also think of $U(\mathcal{G})$ as the left invariant differential operators on G, and this gives the multiplication on $U(\mathcal{G})$. The comultiplication is dual to restriction to the diagonal on $\mathcal{F}(G \times G)$. For elements $X \in \mathcal{G}$ this translates into

$$\Delta(X) = X \otimes 1 + 1 \otimes X, \quad X \in \mathcal{G} \tag{3.7.1}$$

and extends by multiplication to $U(\mathcal{G})$. So the comultiplication is cocommutative, while the multiplication is non trivial and determined by the Lie algebra structure of $\mathcal{G}$. This is, of course, dual to the situation in $\mathcal{F}(G)$ where the multiplication is commutative but the comultiplication carries the information relating to the group structure.

For a general Lie algebra, $\mathcal{G}$ over an arbitrary field,k, there is an abstract definition of $U(\mathcal{G})$ as an associative algebra. It is determined as the solution to a "universal mapping problem": For any linear map ρ of $\mathcal{G}$ into an associative algebra A satisfying

$$\rho(X)\rho(Y) - \rho(Y)\rho(X) = \rho([X,Y]) \tag{3.7.2}$$

there is a unique algebra homomorphism, also denoted by ρ, of $U(\mathcal{G})$ into A. This determines $U(\mathcal{G})$ uniquely up to isomorphism, and $U(\mathcal{G})$ can be constructed as a quotient of the tensor algebra $T(\mathcal{G})$. If k is of characteristic zero, as we shall assume from now on, there is a natural embedding of $\mathcal{G}$ into $U(\mathcal{G})$, so we may identify $\mathcal{G}$ as a subspace of $U(\mathcal{G})$, and $U(\mathcal{G})$ is generated by

$\mathcal{G}$ as an algebra. The map given by (3.7.1) clearly satisfies (3.7.2) and hence extends to an algebra homomorphism

$$\Delta : U(\mathcal{G}) \to U(\mathcal{G}) \otimes U(\mathcal{G}).$$

The map $\epsilon : \mathcal{G} \to k$ given by $\epsilon(X) = 0 \ \ \forall X$ clearly satisfies (3.7.2) and hence extends to an algebra homomorphism, also denoted by ϵ. If follows immediately from (3.7.1) that

$$(\epsilon \otimes id)(\Delta a) = (id \otimes \epsilon)(\Delta a) = a$$

for $a = X \in \mathcal{G}$ and hence by multiplication for all $a \in U(\mathcal{G})$. So $U(\mathcal{G})$ is an associative, coassociative and cocommutative bialgebra.

The map
$$T : X \mapsto -X$$

satisfies (3.7.2) as a map from $\mathcal{G}$ to the opposite algebra, $U(\mathcal{G})^{op}$. Thus T extends as an algebra antihomomorphism

$$T : U(\mathcal{G}) \to U(\mathcal{G}), \quad T(ab) = T(b)T(a)$$

so

$$T(X_1 \cdots X_n) = (-1)^n X_n X_{n-1} \cdots X_1. \tag{3.7.3}$$

Clearly
$$[m \circ (T \otimes id) \circ \Delta](a) = [m \circ (id \otimes T) \circ \Delta](a) = 0$$

for $a = X \in \mathcal{G}$ and hence by multiplication for all products of X's. Hence T is an antipode for the bialgebra $U(\mathcal{G})$. We will refer to the Hopf algebra $(U(\mathcal{G}), m, \Delta, T)$ simply by $U(\mathcal{G})$. It will play the central role in Chapters 10-13. Clearly every element of $\mathcal{G}$ is primitive, and it is known that the converse is true, [182],

$$\text{Prim}(U(\mathcal{G})) = \mathcal{G}. \tag{3.7.4}$$

3.8 The shuffle algebra.

In this section and the next, we want to describe some ideas of Chen that appear in an important series of articles [228] and the references listed there. The results of this section are based on an important paper of Ree,[1008]. We begin with some preliminaries. Recall that the primitive elements of an associative and coassociative bialgebra, B, are those elements which satisfy

$$\Delta(X) = X \otimes 1 + 1 \otimes X$$

and that the set, Prim(B), of primitive elements is a Lie algebra, i.e. is closed under commutator bracket. An element $b \in B$ is called *grouplike* if

$$\Delta(b) = b \otimes b. \tag{3.8.1}$$

Clearly the set of grouplike elements is closed under multiplication. Also,

$$[(\epsilon \otimes id) \circ \Delta](b) = \epsilon(b) \otimes b$$

if b is grouplike, but

$$(\epsilon \otimes id) \circ \Delta$$

is just the standard identification, $b \mapsto 1 \otimes b$ of B with $k \otimes B$. Thus

$$\epsilon(b) = 1 \tag{3.8.2}$$

if b is grouplike. If B is a Hopf algebra with antipode T then

$$\begin{aligned} T(b)b &= m(T \otimes id)\Delta(b) \\ &= \epsilon(b) \\ &= 1 \end{aligned}$$

if b is grouplike. So in a Hopf algebra the grouplike elements form a group with

$$b^{-1} = T(b).$$

We will denote the set of grouplike elements of B by Group(B). We would like to be able to say that

$$\exp : \text{Prim}(B) \to \text{Group}(B)$$

where the exponential map is defined by its usual series:

$$\exp(X) = 1 + X + \frac{1}{2!}X^2 + \frac{1}{3!}X^3 + \cdots.$$

The problem, of course, is that the series need not converge. For example, if $B = U(\mathcal{G})$ is the universal enveloping algebra of a Lie algebra, $\mathcal{G}$, such an infinite sum makes no sense. But suppose we were in a situation where series such as the exponential series converge. Then we could argue as follows: For any polynomial with coefficients in k, the consistency of multiplication and comultiplication implies that $\Delta(P(X)) = P(\Delta(X))$ for any $X \in B$.

Assuming that we can also apply this to the convergent power series for the exponential, we have, for X primitive,

$$\begin{aligned}\Delta(\exp(X)) &= \exp(\Delta(X)) \\ &= \exp(X\otimes 1 + 1\otimes X) \\ &= \exp(X\otimes 1)\cdot\exp(1\otimes X) \\ &= (\exp X\otimes 1)\cdot(1\otimes\exp X) \\ &= \exp X\otimes\exp X\end{aligned}$$

so

$$\exp X \in \mathrm{Grou}(B).$$

In passing from the second line to the third we have use the fact that $X\otimes 1$ commutes with $1\otimes X$ and then that the exponential of the sum of two commuting elements is the product of their exponentials.

A similar type argument shows that the logarithm of a group like element is primitive, where $\log(1+x)$ is defined by its power series. One uses the fact that the logarithm of the product of two commuting elements is the sum of their logarithms.

Here is a situation where we will have no trouble with convergence: Let $\mathbf{I}$ be a collection of symbols. In what follows I, J etc. will denote finite ordered sequences of elements of $\mathbf{I}$. When we write $I\subset J$ we mean that I appears as an ordered subset of J . If $\#J$ denotes the number of elements of J and $1\le k\le \#J$ then J_k denotes the sequence of length k consisting of the first k elements of J and J^k denotes the sequence of length $\#J-k$ consisting of the last $\#J-k$ elements of J . Thus if I consists of the positive integers then

$$(4,3,2,2,5,6)_4=(4,3,2,2) \text{ and } (4,3,2,2,5,6)^4=(5,6).$$

If J and K are sequences we let (J,K) denote the sequence of length $\#J+\#K$ whose first $\#J$ elements constitute J and whose last $\#K$ elements consist of K . We consider the k module $C=C(\mathbf{I})$ generated by 1 and elements Y_I where I ranges over all finite ordered sequences of elements of $\mathbf{I}$. We define a comultiplication $\Delta: C\to C\otimes C$ by $\Delta 1 = 1\otimes 1$ and

$$\Delta Y_I = \sum_{(J,K)=I} Y_J\otimes Y_K = \sum_{k=0}^{\#I} Y_{I_k}\otimes Y^{I^k}. \qquad (3.8.3)$$

Here $Y_\emptyset$ corresponding to subscript 0 or superscript $\#I$ is taken as 1. Thus

$$\Delta Y_{(1,2,3)} = Y_{(1,2,3)} \otimes 1 + Y_{(1,2)} \otimes Y_{(3)} + Y_{(1)} \otimes Y_{(2,3)} + 1 \otimes Y_{(1,2,3)}$$

for example. Notice that Δ is coassociative as

$$(\Delta \otimes id) \circ \Delta(Y_I) = (id \otimes \Delta) \circ \Delta(Y_I) = \sum_{(J,K,L)=I} Y_J \otimes Y_K \otimes Y_L.$$

The counit $\epsilon : C \to k$ is defined by $\epsilon(1) = 1$ and $\epsilon(Y_I) = 0, I \neq \emptyset$. It clearly satisfies the axioms for a counit, so C is an associative coalgebra.

Before defining the multiplication and the antipode on C and showing that it is a Hopf algebra, we study some immediate consequences of the coalgebra structure. Let F denote the dual space to C. So F consists of all infinite series of the form

$$\sum a_I X^I$$

where the X^I are dual to the Y_I . Then F inherits a multiplication given by

$$\langle X^J \cdot X^K, Y_I \rangle = \langle X^J \otimes X^K, \Delta Y_I \rangle.$$

This last expression is 0 unless $(J, K) = I$ and is 1 if $(J, K) = I$ as follows from the definition of the comultiplication on C. Thus

$$X^J \cdot X^K = X^{(J,K)}.$$

So if $J = (p, q, r, ..., s)$ then $X^J = X^p \cdot X^q \cdot X^r \cdots X^s$ where we write X^p instead of $X^{(p)}$ etc.. In other words, F is the ring of noncommutative formal power series in the variables indexed by I.

Let $A = A(\mathbf{I})$ be the ring of non-commutative polynomials in the variable X^p, $p \in \mathbf{I}$, so that F is the completion of A in the Krull (formal power series) topology. We can also think of A as the tensor algebra $A = T(V)$ where V is the vector space with basis $\{X^p\}$. It follows from the defining properties of the universal enveloping algebra that $A = U(\mathcal{G})$ where $\mathcal{G}$ if the free Lie algebra generated by the X^p. Hence A is a Hopf algebra. Since F is the completion of A it inherits a Hopf algebra structure. In particular, if we let $\mathcal{L}$ denote the completion of $\mathcal{G}$ in F, then the Lie algebra $\mathcal{L} = \text{Prim}(F)$. In F there is no problem about convergence of power series such as $\exp a$ or $\ln(1 + a)$ so long as $a \in m$, the maximal ideal of F. In particular the argument we gave at the beginning of this section is legitimate and we have

Proposition 3.8.1 *The map* $\exp : \mathit{Prim}(F) \to \mathit{Group}(F)$ *is a bijection.*

The comultiplication on F induces a multiplication on C by the formula

$$\langle X^K, \mu(Y_I \otimes Y_J) \rangle = \langle \Delta X^K, Y_I \otimes Y_J \rangle. \tag{3.8.4}$$

To see what this is more explicitly, recall that the comultiplication on F is generated multiplicatively by

$$\Delta X^p = X^p \otimes 1 + 1 \otimes X^p.$$

For example, if $K = (i, j, k)$ then

$$\begin{aligned}
\Delta(X^K) &= \Delta X^i \Delta X^j \Delta X^k \\
&= (X^i \otimes 1 + 1 \otimes X^i)(X^j \otimes 1 + 1 \otimes X^j)(X^k \otimes 1 + 1 \otimes X^k) \\
&= X^i X^j X^k \otimes 1 + X^i X^j \otimes X^k + X^i X^k \otimes X^j + X^j X^k \otimes X^i \\
&\quad\; X^i \otimes X^j X^k + X^j \otimes X^i X^k + X^k \otimes X^i X^j + 1 \otimes X^i X^j X^k.
\end{aligned}$$

If we look at

$$I = (p), \quad J = (q, r)$$

then we get a contribution of 1 to (3.8.4) only if K is such that p is one of the three letters i, j or k and the remaining two are q and r in the correct order. Otherwise the contribution will be zero. In other words, we get the non-zero contribution if and only if K is a riffle shuffle of (I, J). (A riffle shuffle with a cut at k is a permutation $\sigma \in S_n$ such that $\sigma(x) < \sigma(y)$ if $x, y \leq k$ or $k < x < y \leq n$.) So we define

$$\mu : C \otimes C \to C$$

by

$$\mu(Y_I \otimes Y_J) = \sum_{\sigma(I,J)=K} Y_K \tag{3.8.5}$$

where the sum is over all riffle shuffles. From general principles we know that Δ and μ are compatible since they are the duals of the compatible multiplication and comultiplication on F.

It is interesting to study the composition $\mu \circ \Delta : C \to C$. If we think of Y_I as a stack of cards labeled by the elements of I, then Δ can be thought of as the direct sum of all possible cuts of the deck. Applying μ then adds up all possible shuffles. Thus the symmetric element

$$\sum_{\sigma \in S_{\#I}} Y_{\sigma I}$$

is clearly an eigenvector of $\mu \circ \Delta$, since every term in the expansion of $\mu \circ \Delta$ into a sum of riffle shuffles leaves this completely symmetric element invariant. The eigenvalue will be the number of summands and this equals $2^{\#I}$ since the position, k, of the cut (in the expression for $\Delta = \Delta_C$) ranges over $0 \leq k \leq I$ and there are

$$\binom{\#I}{k}$$

choices of riffle shuffle with cut at k.

In fact, the operator $\mu \circ \Delta$ is completely reducible and all its eigenvalues of are powers of two. To see this, observe that the transpose of $\mu \circ \Delta$ is

$$m \circ \Delta, \quad \Delta = \Delta_F : F \to F \otimes F, \quad m : F \otimes F \to F,$$

where, with apologies for the notation, $\Delta = \mu^*$ is the comultiplication in F and $m = \Delta_C^*$ is the multiplication. The assertion about the complete reducibility of $\mu \circ \Delta_C$ is an immediate consequence of the following proposition, valid for any Lie algebra.

Proposition 3.8.2 *Let $\mathcal{G}$ be a Lie algebra over a field of characteristic zero. Let*

$$W_n : S^n(\mathcal{G}) \to U(\mathcal{G})$$

denote the completely symmetric (Weyl) embedding, so that we have the vector space identification

$$U(\mathcal{G}) = \bigoplus W_n(S^n(\mathcal{G})). \tag{3.8.6}$$

Then (3.8.6) is a decomposition of $U(\mathcal{G})$ into eigenspaces of $m \circ \Delta$ with eigenvalues 2^n on $W_n(S^n(\mathcal{G}))$.

Indeed, in terms of a Poincaré Birkhoff Witt basis, $\{X^K\}$, the space $W_n(S^n(\mathcal{G}))$ is spanned by

$$\sum_{\sigma \in S_n} X^{\sigma K}, x, \quad \#K = n.$$

But

$$(m \circ \Delta) X^K = \sum_{\tau \text{ a shuffle}} X^{\tau K}.$$

Applying the corresponding formula to each of the $X^{\sigma K}$ and using the fact that there are 2^n shuffles of a deck of size n proves the proposition. We will need this proposition in Chapter 11.

As an aside, we observe that if the elements of I are distinct, then on the space spanned by the

$$\{Y_{\sigma I}\}, \quad \sigma \in S_{\# I}$$

the totally symmetric element is the unique eigenvector with eigenvalue $2^{\# I}$, the space being spanned by eigenvectors, with remaining eigenvalues being $2^r, 1 \le r < \# I$. Indeed, the totally symmetric element, $\sum Y_{\sigma I}$ is characterized (up to scalar multiple) as being orthogonal to $W_k(S^k(\mathcal{L})), k \neq n$. To illustrate, consider the case where $I = (1,2,3)$. In F, we need only take into account expressions which involve sums of products of X^1, X^2 and X^3 in various orders. This six dimensional space is spanned by

$$\begin{aligned} [X_1,[X_2,X_3]],[[X_1,X_2].X_3] &\in W_1(S^1(\mathcal{L})) \\ X^1[X^2,X^3]+[X^2,X^3]X^1, & \\ X^2[X^3,X^1]+[X^3,X^1]X^2, & \\ X^3[X^1,X^2]+[X^1,X^2]X^3 &\in W_2(S^2(\mathcal{L})) \\ \sum_{\sigma\in S_3} X^{\sigma I} &\in W_3(S^3(\mathcal{L})). \end{aligned}$$

Clearly $\sum Y_{\sigma I}$ is orthogonal to all these elements except the last.

The operator

$$\frac{1}{2^n}\mu \circ \Delta, \quad n = \# I$$

acts on the space spanned by the Y_σ as an element of the group algebra of S_n with positive coefficients supported on the shuffles, and whose coefficients add up to one. In other words, it is a probability measure concentrated on the shuffles. As an operator, its maximal eigenvalue = 1, and all other eigenvalues are positive and strictly less than 1. The assertion above about the uniqueness of the maximal eigenvector implies that repeated shuffling, i.e. applying powers of the operator, converges to projection onto one dimensional space of totally symmetric elements. Expressed in terms of probability measures on S_n, it says that repeated convolution converges to the uniform measure, i.e. to complete randomness. For a full discussion of more realistic shuffling paradigms see [120] and the references given there.

The antipode, S, on C, is defined by $S1 = 1$ and

$$SY_I = (-1)^{\# I} Y_{I_{\text{rev}}}$$

where I_{rev} is I with the order of its elements reversed. So

$$SY_{(1,2,3,4)} = Y_{(4,3,2,1)},$$

for example. This is the dual of the antipode, T, on F given by

$$T(X^I) = (-1)^{\#I} X^{I_{\mathrm{rev}}}.$$

the is the standard antipode for the universal enveloping algebra, generated from $TX^p = -X^p$. Since $S = T^*$ it follows that S is an antipode for C. The fact that S is an antipode, i.e. that

$$[\mu \circ (S \otimes id) \circ \Delta] Y_I = 0, \quad I \neq \emptyset.$$

contains nontrivial combinatoric information. Here is a sample of how a direct computation of the cancelation involved in the above equation goes, where we have dropped the commas and parentheses on the subscripts to save space:

$$\Delta Y_{1234} = Y_{1234} \otimes 1 + Y_{123} \otimes Y_4 + Y_{12} \otimes Y_{34} + Y_1 \otimes Y_{234} + 1 \otimes Y_{1234}$$

so

$$(S \otimes id) \circ \Delta Y_{1234} = Y_{4321} \otimes 1 - Y_{321} \otimes Y_4 + Y_{21} \otimes Y_{34} - Y_1 \otimes Y_{234} + 1 \otimes Y_{1234}.$$

Hence $[\mu \circ (S \otimes id) \circ \Delta](Y_{1234})$ is given by

$$\begin{array}{llllll} Y_{4321} & + & & & & \\ -Y_{4321} & -Y_{3421} & -Y_{3241} & -Y_{3214} & & \\ +Y_{2134} & +Y_{3421} & +Y_{3241} & +Y_{3214}+ & Y_{2341}+ & Y_{2314} \\ -Y_{2134} & -Y_{1234} & & & -Y_{2341} & -Y_{2314} \\ & & +Y_{1234} & & & \\ & & = 0. & & & \end{array}$$

When **I** consists of a single element, i.e. when $\#\mathbf{I} = 1$, then the Hopf algebra C is called the algebra of *divided powers*.

3.9 Chen's iterated integrals.

Chen, in his series of remarkable papers, associates to piecewise differentiable paths, α in $\mathbf{R}^n$ an element $\theta(\alpha) \in \mathrm{Group}F$ where $\mathbf{I} = \{1, 2, \ldots, n\}$

This assignment carries composition of paths into multiplication in the group, Group(F). Here is how it works: Let $f_i, i \in \mathbf{I}$ be a collection of (real or complex) piecewise continuous functions defined on some interval containing the points a, b, s etc. Define the iterated integral

$$\int_a^b J(f)$$

inductively as follows: If $\#J = 1$ and J consists of the single index, j, then

$$\int_a^b J(f)dt = \int_a^b f_j(t)dt,$$

the usual integral. If the iterated integral has been defined for all sequences of length n, and $\#J = n+1$ with $J = (K, j)$ and $\#K = n$ (so that the iterated integral for K is defined) then define

$$\int_a^b J(f)dt = \int_a^b \int_a^\tau K(f)dt f_j(\tau)d\tau. \tag{3.9.1}$$

We then have Chen's lemma which asserts that

$$\int_a^s I(f)dt = \sum_{(J,K)=I} \int_a^b J(f)dt \int_b^s K(f)dt \tag{3.9.2}$$

it being understood that, in the sum on the right, the integrals corresponding to empty K or J are just 1 . So if $\#I = 1$ this equation says that the ordinary integral from a to b plus the ordinary integral from b to s equals that ordinary integral from a to s. So (3.9.2) is clearly true for $\#I = 1$. It then follows by induction on $\#I$ by differentiation: the left hand side of (3.9.2) applied to (I, p) is

$$\int_a^s \int_a^\tau I(f)(t)dt f_p(\tau)d\tau$$

whose derivative is

$$f(s) \int_a^s I(f)(t)dt.$$

The right hand side of (3.9.2) when applied to (I, p) will have one term which is not a function of s, namely the term

$$\int_a^b \int_a^\tau I(f)(t)dt f_p(\tau)d\tau.$$

The sum of the remaining terms, when differentiated with respect to s give $f(s)$ times the right hand side of (3.9.2). Hence (3.9.2) is proved by induction.

In particular, consider a piecewise differentiable path, α, in $\mathbf{R}^n$, and let

$$f_i(t) = x_i'(t)$$

where the x_i are the coordinates. Then iterated integration assigns a number, call it $Y_I(\alpha)$, to each piecewise differentiable path. If α and β are two paths with the end point of α coinciding with the initial point of β, let $\gamma = \alpha\beta$ denote the combined path. Let $\theta(\alpha)$ be the linear function on C which assigns the value $Y_I(\alpha)$ to each Y_I . Thus

$$\theta(\alpha) \in F$$

and Chen's lemma, equation (3.9.2), implies that

$$\theta(\alpha\beta) = \theta(\alpha)\theta(\beta). \tag{3.9.3}$$

where the multiplication on the right is in F.

We wish to show that the image of θ lies in Group(F). If α is the straight line segment joining the origin to $a = (a_1, \ldots, a_n)$, it follows immediately from the definition of the iterated integral that

$$\theta(\alpha) = \exp(a_1 X_1 + \cdots a_n X_n).$$

From this and Proposition 3.8.1 it follows that $\theta(\alpha) \in$ Group(F) for this case. Since θ is defined in terms of the derivatives of the path, it is invariant under translation. Since Group(F) is a group, we know it for all broken line segments. By continuity (relative to an appropriate topology on F), it follows for all piecewise differentiable paths. So θ defines a homomorphism from paths under composition into the group like elements of F. This is Chen's generalization of the Campbell-Hausdorff theorem.

Suppose that we are given functions $f_1, \ldots f_n$ and consider their indefinite integrals, solutions to

$$x_i' = f_i,$$

giving a family of curves $\alpha(s) : [0, s] \to \mathbf{R}^n$. Then

$$G(s) = \theta(\alpha(s))$$

is a curve in Group(F). It follows from Chen's lemma that

$$G'(s) = G(s)[X_1 f_1(s) + \cdots + X_n f_n(s)].$$

Thus Chen's iterated integral provides a solution to this differential equation with values in F. In Chapter 12 we will want to consider such differential equations. The convention we follow there will write the differential equation in the form

$$G' = (f_1 X_1 + \ldots + f_n X_n)G. \tag{3.9.4}$$

To accommodate this convention we need only modify our inductive definition of the iterated integral by passing from K to (j, K) instead of putting the j in at the right, i.e. (K, j), as we did above.

Chapter 4

Representations of S_n

4.1 PSH algebras.

In this chapter we give an introduction to the Hopf algebra approach to the representation theory of the finite classical groups. For the case of the symmetric groups, this theory was developed independently by Zelevinski [1263] and Liulevicius [795] and extended by by Zelevinski to include the representation theory of general linear groups over a finite field. The material in this chapter will not be used in the sequel, and so the reader who wants to concentrate on the main theme can proceed directly to Chapter 5. But the Zelevinski theory gives a powerful application of many of the ideas introduced so far, and we thought that it is appropriate to include an introduction to this theory here. For the detailed presentation of this theory, we refer to Zelevinski's beautiful book, [1263].

In this chapter we will consider graded bialgebras $B = \bigoplus_{n\in\mathbf{Z}^+} B^n$ over a ring K for which $B^0 = K$, and

$$B^{2m+1} = (0)$$

for all m. In other words, B has only even components. To avoid having to carry the factors of 2 we shall set

$$R_n = B^{2n}$$

and

$$R = \bigoplus_{n\in\mathbf{Z}^+} R_n$$

with

$$m : R_k \otimes R_l \to R_{k+l}, \quad \Delta : R_n \to \bigoplus R_k \otimes R_l.$$

For most of this chapter, $K = \mathbf{Z}$ and R comes equipped with a preferred $\mathbf{Z}$ basis of homogeneous generators.

The example to bear in mind is: $R_0 = \mathbf{Z}$ and, for $n > 0$, to let R_n be the space of integer combinations of (equivalence clases of) irreducible representations of the group S_n. The (equivalence classes of the) irreducible representations are the preferred $\mathbf{Z}$ basis.

In this example, the multiplication $m : R_k \otimes R_l \to R_n$ $(n = k + l)$ is determined by

$$m(\rho_k \otimes \rho_l) = (\rho_k \otimes \rho_l) \uparrow$$

where $S_k \times S_l$ is regarded as the subgroup of S_n in the obvious way, where $(\rho_k \otimes \rho_l)$ is regarded as a representation of $S_k \times S_l$ and where $\uparrow$ denotes induction from $S_k \times S_l$ to S_n.

The comultiplication is the map

$$\Delta : R_n \to \bigoplus_{k+l=n} R_k \otimes R_l$$

where the k, l component of this map, call it $\Delta_{k,l} : R_n \to R_k \otimes R_l$, is restriction of the representation from S_n to $S_k \times S_l$.

The fact the multiplication and comultiplication are consistent is a consequence of the Mackey double coset theorem which describes the operation of induction from one subgroup followed by restriction to another. In order not to interrupt the flow, we postpone the details, which are quite elementary, to the end of the chapter.

Let x_n denote the (equivalence class of the) trivial representation of S_n and y_n the sign representation, for $n \geq 2$, with $y_1 = x_1$ and $x_0 = y_0 = 1$. Then clearly

$$\Delta x_n = \bigoplus_{k+l=n} x_k \otimes x_l, \quad \Delta y_n = \bigoplus_{k+l=n} y_k \otimes y_l. \tag{4.1.1}$$

The multiplication for the example of the symmetric groups representations has the property that both the multiplication and the comultiplication are expressible in terms of the preferred basis by non-negative integers. That is, if $\{\omega_i\}$ is the preferred basis, we have

$$m(\omega_i \otimes \omega_j) = \sum m_{ij}^k \omega_k, \quad \Delta \omega_k = \sum d_k^{ij} \omega_i \otimes \omega_j, \tag{4.1.2}$$

where the m_{ij}^k and the d_k^{ij} are non-negative integers. The preferred basis induces a $\mathbf{Z}$ valued symmetric bilinear positive definite form on R by declaring the $\{\omega_i\}$ to be an orthonormal basis. Suppose also that R is self adjoint relative to this basis, so that $m_{ij}^k = d_k^{ij}$. Such an algebra is called by Zelevinsky *a Positive Self adjoint Hopf algebra*, or PSH algebra for short.

In the example of the representation theory of the S_n, the symmetric $\mathbf{Z}$ valued bilinear form determined by the generators is clearly given by

$$\langle x, y\rangle = 0 \text{ if } deg\ x \neq deg\ y$$

while

$$\langle x, y\rangle = \dim \ \mathrm{Hom}_{S_n}(x, y)$$

if x and y are (equivalence classes of) representations of S_n, as follows from Schur's lemma. Then the selfadjointness:

$$\langle m(x \otimes y), z\rangle = \langle x \otimes y, \Delta(z)\rangle$$

is an immediate consequence of the Frobenius reciprocity theorem.

Notice that for the symmetric group case, the element $p = x_1$ is primitive:

$$\Delta x_1 = x_1 \otimes 1 + 1 \otimes x_1.$$

But no other basis element is primitive, since any representation of S_n restricts to *some* representation of $S_k \times S_l$ if $k + l = n$.

In the symmetric group case every element of our basis which lies in R_n occurs with a positive coefficient in the decomposition of $p^n = x_1^n$. This is because x_1^n is the (equivalence class of the) regular representation of S_n, and we know that every representation occurs in the regular representation. But in fact, this is already a formal consequence of the axioms in the sense of the following

Proposition 4.1.1 *Let R be a PSH algebra with exactly one of the basis elements primitive. Call this element p. Then every basis element occurs with a positive coefficient in the decomposition of p^n for some n.*

Proof. Let ω be a basis element. If ω is primitive, it must be equal to p by hypothesis. If ω is not primitive, it can not be orthogonal to I^2 by (3.5.1). So there must be basis elements ω_1 and ω_2 so that $\langle \omega, \omega_1 \cdot \omega_2\rangle \neq 0$. But our hypotheses about the positive integer character of the multiplication then imply that ω occurs with a positive coefficient in the decomposition of $\omega_1 \cdot \omega_2$, and both these elements have smaller degree than ω. Hence the proposition follows by induction on the degree.

Suppose that R is a PSH algebra with exactly one primitive basis element, p. The proposition implies that the degree of any basis element must be a multiple of the degree of p. By dividing all the degrees by deg p if necessary, we may assume without loss of generality that deg $p = 1$. The essence of the Zelevinski theory is that (up to isomorphism) there exists exactly one PSH algebra with a single primitive basis element of degree one, and that is the example provided by the representation theory of the symmetric groups. This can be regarded as a (deep) variation on the theorem of Hopf that we proved in the preceding chapter, saying that over a *field* the only evenly graded bialgebra with a single generator is $k[x]$.

Furthermore the entire theory of the representations of the symmetric group, including the Littlewood Richardson rule becomes much more transparent when formulated in terms of this abstract structure. Finally, this entire approach then applies immediately to give the corresponding theorems for the representation theory of the general linear groups over finite fields.

4.2 PSH algebras with one generator.

So for the rest of this section and next, $R = \bigoplus R_n$ will be a PSH algebra with exactly one primitive basis element, $p = x_1$ of degree 1. We claim that p^2 is the sum of exactly two distinct basis elements, call them x_2 and y_2:

$$p^2 = x_2 + y_2. \tag{4.2.1}$$

According to the proposition in the preceding section, once we establish (4.2.1) it will follow that x_2 and y_2 are the only basis elements of degree two. To prove (4.2.1) it is enough to prove that

$$\langle p^2, p^2 \rangle = 2,$$

since $\langle p^2, p^2 \rangle$ is the sum of the squares of the expansion of p^2 in terms of basis elements. But, p^* is a derivation since p is primitive so

$$\begin{aligned} \langle p^2, p^2 \rangle &= \langle p, p^*(p^2) \rangle \\ &= 2\langle p, p \rangle \\ &= 2 \end{aligned}$$

since p is a basis element.

Proposition 4.2.1 *For every $n \geq 2$ there exist unique basis elements, x_n and y_n characterized by*

$$y_2^* x_n = 0, \qquad x_2^* y_n = 0. \tag{4.2.2}$$

Furthermore, for $0 \leq k \leq n$ we have

$$x_k^* x_n = x_{n-k}, \qquad y_k^* y_n = y_{n-k}. \tag{4.2.3}$$

(Here we have defined $x_0 = y_0 = 1$ and $x_1 = y_1 = p$.) If ω is any basis element other than $x_0, \ldots, x_n$ then $\omega^* x_n = 0$, and similarly, if ω is any basis element other than $y_0, \ldots, y_n$ then $\omega^* y_n = 0$. In particular,

$$y_k^* x_n = 0, \qquad x_k^* y_n = 0. \tag{4.2.4}$$

The Δx_n and Δy_n are given by (4.1.1).

Proof. We know the proposition for the case $n = 2$. So assume by induction that we know it up to values less than n.

As a first step, we claim that a basis element in R_n can satisfy the first equation in (4.2.2) only if it occurs with a positive coefficient in the decomposition of px_{n-1}. Indeed, assuming the first equation in (4.2.2), we have

$$y_2^* p^* x_n = p^* y_2^* x_n = 0,$$

so $p^* x_n$ must be a non-negative integer multiple of x_{n-1}, by the uniqueness of x_{n-1} and the fact that p^* and y_2^* have non-negative integer matrix entries with respect to the preferred basis.

On the other hand, $\langle p^{n-1}, p^* x_n \rangle = \langle p^n, x_n \rangle \neq 0$ since every basis element of R_n occurs with a positive coefficient in the decomposition of p^n. Thus $\langle px_{n-1}, x_n \rangle = \langle x_{n-1}, p^* x_n \rangle > 0$.

Now

$$\begin{aligned}
\langle px_{n-1}, px_{n-1} \rangle &= \langle x_{n-1}, p^*(px_{n-1}) \rangle \\
&= \langle x_{n-1}, x_{n-1} \rangle + \langle x_{n-1}, p(p^* x_{n-1}) \rangle \\
&= 1 + \langle x_{n-1}, px_{n-2} \rangle \\
&= 1 + \langle p^* x_{n-1}, x_{n-2} \rangle \\
&= 1 + \langle x_{n-2}, x_{n-2} \rangle \\
&= 2.
\end{aligned}$$

So px_{n-1} is the sum of exactly two basis elements. We will now show that exactly one of these two basis elements is mapped into zero by y_2^*. This will prove the first assertion in the proposition.

For this purpose observe that

$$\langle \Delta x_2, p\otimes p\rangle = \langle \Delta_+ x_2, p\otimes p\rangle = \langle x_2, p^2\rangle = 1,$$

since $p^2 = x_2 + y_2$. Thus $\Delta_+ x_2 = p\otimes p$ and similarly $\Delta_+ y_2 = p\otimes p$. So

$$\Delta y_2 = y_2\otimes 1 + p\otimes p + 1\otimes y_2$$

and so by (3.5.2)

$$\begin{aligned} y_2^*(px_{n-1}) &= (y_2^*p)x_{n-1} + (p^*p)(p^*x_{n-1}) + p(y_2^*x_{n-1}) \\ &= x_{n-2} \end{aligned}$$

because $(y_2^*p) = 0$ as p is of degree 1 and y_2 is of degree 2, because $y_2^*x_{n-1} = 0$ and because $p^*x_{n-1} = x_1^*x_{n-1} = x_{n-2}$ is part of our induction hypothesis (4.2.3). Now px_{n-1} is the sum of two basis elements, and applying y_2^* to it gives the single basis element, x_{n-2}. Hence exactly one of the two summands in px_{n-1} gets sent into 0 under y_2^* and we call this element x_n. The other element is mapped onto x_{n-2}. Similarly for the second equation in (4.2.2). This completes the proof of the first statement of our proposition.

We have also proved that

$$p^*x_n = x_{n-1}, \quad p^*y_n = y_{n-1}.$$

where the conclusion for y_n follows from exactly the same argument. Hence, by induction,

$$(p^k)^*x_n = (p^*)^k x_n = x_{k-n}.$$

If we decompose p^k as a sum of basis elements, this last equation implies that exactly one of them has the property that $\omega^*x_n = x_{n-k}$, all the rest send x_n to 0.

Since every basis element of R_k occurs in the decomposition of p^k, we see that exactly one basis element satisfies $\omega^*x_n = x_{n-k}$, all the others satisfy $\omega^*x_n = 0$. We must show that this preferred element is x_k. For this it suffices to show that $x_k^*x_n \neq 0$. But

$$(p^{n-k})^*(x_k^*(x_n)) = x_k^*((p^{n-k})^*(x_n)) = x_k^*(x_k) = 1,$$

proving (4.2.3) and the assertion that $\omega \neq x_k \in R_k$ satisfies $\omega^*x_n = 0$.

Since $y_2^* x_n = 0$ while $y_2^* y_n = y_{n-2} \neq 0$ we conclude that $x_n \neq y_n$ for any $n \geq 2$ completing the proof of (4.2.4).

Finally,

$$\langle \Delta(x_n), u \otimes v \rangle = \langle x_n, uv \rangle = \langle u^* x_n, v \rangle,$$

so (4.1.1) follows from the preceding assertions in the proposition. QED

The rest of this section is devoted to the proof of

Theorem 4.2.1 *Let R be a PSH algebra with one primitive basis element, and let $\{x_n\}$ and $\{y_n\}$ be as above. Then, as a ring, R is the polynomial ring $R = \mathbf{Z}[x_1, x_2, \ldots]$ in the infinitely many indeterminates $\{x_n\}$. It is also a polynomial ring $R = \mathbf{Z}[y_1, y_2 \ldots]$ in the $\{y_n\}$, and the relation*

$$\sum_{k=0}^{n} (-1)^k x_k y_{n-k} = 0 \tag{4.2.5}$$

holds. The comultiplication is given by (4.1.1). Thus R is determined up to isomorphism.

The unique non trivial automorphism, t, of R (preserving positivity) is determined by

$$t x_n = y_n, \quad t y_n = x_n, \tag{4.2.6}$$

and the antipode of R is given by

$$T x_n = (-1)^n y_n, \quad T y_n = (-1)^n x_n. \tag{4.2.7}$$

Remarks.

1. Recall from the last chapter that every graded associative and coassociative bialgebra has an antipode which is unique within the class of graded maps. This is the meaning of the word "unique" in the theorem. **2.**If we introduce generating functions (formal power series in one variable) with coefficients in R, we can write the relations (4.2.5) as

$$X(\xi) = \frac{1}{Y(-\xi)} \tag{4.2.8}$$

where

$$X(\xi) = \sum_k x_k \xi^k, \quad Y(\xi) = \sum_k y_k \xi^k.$$

3. The relations (4.2.5) follow immediately from (4.2.7) and (4.1.1). Indeed, the definition of antipode requires that

$$m \circ (id \otimes T) \circ \Delta(u) = 0, \quad \forall u \in I,$$

in particular for $u = x_n$. But applied to x_n with (4.1.1) to compute Δx_n and using (4.2.7) gives (4.2.5).

So our first task will be to prove (4.2.7). For any primitive element,r, the definition of T implies that $Tr = -r$. Indeed

$$\Delta(r) = r \otimes 1 + 1 \otimes r$$

so

$$(m \circ (1 \otimes T) \circ \Delta)(r) = r + Tr = u(\epsilon(r)) = 0$$

since $\epsilon(r) = 0$ as r is of positive degree.

So if we define t by $t_n = (-1)^n T_n$ we see that $t(p) = p$ where p is our unique basic element. From Proposition 3.6.1 we know that T and hence t is an algebra morphism,

$$t(p^n) = p^n.$$

Now an element, $\omega \in R_n$ is a basis element if and only if $\langle \omega, \omega \rangle = 1$ and $\langle \omega, p^n \rangle > 0$. From the above it follows that if ω satisfies these two equations, so does $t(\omega)$. In other words, t is an automorphism of R as a PSH algebra.

Since x_2 has positive degree, $m \circ (id \otimes T) \circ \Delta\ (x_2) = 0$. But

$$\Delta x_2 = x_2 \otimes 1 + p \otimes p + 1 \otimes x_2$$

and $p \cdot Tp = -p^2 = -(x_2 + y_2)$. We conclude that

$$t(x_2) = T(x_2) = y_2, \quad t(y_2) = T(y_2) = x_2.$$

Now let u, v, w be elements of R. We have

$$\begin{aligned} \langle v, t(u)^* w \rangle &= \langle vt(u), w \rangle \\ &= \langle t(v)u, t(w) \rangle \\ &= \langle t(v), u^* t(w) \rangle \\ &= \langle v, t[u^* t(w)] \rangle. \end{aligned}$$

In other words, (since $t = t^{-1}$)

$$(t(u))^* = t \circ u^* \circ t^{-1}.$$

In particular,

$$(x_2^* \circ t)(x_n) = t \circ y_2^* x_n = 0,$$

so $t(x_n) = y_n$ and $t(y_n) = x_n$.

Notice that if a is any automorphism of R as a PSH algebra, we must have $a(p) = p$ and either $a(x_2) = y_2$, $a(y_2) = x_2$ or $a(x_2) = x_2$, $a(y_2) = y_2$. Hence, by the above argument, either $a(x_n) = y_n, a(y_n) = x_n$ for all n or $a(x_n) = x_n$, $a(y_n) = y_n$ for all n. Once we know that $R = \mathbf{Z}[x_1, x_2, \ldots]$ these two alternatives imply that either $a = t$ or $a = id$.

Notice also that we can use (4.2.5) to recursively express all the y's in terms of the x's or vice versa. So what remains is to prove that the monomials in the x's form a basis of R over $\mathbf{Z}$. We will write the monomial $x_1^{m_1} x_2^{m_2} \cdots x_k^{m_k}$ as x_λ where $\lambda = 1^{m_1} 2^{m_2} \cdots k^{m_k}$ is the partition of $|\lambda| = m_1 + 2m_2 + \cdots + km_k$ with m_1 parts of size 1, m_2 parts of size 2 etc. We use the notation $l(\lambda) = m_1 + m_2 + \cdots + m_k$ for the length of the partition. We wish to show that the x_λ with $|\lambda| = n$ form a $\mathbf{Z}$ basis of R_n.

Let M be a $\mathbf{Z}$ module with a preferred basis, and let $\langle , \rangle$ be the corresponding $\mathbf{Z}$-valued symmetric bilinear form.

Lemma 4.2.1 *The elements $e_1, \cdots e_p$ of M form a basis for a $\mathbf{Z}$ submodule spanned by a subset of the preferred basis if and only if*

$$det(\langle e_i, e_j \rangle)_{i.j=1,\ldots,p} = 1. \tag{4.2.9}$$

Proof. Suppose that $\omega_1, \ldots, \omega_p$ be the preferred basis elements which span the submodule. Then the change of basis matrix, A, has integer entries and as does its inverse, so $\det A = \pm 1$. So $\det(\langle e_i, e_j \rangle = \det AA^\dagger = (\det A)^2 = 1$.

Conversely, suppose that (4.2.9) holds, so the e_i are linearly independent. Let N be the submodule they generate, and $N^\perp$ the orthogonal complement of N in M. Given any $x \in M$, there will be a unique solution in integers to the equations

$$\langle \sum a_i e_i, e_j \rangle = \langle x, e_j \rangle.$$

This means that we have the direct sum decomposition

$$M = N \oplus N^\perp.$$

We must show that N is spanned by the elements of the preferred basis that it contains. So let ω be any basis element. We have

$$\begin{aligned} 1 &= \langle \omega, \omega \rangle \\ &= \langle \omega_N + \omega_{N^\perp}, \omega_N + \omega_{N^\perp} \rangle \\ &= \langle \omega_N, \omega_N \rangle + \langle \omega_{N^\perp}, \omega_{N^\perp} \rangle. \end{aligned}$$

So every basis element either belongs to N or to $N^\perp$. QED

To complete the proof of the theorem it is enough to prove that

$$\det(\langle x_\mu, x_\lambda \rangle) = 1 \qquad (4.2.10)$$

as μ and λ rangle over all partitions of n. Indeed, by the lemma they span a submodule, and if they don't span the whole of R_n there must be a basis element orthogonal to all the x_λ. But taking $\lambda = 1^n$ we get $x_\lambda = x_1^n = p^n$, and we have already proved that $\langle w, p^n \rangle > 0$ for every basis element in R_n. So we must prove (4.2.10).

We have proved that the y_j can be expressed in terms of the x_i, and hence the y_μ can be expressed as linear combinations of the x_λ. Let $\mu^\sim$ denote the partition dual to μ. Then there are integers $a_{\nu\lambda}$ such that

$$y_{\nu^\sim} = \sum a_{\nu\lambda} x_\lambda,$$

and so

$$\det(\langle x_\lambda, y_{\mu^\sim} \rangle) = \det(\langle x_\mu, x_\lambda \rangle) \det(a_{\nu\mu}).$$

We will prove that the matrix entering into the determinant on the left is upper triangular with ones on the diagonal, when we order the partitions lexicographically. This means that the product of the two determinants on the right (which are integers) is one. So each determinant is ± 1. But the determinant of a matrix of scalar products is non-negative (it is the length squared of the corresponding exterior product). Hence we will be finished once we prove

Proposition 4.2.2 *Let λ and μ be two partitions of n. If μ is lexicographically higher than $\lambda^\sim$ then*

$$y_\mu^* x_\lambda = 0,$$

and

$$y_{\lambda^\sim}^* x_\lambda = 1.$$

Recall that $y_j^* x_k = 0$ if $j > 1$ and $y_1^* x_k = x_{k-1}$. So $y_j^* x_\lambda = 0$ if $j > l(\lambda)$ while $y_{l(\lambda)}^* x_\lambda$ replaces each factor, x_k in x_λ by x_{k-1}. This is just a repeated application of (3.5.2) and (4.1.1). To illustrate, $y_5^*(x_4 x_3^2 x_1) = 0$ since $x_4 x_3^2 x_1$ has only four factors, while $y_4^* x_4 x_3^2 x_1 = x_3 x_2^2$. In terms of diagrams, $y_j^* x_\lambda = 0$ if j is bigger than the number of rows, l of λ, while $y_l x_\lambda = x_{\lambda^\leftarrow}$ where the diagram of $\lambda^\leftarrow$ is obtained from the diagram of λ by removing the leftmost column. So proceeding inductively, using the rows of μ to peel off the columns of λ (or get zero) proves the proposition, and completes the proof of the theorem.

As a corollary to the method of proof we can determine all the primitive elements of R:

Proposition 4.2.3 *For every $n \geq 1$ there exists a unique primitive element, $z_n \in R_n$ satisfying*

$$\langle z_n, x_n \rangle = 1.$$

The set of primitive elements in R_n consists of all integer multiples of z_n.

Proof. By (3.5.1) and the theorem, we know that an element of R_n is primitive if and only if it is orthogonal to all the $x_\lambda, |\lambda| = n$ except x_n. Any such element will be an integer multiple of the one with the least positive value of $\langle z, x_n \rangle$. Since the determinant in (4.2.10) is 1, this least possible value is 1. QED

4.3 Relations between x, y and z.

We introduce the notation z_λ for monomials in $\{z_1, z_2, \ldots\}$ similar to x_λ and y_λ and the generating function

$$Z(\xi) = \sum_{n \geq 0} z_{n+1} \xi^n.$$

We have already remarked that

$$X(\xi) = \frac{1}{Y(-\xi)}.$$

We claim that

$$Z(\xi) = \frac{d}{d\xi} \ln X(\xi) = \frac{\frac{d}{d\xi} X(\xi)}{X(\xi)}. \tag{4.3.1}$$

Proof. The power series on the right of (4.3.1) makes sense since the constant term in X is 1. Let us temporarily denote it by

$$W(\xi) = \sum w_{n+1} \xi^n.$$

So $w_n \in R_n$. We wish to prove that w_n is primitive and that $\langle w_n, x_n \rangle = 1$. We prove each of these assertions by using the following lemma whose proof is obvious:

Lemma 4.3.1 *If $\phi : A \to B$ is a ring homomorphism, the induced homomorphism of the power series rings $A[[\xi]] \to B[[\xi]]$ (also denoted by ϕ) commutes with the operation of logarithmic derivative.*

To prove the primitivity of the w_n we will apply the lemma to $\Delta : R \to R \otimes R$. If $U = \sum u_n \xi^n \in R[[\xi]]$, we let U_1 and U_2 denote the elements of $(R \otimes R)[[\xi]]$ given by

$$U_1(\xi) = \sum (u_n \otimes 1)\xi^n, \quad U_2(\xi) = \sum (1 \otimes u_n)\xi^n.$$

So, in this notation we can write (4.1.1) as

$$\Delta X(\xi) = X_1(\xi)X_2(\xi), \quad \Delta Y(\xi) = Y_1(\xi)Y_2(\xi). \tag{4.3.2}$$

So

$$\begin{aligned} \Delta W(\xi) &= \Delta \frac{d}{d\xi} \ln X(\xi) \\ &= \frac{d}{d\xi} \ln \Delta X(\xi) \\ &= \frac{d}{d\xi} \ln(X_1(\xi)X_2(\xi)) \\ &= \frac{d}{d\xi} [\ln X_1(\xi) + \ln X_2(\xi)] \\ &= W_1(\xi) + W_2(\xi), \end{aligned}$$

so all the coefficients of W are primitive.

Next consider the map $\delta_x : R \to \mathbf{Z}$ defined on each R_n by

$$\delta_{x|R_n} = x_n^* = \langle x_n, \cdot \rangle \tag{4.3.3}$$

and extended by linearity. According to (4.1.1) and (3.5.2), if $a \in R_k, b \in R_l$, $k + l = n$ we have

$$x_n^*(ab) = \sum_{i+j=n} x_i^* a \cdot x_j^* b = x_k^* a \cdot x_l^* b,$$

so we see that

$$\delta_x(ab) = \delta_x(a)\delta_x(b).$$

In other words $\delta_x : R \to \mathbf{Z}$ is a ring homomorphism. Furthermore, by definition

$$\delta_x X(\xi) = \sum \xi^n = \frac{1}{1 - \xi}.$$

Therefore

$$\delta_x W(\xi) = \frac{d}{d\xi} \ln[(1 - \xi)^{-1}] = \frac{1}{1 - \xi}.$$

This says that $\langle x_n, w_n\rangle = 1$ for all n, in other words $W = Z$.

For any primitive element, r, in any Hopf algebra we have $Tr = -r$ by the definition of the antipode, as we have already remarked. So $t(z_n) = (-1)^{n-1} z_n$ and we have the formulas

$$t(x_\lambda) = y_\lambda, \quad t(y_\lambda) = x_\lambda, \quad t(z_\lambda) = (-1)^{|\lambda| - l(\lambda)} z_\lambda. \tag{4.3.4}$$

(Recall that $(-1)^{|\lambda|-l(\lambda)}$ is the sign of the permutation with cyclic structure λ.)

Since the z_n are primitive, the operators z_n^* are derivations. We claim that they act on generators by

$$z_k^* x_n = x_{n-k} \quad z_k^* y_n = (-1)^{n-1} y_{n-k}. \tag{4.3.5}$$

As the second of these equations follows from the first by the application of t, it suffices to prove the first. It follows from Proposition 12 that for any $u \in R_k$, $u^* x_n = c x_{n-k}$. We must show that $c = 1$ for $u = z_k$. But

$$c = x_{n-k}^* z_k^* x_n = z_k^* x_{n-k}^* x_n = z_k^* x_k = 1.$$

It follows from (4.3.5) and the fact that z_n^* is a derivation that

$$\langle z_n, x_\lambda\rangle = \langle 1, z_n^* x_\lambda\rangle = 0,$$

unless $\lambda = (n)$, so if we expand z_n in terms of the basis $\{x_\lambda\}$:

$$z_n = \sum_{|\lambda|=n} a_\lambda x_\lambda,$$

then $\langle z_n, z_n\rangle$ is $a_{(n)}$, the coefficient of x_n in this expansion. Comparing the coefficient of

$$\sum z_n \xi^{n-1} = (\sum n x_n \xi^{n-1})(1 + x_1\xi + \cdots)^{-1}$$

shows that this coefficient is n. Thus

$$\langle z_n, z_n\rangle = n. \tag{4.3.6}$$

We also see from this argument that we can express x_λ in terms of the z_λ if we allow division by integers, i.e *the* $\{z_\lambda\}$ *form a basis of* $R \otimes \mathbf{Q}$.

To understand this assertion in terms of the model for R consisting of the representation theory of the symmetric groups, observe that $R_n \otimes \mathbf{Q}$ in this case is the space of class functions (functions constant on conjugacy classes) with values in $\mathbf{Q}$ on S_n. Under this correspondence between representations and their characters,the maps $\Delta_{k,l}$ become the restriction of functions from S_n to $S_k \times S_l$. The function which is one on the cycles of length n and zero on all other cycles clearly has the property that it goes to zero under $\Delta_{k,l}$ if $0 < k < n$ and so is primitive. They form a system of generators for $R \otimes \mathbf{Q}$.

Also notice that if $k < n$, then $\langle a, x_k^* z_n \rangle = \langle a x_k, z_n \rangle = 0$ by (3.5.1), and similarly for y_k. So

$$x_k^* z_n = 0, \quad y_k^* z_n = 0, \qquad 0 < k < n. \tag{4.3.7}$$

Let $\lambda = (\lambda_1, \ldots, \lambda_{l(\lambda)})$ and $\mu = (\mu_1, \ldots, \mu_{l(\mu)})$ be two partitions. We let $M^{\lambda,\mu}$ be the set of all matrices with $l(\lambda)$ rows and $l(\mu)$ columns with non-negative integer entries such that the sum of the $i-$th row equals λ_i and the sum of the $j-$th column equals μ_j. Let $M_{01}^{\lambda,\mu} \subset M^{\lambda,\mu}$ be the subset consisting of matrices whose entries are only zero's or one's. Let $M_\delta^{\lambda,\mu} \subset M^{\lambda,\mu}$ denote the subset consisting of those matrices with exactly one non-zero entry in each row. Then

$$\langle x_\lambda, x_\mu \rangle = \langle y_\lambda, y_\mu \rangle = \# M^{\lambda,\mu}, \tag{4.3.8}$$

while

$$\langle x_\lambda, y_\mu \rangle = \# M_{01}^{\lambda,\mu}, \tag{4.3.9}$$

and

$$\langle x_\mu, z_\lambda \rangle = \# M_\delta^{\lambda,\mu}, \quad \langle y_\mu, z_\lambda \rangle = (-1)^{|\lambda| - l(\lambda)} \# M_\delta^{\lambda,\mu}. \tag{4.3.10}$$

Furthermore $\langle z_\mu, z_\lambda \rangle = 0$ unless $\lambda = \mu$ and

$$\langle z_\lambda, z_\lambda \rangle = 1^{m_1} m_1! 2^{m_2} m_2! \cdots, \tag{4.3.11}$$

where m_k denotes the number of parts of size k in v.

To prove (4.3.8) observe that (4.1.1) and (3.5.2) imply that

$$x_{\lambda_1}^* x_\mu = \sum x_{\mu_1 - k_1} x_{\mu_2 - k_2} \cdots x_{\mu_{l(\mu)} - k_{l(\mu)}}$$

where the sum ranges over all non-negative integers $\{k_i\}$ with $\sum k_i = \lambda_1$. So throwing away the terms with $\mu_i - k_i$ negative, this sum is over all possible first rows of matrices in $M^{\lambda,\mu}$. Successive application of this rule gives (4.3.8). To prove (4.3.9) we proceed the same way except that now the sum occurring in

$$y_{\lambda_1}^* x_\mu = \sum x_{\mu_1 - k_1} x_{\mu_2 - k_2} \cdots x_{\mu_{l(\mu)} - k_{l(\mu)}}$$

is restricted also by $k_i = 0, 1$ since $y_r^* x_n = 0$ if $r > 1$. Since $z_{\lambda_1}^*$ is a derivation, the corresponding sum is limited to where one of the k_i equals λ_1 and all the others vanish. This proves the first equation in (4.3.10) and the second follows by an application of t.

To deal with $\langle z_\lambda, z_\mu \rangle$ we make use of the fact that the z_n are primitive and use the following:

Lemma 4.3.2 *Let $p_1, \ldots, p_q$ be mutually orthogonal primitive elements in a self adjoint bialgebra. Let $u = p_{i_1} p_{i_2} \cdots p_{i_r}$, $v = p_{j_1} p_{j_2} \cdots p_{j_s}$ be products of the primitive elements (repetitions allowed). Then $\langle u, v \rangle = 0$ unless $r = s$ and each primitive element, p_i occurs with the same multiplicity, m_i in the product decompositions of u and v. If this happens then*

$$\langle u, v \rangle = m_1! \langle p_1, p_1 \rangle^{m_1} m_2! \langle p_2, p_2 \rangle^{m_2} \cdots.$$

Proof of Lemma. Since $p_{i_1}^*$ is a derivation, and $\langle p_i, p_j \rangle = 0, i \neq j$ we conclude that

$$p_{i_1}^* v = m_{i_1} v'$$

where $m_i = 0$ if p_i does not occur in the decomposition of v and equal to the multiplicity of p_i if it does occur, and where v' is v with one factor of p_{i_1} deleted. Induction proves the lemma.

Since the z_j with differing j belong to different R_j they are orthogonal, and since $\langle z_n, z_n \rangle = n$, the lemma gives (4.3.11).

4.4 Irreducible elements

If $\mu = \lambda^\sim$, then $M_{01}^{\lambda,\mu}$ consists of matrices of zeros and ones whose i−th row sum equal the size of the i−th row of λ and j−th column sum equals the j−th column of λ. There is clearly only one such matrix, with ones in the positions given by the standard diagram of λ and zeros elsewhere. Thus, by (4.3.9),

$$y_{\lambda^\sim}^* x_\lambda = 1.$$

But the operator $y_{\lambda^\sim}^*$ takes basis elements into non- negative integer combinations of basis elements. Thus the only way that the preceding equation can occur is that there is a unique basis element in the decomposition of x_λ which is not sent to 0 by $y_{\lambda^\sim}^*$, that this element is sent to 1, and that this element occurs with multiplicity one in the decomposition of x_λ. In other words, we have proved the first part of

Proposition 4.4.1 *For any partition, λ there exists a unique basis element, call it $\{\lambda\}$, in $R_{|\lambda|}$ such that $x^*_\lambda\{\lambda\} \neq 0, y^*_{\lambda^\sim}\{\lambda\} \neq 0$. This element satisfies*

$$x^*_\lambda\{\lambda\} = y^*_{\lambda^\sim}\{\lambda\} = 1. \tag{4.4.1}$$

The map $\lambda \mapsto \{\lambda\}$ is bijective and the $\{\lambda\}$ exhaust all the basis elements of R. The automorphism, t, acts on $\{\lambda\}$ by

$$t(\{\lambda\} = \{\lambda^\sim\}.$$

To show that $\lambda \neq \mu \Rightarrow \{\lambda\} \neq \{\mu\}$ we may assume that $\mu^\sim > \lambda^\sim$ in the lexicographic order and hence $y^*_{\mu^\sim} x_\lambda = 0$ so certainly $y^*_{\mu^\sim}\{\lambda\} = 0$ since the entries of $y^*_{\mu^\sim}$ are non-negative with respect to the preferred basis. The rank of R_n over $\mathbf{Z}$ is equal to the number of partitions of n, since the monomials x_λ form a $\mathbf{Z}$ basis. This proves that the $\{\lambda\}$ exhaust the entire basis. The last statement is an immediate consequence of (4.2.6). QED .

For the model of R as representations of the symmetric groups, the basis of R_n consists of the the irreducible representations of S_n. So we have shown that every irreducible representation of S_n corresponds to a diagram λ with $|\lambda| = n$ according to the Proposition. In view of the last equation in the proposition, we shall use the notation λ^t (for transpose) instead of $\lambda^\sim$ so we can write the last equation in the proposition as

$$t(\{\lambda\}) = \{\lambda^t\}. \tag{4.4.2}$$

4.5 Actions on basis elements.

We begin with

Proposition 4.5.1 . *We have*

$$x^*_i\{\lambda\} = \sum_{|\lambda|-|\mu|=i} a_{\lambda,\mu}\{\mu\},$$

where $a(\lambda,\mu) = 1$ if μ can be obtained from λ by removing at most one box from each column of λ, and $a(\lambda,\mu) = 0$ otherwise.

Let us introduce the notation $\mu \Downarrow \lambda$ for the assertion that "μ can be obtained from λ by removing at most one box from each column of λ". So we can write the proposition as

$$x^*_i\{\lambda\} = \sum_{|\lambda|-|\mu|=i,\ \ \mu\Downarrow\lambda} \{\mu\}. \tag{4.5.1}$$

We will also use the notation $\overleftarrow{\lambda}$ for the diagram λ with the left hand column removed (and shifted one unit to the left). Thus

$$\mu \Downarrow \lambda \iff \overleftarrow{\mu} \Downarrow \overleftarrow{\lambda} \;\; \& \;\; l(\mu) \leq l(\lambda) \leq l(\mu) + 1. \tag{4.5.2}$$

Let us temporarily introduce the notation $a_i(\lambda, \mu)$ for the coefficients (as yet uncalculated) that occur on the right hand side of the equation in the proposition. So $a_i(\lambda, \mu) > 0$ if and only if μ occurs with positive coefficient in the decomposition of $x_i^*\{\lambda\}$, which is the same as saying that $\{\lambda\}$ occurs with positive coefficient in the decomposition of $x_i\{\mu\}$.

Lemma 4.5.1 *If $a_i(\lambda, \mu) > 0$ then $l(\mu) \leq l(\lambda) \leq l(\mu) + 1$.*

Proof. If $l > l(\lambda)$ then $y_l^* x_i^* \{\lambda\} = x_i^* y_l^* \{\lambda\} = 0$, since $y_l^* x_\lambda = 0$. As the operator y_l^* has non-negative entries relative to the basis, this implies that $y_l^*\{\mu\} = 0$ if μ occurs with positive coefficient in the decomposition of $x_i^*\{\lambda\}$. So $l(\mu) \leq l(\lambda)$. On the other hand, if $l > l(\mu) + 1$ then

$$y_l^*[(x_i\{\mu\})] = x_i y_l^*\{\mu\} + x_{i-1} y_{l-1}^*\{\mu\} = 0.$$

So if $\{\lambda\}$ occurs with positive coefficient in the decomposition of $x_i\{\mu\}$, then $y_l\{\lambda\} = 0$, so $l(\lambda) \leq l(\mu) + 1$. QED

Lemma 4.5.2 *We have*

$$y_{l(\lambda)}^*\{\lambda\} = \{\overleftarrow{\lambda}\}. \tag{4.5.3}$$

Proof. We know from (4.1.1), (4.2.4) and a repeated application of (3.5.2) that $y_{l(\lambda)}^* x_\lambda = x_{\overleftarrow{\lambda}}$. Since $\{\lambda\}$ is a basis element occurring in the decomposition of x_λ, it follows that any basis element, ω, occurring in the decomposition of $y_{l(\lambda)}^*\{\lambda\}$ must occur in $x_{\overleftarrow{\lambda}}$ hence

$$x_{\overleftarrow{\lambda}}^* \omega \neq 0.$$

But

$$y_{(\overleftarrow{\lambda})^t}^* y_{l(\lambda)}^* \{\lambda\} = y_{\lambda^t}^*\{\lambda\} = 1.$$

So $\{\overleftarrow{\lambda}\}$ is a basis element which does occur with a non zero coefficient in the decomposition of $y_{l(\lambda)}^*\{\lambda\}$. So it is enough to show that $y_{l(\lambda)}^*\{\lambda\}$ is itself a basis element, i.e. that $\langle y_{l(\lambda)}^*\{\lambda\}, y_{l(\lambda)}^*\{\lambda\}\rangle = 1$. Since $x_{\overleftarrow{\lambda}}$ is a positive combination of basis elements, including $y^*_{l(\lambda)}\{\lambda\}$, it is more than enough to prove that

$$\langle x_{\bar{\lambda}}, y^*_{l(\lambda)}\{\lambda\}\rangle = 1,$$

or that

$$x^*_{\bar{\lambda}}\, y^*_{l(\lambda)}\{\lambda\} = 1. \tag{4.5.4}$$

We prove this using the following lemma:

Lemma 4.5.3 *Let p and q be positive integers, and let $v \in R$. Suppose that $x^*_l v = 0$ for all $l > p$ and $y^*_m v = 0$ for all $m > q$. Then*

$$x^*_p y^*_q v = 0,$$

and

$$x^*_{p-1} y^*_q v = x^*_p y^*_{q-1} v.$$

Proof. From the fundamental relation (4.2.5) we have

$$\sum_{k=0}^{n}(-1)^k x^*_k y^*_{n-k} = 0.$$

Applied to v with $p + q = n$ gives the first equation in the lemma while $n = p + q - 1$ gives the second equation. QED

Let $\lambda_1 \geq \lambda_2 \geq \cdots \geq \lambda_{l(\lambda)}$ be the rows of λ and define inductively $v_0 = \{\lambda\}$, $v_1 = x^*_{\lambda_1}\{\lambda\}$, and

$$v_k = x^*_{\lambda_k} v_{k-1}.$$

If we had defined $w_0 = y_{\lambda^t}, w_1 = x^*_{\lambda_1} y_{\lambda^t}$. and

$$w_k = x^*_{\lambda_k} w_{k-1},$$

then

$$w_k = y_{\mu^t_k}$$

where μ_k is the diagram obtained by stripping off the top k rows of λ. In particular,

$$x^*_r w_k = 0, \quad \forall r > \lambda_{k+1}.$$

But since $\{\lambda\}$ is a basis element which occurs with coefficient one in the expansion of y_{λ^t}, it then follows from the non-negativity of all the operators x^*_i that

$$x^*_r v_k = 0, \quad \forall r > \lambda_{k+1}. \tag{4.5.5}$$

We claim that also

$$y_s^* v_k = 0 \quad \forall s > l(\lambda) - k. \tag{4.5.6}$$

Indeed, the case $k = 0$ says that $y_s^*\{\lambda\} = 0$ if $s > l(\lambda)$ which follows from the fact that $y_s^* x_\lambda = 0$ if $s > l(\lambda)$. So suppose that we have proved that

$$y_s^* v_{k-1} = 0, \quad \forall s > l(\lambda) - k + 1.$$

Then $y_s^* v_k = 0$ for $s > l(\lambda) - k + 1$ since y_s^* and $x_{\lambda_k}^*$ commute. So we must show that

$$x_{\lambda_k}^* y_{l(\lambda)-k+1}^* v_{k-1} = 0.$$

But this now follows from the lemma applied to $v = v_{k-1}$.

We now apply the second equation in Lemma 6 to $v = v_k$ to get

$$x_{\lambda_{k+1}-1}^* y_{l(\lambda)-k}^* v_k = y_{l(\lambda)-k-1}^* x_{\lambda_{k+1}}^* v_k = y_{l(\lambda)-k-1}^* v_{k+1}.$$

Applying these inequalities for $k = 0, 1, \ldots, l(\lambda) - 1$ gives

$$\begin{aligned} x_{\overleftarrow{\{\lambda\}}}^* y_{l(\lambda)}^* \{\lambda\} &= v_{l(\lambda)} \\ &= x_\lambda^* \{\lambda\} \\ &= 1, \end{aligned}$$

proving (4.5.4).

We recall that we are temporarily using the notation $a_i(\lambda, \mu)$ for the coefficient of $\{\mu\}$ in the expansion of $x_i^*\{\lambda\}$.

Lemma 4.5.4 *If $l(\lambda) = l(\mu)$ then*

$$a_i(\lambda, \mu) = a_i(\overleftarrow{\lambda}, \overleftarrow{\mu}),$$

while if $l(\lambda) = l(\mu) + 1$ then

$$a_i(\lambda, \mu) = a_{i-1}(\overleftarrow{\lambda}, \overleftarrow{\mu}).$$

Proof. Set $l = l(\lambda)$. Then $y_l^*\{\lambda\} = \{\overleftarrow{\lambda}\}$ so

$$x_i^*\{\overleftarrow{\lambda}\} = x_i^* y_l^* \{\lambda\} = \sum_\mu a_i(\lambda, \mu) y_l^* \{\mu\}.$$

But recall from Lemma 4 that the $a_i(\lambda, \mu) = 0$ unless $l(\mu) \leq l(\lambda)$ while $y_l^*\{\mu\} = 0$ if $l > l(\mu)$ so

$$x_i^*\{\overleftarrow{\lambda}\} = \sum_{l(\mu)=l(\lambda)} a_i(\lambda, \mu)\{\overleftarrow{\mu}\},$$

which gives the first part of the lemma. Now let $l = l(\mu) + 1$. Then

$$y_l^*(x_i\{\mu\}) = x_{i-1} \cdot \{\overleftarrow{\mu}\}.$$

So

$$x_{i-1}\{\overleftarrow{\mu}\} = \sum_\lambda a_i(\lambda, \mu) \cdot y_l^*\{\lambda\}.$$

From Lemma 4 we know that if $a_i(\lambda, \mu) > 0$ then $l(\lambda) \leq l$, while $y_l^*\{\lambda\} = 0$ if $l(\lambda) < l$. So

$$x_{i-1} \cdot \{\overleftarrow{\mu}\} = \sum_{l(\lambda)=l} a_i(\lambda, \mu) \cdot y_l^*\{\lambda\}.$$

This proves the second part of the lemma.

But now we can complete the proof of the proposition: If $a_i(\lambda, \mu) > 0$ then either $l(\lambda) = l(\mu)$ which means that no boxes have been removed from the first column of λ, and then $a_i(\lambda, \mu) = a_i(\overleftarrow{\lambda}, \overleftarrow{\mu})$ which, by induction we may assume is given by the proposition, or $l(\lambda) = l(\mu) + 1$ which means that a box has been removed from the first column of λ to obtain μ. But then $a_i(\lambda, \mu) = a_{i-1}(\overleftarrow{\lambda}, \overleftarrow{\mu})$ and once again the proposition follows by induction, since we have used up one box that we have removed from the first column.

Let $\mu \Leftarrow \lambda$ mean that μ is obtained from λ by removing at most one box from each row. Applying the automorphism, t, to (4.5.1) gives

$$y_i^*\{\lambda\} = \sum_{|\lambda|-|\mu|=i,\ \mu \Leftarrow \lambda} \{\mu\}. \tag{4.5.7}$$

Introduce the operators

$$X^* = \sum_{i=0}^{\infty} x_i^*,\ Y^* = \sum_{i=0}^{\infty} y_i^*. \tag{4.5.8}$$

Because

$$x_n^*(ab) = \sum_{k+l=n} x_k^*(a) x_l^*(b)$$

it follows that X^*, and similarly Y^* are algebra homomorphisms.

Lemma 4.5.5 *For all $n \geq 1$ the maps $X^* - 1$ and $Y^* - 1$ are injections of R_n into $\bigoplus_{k<n} R_k$.*

This is because the map $\bigoplus_{k<n} R_k \to R_n$ which is given by multiplication by x_k on R_{n-k} is a surjection. (Similarly for y.) We also have the formulas

$$(X^* - 1)\{\lambda\} = \sum_{\mu \Downarrow \lambda,\ \mu \neq \lambda} \{\mu\} \tag{4.5.9}$$

and

$$(Y^* - 1)\{\lambda\} = \sum_{\mu \Leftarrow \lambda,\ \mu \neq \lambda} \{\mu\}. \tag{4.5.10}$$

4.6 The Littlewood Richardson rule.

We wish to compute the expression for m in terms of the irreducible basis. That is, we wish to compute the coefficients, $g_{\lambda,\mu.\nu}$ in

$$\{\mu\} \cdot \{\nu\} = \sum_{\lambda} g_{\lambda,\mu.\nu}\{\lambda\}. \tag{4.6.1}$$

So

$$g_{\lambda,\mu.\nu} = \langle \{\mu\} \cdot \{\nu\}, \{\lambda\} \rangle.$$

If we define

$$[\lambda \setminus \mu] = \{\mu\}^*\{\lambda\}$$

then

$$g_{\lambda,\mu.\nu} = \langle \{\nu\}, [\lambda \setminus \mu] \rangle. \tag{4.6.2}$$

Also, by selfadjointness,

$$\Delta\lambda = \sum g_{\lambda,\mu.\nu}\{\mu\} \otimes \{\nu\}. \tag{4.6.3}$$

We will see that the equations (4.5.9) and (4.5.10) give a recursive computation of the $g_{\lambda,\mu.\nu}$ starting from the obvious fact that

$$g_{\mu,\mu.\emptyset} = 1, \quad g_{\mu,\mu.\nu} = 0\ \forall \nu \neq \emptyset. \tag{4.6.4}$$

As a beginning and as an illustration let us prove

Proposition 4.6.1 *The element $[\lambda \setminus \mu]$ is non-zero if and only if $\lambda \supset \mu$ in which case it depends only on the shape of the set $\lambda \setminus \mu$, i.e. it is invariant under a shift of λ and μ by a lattice vector.*

Proof. We have

$$\begin{aligned}(X^* - 1)[\lambda \setminus \mu] &= (X^* - 1)\{\mu\}^*\{\lambda\} \\ &= \{\mu\}^*(X^* - 1)\{\lambda\} \\ &= \{\mu\}^* \sum \{\lambda'\} \\ &= \sum [\lambda' \setminus \mu],\end{aligned}$$

where the sum ranges over all $\lambda' \Downarrow \lambda$, $\lambda' \neq \lambda$. By induction (on, say, $|\lambda|$) we may assume that the proposition is true for all the λ' occurring in the sum. By the injectivity of $X^* - 1$ on $R_n, n > 0$, this proves the proposition except when $|\mu| = |\lambda|$. But in this case we already know that the $\{\lambda\}$'s form and orthonormal base. QED

Now let us apply the operator $Y^* - 1$ to both sides of the equation

$$\{\mu\}^*\{\lambda\} = \sum g_{\lambda,\mu.\nu}\{\nu\}.$$

On the left we get

$$\{\mu\}^*(Y^* - 1)\{\lambda\} = \sum g_{\lambda',\mu.\nu}\ \{\nu\},$$

where the sum is over all $\lambda' \Leftarrow \lambda, \lambda' \neq \lambda$ and over all ν. On the right we get

$$\sum g_{\lambda,\mu.\nu} \sum \{\nu'\},$$

where the sum is over all $\nu' \Leftarrow \nu, \nu' \neq \nu$. So we get the equation

$$\sum g_{\lambda',\mu.\nu}\ \{\nu\} = \sum g_{\lambda,\mu.\nu} \sum \{\nu'\}. \tag{4.6.5}$$

We claim that (4.6.5) can be used to determine the coefficients $g_{\lambda,\mu.\nu}$ recursively, using the proposition and starting from (4.6.4). For example, suppose that $|\lambda| = |\mu| + 1$, the first case not covered by the proposition or by (4.6.4). So $\mu^*\lambda \in R_1$ and hence

$$\mu^*\lambda = a\{(1)\}$$

is some multiple of unique partition, $\{\nu\} = \{(1)\}$ of 1. The question is to determine the coefficient, a. By the proposition, the only possibility for λ' with $g_{\lambda',\mu.\nu} \neq 0$ is $\lambda' = \mu$ in which case the left hand side of (4.6.5) is

$$\{\emptyset\}.$$

The right hand side becomes $\{\emptyset\}$. We conclude that

$$a = 1.$$

This result is of sufficent importance to have a name of its own in the theory of the symmetric group where it is known as the *Branching rule*, cf. [1069] p.76:

If λ is a partition, then an *inner corner* of λ is a node (i, j) in the diagram of λ whose removal leaves a legitimate diagram of a partition. Any partition so obtained will be denoted by λ^-. An *outer corner* of a partition μ is a node (i, j) not in the diagram of μ whose addition produces a legitimate diagram of a partition. Any diagram so obtained will be denoted by μ^+.

Recall that in the symmetric group interpretation, $\Delta_{n-1,1}$ is just the restriction, $\downarrow_{S_{n-1}}$, from S_n to $S_{n-1} = S_{n-1} \times S_1$ while the multiplication $m : R_n \times R_1 \to R_{n+1}$ is induction, $\uparrow^{S_{n+1}}$ from S_n to S_{n+1}. Thus the result $a = 1$ above can be stated as

Proposition 4.6.2 Branching rule. *If λ is a partition of n then*

$$\{\lambda\} \downarrow_{S_{n-1}} = \bigoplus_{\lambda^-} \{\lambda^-\}$$

and

$$\{\lambda\} \uparrow^{S_{n+1}} = \bigoplus_{\lambda+} \{\lambda^+\}.$$

Let us examine what (4.6.5) says when $|\lambda| = |\mu| + 2$. So $\mu^*\lambda \in R_2$ and so

$$\{\mu\}^*\{\lambda\} = x\{(2)\} + y\{(1,1)\}$$

and we must determine the coefficients

$$x = g_{\lambda,\mu.(2)} \quad \text{and} \quad y = g_{\lambda,\mu.(1,1)}.$$

The right hand side of (4.6.5) is

$$(x + y)\{(1)\} + y\{\emptyset\}$$

since we can remove only one box from the diagram of (2) which has a single row, while we can remove one or two boxes from the diagram of (1, 1).

Now let us examine the possibilities for the left hand side: If the two additional boxes occurring in the diagram of λ do not lie in the same row or column (in other words if λ is obtained from μ by filling in two outer corners) then there are three possibilities for λ' as we can remove either or both additional boxes. Hence the left hand side of (4.6.5) is

$$\{\emptyset\} + 2\{(1)\}$$

and so

$$x = y = 1.$$

If both additional boxes lie in the same row, then there is only one λ', obtained by removing the right hand box from λ. So the left hand side of (4.6.5) is

$$\{(1)\}$$

and so

$$x = 1, \quad y = 0.$$

If both additional boxes lie in the same column, then there are two possibilities for λ': we can remove the bottom box or we can remove both boxes. So the left hand side of is

$$\{\emptyset\} + \{(1)\}$$

and hence

$$x = 0, \quad y = 1.$$

This should give the flavor of how (4.6.5) can be used recursively to determine the $g_{\lambda,\mu.\nu}$. For reformulation of this recursion formula in terms of other combinatorial statements, we refer to standard texts such as [575],[1069] or [1263].

4.7 Clebsch Gordan decompositions.

The Littlewood Richardson coefficients, $g_{\lambda,\mu.\nu}$ play an important role in the representation theory of the unitary groups. Essentially, they give the decomposition of the tensor product of any two irreducible representations of $U(d)$. It is worth briefly recalling how this goes, referring the reader to the standard texts for details, cf. for example [1132].

Let V be a finite dimensional vector space over the complex numbers and let $Gl(V)$ denote the group of invertible linear transformations of V. We let $V^{\otimes n}$ denote the n-fold tensor product of V with itself, so

$$V^{\otimes n} = V \otimes \cdots \otimes V \quad n \text{ times}.$$

The group $Gl(V)$ has a natural representation on $V^{\otimes n}$ where $A \in Gl(V)$ acts via

$$A(v_1 \otimes \cdots \otimes v_n) = Av_1 \otimes \cdots Av_n.$$

The symmetric group S_n acts via permuting the tensor positions, so $\pi \in S_n$ acts via

$$\pi(v_1 \otimes v_2 \otimes \cdots \otimes v_n) = v_{\pi^{-1}1} \otimes v_{\pi^{-1}2} \otimes \cdots \otimes v_{\pi^{-1}n}.$$

It is obvious that the action of every $\pi \in S_n$ commutes with the action of every $A \in Gl(V)$. What is not so obvious but nevertheless true is that the algebras generated by these two groups are mutual centralizers. That is, if $Op[S_n]$ denotes the algebra spanned by the π's and $Op[Gl(V)]$ the algebra spanned by the A's inside $\mathrm{End}(V^{\otimes n})$, then $Op[Gl(V)]$ is the centralizer of $Op[S_n]$ and vice versa. This has the implication that when we decompose $V^{\otimes n}$ into irreducibles under $Gl(V) \times S_n$ we get a decomposition

$$V^{\otimes n} = \bigoplus_{\lambda} U_\lambda \otimes \{\lambda\}.$$

Here the sum is over all partitions, λ , of n and $\{\lambda\}$ denotes the irreducible representation of S_n given by λ. Each U_λ is an irreducible representation of $Gl(V)$, and the representations corresponding to different λ's are inequivalent, except that $U_\lambda = (0)$ if $l(\lambda) > d$ where $d = \dim V$. If $V = \mathbf{C}^d$ and we restrict the representation U_λ to the unitary subgroup, $U(d)$, it remains irreducible, and,up to multiplying by a character of the center, S^1, these irreducibles exhaust all the irreducible representations of $U(d)$. This, in a nutshell, is the Schur-Weyl theory of the representations of the unitary groups.

Now let $n = k + l$ and consider $V^{\otimes n} = V^{\otimes k} \otimes V^{\otimes l}$. Restricting from S_n to $S_k \times S_l$ we have

$$\begin{aligned} V^{\otimes n} &= \bigoplus U_\lambda \otimes \Delta_{k,l}\{\lambda\} \\ &= \bigoplus U_\lambda \otimes \left(\bigoplus g_{\lambda,\mu.\nu}\{\mu\} \otimes \{\nu\}\right) \\ &= V^{\otimes k} \otimes V^{\otimes l} \\ &= \left(\bigoplus U_\mu \otimes \{\mu\}\right) \otimes \left(\bigoplus U_\nu \otimes \{\nu\}\right) \\ &= \bigoplus (U_\mu \otimes U_\nu) \otimes \{\mu\} \otimes \{\nu\}. \end{aligned}$$

Comparing the "coefficients" of the irreducible representation $\{\mu\} \otimes \{\nu\}$ of $S_k \times S_l$ shows that

$$U_\mu \otimes U_\nu = \bigoplus g_{\lambda,\mu.\nu} U_\lambda. \tag{4.7.1}$$

If we allow ourselves to pass to $d = \infty$ then we can express (4.7.1) as saying that the multiplication on the Grothendieck ring $R[U(\infty)]$ is dual to the comulptiplication on our Hopf algebra, R.

It is also known that the Litlewood-Richardson coefficients determine the Clebsch Gordan decomposition of S_n, [575]. By this we mean the following: Recall that R_n denote the Grothendieck ring of S_n. Tensor product induces a multiplication on R_n, which we will denote by $\times$. So

$$\times : R_n \otimes R_n \to R_n.$$

Then we are saying that the Hopf algebra multiplication (or comultiplication) on R determines the product $\times$ on R_n. Here is a beautiful proof of this fact, together with an explicit formula relating the two types of multiplication due to Liulevicius, [795](see also [237]): Recall that x_n corresponds to the trivial representation, and hence

$$x_n \times c = c \quad \forall c \in R_n. \tag{4.7.2}$$

Every other element of R_n can be written as a sum of products of lower degree elements.

Proposition 4.7.1 Liulevicius. *If $m(a,b) \in R_n$ with $a \in R_k$, then for any $c \in R_n$ we have*

$$m(a,b)\times c = \sum_{deg\ c_{(1)}=k} m(a\times c_{(1)}, b\times c_{(2)}) \tag{4.7.3}$$

where

$$\Delta c = \sum c_{(1)} \otimes c_{(2)}$$

but the sum in (4.7.3) is only over those terms where deg $c_{(1)} = k =$ deg a. In other words,

$$\times \circ (m \otimes id) = m \circ (\times \otimes \times) \circ s_{23} \circ (id \otimes \Delta_{k,l}) \tag{4.7.4}$$

as maps from

$$R_k \otimes R_l \otimes R_n \longrightarrow R_n.$$

Proof. It is enough to prove the formula for the case that a is (the equivalence class of) a representation of S_k, b of S_l and c of S_n. Frobenius reciprocity implies that for any representation, r of a subgroup H of G and any representation t of G we have

$$r \uparrow \otimes t = (r \otimes t \downarrow) \uparrow .$$

Applied to $H = S_k \times S_l$, $G = S_n, r = a \otimes b$, $t = c$ we get

$$\begin{aligned} m(a,b)\times c &= (a \otimes b) \uparrow \otimes c \\ &= ((a \otimes b)\times c \downarrow) \uparrow \\ &= (a \otimes b)\times \Delta_{k,l} c) \uparrow \\ &= \sum m(a\times c_{(1)}, b\times c_{(2)}) \end{aligned}$$

where the sum is as in the proposition. QED.

The application of formula (4.7.3) is somewhat complicated, especially as it involves expressing a given representation in terms of lower order elements. Here is a case where the formula takes on an especially simple form. Suppose we take $a = y_k, b = x_{n-k}$. (Notice that, as a representation, this is just the $k-$th exterior power of the permutation representation.) Tensoring by y_k is just the automorphism τ on R_k. Thus, cf. see [237],

Proposition 4.7.2 *The operation of tensoring by the $k-$th exterior power of the permutation representation is*

$$m \circ (\tau \otimes id) \circ \Delta_{k,l}.$$

It is easy to show that the exterior powers of the permutation representation generate the ring R_n (under the multiplication $\times$). Hence if we know the expression of a representation in terms of the exterior powers of the permutation representation, we can use this Proposition 4.7.2 instead of (4.7.3).

4.8 Symmetric functions.

Let A be a commutative ring and $\xi = \{\xi_1, \xi_2, \xi_3, \ldots\}$ a countable set of variables. We let $A[[\xi]]$ denote the corresponding power series ring. For every n there is a natural action of S_n on $A[[\xi]]$ by

$$\pi f(\xi_1, \xi_2, \xi_3, \ldots = f(\xi_{\pi 1}, \xi_{\pi 2}, \xi_{\pi 3}, \ldots)$$

where $\pi i = i$ for $i > n$. The monomial

$$\xi_{i_1}^{\lambda_1} \xi_{i_2}^{\lambda_2} \xi_{i_3}^{\lambda_3} \cdots \xi_{i_l}^{\lambda_l}$$

is said to have exponent λ and has degree n if $|\lambda| = n$. Clearly the *monomial symmetric functions*

$$u_\lambda = u_\lambda(\xi) = \sum \xi_{i_1}^{\lambda_1} \xi_{i_2}^{\lambda_2} \xi_{i_3}^{\lambda_3} \cdots \xi_{i_l}^{\lambda_l}$$

where the sum is over all distinct monomials with exponent λ is invariant under all the S_n. The ring $S[A]$ is defined to be the submodule of $A[[\xi]]$ generated by the u_λ. It is clear that the consistency conditions are satisfied.

It follows from the definitions that for $v \in R_n$,

$$p(v) = \sum x_\lambda^*(v) u_\lambda = \sum \langle x_\lambda, v \rangle u_\lambda$$

where the sum ranges over all partitions of n. In particular, we have

$$\begin{aligned} p(x_n) &= \sum_{|\lambda|=n} u_\lambda \quad \text{sum of all monomials of degree } n \\ p(y_n) &= u_{1^n} \quad \text{elementary symmetric polynomial of degree } n \\ p(z_n) &= u_{(n)} \quad \text{sum of } n\text{-th powers} \end{aligned}$$

It is easy to check that $S[A]$ is closed under multiplication, i.e. that the product of two monomial symmetric functions is a finite sum of monomial symmetric functions. So $S[A]$ is indeed a ring.

Clearly if $\phi : A \to A'$ is a ring homomorphism, then we get an induced homomorphism from $S[A]$ to $S[A']$. In this section we sketch a construction, due to Bernstein, which gives an algebra isomorphism of our R with $S[\mathbf{Z}]$. This then puts a Hopf algebra structure on the ring of symmetric functions, $S[\mathbf{Z}]$. For details we refer [1263]. Bernstein defines a homomorphism $P : R \to S[R]$ and then the homomorphism

$$\delta_x : R \to \mathbf{Z}$$

given by (4.3.3) induces a homomorphism

$$p = \delta_x \circ P : R \to S[\mathbf{Z}]$$

which is is proved to be an isomorphism. In fact, let $S[A, N]$ denote the ring of symmetric polynomials in finitely many variables $\xi_1, \ldots, \xi_N$. There is an obvious homomorphism

$$\psi_N : S[A, N'] \to S[A, N], \quad N' > N$$

obtained by sending $\xi_i \to 0, i > N$. Clearly

$$S[A] = S[A, \infty] = \varprojlim_N S[A, N]$$

identifies $S[A]$ as the inverse limit of the $S[A,N]$. So we will construct a homomorphism

$$P_N : R \to S[A,N]$$

consistent with the ψ so as to obtain the desired P.

Let $Q : R \to R[\xi]$ be the map which sends $v \in R_n$ into $v\xi^n$. More generally, let

$$Q : R \otimes R \otimes \cdots \otimes R \to R[\xi_1, \ldots \xi_N]$$

given by

$$v_1 \otimes \cdots \otimes v_N \mapsto v_1 v_2 \cdots v_N \xi_1^{n_1} \xi_2^{n_2} \cdots \xi_n^{n_N}, \; v_i \in R_{n_i} >$$

The map Q is clearly an algebra homomorphism. The map P_N is defined as the composite

$$P_N = M \circ Q \circ \Delta^N$$

where

$$\Delta^N : R \to R \otimes R \otimes \cdots \otimes R \quad N \text{ times}$$

is the $N-$ th iterate of Δ. By the associativity of the comultiplication this is well defined and by the cocommutativity the image of P_N lies in $S[A,N]$. The consistency condition is obvious. where Q is the tensor product of N copies

4.9 Representations of $Gl(n,q)$.

Let us fix a finite field $\mathbf{F}_q$ and let G_n denote the group of invertible $n \times n$ matrices over the field $\mathbf{F}_q$. Let $R_n = R_n(q)$ denote space of $\mathbf{Z}$ linear combinations of equivalence classes of irreducible representations of G_n. We have a natural embedding of $G_k \times G_l \to G_n$, $n = k + l$ Let $U_{k,l}$ be the subgroup of G_n consisting of unipotent upper triangular matrices

$$u = (u_{ij})$$

such that u_{ij} can be non-zero only if $i = j$ or

$$i \leq k < j.$$

In other words, the non-zero elements of u off the diagonal lie in the upper right hand block left over after removing the blocks corresponding to G_k and

G_l along the diagonal. Clearly $U_{k,l}$ normalizes $G_K \times G_l$. We now want to define a multiplication

$$m : R_k \otimes R_l \to R_n$$

and a comultiplication

$$\Delta_{k,l} : R_n \to R_k \otimes R_l.$$

Let r be any representation of $G_k \times G_l$. Extend it to the group

$$B_{k,l} = U_{k,l} \cdot (G_k \times G_l),$$

consisting of all block upper matrices by letting $U_{k,l}$ act trivially. Since $U_{k,l}$ normalizes $G_k \times G_l$, this makes sense, and let us denote this extended representation by r^e. We can then construct the induced representation $r^e \uparrow$, induced from $B_{k,l}$ to G_n. Applied to $r = v \otimes w,\ v \in R_k,\ w \in R_l$ gives the multiplication map.

To define the comultiplication, let r be a representation of G_n on a vector space, V. Let

$$V_{k,l} = V_{U_{k,l}} = \{v \in V | r(u)v = v \forall u \in U_{k,l}\}$$

be the subspace of $U_{k,l}$ fixed vectors. Since $U_{k,l}$ normalizes $G_k \times G_l$, the subspace $V_{k,l}$ is invariant under $G_k \times G_l$, and hence we consider the restriction,

$$V_{k,l} \downarrow_{G_k \times G_l}$$

whose equivalence class will be an element of $R_k \otimes R_l$. This defines $\Delta_{k,l}$ and hence the comultiplication

$$\Delta : R = R(q) = \bigoplus R_n(q) \to R \otimes R.$$

It is straightforward to check that the axioms for a PSH algebra are satisfied, as in the case of the symmetric group. A basic (i.e. irreducible) primitive element of $R(q)$ is called the (equivalence class of) a *cuspidal* representation. So a representation is cuspidal if and only if it contains no non-zero vectors which are invariant under any non-trivial subgroup $U_{k,l}$. In contrast to the representation theory of S_n where there was only one basic primitive element, there is now a whole family of them. But Zelevinsky proves that any PSH algebra can be decomposed into the tensor product of PSH algebras with only one basic primitive element.

More precisely, let BP denote the set of basic primitive elements in a PSH algebra, R. For any $p \in BP$ let $\Omega(p)$ denote the set of basic elements ω such that

$$\langle \omega, p^n \rangle \neq 0, \quad \text{for some } n \geq 0.$$

Let $R(p)$ denote the space of $\mathbf{Z}$ linear combinations of the $\omega \in \Omega(p)$:

$$R(p) = \bigoplus_{\omega \in \Omega(p)} \mathbf{Z}\omega.$$

Then Zelevinky proves, [1263] p.22

Proposition 4.9.1 *Each $R(p)$ is a PSH subalgebra of R with basic elements $\Omega(p)$ and p is the unique primitive element of $R(p)$. Furthermore,*

$$R \sim \bigotimes_{p \in BP} R(p)$$

is an isomorphism of PSH algebras.

In particular, up to shift in degree, each $R(p)$ is isomorphic to our basic example of the representation theory of the symmetric group, and so the methods of this chapter can be applied to the study of the representations of $Gl(n,q)$.

4.10 The compatibility of m and Δ.

To prove compatibility of multiplication and comultiplication in R we need to use Mackey's double coset theorem.

Theorem 4.10.1 *Let G be a finite group, H, K subgroups and $\rho : H \to Gl_k(V)$ a finite dimensional representation. Let $\{x_1, \ldots, x_n\}$ be a set of representatives of the K-H double cosets and ρ^{x_i} the composition of ρ with conjugation by x_i^{-1} defined on the group $x_i H x_i^{-1}$, then*

$$\left(\rho \uparrow_H^G\right) \downarrow_K^G = \bigoplus_{i=1}^{n} \left(\rho \downarrow_{H \cap x_i^{-1} K x_i^H}\right)^{x_i} \uparrow_{x_i H x_i^{-1} \cap K}^{K}.$$

Proof. The vector space carrying the representation on the left side is the space of sections of the homogeneous vector bundle $G \times_\rho V$ over G/H, or equivalently functions on G with values in V satisfying

$$f(gh) = \rho(h)^{-1} f(g) \quad \text{for } h \in H. \tag{4.10.1}$$

The action of K is given by

$$(\hat{\rho}(k)f)(g) = f(k^{-1}g). \tag{4.10.2}$$

To prove the decomposition of W given in the theorem, we decompose G/H into K orbits, which correspond to the double cosets $X_i = Kx_iH$, $i = 1, \dots n$. Over each X_i we have a homogeneous bundle with respect to the K-action. The isotropy group of the coset x_iH is the subgroup $K_i = K \cap x_iHx_i^{-1}$ and the isotropy representation is $\rho^{x_i}|_{K_i}$, since for $k \in K_i$ and f satisfying (1) we have

$$\begin{aligned} f(kx_i) &= f(x_i(x_i^{-1}kx_i)) \\ &= \rho(x_i^{-1}kx_i)^{-1}f(x_i) \\ &= \rho^{x_i}(k)^{-1}f(x_i). \end{aligned}$$

The representation of K on the sections of the homogeneous bundle over X_i is given by inducing up from this isotropy representation.

$$W_i = \left(\rho \downarrow^{H}_{H\cap x_i^{-1}Kx_i}\right)^{x_i} \downarrow^{K}_{K_i}. \tag{4.10.3}$$

Since the space W is a direct sum of the subspaces W_i of sections with support in X_i, we have proved the theorem. QED.

Now we apply this theorem to

$$G = S_n \quad H = S_k \times S_{n-k} \quad K = S_\ell \times S_{n-\ell}.$$

The map

$$\sigma H \to (\{\sigma(1), \dots \sigma(k)\}, \{\sigma(k+1), \dots \sigma(n)\})$$

is a bijection of the right H cosets with partitions of n into two subsets of k and $n-k$ elements. The cosets σH and τH are in the same orbit under the action of $S_\ell \times S_{n-\ell}$ on the left if and only if

$$\#(\{\sigma(1), \dots \sigma(k)\} \cap \{1, \dots \ell\}) = \#(\{\tau(1), \dots \tau(k)\} \cap \{1, \dots \ell\}).$$

Thus the partitions into four subsets

$$\{1, \dots a\}\{a+1, \dots k\}\{k+1, \dots k+\ell-a\}\{k+\ell-a+1, \dots n\}$$

for $0 \le a \le \min\{k, \ell\}$ correspond to a complete set of representative x_a of K-H double cosets. The passage from $H \cap x_a^{-1}Kx_a$ to $x_aHx_a^{-1} \cap K$ corresponds to the twist map $s : (S_a \times S_b) \times (S_c \times S_d) \to (S_a \times S_c) \times (S_b \times S_d)$ giving:

$$\begin{aligned}&\left(\rho_k \otimes \rho_{n-k} \uparrow_{S_k \times S_{n-k}}\right) \downarrow_{S_\ell \times S_{n-\ell}} \\ = \bigoplus\nolimits_{0 \le a \le \min\{k,\ell\}} &\left(\left((\rho_k \otimes \rho_{n-k}) \downarrow^{S_k \times S_{n-k}}_{(S_a \times S_b) \times (S_c \times S_d)}\right) \circ s\right) \uparrow^{S_\ell \times S_{n-\ell}}_{S_a \times S_c) \times (S_b \times S_d)} .\end{aligned}$$

In terms of the bialgebra structure this is the desired equality

$$(m \otimes m) \circ s \circ (\Delta' \otimes \Delta)' = \Delta_{\ell, n-\ell} \circ m$$

where Δ' is the component of the comultiplication which contributes terms in $R_\ell \otimes R_{n-\ell}$ after multiplication.

Chapter 5

The Stasheff polyhedra

5.1 Associativity laws.

We return to the subject of H-spaces as introduced in Chapter 3. In the definition of an H-space, it is required that there be a homotopy identity but no assumption is made concerning associativity. For example, among the spheres of positive dimension, the theorem of Adams says that only S^1, S^3, and S^7 are H- spaces, without any requirement of associativity. Now S^1 and S^3 can be given the structure of Lie groups.

The "natural" multiplication on S^7, regarding it as the set of unit octonians, is not associative. So one can ask the following question: Suppose we weaken the condition of associativity. For example, suppose we consider the two maps of $X^3 \to X$:

$$\{x, y, z\} \mapsto (xy)z \qquad \text{and } \{x, y, z\} \mapsto x(yz).$$

We say that the multiplication satisfies the condition A_3 (homotopy associativity for three multiplicands) if these two maps are homotopic to one another. Does S^7 satisfy A_3? It turns out that the answer to this question is "no".

But the question gives rise to a whole series of questions: Let us reformulate condition A_3 as follows: Consider the interval, which we denote by K_3.

$$K_3 \qquad\qquad (xy)z \circ\!\!-\!\!-\!\!-\!\!-\!\!-\!\!-\!\!\circ\, x(yz)$$

where we label the two vertices by the placement of the parentheses. Then condition A_3 asserts that there exists a map of $K_3 \times X^3 \to X$ which coincides

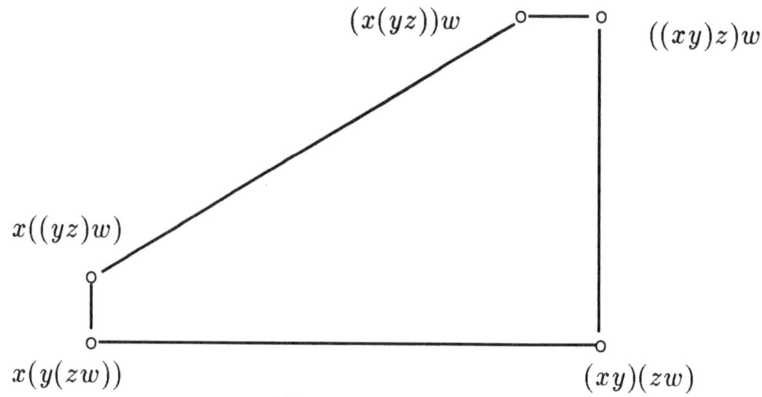

Figure 5.1:

with the designated maps at the end points. This is, of course, much weaker than associativity. But, as we shall see in a moment, this condition is also much weaker than the condition "X is homotopic to a space Y which is an H space with an associative multiplication".

To see why this is so, we have to introduce the Stasheff polyhedra $K_n(n \geq 2)$ and the corresponding conditions, A_n. We define K_2 to consist of a point. We have defined K_3 above. We draw K_4 as in Figure 5.1
with the understanding that the vertices of K_4 consist of all possible ways of inserting two sets of brackets among the four letters x, y, z and w. Two vertices are connected by and edge if they can be obtained from one another by moving one set of brackets. This edge can be labelled by the unchanged set of brackets. For example, the edge joining $x((yz)w)$ to $(x(yz))w$ can be labelled by $x(yz)w$.

In general, we will want K_n to be an $n-2$ dimensional cell. We will think of K_n as being labelled by n letters $x_1x_2\ldots x_n$. We want each of the faces of K_n to correspond introducing one set of brackets $x_1x_2\ldots(x_k\ldots x_{k+s-1})\ldots x_n$ where

$$2 \leq s \leq n-1 \quad \text{and } 1 \leq k \leq n-s-1. \tag{5.1.1}$$

We want to think of this insertion of brackets as requiring that the next insertion of brackets be either entirely within the block $(x_k\ldots x_{k+s-1})$ or

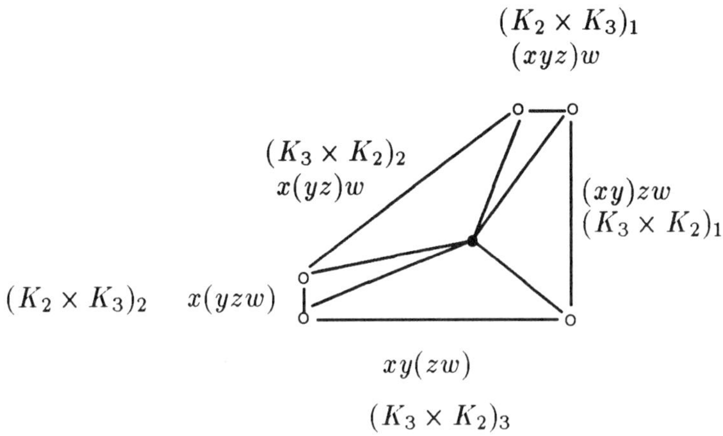

Figure 5.2:

treat this block as a unit. In other words, the face corresponding to the choice of s look like $K_{n-s+1} \times K_s$. So we inductively define

$$K_n = \quad \text{cone on } \partial K_n = \cup (K_{n-s+1} \times K_s)_k,$$

where s and k are restricted as above, and we think of $(K_{n-s+1} \times K_s)_k$ as corresponding to the insertion of brackets $x_1 x_2 \ldots (x_k \ldots x_{k+s-1}) \ldots x_n$.

For example, in Figure 5.2 we have drawn the pentagon K_4 as a union of triangles, with each triangle having one of the edges of K_4 as a base. The edge labeled $(xyz)w$ is also labeled $(K_2 \times K_3)_1$. The meaning of the K_2 in this labeling is that once we have treated (xyz) as a single symbol, then we are left with two symbols, (xyz) and w. The meaning of the K_3 in this labelling is that there are three symbols, x, y and z inside the parentheses. The meaning of the subscript 1 is that the first symbol inside the parentheses is $x = x_1$.

Of course, in this example, where there are only four symbols, x, y, z, and w to start with, the only choice for s is either $s = 2$ or $s = 3$, so all the faces are either $K_2 \times K_3$ or $K_3 \times K_2$. Since K_2 is a point and K_3 is an interval, all of these faces are edges. For $n > 4$ the possibilities become more complicated.

We say that the multiplication *satisfies the condition* A_n if there exist maps

$$M_i : K_i \times X^i \to , 2 \leq i \leq n$$

such that

(i)M_2 is the H space multiplication
and

(ii) the M_i are compatible in the sense that the restriction $M_i|\partial K_i \times X^i$ is given on the component $(K_{i-s+1} \times K_s)_k \times X^i$ of the boundary as

$$\begin{aligned} &M_i(\alpha, \beta, x_1, \ldots, x_i) \\ &= M_{i-s+1}(\alpha, x_1, \ldots, x_{k-1}, M_s(\beta, x_k, \ldots, x_{k+s-1}), x_{k+s}, \ldots, x_i), \end{aligned}$$

$(\alpha, \beta) \in K_{i-s+1} \times K_s,$

Intuitively speaking, the idea is that M_n restricted to each vertex of $K_n \times X^n$ gives a map of $X^n \to X$ corresponding to a definite choice of $n-2$ brackets; restricted to each edge it gives a homotopy between two "adjacent" choices of brackets, restricted to each two dimensional face it gives a homotopy between these homotopies, etc.

If the multiplication $m = M_2$ is actually associative, then there is a well defined map, $m_n : X^n \to$ since any grouping of brackets gives the same answer. So define $M_n(\alpha, x_{1,\ldots,x_n}) = m_n(x_{1,\ldots,x_n})$ and the above conditions are automatically satisfied by the associative law. Thus an associative multiplication satisfies all the A_n.

5.2 Loop spaces.

Suppose that Y is an associative H space with multiplication μ and suppose that

$$f : X \to Y$$

and

$$g : Y \to X$$

are maps which are homotopy inverses. Then suppose that we define the multiplication on X by

$$m = g \circ \mu \circ (f \times f).$$

We claim that m also satisfies A_n for all n.

We sketch the proof for A_3 and A_4, the general idea then being clear. Let h be a homotopy between $f \circ g$ and id_Y. Then

$$\begin{aligned} m(m(\cdot,\cdot),\cdot) &= g \circ \mu \circ [(f \circ g \circ \mu \circ (f \times f)) \times f] \\ &\overset{h}{\sim} g \circ \mu \circ [(\mu \circ (f \times f)) \times f] = g \circ \mu_3 \circ (f \times f \times f) \end{aligned}$$

where $\mu_3 : Y^3 \to Y$ is the triple product corresponding to the associative multiplication μ on Y.

Similarly, $m(\cdot, m(\cdot,\cdot))$ is also homotopic to $g \circ \mu_3 \circ (f \times f \times f)$. So if we denote $g \circ \mu_3 \circ (f \times f \times f)$ symbolically by the triple product xyz then we can join these to homotopies together to get a homotopy

$$(xy)z \circ \text{———} \overset{xyz}{\bullet} \text{———} \circ\, x(yz)\ ,$$

This verifies A_3.

To verify A_4, consider two adjacent edges of the pentagon, for example $x(yz)w$ and $(xyz)w$. By the above procedure, we have associated a homotopy to each of these edges, where the midpoint of each edge is given by a definite map:

$$x(yz)w \text{ is the map } g \circ \mu_3 \circ (f \times f \circ g \circ \mu \circ (f \times f) \times f)$$

and

$$(xyz)w \text{ is the map } g \circ \mu \circ (f \circ g \circ \mu_3 \circ (f \times f \times f) \times f).$$

Applying the homotopy h to the $f \circ g$ makes each of these maps homotopic to the map

$$g \circ \mu_4 \circ (f \times f \times f \times f) \text{ which we denote by } xyzw$$

and which we associate to the apex of the cone drawn above.

Each of these homotopies can be thought of as being a map of the segment from the apex of the cone to the midpoint of the segment. But it is clear that this map extends to a continuous map of the entire cone K_4 : the removal of any pair of parentheses corresponds to some homotopy h, and each line joining the apex to the boundary is associated to differing parametrizations of the various h's that occur in removing different pairs of parentheses.

There is a triangulation of K_4 whose vertices consist of the vertices of K_4 together with the midpoints of the segments and the apex. In each triangle we have a two parameter homotopy depending on barycentric coordinates. For example along the line joining the apex to the common vertex $(x(yz))w$

of the two edges we have been discussing, we would associate the "simultaneous" homotopy

$$g \circ \mu \circ (h \circ \mu \circ (f \times h \circ \mu \circ (f \times f)) \times f).$$

An important application of the above result is to the case of loop spaces. Let M be a connected topological space with base point, and let $X = \Omega M$ be the space of loops on M. So an element of X is a map $g : [0,1] \to M$ with $g(0) = g(1) = *$, the base point of M. On the other hand, let $Y = \Theta M$ be the space of maps of all intervals $[0,r] \to M$ where $r \geq 0$, and the end points are mapped to $*$. The space Y has an associative multiplication: If $\alpha : [0,r] \to M$ and $\beta : [0,s] \to M$ then $\mu(\alpha,\beta) : [0,r+s] \to M$ is defined by

$$\mu(\alpha,\beta)(t) \begin{array}{ll} = \alpha(t) & 0 \leq t \leq r \\ \\ = \beta(t) & r \leq t \leq r+s. \end{array}$$

The multiplication on Y is clearly associative. We have an inclusion map $f : X \to Y$. We also have a retraction map $g : Y \to X$, given by

$$(g(\gamma))(u) = \gamma(ru) \ := \gamma_r(u)$$

if $\gamma \in Y, \ \ \gamma : [0,r] \to M$. The homotopy h between $f \circ g$ and id can be taken to be

$$h(t,\gamma)(u) = \gamma([(1-t)r + t]u).$$

Thus loop spaces are examples of A_∞ spaces. In [1123], Stasheff shows how to associate to every A_∞ space a classifying space, generalizing the construction of the classifying spaces for Lie groups. As part of his work, he shows that the spaces K_n are all cells and studies the structure of the boundary operators associated with these cells.

5.3 The polyhedron K_5.

In Figure 5.3 we draw the graph which is the one-skeleton of the Stasheff polyhedron, K_5. The "faces" of this graph including the bounding pentagon are three quadrilaterals and six pentagons. This is the boundary of K_5. Recall that the vertices are all ways of inserting three pairs of parentheses among five letters which we shall choose as $ABCDE$. We will also orient this graph with an interpretation we will soon provide. For the moment ignore the arrows at the ends of the edges.

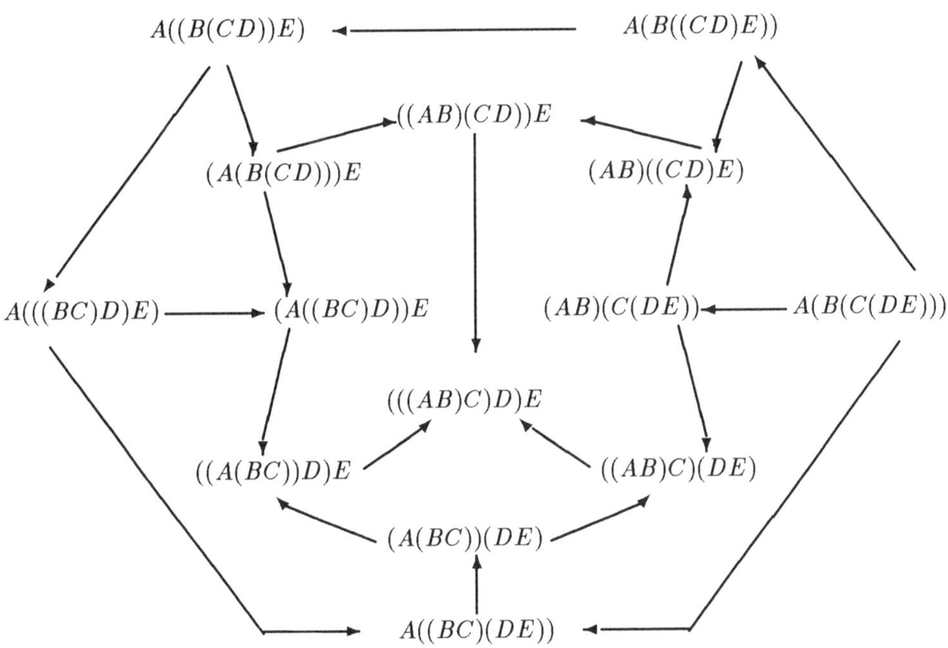

A((B(CD))E)
A(B((CD)E))
((AB)(CD))E
(A(B(CD)))E
(AB)((CD)E)
A(((BC)D)E)
(A((BC)D))E
(AB)(C(DE))
A(B(C(DE)))
(((AB)C)D)E
((A(BC))D)E
((AB)C)(DE)
(A(BC))(DE)
A((BC)(DE))

Figure 5.3:

Notice that the faces are all either pentagons, corresponding to $K_2 \times K_4$ and $K_4 \times K_2$, or quadrilaterals, corresponding to $K_3 \times K_3$, as required. The arrows corresponding to a move of the form a : $\_(\_\ \_) \rightsquigarrow (\_\ \_)\ \_$ where each $\_$ stands for a letter or a block of letters.

For example, the arrow going from $(A(BC))(DE)$ to $((AB)C)(DE)$ near the bottom of the diagram means that we have applied this move to the letters ABC leaving (DE) unchanged. We can denote this move by $a \otimes (1 \otimes 1)$ or even more precisely by $a_{A,B,C} \otimes (1 \otimes 1)$ where the $\otimes(1 \otimes 1)$ indicates that we are not touching the (DE). Similarly, the arrow coming diagonally down from the upper left hand corner going from $A(B(CD))E$ to $(A(B(CD)))E$ is a move of the type $\_(\_\ \_) \rightsquigarrow (\_\ \_)\ \_$ where the first $\_$ stands for A, the second $\_$ stands for $(B(CD))$ and the third $\_$ stands for E. We would denote this move by a, or more precisely by $a_{A,(B(CD)),E}$.

Clearly every two adjacent vertices are connected by a move of the type a (with possibly some $1'$s tensored in) or, if we go in the opposite direction, by the inverse move. In Figure 5.4 we have drawn the same graph as in Figure 5.3 with the moves labelled, but with the subscripts on the a's omitted.

5.4 Closed paths.

We can now ask the following combinatorial question: how do we describe the closed paths on the graph which is the one skeleton of K_n, that is, the graph whose vertices are the vertices of K_n and whose edges are the edges of K_n. It is clear that every edge corresponds to a move of the form a with some 1's tensored in. So a closed path is a succession of such a's and their inverses. From the Figure 5.3 it is clear that for $n = 5$, the two dimensional skeleton of K_5 can be decomposed into pentagons and quadrilaterals. The fact that K_5 is a cell then implies that every closed path on K_5 can be written as a succession of quadrilaterals and pentagons. In fact, by filling in the outside pentagon in Figure 5.3 from underneath, it is clear that the boundary of K_5 is topologically a sphere.

In general, the two dimensional skeleton of any K_n for $n > 4$ is a union of faces which are either pentagons or quadrilaterals. Indeed, any such face can be described by having $n-4$ parenthesis pairs for n letters. By construction, it will be the product of $n-3$ factors of the type K_{s_i} where $\sum(s_i - 2) = 2$. So at most all but two of the s_i must equal 2, and either two of them equal 3 in which case the face is the product of ($n-5$ factors of type K_2 with) $K_3 \times K_3$ and the face is a quadrilateral, or it is the product of $n-4$ copies of

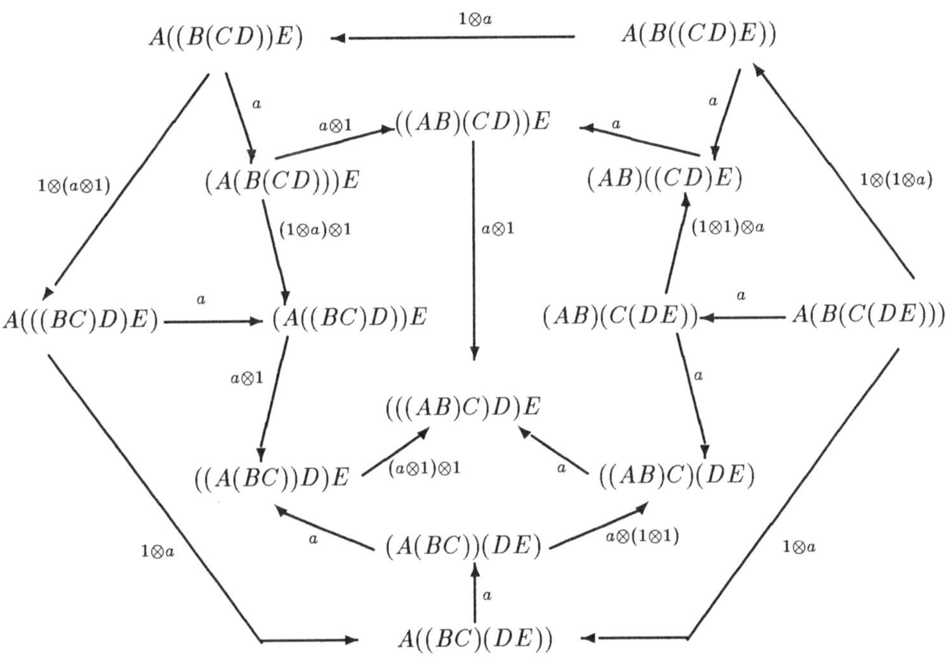
A((B(CD))E)
1⊗a
A(B((CD)E))
a
a⊗1
((AB)(CD))E
a
a
1⊗(a⊗1)
(A(B(CD)))E
(AB)((CD)E)
1⊗(1⊗a)
(1⊗a)⊗1
a⊗1
(1⊗1)⊗a
A(((BC)D)E)
a
(A((BC)D))E
(AB)(C(DE))
a
A(B(C(DE)))
a⊗1
a
(((AB)C)D)E
((A(BC))D)E
(a⊗1)⊗1
a
((AB)C)(DE)
a
(A(BC))(DE)
a⊗(1⊗1)
1⊗a
1⊗a
a
A((BC)(DE))

Figure 5.4:

K_2 with K_4 in which case it is a pentagon. We shall call the paths around the boundaries of these faces the quadrilateral and pentagonal paths. The following important theorem is the a consequence of Stasheff's theorem that K_n is a cell, cf. [1123]

Theorem 5.4.1 *The quadrilateral and pentagonal paths generate the closed paths on K_n.*

In [824] Mac Lane gave a far reaching interpretation of this theorem in terms of category theory, which we will explain in the next Chapter. Hence this theorem should be known as the *Stasheff-Mac Lane coherence theorem* for associativity. We will explain the meaning of the term "coherence" in the next chapter. A direct combinatorial proof of this theorem was given by Mac Lane in [824] and can be found in his text [828].

5.5 Free magmas.

In Figure 5.3, the letters $ABCDE$ are just place holders. Instead of writing, for example, $(AB)(C(DE))$ we should, from a strictly logical point of view write

$$(\_\_)(\_(\_\_)).$$

All that matters is the location of the parentheses.

Recall (Bourbaki's notation) that a *magma* is a set, S, with multiplication

$$S \times S \to S$$

but with no further conditions. So no identity or associative law is a assumed. A *free magma*, $F(Q)$, on a set of generators, Q, is then the magma generated by Q with no relations whatsoever between the expressions. Since there are no relations, $F(Q)$ is graded: every element in the free magma generated by Q has a "length" n, the number of elements of Q entering into its expression.

Suppose we consider the free magma generated by a single element, _ . Then the elements of degree 1,2,3,4 of the free magma generated by _ are listed as follows:

$$
\begin{array}{cl}
\_ & F_1(\{\_\}) \\
\_\,\_ & F_2(\{\_\}) \\
\_(\_\,\_),\ (\_\,\_)\_ & F_3(\{\_\}) \\
\_(\_(\_\,\_)),\ \_((\_\,\_)\_),\ (\_\,\_)(\_\,\_), & \\
(\_(\_\,\_))\_,\ ((\_\,\_)\_)\_ & F_4(\{\_\})
\end{array}
$$

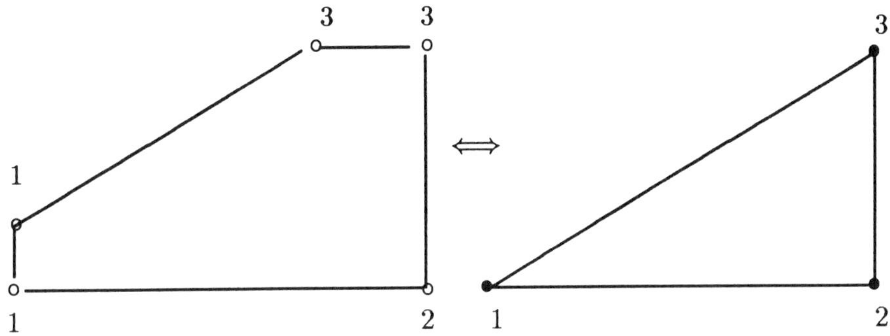

Figure 5.5:

It is clear we have an identification of $F_n(\{\_\})$ with $Vert(K_n)$, the set of vertices of K_n for $n \geq 2$. Thus $\{\_\} \cup \bigcup_n Vert(K_n)$ can be identified with the free magma, $F(\{\_\})$, on one generator, and hence

$$X \cup \bigcup_n Vert(K_n) \times X^n$$

can be identified with the free magma, $F(X)$, generated by the set X. In fact, it is precisely the space $\bigcup_n K_n \times X^n$ with appropriate identifications coming from the multiplication given on X which is Stasheff's universal bundle.

Associated to any element, b, of $F_n(\{\_\})(n \geq 2)$ there is a unique integer i such that $b = cd$ where $c \in F_i(\{\_\})$ and $d \in F_{n-i}(\{\_\}), 1 \leq i \leq n-1$. We will want to think of i as ranging over the vertices of the $(n-2)$ dimensional simplex, Σ_{n-2}. If we think of K_n as a polyhedron, then we have a map from the vertices of K_n to the vertices of the simplex,which then extends to a map from K_n to Σ_{n-2}. We draw the picture for $n = 4$ relating the pentagon to the triangle in Figure 5.5.

Conversely, we can regard the pentagon as coming from the triangle where the vertices 1 and 3 have been "blown up" into edges. This point of view will be important from the standpoint of the completion of moduli space and from the standpoint of symplectic geometry. In symplectic geometry, the triangle is the image of the moment map for the standard action of the two dimensional torus on complex projective two space, with the vertices corresponding to the fixed points. Truncating the triangle at two vertices corresponds to blowing up two of these fixed points, [499]. The connection of this remark with associativity laws will become apparent when we discuss

Drinfel'd proof of the existence of quantum deformations of the universal enveloping algebra using the Knizhnik-Zamolodchikov equations, cf. Chapter 12.

We now describe the Stasheff polyhedron, K_5, with its fourteen vertices, six pentagonal faces and three quadrilateral faces as a truncated simplex in three dimensions: The simplex has four vertices, labelled by the number of elements included in the leftmost parenthesis. (We don't draw the parenthesis containing just one element.) Thus, including just the outermost parentheses, we have the labelling

$$\begin{aligned} 1 &= \bullet(\bullet\bullet\bullet\bullet) \\ 2 &= (\bullet\bullet)(\bullet\bullet\bullet) \\ 3 &= (\bullet\bullet\bullet)(\bullet\bullet) \\ 4 &= (\bullet\bullet\bullet\bullet)\bullet \end{aligned}$$

Eventually, when we insert all the parentheses, 1 will correspond to five vertices of K_5 corresponding to the insertion of two parentheses inside the outer ones. The vertices 2 and 3 correspond to two vertices each of K_5 and 4 to five. Our truncations will consist of two vertex truncations, at the vertices 1 and 4, and three edge truncations, along the edges $\overline{12}, \overline{14}$ and $\overline{34}$. To visualize this, let us first draw the simplex and the positions of the vertex truncations:

These two truncations at the vertices converts two of the four triangular faces into pentagons, two into quadrilaterals and introduces two new triangular faces. Next we draw the simplex truncated at the vertices 1 and 4, just before the three edge truncations have been performed. This truncation converts the triangles situated at 1 and at 4 into pentagons, the two quadrilaterals into pentagons, the vertices 2 and 3 into edges, and the the remaining portions of the edges $\overline{12}, \overline{34}$ and $\overline{14}$ into quadrilaterals.

All fourteen vertices of K_5 are visible in Figure 5.??, and are labeled as

$$\begin{aligned} 1a &= \bullet(\bullet(\bullet(\bullet\bullet))) \\ 1a' &= \bullet(\bullet((\bullet\bullet)\bullet)) \\ 1b &= \bullet((\bullet(\bullet\bullet))\bullet) \\ 1b' &= \bullet(((\bullet\bullet)\bullet)\bullet) \\ 1c &= \bullet((\bullet\bullet)(\bullet\bullet)) \\ 2a &= (\bullet\bullet)(\bullet(\bullet\bullet)) \\ 2a' &= (\bullet\bullet)((\bullet\bullet)\bullet) \end{aligned}$$

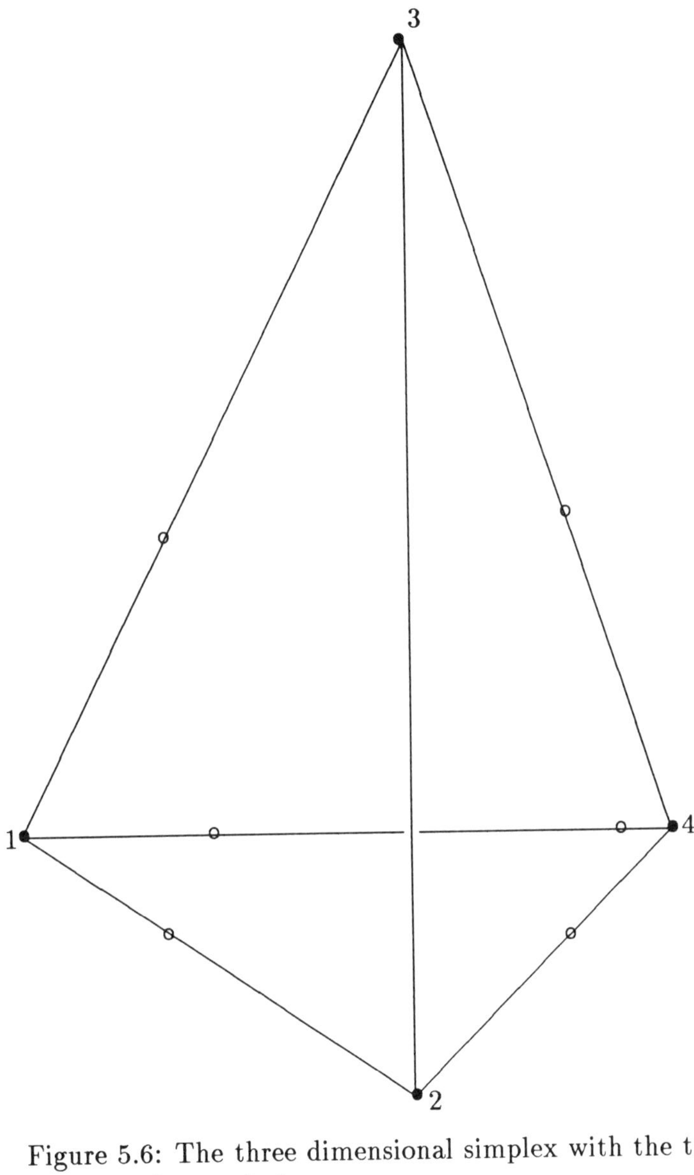

Figure 5.6: The three dimensional simplex with the truncations at the vertices 1 and 4 marked.

$$
\begin{aligned}
3a &= ((\bullet\bullet)\bullet)(\bullet\bullet) \\
3a' &= (\bullet(\bullet\bullet))(\bullet\bullet) \\
4a &= (((\bullet\bullet)\bullet)\bullet)\bullet \\
4a' &= ((\bullet(\bullet\bullet))\bullet)\bullet \\
4b &= (\bullet(\bullet(\bullet\bullet)))\bullet \\
4b' &= (\bullet((\bullet\bullet)\bullet))\bullet \\
4c &= ((\bullet\bullet)(\bullet\bullet))\bullet
\end{aligned}
$$

In general, it can be proved by induction that K_{n+2} can be obtained as a truncation (i.e. an intersection with half spaces) of the n-simplex and hence can be realized as a convex body. In particular it is homeomorphic to the ball and hence is simply connected. Here is a sketch of the inductive argument:

It will be convenient to rotate the the simplex so that now the last vertex is at the apex. So let $\Sigma_{n-2} = [v_1, v_2, \cdots v_{n-1}]$ where we think of v_i as corresponding to the partition $(i, n-i)$ of n coming from the last parenthesis in a non-associative product of n elements. Using lexicographical ordering v_{n-1} is maximal and v_1 is minimal. Truncate Σ_{n-2} at the vertex v_{n-1} by a plane parallel to the face $[v_1, \cdots, v_{n-2}]$ and label the new vertices of the truncation by $w_1, w_2, \cdots w_{n-2}$ with w_i lying on the edge connecting v_{n-1} to v_i and corresponding to the partition $(i, n-i-1, 1)$. The resulting geometric object is a cylinder over Σ_{n-3}. By induction we know how to truncate the $n-3$ simplex to get K_{n-1}. Extend this truncation to the cylinder. The next step is to truncate at the minimal vertex, however to make sure that the procedure is consistent, we need to consider the $n-3$ simplex spanned by $w_1, v_2, \cdots, v_{n-2}$. Truncate at v_1 by a plane parallel to $[w_1, v_2, \cdots, v_{n-2}]$. Label the new vertices $u_1, u_2, \cdots, u_{n-2}$, with u_i corresponding to the partition $(1, n-i-1, i)$. In this case u_1 is maximal and u_{n-2} minimal and for $i > 1$ the vertices u_i and v_{n-i} lie on the same edge. Extend the truncation of Σ_{n-3} which gives K_{n-1} to this cylinder as before, EXCEPT for the truncation at u_1. Along the edge connecting u_1 and w_1 we already have done the necessary truncation at the first stage of the procedure. The truncation at vertex w_1 corresponded to the minimal vertex in a Σ_{n-3}. The vertex u_1 is maximal relative to the lexicographical ordering, but the truncation rule is the same by symmetry which reverses order of partitions and non-associative products.

Proposition 5.5.1 *The truncation procedure just described yields a convex*

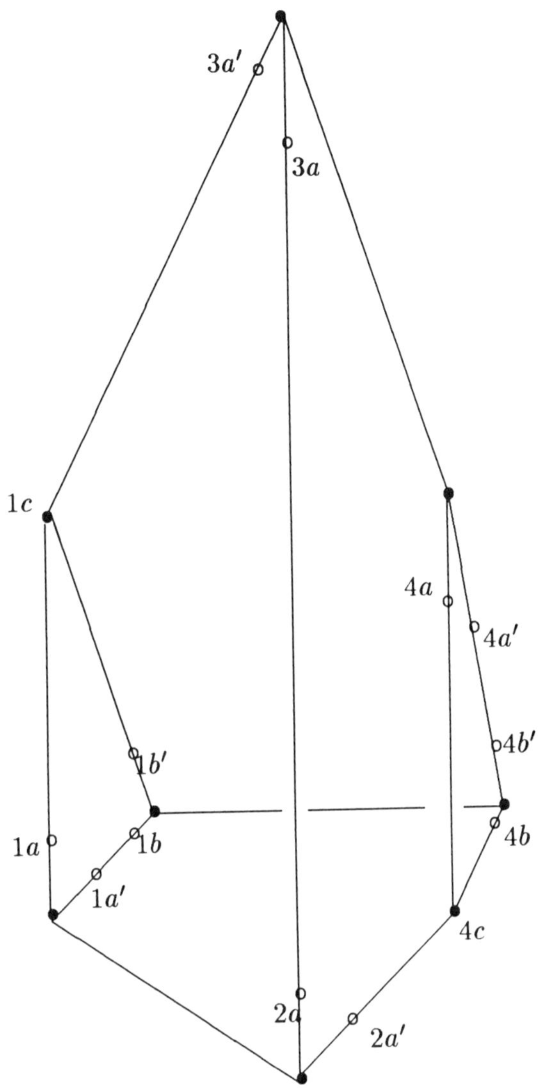

Figure 5.7: The simplex truncated at two vertices and the three edge truncations indicated. When we make these three truncations we obtain K_5. The vertices of K_5 are labelled according to the code indicated in the text.

polyhedron whose vertices and edges form a graph with nodes corresponding bijectively to the nonasssociative words on n elements and whose edges correspond to associativity operations on a triple product.

Proof. The proof is by induction just like the construction itself. For the purpose of the proof we call the graph described in the statement K_n with vertices V_n and edges E_n. To begin, $K_3 = \Sigma_1$, with one edge and two vertices. Any nonassociative word determines a partition $(i, n-i)$, and we can express

$$V_n = \cup_i V_i \times V_{n-i}.$$

Label the vertices in Σ_{n-2} by the partitions. We can consider any vertex in the truncation at $(n-1, 1)$ as defining a refinement $(i, n-i-1, 1)$ connecting this to the vertex of Σ_{n-2} labelled $(i, n-i)$ represents the general associativity between subwords of length $i, n-i-1, 1$. Assume by induction that truncating Σ_{n-3} gives a polyhedron with one skeleton K_{n-1}. If we carry out the product truncation along the cylinder over Σ_{n-3} we get all associativities between words whose corresponding fourfold partitions are of type $\mathbf{A} = \{((i,j,k),1)\}$ or of type $\mathbf{B} = \{((i,j),(k,1))\}$ with $i+j+k = n-1$. Any associativity between the terms i, j, k in $\mathbf{A}$ occurs as an edge in the K_{n-1} given by truncating the Σ_{n-3} at "top"of the cylinder, i.e., the vertex $(n-1, 1)$. Any associativity between terms in $\mathbf{B}$ occurs in the truncation induced on the face $[(n-2,2),(n-3,3)\cdots,(1,n-1)]$ of Σ_{n-3} which is the "bottom" of the cylinder. The associativities between terms of type $\mathbf{A}$ and terms of type bfB are the "vertical" edges connecting vertices in the "top" to vertices in the "bottom" of the cylinder. This gives all the associativities (edges) between vertices representing words which determine fourfold partitions of type $\mathbf{A}$ or $\mathbf{B}$. The remaining vertices of K_n correspond to fourfold partitions of the type $\mathbf{C} = \{((1,i),(j,k))\}$ and $\mathbf{D} = \{(1,(i,j,k))\}$. The associativities between these are constructed by the truncation at vertex $(1, n-1)$ followed by truncating the resulting Σ_{n-3} and extending to the cylinder. All the associativities between vertices both of which correspond to the partitions in $\mathbf{C} \cap \mathbf{B}$ are described by either of the cylinders, so it is enough to truncate only once along the edge connecting the vertex of the "top" truncation labelled $(1, n-1, 1)$ to $(1, n-1)$.

Chapter 6

Mac Lane's coherence

6.1 Monoidal categories.

We say that the category $\mathcal{C}$ has a *multiplication* or a *tensor product* if we are given a covariant bifunctor

$$\mathcal{C} \times \mathcal{C} \to \mathcal{C}.$$

The bifunctor will be denoted by placing the symbol $\otimes$ between the arguments,that is by writing the functor as

$$X \times Y \rightsquigarrow X \otimes Y.$$

So to say that we have a multiplication means that

1. Two every $X, Y \in Ob(\mathcal{C})$ there is assigned an object $X \otimes Y$ of $\mathcal{C}$.

2. To each pair of morphisms $f : X \to X', g : \mathbf{Y} \to \mathbf{Y}'$ of $\mathcal{C}$ there is assigned a morphism $f \otimes g : X \otimes Y \to X' \otimes Y'$.

3. If $f' : X' \to X''$ and $g' : Y' \to Y''$ then $(f' \otimes g') \circ (f \otimes g) = f'f \otimes g'g$ and

4. $id_X \otimes id_Y = id_{X \otimes Y}$.

We can think of $\otimes$ as a sort of multiplication, thus making $Ob(\mathcal{C})$ into a magma. Thus to every $w \in Vert(K_n)$ and every n objects $X_1, \ldots, X_n$ we get a functor

$$w : (X_1, \ldots, X_n) \rightsquigarrow w(X_1, \ldots, X_n).$$

For example, if $n = 3$ we have the two functors

$$(X, Y, Z) \rightsquigarrow X \otimes (Y \otimes Z) \quad \text{and} \quad (X, Y, Z) \rightsquigarrow (X \otimes Y) \otimes Z.$$

Without further assumptions, there need be no relation between these functors. The best we can do is to *postulate* an identification between them. More precisely, we assume that there is a natural transformation, a, between the two functors $\cdot \otimes (\cdot \otimes \cdot)$ and $(\cdot \otimes \cdot) \otimes \cdot$ from $\mathcal{C} \times \mathcal{C} \times \mathcal{C} \to \mathcal{C}$ which is an isomorphism for every three objects. So what this means is that to every three objects X, Y, and Z of $\mathcal{C}$ we are given a morphism

$$a_{X,Y,Z} : X \otimes (Y \otimes Z) \to (X \otimes Y) \otimes Z$$

which has a two sided inverse and which satisfies the relevant commutative diagram with respect to morphisms. The natural transformation a is called an *associativity* or an *associativity constraint*

But now we are faced with the problem of *coherence* which is the following: Each edge of the Stasheff polyhedron K_n will correspond to an application of a (possibly tensored with some id's), or, if traversed in the opposite direction to the inverse natural transformation. For example, in the diagram for K_5 in Figure 5.3, we should insert $\otimes$ signs between the various brackets, so the bottom element should be labeled $A \otimes ((B \otimes C) \otimes (D \otimes E))$ and the element immediately above it should be labeled as $(A \otimes (B \otimes C)) \otimes (D \otimes E)$. Then the vertical arrow a connecting these two is the morphism $a_{A,B\otimes C,D\otimes E}$. We can go from one vertex of K_n to another along different paths. Each such path gives rise by composition to a natural transformation. We say that a is coherent if all these natural transformations are the same. For example, going from the bottom left hand corner of the pentagon in Figure 5.1 to the top right hand corner amounts to moving all the parentheses from the extreme right to the extreme left: the bottom left hand corner is $X \otimes (Y \otimes (Z \otimes W))$ and the upper right hand corner is $((X \otimes Y) \otimes Z) \otimes W$. Moving along the upper path is the composition

$$\begin{aligned} X \otimes (Y \otimes (Z \otimes W)) &\stackrel{id \otimes a_{Y,Z,W}}{\longrightarrow} X \otimes ((Y \otimes Z) \otimes W) \\ &\stackrel{a_{X,Y\otimes Z,W}}{\longrightarrow} (X \otimes (Y \otimes Z) \otimes W \\ &\stackrel{a_{X,Y,Z} \otimes id}{\longrightarrow} ((X \otimes Y) \otimes Z) \otimes W \end{aligned}$$

while moving along the bottom path is the composition

$$X \otimes (Y \otimes (Z \otimes W)) \stackrel{a_{X,Y,Z\otimes W}}{\longrightarrow} (X \otimes Y) \otimes (Z \otimes W) \stackrel{a_{X\otimes Y,Z,W}}{\longrightarrow} ((X \otimes Y) \otimes Z) \otimes W.$$

The equality of these two compositions is the celebrated Mac Lane pentagon condition

$$a_{X\otimes Y,Z,W} \circ a_{X,Y,Z\otimes W} = (a_{X,Y,Z} \otimes id) \circ (a_{X,Y\otimes Z,W}) \circ (id \otimes a_{Y,Z,W}). \quad (6.1.1)$$

So Mac Lane's pentagon condition asserts that the morphism given as the composition of the two morphisms on the left of (6.1.1) is equal to the morphism given by the triple composition on the right. Mac Lane's coherence theorem (for associativity) asserts that if a satisfies (6.1.1) then it is coherent.

Basically, as we have said, this theorem amounts to the assertion that the closed paths on the one skeleton of K_n are generated by the pentagons and quadrilaterals. (The quadrilaterals correspond to the naturality property of a.) From the point of the previous chapter, this coherence result is an immediate consequence of the fact that the Stasheff polyhedra, K_n are simply connected. So the construction by truncation that we gave at the end of the last chapter gives a proof of the coherence theorem. We refer to Mac Lane's original paper [824] or book [828] for a detailed alternative proof of this theorem which follows along the lines of the classical proof of the Jordan Holder theorem.

A category $\mathcal{C}$ is called *monoidal* if it has a multiplication $\otimes$ with an associativity constraint, a, satisfying the Mac Lane pentagon conditions (and hence coherent) and if, in addition, there is a preferred object $I \in \mathcal{C}$ called the *unit* object together with natural isomorphisms

$$l = l_A : I \otimes A \to A, \quad r = r_A : A \otimes I \to A$$

such that, in addition to the pentagon condition on a, the diagrams in Figure 6.1 commute.

(In these diagrams, ϕ denotes a^{-1}. There is an unfortunate but understandable inconsistency in the notation in the literature, as to whether the fundamental associativity should go from $(X \otimes Y) \otimes Z \to X \otimes (Y \otimes Z$ or vice versa. Mac Lane uses the symbol a for associatvity, and uses this to mean that the parentheses are moved to the left. Drinfeld uses the symbol ϕ to mean the operation of moving the parentheses to the right. So to adjust to the literature,we will abide by both conventions we will use the symbol ϕ to denote the associativity that moves the parentheses to the right.)

In fact, we leave as an exercise the fact (see [643] or [596]) that commutativity of the bottom triangle (and the naturality of l and r) imply the commutativity of the other two.

A monoidal category is called *strict* when all the $\phi_{A,B,C}, l_A$ and r_A are identity morphisms.

Let $\mathcal{C}$ and $\mathcal{D}$ be monoidal categories, and let $F : \mathcal{C} \to \mathcal{D}$ be a functor. We then have two functors from $\mathcal{C} \times \mathcal{C}$ to $\mathcal{D}$, namely

$$F \otimes F = F \otimes_{\mathcal{D}} F : (A, B) \mapsto F(A) \otimes_{\mathcal{D}} F(B)$$

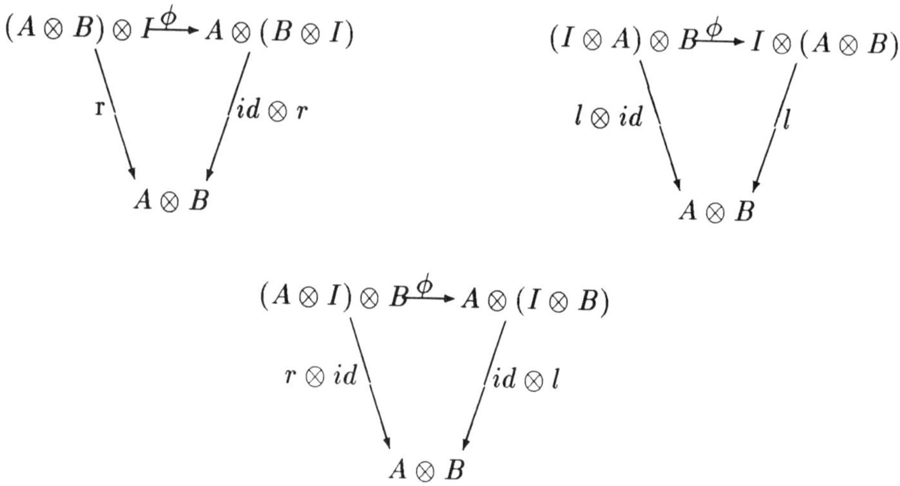

Figure 6.1: The coherence conditions on the unit. Actually, the lower condition suffices.

and

$$F \circ \otimes = F \circ \otimes_{\mathcal{C}} : (A, B) \mapsto F(A \otimes_{\mathcal{C}} B).$$

A *monoidal functor* from $\mathcal{C}$ to $\mathcal{D}$ is a triple (F, b, f) where $F : \mathcal{C} \to \mathcal{D}$ is a functor, where b is a natural transformation between the two functors listed above, where $f \in \mathrm{Hom}_{\mathcal{D}}(I_{\mathcal{D}}, F(I_{\mathcal{C}}))$ is an isomorphism, and such that the diagrams in Figure 6.2 all commute.

A monoidal functor is called *strict* if f and all the $b_{A,B}$ are identity morphisms.

One of the consequences of the MacLane coherence theorem is that any category, $\mathcal{C}$, is equivalent via a monoidal functor to a strict monoidal category. Roughly speaking, just define a new category built out of the objects of but with concatenations of tensor products having all the parentheses moved all the way to the left. Here is a direct proof, due to Joyal and Street, cf. [600] of this consequence of the Mac Lane coherence theorem which does not use the results of Chapter 5 about the convexity of the Stasheff polyhedra, but does use the existence of the unit object.

Proposition 6.1.1 *Any monoidal category is equivalent via a monoidal transformation to a strict monoidal category.*

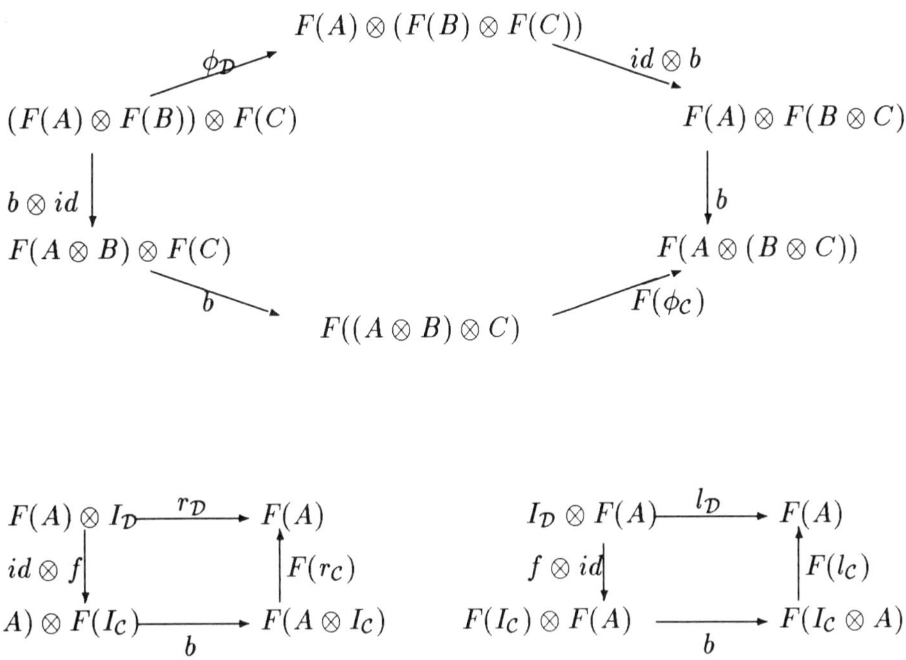

Figure 6.2: Commutativity diagrams for a monoidal functor.

Proof.(Joyal and Street,) Starting with a monoidal category, $\mathcal{C}$, let $\mathcal{E} = \mathcal{E}_{\mathcal{C}}$ be a new category constructed as follows: An obejct of $\mathcal{E}$ is a pair (E, ρ) consisting of a functor:$E : \mathcal{C} \to \mathcal{C}$ and a natural isomorphism ρ, between the two functors

$$(A, B) \mapsto E(A) \otimes B \quad \text{and} \quad (A, B) \mapsto E(A \otimes B)$$

from $\mathcal{C} \times \mathcal{C} \to \mathcal{C}$. So for each $A, B \in \mathrm{Ob}(\mathcal{C})$

$$\rho_{A,B} \in \mathrm{Hom}_{\mathcal{C}}(E(A) \otimes B, E(A \otimes B))$$

is an isomorphism.

A morphism, or arrow, $\theta : (E, \rho) \to (E', \rho')$ in the category $\mathcal{E}$ is a natural transformation from the functor E to the functor E' consistent with ρ and ρ' in the sense that

$$\theta_{A \otimes B} \circ \rho_{A,B} = \rho'_{A,B} \circ (\theta_A \otimes id) \quad \forall A, B.$$

If $\theta : (E, \rho) \to (E', \rho')$ and $\theta' : (E', \rho') \to (E'', \rho'')$ both satisfy the consistency condition, then so does $\theta' \circ \theta : (E, \rho) \to (E'', \rho'')$. So the composition of natural transformations gives the composition of morphisms in $\mathcal{E}$. We now define a tensor product on $\mathcal{E}$: On objects it is given by

$$(E, \rho) \otimes (F, \sigma) = (E \circ F, \tau)$$

where $\tau_{A,B}$ is the composite morphism

$$(E(F(A)) \otimes B \overset{\rho_{F(A),B}}{\longrightarrow} E(F(A) \otimes B) \overset{E(\sigma_{A,B})}{\longrightarrow} E(F(A \otimes B)). \tag{6.1.2}$$

On the level of arrows, if $\theta : (E, \rho) \to (E', \rho')$ and $\psi : (F, \sigma) \to (F', \sigma')$ then

$$\theta \otimes \psi : E \circ F \to E' \circ F'$$

is given as

$$\theta \otimes \psi = \theta \circ E(\psi).$$

Since composition is an associative operation, we see that $\mathcal{E}$ is a strictly monoidal category with unit object

$$(id, id).$$

There is a functor $L : \mathcal{C} \to \mathcal{E}$ given by

$$L(X)(\bullet) = (X \otimes \bullet, \phi_{X,\bullet,\bullet}), \quad L(f) = f \otimes id. \tag{6.1.3}$$

Lemma 6.1.1 *The functor $L : \mathcal{C} \to \mathcal{E}_{\mathcal{C}}$ is a fully faithful monoidal functor.*

We recall that the words "fully faithful" mean that for every pair of objects $X, Y \in \mathcal{C}$ the functor L gives an isomorphism

$$\mathrm{Hom}_{\mathcal{C}}(X,Y) \cong \mathrm{Hom}_{\mathcal{E}}(L(X), L(Y)).$$

Since the image of $\mathcal{C}$ is closed under tensor product, the lemma implies Proposition 6.1, since we may take the image of L as the strict monoidal category.

Proof.(Joyal and Street.) Suppose that $h \in \mathrm{Hom}_{\mathcal{C}}(X,Y)$. Then $L(h) = h \otimes id$. Clearly $h \otimes id = h' \otimes id$ implies that $h = h'$, because of our axioms about the unit element.So L is injective on morphisms. To show that it is surjective, suppose that $\theta : L(X) = X \otimes I \to Y \otimes I = L(Y)$. Define $h = h(\theta) : X \to Y$ as the composite

$$X \xrightarrow{r_X} X \otimes I \xrightarrow{\theta_I} Y \otimes I \xrightarrow{r_Y^{-1}} Y.$$

We claim that

$$\theta = L(h) = h \otimes id.$$

Indeed, this follows from the commutativity of the diagram

$$\begin{array}{ccccccc}
X \otimes A & \xrightarrow{r_X^{-1}} & (X \otimes I) \otimes A & \xrightarrow{\phi_{X,I,A}} & X \otimes (I \otimes A) & \xrightarrow{id_X \otimes l_A} & X \otimes A \\
\downarrow{\scriptstyle h \otimes id_A} & & \downarrow{\scriptstyle \theta_I \otimes id_A} & & \downarrow{\scriptstyle \theta_{id \otimes A}} & & \downarrow{\scriptstyle \theta_A} \\
Y \otimes A & \xrightarrow[r_Y^{-1}]{} & (Y \otimes I) \otimes A & \xrightarrow[\phi_{Y,I,A}]{} & Y \otimes (I \otimes A) & \xrightarrow[id_Y \otimes l_A]{} & Y \otimes A
\end{array}$$

and the commutativity of the triangle for the the unit. So L is fully faithful.

We must now show that L is a monoidal functor. So we must first find an equivalence between the functors

$$L \otimes L \quad \text{and} \quad L \circ \otimes.$$

Now

$$\begin{aligned}
(L \otimes L)(X,Y) &= (X \otimes \bullet, \phi_{X,\bullet,\bullet}) \otimes (Y \otimes \bullet, \phi_{Y,\bullet,\bullet}) \\
&= (X \otimes (Y \otimes \bullet), (X \otimes \phi_{Y,\bullet,\bullet}) \circ \phi_{X,Y \otimes \bullet,\bullet})),
\end{aligned}$$

where the expression after the comma is the composite natural transformation as required by (6.1.2); while

$$[L \circ \otimes](X, Y) = ((X \otimes Y) \otimes \bullet, \phi_{X \otimes Y, \bullet, \bullet}).$$

We claim that composition on the left with $\phi^{-1}_{X,Y}$ is the desired morphism between these two functors. To see this, observe that on the level of functors, we have

$$\phi^{-1}_{X,Y,\bullet} : X \otimes (Y \otimes \bullet) \longrightarrow (X \otimes Y) \otimes \bullet.$$

On the level of natural transformations,

$$\phi^{-1}_{X,Y,\bullet} \circ X \otimes \phi_{Y,\bullet,\bullet}) \circ \phi_{X,Y \otimes \bullet, \bullet}))$$

when applied to (A, B) is the composite

$$(X \otimes (Y \otimes A) \otimes B) \longrightarrow X \otimes ((Y \otimes A) \otimes B) \longrightarrow X \otimes (Y \otimes (A \otimes B)) \longrightarrow (X \otimes Y) \otimes (A \otimes B)$$

This is a path of three steps around the pentagon. By the pentagon identity, this is equal to the two step path

$$(X \otimes (Y \otimes A)) \otimes B \longrightarrow ((X \otimes Y) \otimes A) \otimes B \longrightarrow (X \otimes Y) \otimes (A \otimes B).$$

So composition on the left with $\phi^{-1}_{X,Y,\bullet}$ is indeed the desired

$$b_{X,Y} : L(X) \otimes L(Y) \longrightarrow L(X \otimes Y).$$

To check the upper consistency diagram in Figure 6.2, observe that the associativity occurring in the upper left is an identity, since $\mathcal{E}$ is a strict monoidal category. The diagram, in fact reduces exactly to the pentagon constraint. The quadrilateral consistency conditions in Figure 6.2 become triangle conditions, which are just two of the three axioms on the unit. QED

As we indicated above, Proposition 6.1 is a consequence of the Mac Lane coherence theorem. With a little more work, [600], one can actually deduce the full Mac Lane coherence theorem from the proposition.

Proposition 6.1 says, roughly speaking, that, in a certain sense, we can replace a monoidal category by a strictly monoidal category. But we should treat this formulation with some care. When we deal with more than one category at the same time, the actual associativities become very important. For example, one of the key results of Chapter 11 will assert that when we deform the universal enveloping algebra of a simple Lie algebra, and hence its category of representations, all of the essential properties of the deformation can be reduced to a deformation of the associativity constraint.

Here is an example of a monoidal category : Let $Ob(\mathcal{M}_n)$ consist of oriented $n-1$ dimensional manifolds. If X and Y are $n-1$ dimensional manifolds, then an element of $\mathrm{Hom}(X,Y)$ is an oriented n - dimensional manifold whose oriented boundary is identified with $Y \cup X^-$ where X^- is X with the opposite orientation. Here we identify two such M if they are diffeomorphic with a diffeomorphism consistent with the identifications along the boundary. If $f \in \mathrm{Hom}(X,Y)$ and $g \in \mathrm{Hom}\,(Y,Z)$ then $g \circ f$ is defined to be the union of the manifolds g and f along Y. For any X, the element id_X is taken to be $X \times [0,1]$ with the standard identifications on the boundary. (The tubular neighborhood theorem guarantees that this satisfies the requirements for id_X.) We define the tensor product to be disjoint union (on objects and on morphisms). The element 1 is then the empty set, $1 = \emptyset$. A *topological quantum field theory* is a functor from $\mathcal{M}_n$ to **Vec** satisfying some additional conditions, cf. [56] [143]

6.2 Braided categories.

There are various other properties of **Vec** that we will want to mimic by axiomatization for general categories. For example, for vector spaces we have a natural identification of $X \otimes Y$ with $Y \otimes X$. In a general category with product, we will *postulate* the existence of a natural transformation, s, between the two functors $(X,Y) \rightsquigarrow X \otimes Y$ and $Y \otimes X$ from $\mathcal{C} \times \mathcal{C} \to \mathcal{C}$. Thus we assume that for every pair of objects X and Y there is a (two sided invertible) morphism

$$s_{X,Y} : X \otimes Y \to Y \otimes X.$$

(Mac Lane added the requirement that $s^2 = id$. But recent investigations have shown that it is better to drop or weaken this requirement.)

Given three objects, X, Y and Z, there are two different paths going from $X \otimes (Y \otimes Z)$ to $(Z \otimes X) \otimes Y$ following the right or left hand paths in the hexagon of Figure 6.1. We want these to compositions to yield the same morphism.

In other words, we impose the requirement

$$a_{Z,X,Y} \circ s_{X \otimes Y,Z} \circ a_{X,Y,Z} = (s_{X,z} \otimes id) \circ a_{X,Z,Y} \circ (id \otimes s_{Y,Z}). \qquad (6.2.1)$$

Of course, if $s_{X,Y} : X \otimes Y \to Y \otimes X$ then $(s_{Y,X})^{-1}$ also goes from $X \otimes Y$ to $Y \otimes X$ and hence can serve as another s. It turns out that it is necessary

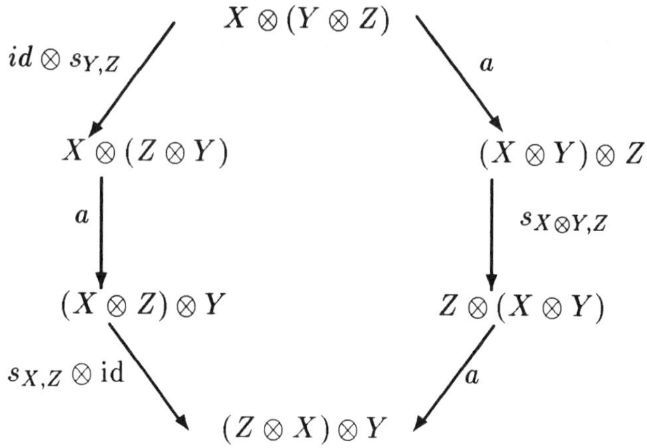

Figure 6.3:

to impose the hexagon condition for this morphism as well, that is it is necessary to impose the hexagon condition on it as well, that is to require that

$$\begin{aligned} &a_{Z,X,Y} \circ (s_{Z,X\otimes Y})^{-1} \circ a_{X,Y,Z} \\ &= ((s_{X,Z})^{-1} \otimes id) \circ a_{X,Z,Y} \circ (id \otimes (s_{Z,Y})^{-1}) . \end{aligned} \tag{6.2.2}$$

A monoidal category having an s satisfying (6.2.1) and (6.2.2), with s compatible with the identifications of $1 \otimes X$ and $X \otimes 1$ with X is called a *braided* category.

In [824], Mac Lane proves a coherence theorem under the additional hypothesis that $s^2 = 1$ which, roughly speaking, says that if the pentagon and hexagon conditions are satisfied, then any two natural morphisms relating the taking of tensor products in one order (and choice of parentheses) to another are the same. A recent proof of this theorem of MacLane was given along geometric lines by Kapranov, [608], similar to the geometric proof using the Stasheff polyhedra that we gave in the preceding chapter. Kapranov constructs a simply connected cell complex whose vertices are the various possible orders and placements of parentheses in an $n-$fold tensor product, and whose edges are the symmetries and associativites.

A generalization of this theorem to braided categories was proved by Joyal and Street, cf.[596]. Here is a good way to think of the Joyal Street coherence theorem suggested by Majid [862]. Introduce notation for the commutativity morphisms

$$s_{X\otimes Y} : X \otimes Y \to Y \otimes X \quad \text{and} \quad s^{-1}_{Y,X} : X \otimes Y \to Y \otimes X$$

as actual braids, as indicated in the figure:

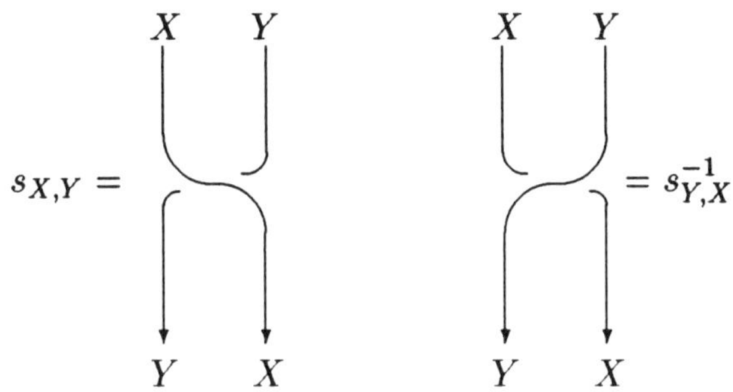

With this notation, the first hexagon identity can be drawn as

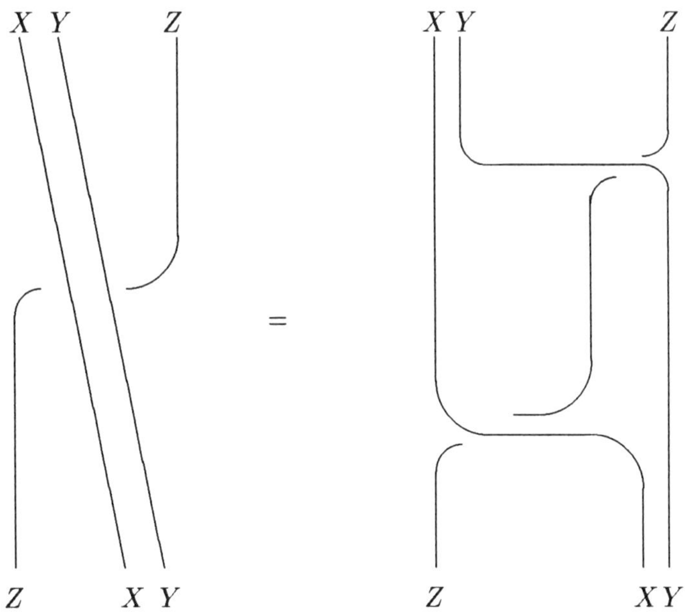

The doubled line on the left is construed as referring to the composite object $X \otimes Y$, and the overcrossing of this double line with the curve joining the Z's is thus $s_{X\otimes Y,Z}$. On the right, the motion of the Y line away from X and towards Z is taken as the associativity

$$(X \otimes Y) \otimes Z \to X \otimes (Y \otimes Z).$$

The upper crossing of the Y line with the Z line then corresponds to

$$id \otimes s_{Y,Z} : X \otimes (Y \otimes Z) \to X \otimes (Z \otimes Y).$$

The motion of the Z line towards the X line corresponds to the associativity

$$X \otimes (Z \otimes Y) \to (X \otimes Z) \otimes Y.$$

The crossing of the Z line and the X line corresponds to

$$s_{X\otimes Z} \otimes id : (X \otimes Z) \otimes Y \to (Z \otimes X) \otimes Y.$$

Finally the motion of the X line towards the Y corresponds to the associativity

$$(Z \otimes X) \otimes Y \to Z \otimes (X \otimes Y).$$

Thus the right hand side of the equation corresponds to a five step trip around the hexagon, and the equation represents the first hexagon identity. The second hexagon identity can be represented by a similar diagram.

Suppose we are in a strictly monoidal category, so that all associativities are identities, and hence all the sliding motion in the preceding diagram corresponds to the identity transformation. For strictly monoidal braided categories the Joyal Street coherence theorem can be thought of as follows: If a series of morphisms built out of associativities and commutativities correspond to the same braid, then they compose to the same morphism.

For the case of a general braided monoidal category we must modify the above statement a little: For example, we might want to think of

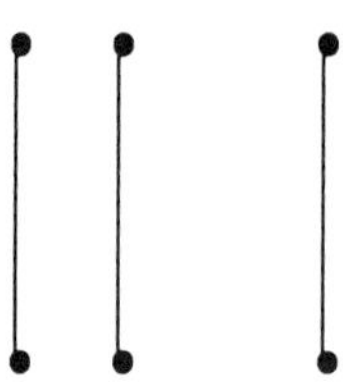

and

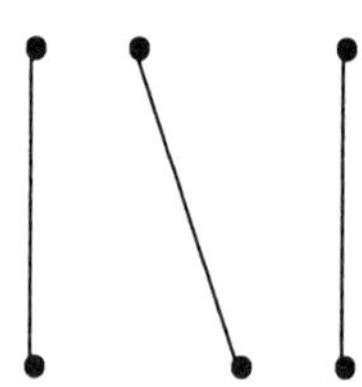

as representing the same braid. But the figure on the left corresponds to $id \in \text{Hom}((X \otimes Y) \otimes Z, (X \otimes Y), \otimes Z)$ while the figure on the right is an associativity and lies in $\text{Hom}(X \otimes Y) \otimes Z, X \otimes (Y \otimes Z))$. So we want to consider only compositions which start from the same initial expression, in and the same final expression, fin. Here in and fin are expressions that can be written as tensor product formulas of objects in a category (and with the identity, I) with various placements of the parentheses. We can then formulate one version of the Joyal Street coherence theorem as follows: Suppose that we have two morphisms between built out of a succession of of ϕ, r, l and s, both morphisms between the same initial and final expression. Then

Proposition 6.2.1 Joyal and Street. *To check whether two morphisms built out of ϕ, r, l, s are the same in all braided monoidal categories it suffices to check that the underlying braids are the same.*

For example, consider the two morphisms indicated by the diagrams in Figure 6. Reading from the top down, both diagrams represent morphisms in

$$\text{Hom}((A \otimes B) \otimes C, (C \otimes (B \otimes A)).$$

The diagrams on the left and right in Figure 6.3 correspond to the compositions

$$\begin{array}{ccccccccccc} (AB)C & \xrightarrow{\phi} & A(BC) & \xrightarrow{id\otimes s} & A(CB) & \xrightarrow{\phi^{-1}} & (AC)B & \xrightarrow{s\otimes id} & (CA)B & \xrightarrow{\phi} & C(AB) \\ & & & & & & & & & & \downarrow{\scriptstyle id\otimes s} \\ & & & & & & & & & & C(BA) \end{array}$$

and

$$\begin{array}{ccccccccccc} (AB)C & & & & & & & & & & \\ {\scriptstyle s\otimes id}\downarrow & & & & & & & & & & \\ (BA)C & \xrightarrow{\phi} & B(AC) & \xrightarrow{id\otimes s} & B(CA) & \xrightarrow{\phi^{-1}} & (BC)A & \xrightarrow{s\otimes id} & (CB)A & \xrightarrow{\phi} & C(BA) \end{array} \tag{6.2.3}$$

where we have omitted the tensor product sign between objects so that the formulas can fit on the page. The braids on both sides in Figure 6.3 are equal as can be seen by sliding the strings across. In fact, the equality of the braids drawn on both sides is exactly the famous braid relation

$$\tau_2\tau_1\tau_2 = \tau_1\tau_2\tau_1. \tag{6.2.4}$$

The Joyal Street coherence theorem asserts that both compositions in (6.2.3) are equal. This equality is one of the most important equations in this book. It is the categorical version of the celebrated *Yang-Baxter equation* whose origins in statistical mechanics we will explain in the next chapter.

But we can give a direct proof of the equality of the two compositions in (6.2.3). Indeed, each of the horizontal rows is a five step trip around a hexagon, and hence can be replaced by the single remaining step, in view of the hexagon conditions. Thus first horizontal row can be replaced by

$$s_{A\otimes B,C} : (A \otimes B) \otimes C \longrightarrow C \otimes (A \otimes B)$$

and the second horizontal row can be replaced by

$$s_{B\otimes A,C} : (B \otimes A) \otimes C \longrightarrow C \otimes (B \otimes A).$$

Thus the Yang Baxter equation, the equality of both compositions in (6.2.3), boils down to the equation

$$(id \otimes s) \circ s = s \circ (s \otimes id)$$

which is a consequence of the naturality of s.

In fact, we have essentially just given the entire proof of the Joyal Street coherence theorem, since the braid relations are the key ingredient in the equality of braids. We will present a more formal version of the coherence together with an indication of the proof in the next section.

6.3 The category of braids.

In order to give a more precise formulation to the Joyal Street coherence theorem, we introduce the following

Definition. Let $\mathcal{C}$ and $\mathcal{D}$ be braided monoidal categories and let (F, c, f) be a monoidal functor from $\mathcal{C}$ to $\mathcal{D}$. We say that (F, c, f) is *braided* if

$$\begin{array}{ccc} F(A) \otimes F(B) & \xrightarrow{c} & F(A \otimes B) \\ {\scriptstyle s}\downarrow & & \downarrow{\scriptstyle F(s)} \\ F(A) \otimes F(B) & \underset{c}{\longrightarrow} & F(B \otimes A) \end{array} \tag{6.3.1}$$

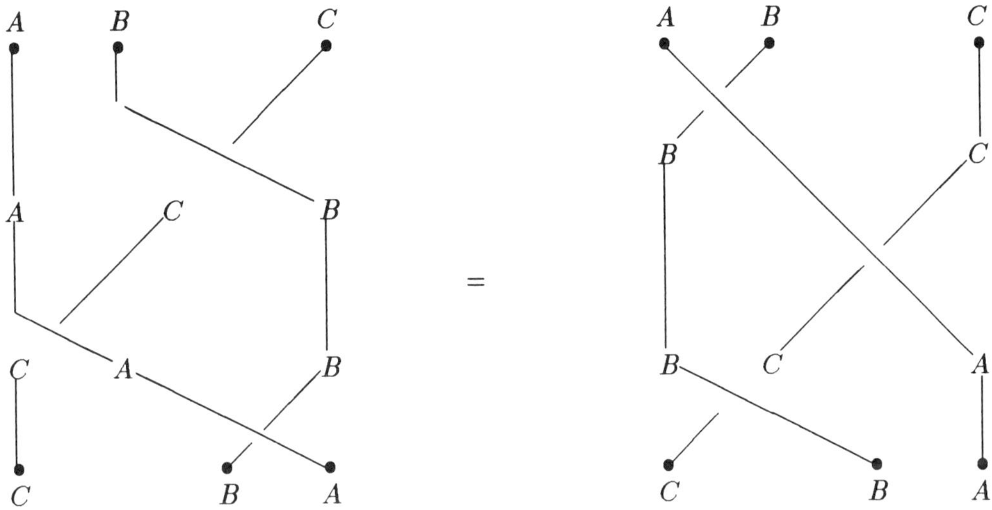

Figure 6.4: The categorical version of the Yang Baxter equation

commutes.

For example, suppose that (F, c, f) is a monoidal equivalence from a braided monoidal category, $\mathcal{C}$ to a monoidal category $\mathcal{D}$. Then the braiding on on $\mathcal{C}$ transports along F to a unique braiding on $\mathcal{D}$ so that (F, c, f) is braided. In particular, let $\mathcal{C}$ be a braided monoidal category and let $\mathbf{st}\mathcal{C}$ be a strict monoidal category equivalent to C. We know from Proposition 6.1 that such a category exists. Then $\mathbf{st}\mathcal{C}$ inherits a braiding. So starting with any braided monoidal category we can produce a strictly monoidal braided category (unique up to braided monoidal equivalence).

The basic braided monoidal category is the category of braids. It is defined as follows. The objects in $\mathcal{B}$ are the non-negative integers $0, 1, 2...$ with

$$\mathrm{Hom}(m, n) = \emptyset, \quad m \neq n$$

with Hom(0, 0) the trivial group and with

$$\mathrm{Hom}(n, n) = B_n$$

where B_n denotes the Artin braid group [48] and [49]. We briefly recall their definition. Let C_n denote the space of subsets of cardinality n of $\mathbf{R}^2$.

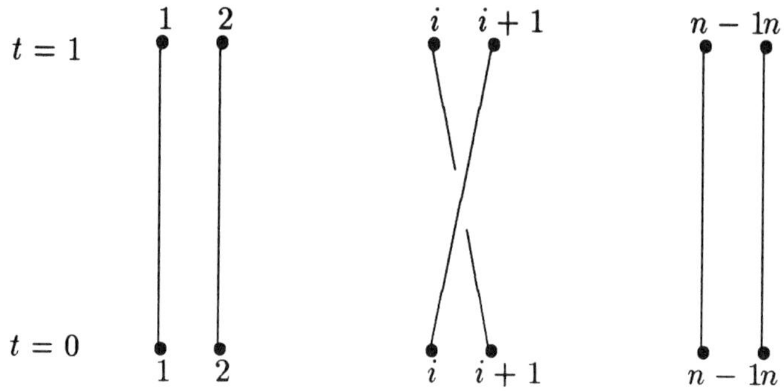

Figure 6.5: The i−th generator, τ_i, of B_n

Then B_n is the fundamental group of C_n. Usually the set of n collinear points $\{(1,0),(2,0),\ldots,(n,0)\}$ along the x−axis is taken as base point and is denoted by $[1,2,\ldots,n]$. A loop based at $[1,2,\ldots,n]$ can be described by its graph in $I \times \mathbf{R}^2$ where I denotes the unit interval, and so is a union of n curves in $\mathbf{R}^3$ called strings. The k−th curve joins the point $(0,k,0)$ to the point $(1,\sigma(k),0)$ where $\sigma \in S_n$ is called the permutation associated to the braid. Usually, this union of curves is drawn by projection (along the y−axis) with t− direction increasing vertically, and the x− axis horizontal. Let τ_i be the braid depicted in Figure 6.4.

Then a presentation of the braid group, B_n is given by the generators $\tau_1,\ldots,\tau_{n-1}$ and the relations

$$\tau_i\tau_{i+1}\tau_i = \tau_{i+1}\tau_i\tau_{i+1} \qquad 1 \leq i \leq n-2 \tag{6.3.2}$$

$$\tau_i\tau_j = \tau_j\tau_i \qquad 1 \leq i < j-1 \leq n-2. \tag{6.3.3}$$

see [49]. The category $\mathcal{B}$ is equipped with a strict monoidal structure given by the juxtaposition (side by side) of two braids: given a braid $b_m \in B_m$ and a braid $b_n \in B_n$ we get a braid with $m+n$ strings by using b_m as strings attaching the first m points and b_n on the last n points. Algebraically, this can be described as the homomorphism of $B_m \times B_n \to B_{m+n}$ determined by

$$(\tau_i, \tau_j) \mapsto \tau_i\tau_{m+j}.$$

The braiding on $\mathcal{B}$ must be a map of B_{m+n} into itself (for each m and

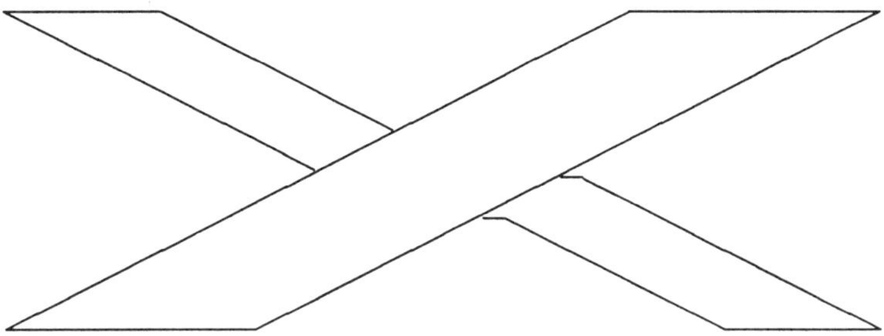

Figure 6.6: The braiding on the category $\mathcal{B}$.

n) which sends

$$\tau_i \tau_{m+j} \mapsto \tau_j \tau_{n+i}.$$

It is given by multiplication in B_{m+n} by the braid which is drawn by straight line overcrossings (in order) of the first m positions over the last n positions. Geometrically it can be depicted as in Figure 6,4

The inverse braiding is then given by an undercrossing instead of an overcrossing. The functoriality (i.e. natural transformation property) of this braiding can b visualized diagrammatically as sliding down the shute as in Figure 6.5.

Similarly, the first hexagon identity is proved geometrically by Figure 6.6.

Let $\mathcal{C}$ and $\mathcal{D}$ be two braided categories. We can consider the category whose objects are braided monoidal functors from $\mathcal{C}$ to $\mathcal{D}$ and whose arrows are monoidal natural transformations. Let $\mathbf{BM}(\mathcal{C}, \mathcal{D})$ denote this category. Then

Proposition 6.3.1 Joyal and Street. *For every braided monoidal category, $\mathcal{C}$, evaluation at $1 \in \mathcal{B}$ is an equivalence of categories*

$$\mathbf{BM}(\mathcal{B}, \mathcal{C}) \cong \mathcal{C}.$$

Indication of proof.(Joyal and Street [596]) By transport of structure, we may replace $\mathcal{C}$ by an equivalent strictly monoidal category, and so it suffices to prove the proposition under the hypothesis that $\mathcal{C}$ is strictly monoidal.

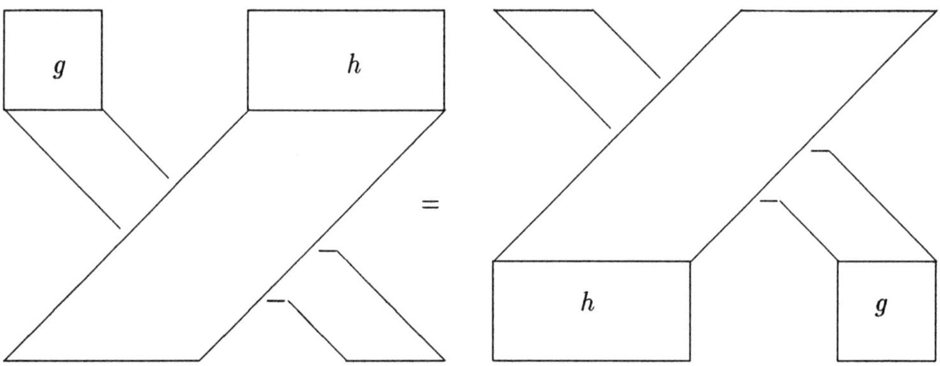

Figure 6.7: Functoriality of the braiding.

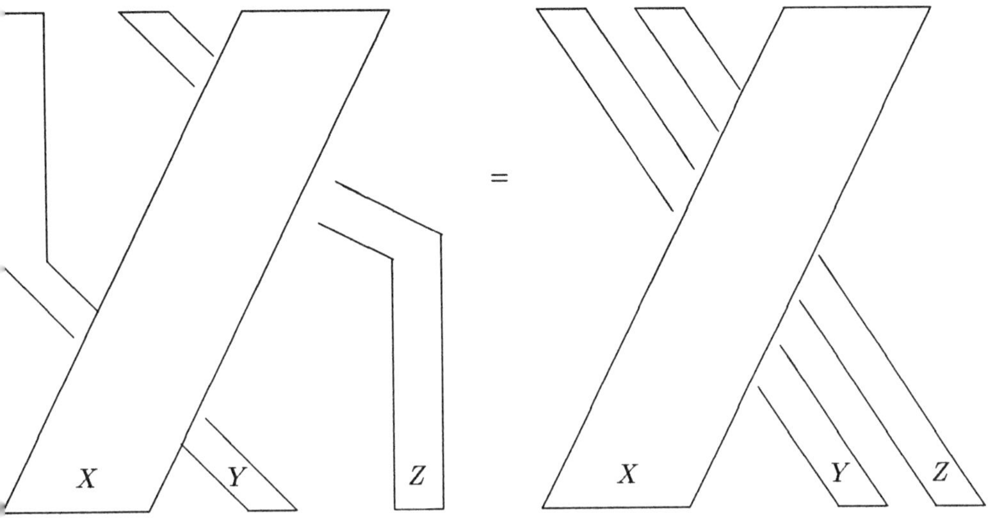

Figure 6.8: A proof of one of the two hexagon identities for the category of braids.

To each object A of $\mathcal{C}$ we describe the unique strict monoidal functor, $F_A : \mathcal{B} \to \mathcal{C}$ with $F_A(1) = A$. This forces

$$F_A(n) = A^{\otimes n}$$

since the tensor product of n copies of 1 in $\mathcal{B}$ is n. Now the braiding in $\mathcal{B}$ on the object $1 \otimes 1 = 2$ is

$$s_{1,1} = \tau_1 \in \mathrm{Hom}(2,2) \quad \text{in} \quad \mathcal{B}.$$

This forces

$$F(\tau_1) = s_{A,A} \in \mathrm{Hom}(A, A) \quad \text{in} \quad \mathcal{C}.$$

But in $\mathcal{B}$ we have the expression

$$\tau_i = id_{i-1} \otimes \tau_1 \otimes \tau_{n-i-1} \in B_n = \mathrm{Hom}(n, n).$$

This forces

$$\begin{aligned} F(\tau_i) = \\ id \otimes s_{A,A} \otimes id \in \\ \in \mathrm{Hom}(A^{\otimes(i-1)} \otimes A \otimes A \otimes A^{\otimes(n-i-1)}, A^{\otimes(i-1)} \otimes A \otimes A \otimes A^{\otimes(n-i-1)}) \\ = \mathrm{Hom}(A^{\otimes n}, A^{\otimes n}). \end{aligned}$$

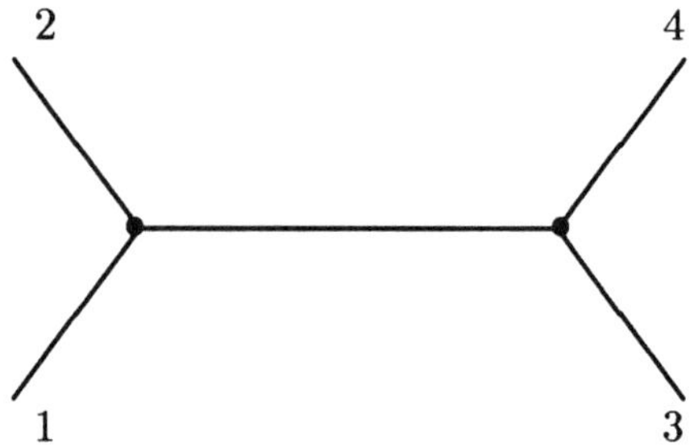

Figure 6.9:

To check that this is a functor one must check that the braid relations are satisfied. but this is precisely what is guaranteed by the categorical Yang Baxter equation that we proved in the preceding section. One then checks in a straightforward manner that this is a braided strict monoidal functor, and that the operation of evaluation at 1 is fully faithful. For details we refer the reader to [596] or [600].

In addition, it is convenient to demand of a category the existence of something analogous to a dual space which then allows a representing object for Hom (X, Y) to be thought of as the object $Y \otimes X^*$. We now interrupt the abstract discussion to explain how some of these ideas arise in physics, the hexagon and associated braid relations in the location of the critical temperature of the Ising model (work of the 1940's) and the pentagon in the dual theories of strong interaction physics of the 1970's and the more recent conformal string theories. We will discuss the relations with dual models in this chapter, and turn to statistical mechanics in the next.

6.4 Duality and crossing.

In the diagrammatic description of particle interactions one may regard a diagram such as Figure 6.8 as indicating that particles 1 and 2 combine to form an intermediate particle which decays into particles 3 and 4.

"Crossing" refers to the assertion that there is a relation between the interaction described by the diagram of Figure 6.8 and the interactions associated to the diagrams in Figures 6.9 and 6.10. In its simplest form the assertion relates the analytic continuation of the amplitudes of these processes when expressed in terms of Lorentz invariant combinations of the four momenta of the interacting particles, the "Mandelstam variables" s, t, and u, cf.[882].

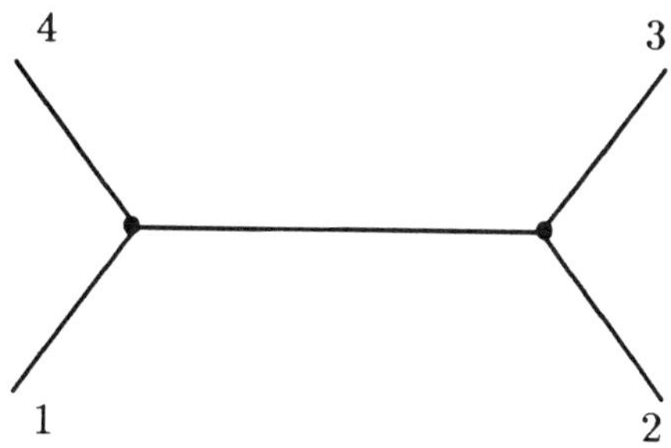

Figure 6.10:

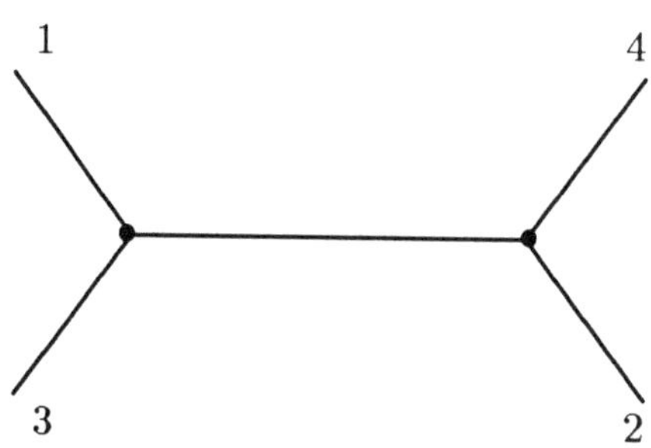

Figure 6.11:

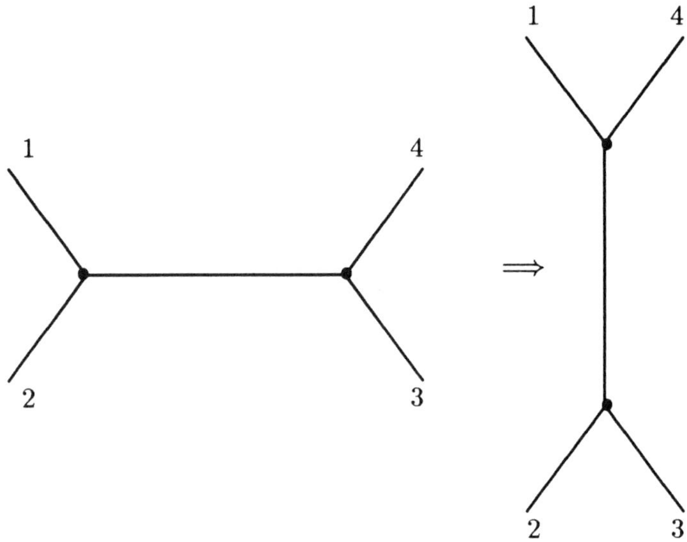

Figure 6.12:

In the 1960's a systematic study of crossing in the context of "dual models" was begun cf. [297]. With the advent of the Veneziano model [1181] and its reinterpretation as a quantum string theory, duality was considered as a basic ingredient in the formulation of strong interaction physics [941], and [1138]. In some of the early papers on dual models [659], [522], [1054] one sees the "crossing transformation" of Figure 6.11 applied to an internal portion of a more complicated graph. Thus, for example, the above transformation, when applied to the graph of Figure 6.12 yields the graph of Figure 6.7.

In 1988 [933],[1175] [1010], [1183] [190] crossing transformations on graphs reemerged in the context of finding constraints on two dimensional rational conformal field theories. Roughly speaking, one associates to each crossing transformation a matrix which is, essentially, the matrix of a monodromy operator on a space of meromorphic sections of a line bundle over a Riemann surface whose knowledge determines the theory. Now one may pass from one graph to another by two different sequences of crossing transformations. Each of these routes will correspond to an expression in the basic matrices. Equating the expression corresponding to two different routes then provides a constraint on the basic matrices. It then becomes important to determine a set of generating relations from which all the others follow. It will turn out

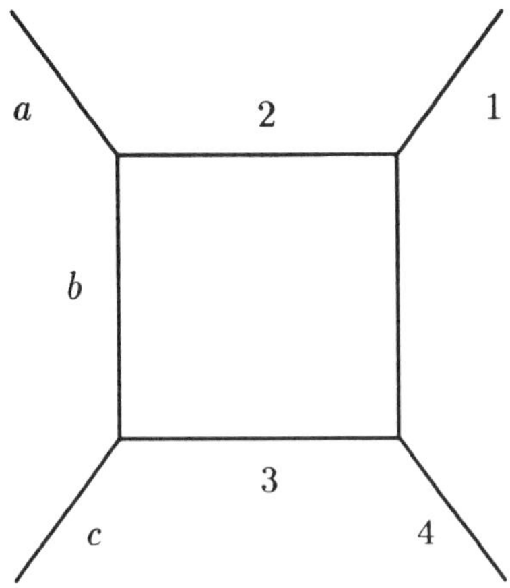

Figure 6.13:

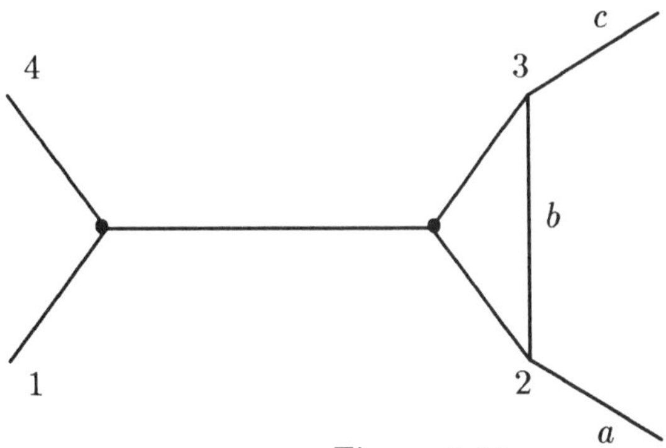

Figure 6.14:

that this question is intimately related to the Mac Lane coherence theorems stated earlier in this Chapter. The results of the next few sections are joint work of Brustein, Ne'eman and Sternberg obtained in the Tel Aviv seminar of the summer of 1988 and published with some delay in [189].

6.5 Constructions on graphs.

In order to avoid confusion arising from differences in terminology between mathematicians and physicists we make the following definitions. A *graph* consists of two sets, E, its set of edges (or one dimensional objects) and N, its set of nodes (or zero dimensional objects). Furthermore to each edge, $e \in E$, is associated a subset $\{n_1, n_2\}$ of N called its boundary nodes. We say that n_1 and n_2 are *incident* to e. (We do not, at this stage exclude the possibility that n_1 might equal n_2 .)

In the physics literature, nodes that are incident to more than one edge are called vertices. But in the mathematical terminology the word vertex refers to all nodes. So we will avoid the use of the word vertex altogether, and speak of internal and external nodes.

We will consider *trivalent* graphs where every node is incident to either three edges (the internal nodes) or to one edge (the external nodes). In physics language this means that we are considering graphs similar to the Feynman diagrams occurring in a ϕ^3 theory.

External edges (those attached to external nodes) will be called *legs*. We can consider two types of constructions:

Joining: We can pick one leg from each of three graphs and join them. So the three graphs of Figure 6.14 join together to yield the single graph of Figure 6.15. (In this figure X, Y and Z stand for various graphs whose internal structure we have not drawn.)

Or, we can pick two legs from one graph and join them to one from another. So the two graphs of Figure 6.16 combine to yield the single graph of Figure 6.17. Or we can join three legs of a single graph.

Crossing. If the graph of Figure 6.8 occurs as a subdiagram of a graph, then

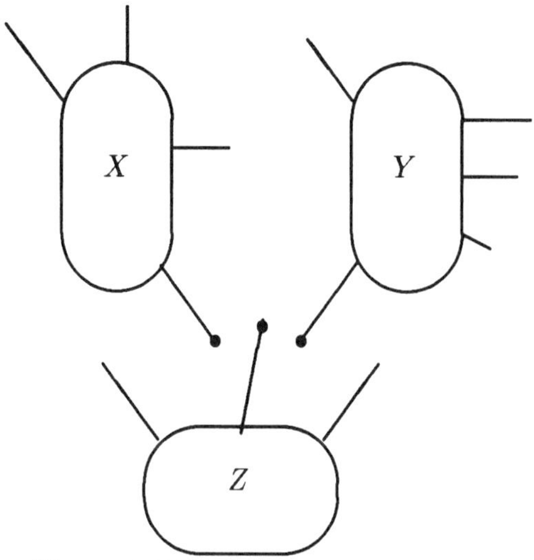

Figure 6.15:

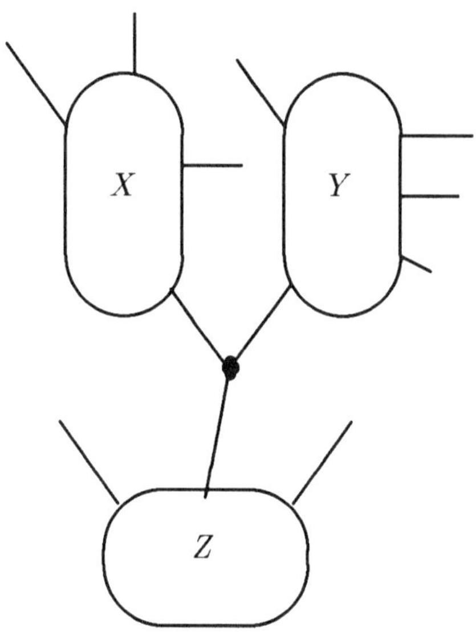

Figure 6.16:

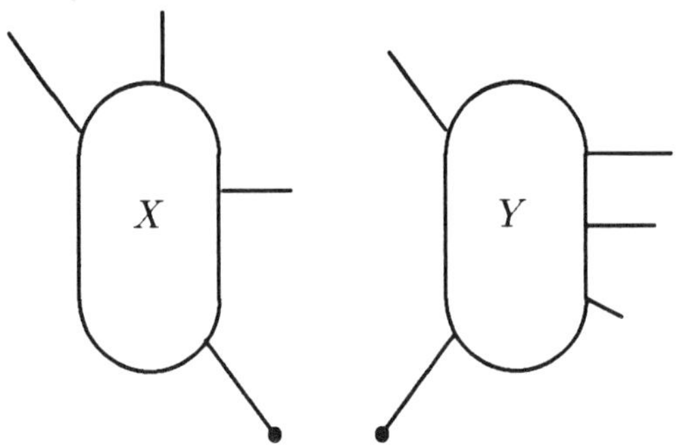

Figure 6.17:

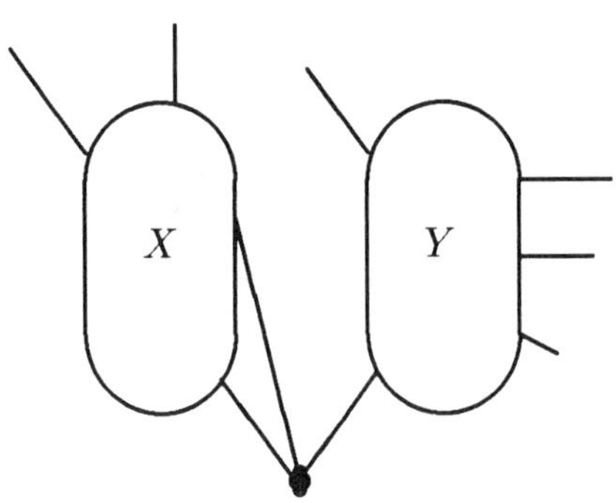

Figure 6.18:

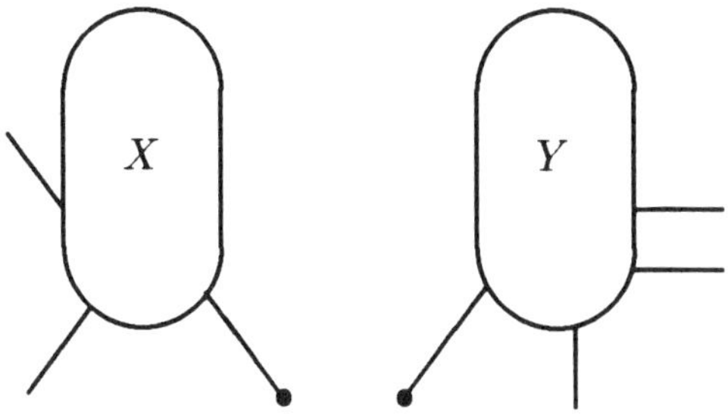

Figure 6.19:

$J(X, Y) =$

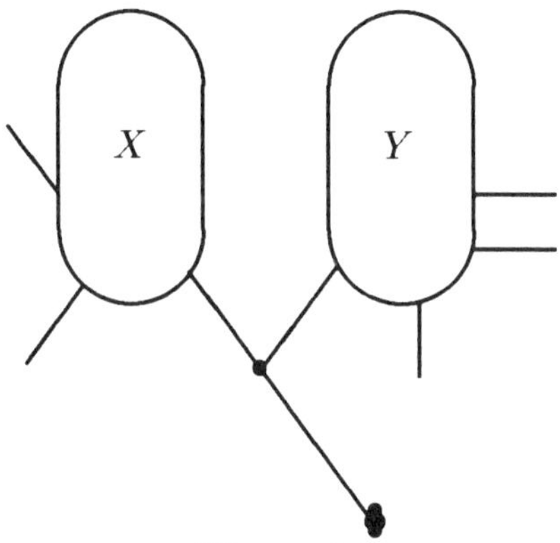

Figure 6.20:

it is replaced by the graph of Figure 6.9 or 6.10 (with all the other connections unchanged). An illustration of this operation was given in Figure 6.13 of the preceding section.

Suppose that we consider connected graphs with one leg with an exterior node marked. At the risk of mixing metaphors, we will call such a graph a rooted graph, and the marked leg the root. Then joining becomes a binary operation, J, on rooted graphs, by joining the two roots to the trivial graph, consisting of a single edge and two nodes, and then marking the remaining node. Thus the two graphs of Figure 6.18 combine to yield the single graph of Figure 6.19.

Notice that the operation J does not introduce any new loops (cycles). In particular, if χ_1 and χ_2 are trees (have no cycles), then $J(\chi_1, \chi_2)$ is again a tree.

Indeed, let n_1 and n_2 denote the number of nodes in χ_1 and χ_2 and e_1

and e_2 the number of edges. Then

$$n = n_1 + n_2$$

and

$$e = e_1 + e_2 + 1$$

where n and e denote the number of nodes and edges in $J(\chi_1, \chi_2)$. So if $n_1 = e_1 + 1$ and $n_2 = e_2 + 1$ then $n = e + 1$.

6.6 Cyclically ordered graphs.

For any finite set, A, let $S(A)$ denote the group of all one to one transformations of A onto itself. So if A has n elements, the group $S(A)$ is isomorphic to the symmetric group $S_n = S(\{1, 2, \ldots, n\})$.

Suppose that A has n elements. By a *cyclic order* on A we mean an element, s, of $S(A)$ which has order n. There are thus $(n-1)!$ different cyclic orders on A.

By a *cyclically ordered* graph we shall mean a graph for which a cyclic order has been chosen at each (internal) node. In some of the more recent literature, a cyclically ordered graph is called a *fat graph.*

When we perform the operation J on two cyclically ordered graphs, X and Y, we must decide which of the two cyclic orders to put on the joining node. If x denotes the root (the marked leg) of X and y denotes the root of Y and j denotes the new leg, we must choose between the cyclic orders (j, x, y) and (j, y, x) at the new node. We will denote the first choice by $J(X, Y)$ and the second choice by $J(Y, X)$. We have thus proved

Proposition 6.6.1 *The join operation defines a nonsymmetric binary operation, J , on cyclically ordered graphs. If X and Y are trees, then so is $J(X, Y)$.*

Let us consider the basic four legged graph of Figure 6.20, which we shall denote by $\times$. Let $G = S(A)$ where A is the set of legs of this graph. So G is isomorphic to S_4 if we label the legs as in Figure 6.8.

There are three unordered labelled graphs corresponding to Figures 6.8, 6.9 and 6.10. The group G acts transitively on this three element set (the action of S_4 on unordered partitions of $\{1, 2, 3, 4\}$ into two unordered subsets).

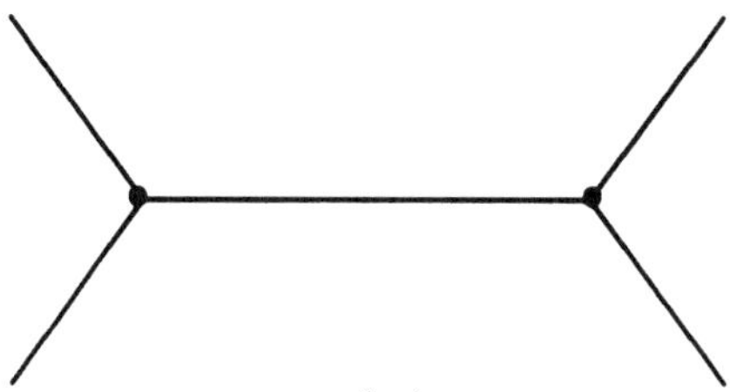

(x)

Figure 6.21:

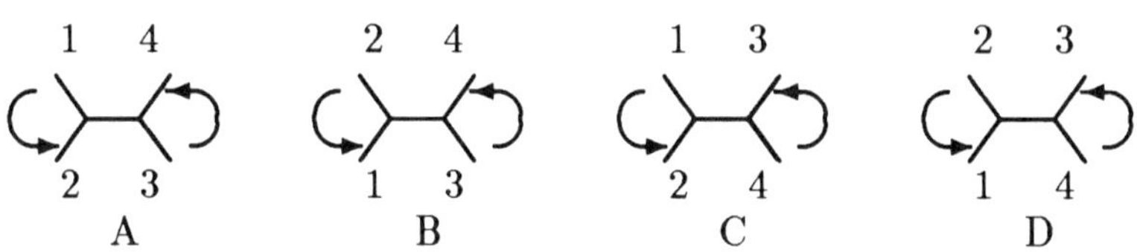

Figure 6.22:

Each of these unordered graphs gives rise to four cyclically ordered graphs. Thus the four graphs in Figure 6.81 are the possible cyclically ordered graphs corresponding to the graph of Figure 6.8.

The group G acts transitively on the twelve element set of all cyclically ordered graphs. The element (13)(24) fixes the graph A in Figure 6.21. (If we were to consider oriented cyclically ordered graphs, where an orientation is chosen on each edge, then (13)(24) would not act trivially and the orbit of G would contain twenty four distinct possibilities.) If a four legged graph, $\times$ is an internal subgraph of some unordered graph X_0, then we have the two graphs Y_0 and Z_0 associated to X_0 by crossing, and G acts transitively on this three element set as before. Let us choose a cyclic order on X_0. We denote the corresponding cyclically ordered graph by X. Conversely, if X is cyclically ordered, we let X_0 denote the corresponding unordered graph. Suppose that X is a cyclically ordered graph, $\times$ is a four legged subdiagram of X and $b \in G(\times)$. We let bX be the cyclically ordered graph given by cyclically ordering the nodes of bX_0 as follows: each node of bX_0 other than the two nodes corresponding to the nodes of $\times$ comes from a unique node of X_0. On each such node we put the old cyclic ordering as in X. At the nodes corresponding to those of the crossing subgraph, $\times$, we put the cyclic ordering given by the action of G as described above.

We can now describe "moves" on cyclically ordered graphs:

Onsager**Move**: For a cyclically ordered graph, X, pick a 4-legged subgraph, $\times$. Then pick an element $b \in G(A(\times))$ and obtain the cyclically oriented graph bX by the procedure given above.

The mathematical problem that then arises is: describe the relations among the moves. That is, describe those sequences of moves that will lead from X back to the same graph X.

We must be precise about what we mean by "the same X ": Notice that the various legs of the graph retain their identity at each move. So after a sequence of k moves, we obtain a graph Y together with an identification of the legs of X with the legs of Y. We say that "Y is the same as X " if there is a one to one map of (the edges and nodes) of X onto (the edges and nodes of) Y which preserves all incidences and cyclic orders, and which reduces to the given identification on the legs.

We can also consider a more restricted class of moves defined as follows: Suppose we consider a four legged graph $\times$. We claim that a cyclic order on the nodes of $\times$ induces a cyclic order on the legs of $\times$ as in Figure 6.22. In other words, the cyclic orders (12m) and (m34) induce the cycle (1234).

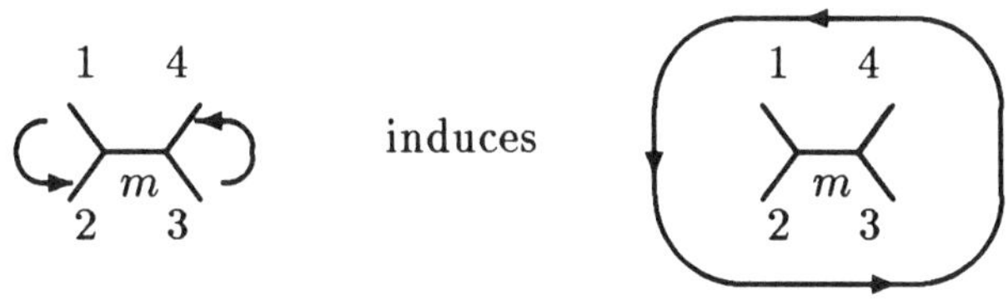

Figure 6.23:

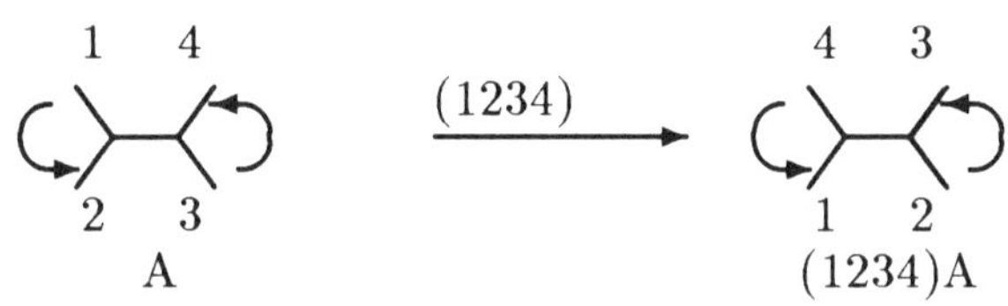

Figure 6.24:

(Notice that if we choose the nodes in the opposite order we get the same cycle, (3412) = (1234).)

Notice that (1234) carries A into the graph (1234)A as drawn in Figure 6.23, and the cycle associated to (1234)A is again (1234).

Notice also that

$$(1234)^2 = (13)(24)$$

which carries A into itself, as is to be expected.

Restricted moves: Let X be any cyclically ordered graph. Each four legged subgraph, $\times$ is cyclically ordered, and hence determines an element, a of $G(\times)$. Apply this preferred element a to X so as to move to aX. Then choose another four legged subgraph etc. These are the restricted moves.

6.7 Restricted moves on rooted trees.

In this section we will show that a cyclically ordered rooted tree is, in fact, just a vertex of K_n, in other words, an element of $F(\{-\})_n$. (We shall also see that a restricted move on a rooted tree is just an associativity operator as in section 6.1 , and hence we can apply Mac Lane's coherence theorem to determine all the relations among restricted moves on cyclically ordered rooted trees.

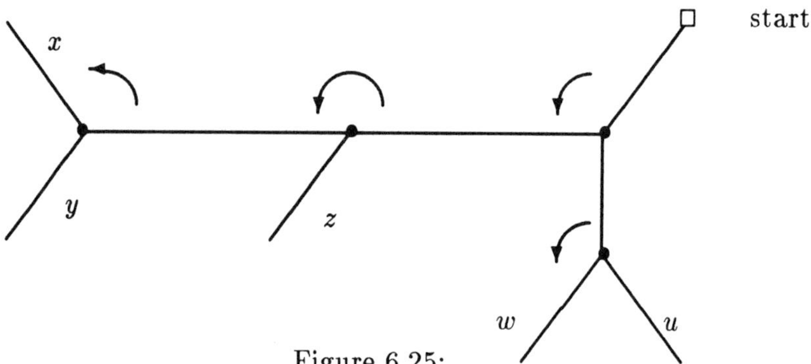

Figure 6.25:

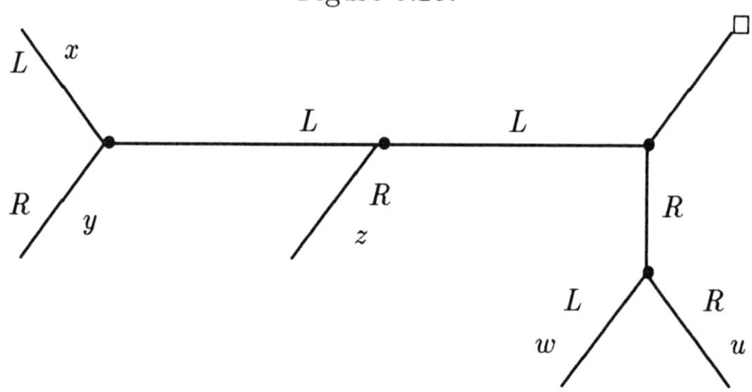

Figure 6.26:

Suppose we are given a rooted tree. We can regard the cyclic order on the nodes as a "sorting procedure" on the remaining legs as follows: Start from the root (the marked leg). At its juncture, mark the first edge encountered in the cyclic order by L and the second edge by R. This is to be interpreted as follows: All legs connected to the edge marked L will be placed to the left of all the legs attached to the edge marked R . If the edge we just marked L is not a leg, continue to its other joining node and repeat the procedure of marking the first edge encountered in the cyclic order by L and the second by R. Do the same for the edge marked R if it is not a leg. Continue. Now consider each interior node as a tensor product and hence each interior edge as a parenthesis system. Thus, for example,the graph of Figure 6.24 is marked as in Figure 6.25 and corresponds to the "tensor product" expression

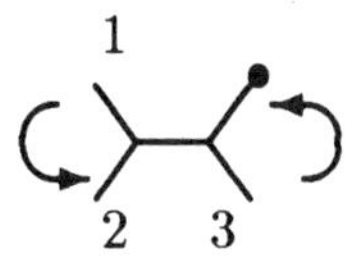

Figure 6.27:

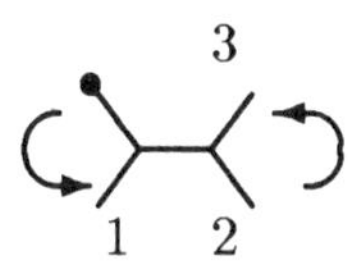

Figure 6.28:

$$((x \otimes y) \otimes z) \otimes (w \otimes u).$$

Here is another way of saying the same thing in the language of the free magma on one generator. Any cyclically ordered rooted tree can be built from smaller trees by the operation J unless it is already the trivial tree consisting of a single edge. So we let this trivial tree be denoted by _ and write

$$X \otimes Y \text{instead of } J(X,Y).$$

Thus we have proved:

Proposition 6.7.1 *There is a bijective correspondence between cyclically ordered rooted trees and elements of the free magma on one generator. The correspondence can be established by representing the tree by successive application of the join operation.*

Now consider the four legged graph of Figure 6.26 where we have marked the leg 4. It corresponds to $(1 \otimes (2 \otimes 3)$. The operator $a = (1234)$ applied to this graph gives the graph of Figure 6.27 which corresponds to $1 \otimes (2 \otimes 3)$.

Thus a^{-1} is an "associativity operator".

If $\times$ is an interior subgraph of a rooted tree , X, one and only one of its legs will lead to the root, since X is a tree, and we may consider this leg of $\times$ as marked. Hence every restricted move on a cyclically ordered rooted tree is an instance of an associativity.

We can now read the relations on restricted moves on rooted trees directly from Mac Lane's theorem. They are generated by relations of two types:

1) a "rectangle identity" or "naturality" which asserts that a restricted move applied to a subtree attached to one of the unmarked legs of $\times$ "commutes" with the restricted move associated to the four legged subdiagram $\times$. Put another way, the moves associated to disjoint subdiagrams commute. and 2) a "pentagon identity" which relates the two paths going from

$$1 \otimes (2 \otimes (3 \otimes 4)) \text{ to} ((1 \otimes 2) \otimes 3) \otimes 4$$

one in two moves and the other in three moves. In our present discussion we would draw Mac Lane's pentagon identity as in Figure 6.28.

A four cycle and any transposition (two cycle) generate S_4. On any rooted tree each four legged subdiagram, $\times$, has a marked edge as we have observed, and hence a unique transposition in $G(A(\times))$, the transposition of the two legs not joined to the marked edge. This transposition clearly corresponds to an instance of commutativity transformation in the tensor product interpretation. Indeed Figure 6.29 corresponds to $(1 \otimes 2) \otimes 3$, and thus interchanging 1 and 2 is a commutativity operator.

Hence we can simply translate Mac Lane's hexagon identities relating associativity and commutativity into the graph language.

Mac Lane's coherence theorem thus asserts that the hexagon, pentagon, and naturality identities determine all the relations on cyclically ordered rooted trees.

Notice, that from the graphical point of view the fundamental binary operation on our free magma is represented by the basic cyclically ordered rooted tree having exactly one vertex and three edges. If we think of this graph as situated in three dimensional Euclidean space, we can "thicken" it (take the normal sphere bundle) so as to obtain a "pair of pants" as in the following figure:

In the conformal field theories in two dimensions, these "pairs of pants" play a basic role.

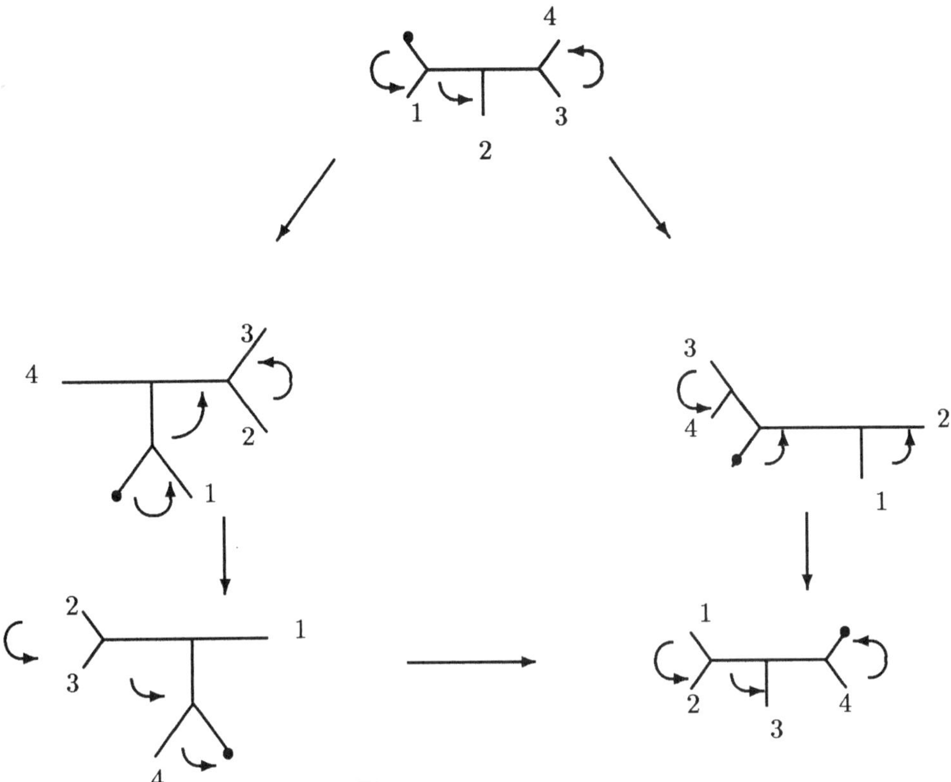

Figure 6.29:

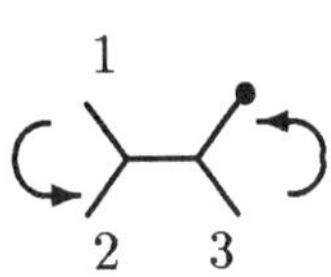

Figure 6.30:

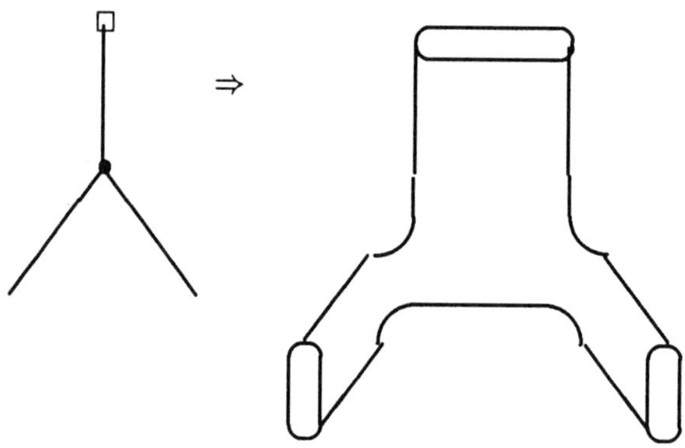

Figure 6.31:

Chapter 7

Lattice statistical mechanics

7.1 Ising models

In this Chapter we discuss the rudiments of statistical mechanics on a lattice. Basically, we just do the bare minimum needed to get to the star triangle relation. The classic on the subject is the book by Baxter, [105], who is one of the main innovators in the field. For a general introduction to the ideas of statistical mechanics and their relation to classical thermodynamics, cf.[74].

We begin with a very sketchy review of discrete statistical mechanics. Suppose we are given a finite set Σ and a function H on Σ called the "energy". Let β be any positive real number. Then we can put a probability measure on Σ by assigning each $\sigma \in \Sigma$ the weight $e^{-\beta H(\sigma)}$. That is, define the partition function to be

$$Z = Z(\beta) = \sum e^{-\beta H(\sigma)}$$

where the sum extends over all $\sigma \in \sum$. Then the probability assigned to σ is

$$p_\beta(\sigma) = e^{-\beta H(\sigma)}/Z(\beta). \tag{7.1.1}$$

The parameter β is related to the absolute temperature by the formula $\beta = 1/kT$ where k is "Boltzmann's constant". So "high temperature" corresponds to small values of β and "low temperature" corresponds to large values of β.

The function

$$\psi = \log Z$$

is called the*Massieu function.* It is the basic function in statistical mechanics.

For example,

$$-\frac{\partial\psi}{\partial\beta} = \sum H(\sigma)p_\beta(\sigma)$$

is the expected value of the energy at the temperature corresponding to β.

In the various *Ising models* we shall consider, the space Σ is described as follows: we are given a finite graph $\Gamma = (E, V)$ where E is the set of edges and V is the set of vertices. The space Σ is the set of maps of V into the two element set $\{1, -1\}$. So a $\sigma \in \Sigma$ assigns the value ± 1 to each vertex of the graph. The elements, σ will sometimes be called "states", or "configurations".

The energy function will be defined as follows. For each edge e there will be an "interaction parameter" $w(e)$, and

$$H(\sigma) = -\sum w(e)\sigma_i\sigma_j. \tag{7.1.2}$$

Here the sum is over all edges of the graph, and $\sigma_i\sigma_j$ is the product of the values of σ at the two end points of the edge, e. (We don't allow loops, so each edge has exactly two end points.)

Usually $w(e)$ will be positive for all e. Then each edge makes a positive contribution to the energy when the values of σ at its ends are different. So the σ with all the values at ends of edges the same will have the lowest energy. From the minus sign in the exponent in (7.1.1) it follows that this state will be most probable, and with increased values of β, that is at lower and lower temperatures, this preference for the more and more correlated σ's becomes more pronounced.

The mathematical problem is to study the behavior of this system at various temperatures, as the graph Γ grows in a regular way. For example, we may want to consider the limit $\psi(\beta, \Gamma)/N$ where $N = N(\Gamma)$ is the number of vertices in Γ, i.e. the Massieu function per unit site. The question is to analyze the values where this limiting function develops singularities. Alternatively, we can consider Γ as having a boundary which recedes to infinity as Γ grows, and the question becomes whether, at some values of β, the imposition of boundary values for σ affects the probability in the interior.

The initial results on these questions, due to Ising, [566] were negative. He considered the "one dimensional case" where Γ is a linear chain of $N-1$ edges and all the $w(e)$ are equal, say $w(e) \equiv 1$. He showed that the Massieu function per unit site does not develop any singularities, and that the limiting probabilities at the interior of the chain are independent of the boundary conditions.

7.2 The Peierls argument.

The first non-trivial result is an ingenious argument due to Peierls in the 1930's which we now describe. We will take Γ to be a square subgraph with N sites of the graph consisting of the lattice of integer points in the plane. We will allow two different values for $w(e)$, the value $w(e) = A$ if e is a horizontal edge and $w(e) = B$ if e is a vertical edge. Let us set $K = \beta A$ and $L = \beta B$. This allows us to get rid of the double negative in the definition of Z and write

$$Z_N = \sum_{\sigma} \exp\left(K\Sigma\sigma_i\sigma_j + L\Sigma\sigma_i\sigma_j\right). \tag{7.2.1}$$

Here the first sum is over all horizontal edges of the square where i and j are the boundary points of each edge and the second sum is over all vertical edges. When $\beta = 0$, all states have equal probability. In particular, imposition of boundary conditions, even for a finite lattice, can have no effect on the interior. The result of Peierls is that for low temperature, i.e. large β, the boundary effects persist even as the boundary recedes to infinity.

Let us consider the case where $K = L$. From general physical arguments one does not expect the existence of a phase transition to depend on the particular choice of boundary conditions. Accordingly, to exhibit the existence of a phase transition we will make a specific choice: we consider only those σ which take on the value $+1$ on the boundary. We wish to prove that there is a critical value, β_c, such that for $\beta > \beta_c$

$$\text{Prob}_\beta\{\sigma(0) = -1\} < \alpha < \frac{1}{2} \tag{7.2.2}$$

where α is independent of N. The expression on the left denotes the probability that the value at the origin is -1 with respect to the probability assignment (7.1.1). So if $\mathcal{E}$ is any *event*, i.e. $\mathcal{E}$ is a subset of the set of states, the probability assigned to $\mathcal{E}$ by (7.1.1) is

$$\text{Prob}_\beta(\mathcal{E}) = \sum_{\sigma\in\mathcal{E}} p_\beta(\sigma). \tag{7.2.3}$$

In (7.2.2) the event $\mathcal{E}$ is that the value of the state at the origin is -1. In other words, $\mathcal{E}$ is the set of all σ with $\sigma(0) = -1$,

$$\mathcal{E} = \{\sigma | \sigma(0) = -1\},$$

and the sum defining the left hand side of (7.2.2) is over this set. The significance of (7.2.2) is that the boundary condition affects the probability

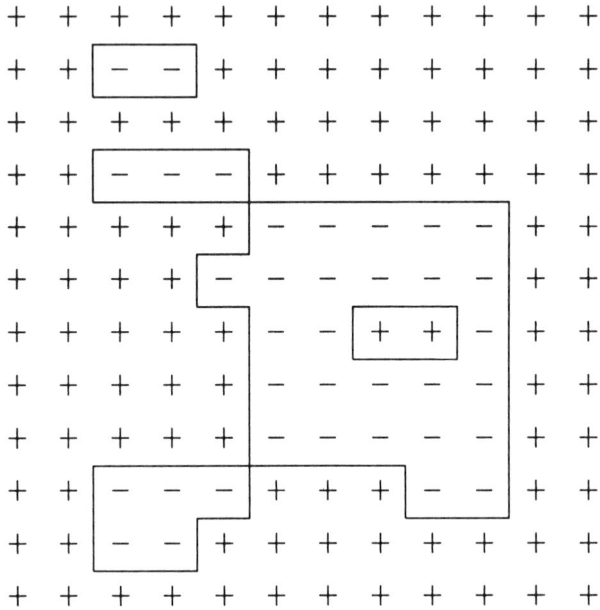

Figure 7.1: A state σ and the corresponding $C(\sigma)$

at an interior point, no matter how far this point is from the boundary, if the temperature is sufficiently small.

Here is Peierls' argument. Define the "dual lattice" to be the square lattice whose vertices are at the centers of the squares of Γ. So each vertex of the dual lattice is surrounded by four vertices of Γ. Any state, σ, defines a series of edges in the dual lattice, namely those edges which subdivide edges of Γ where σ takes on opposite values. Call this subset of edges of the dual lattice, $C(\sigma)$. The number of changes in values at the four vertices of Γ as we go around any vertex in the dual lattice must be even, i.e. $0, 2$, or 4 . Hence the set $C(\sigma)$ must consist of a collection of closed polygons. Conversely $C(\sigma)$ completely determines σ: Starting from the boundary condition $\sigma = +1$ the value of σ changes sign every time we cross a portion of $C(\sigma)$.

Let $|\sigma|$ denote the length $C(\sigma)$. In other words, $|\sigma|$ denotes the number of edges in $C(\sigma)$. Each edge of $C(\sigma)$ corresponds to two adjacent vertices of Γ with opposite values of σ. Hence we can rewrite the partition function (7.2.1) as

$$Z_N = e^{2MK} \sum_{\sigma} e^{-2K|\sigma|} \tag{7.2.4}$$

where M denotes the total number of horizontal = total number of vertical edges in the square (which is essentially equal to N) . Thus

$$p_\beta(\omega) = \frac{e^{-2K|\omega|}}{\sum_\sigma e^{-2K|\sigma|}}. \tag{7.2.5}$$

Let γ be a fixed simple closed polygon in the dual lattice, and let I_γ be the random variable defined by

$$I_\gamma(\sigma) = 1 \text{ if } \gamma \text{ occurs as a component of some} C(\sigma)$$

and

$$I_\gamma(\sigma) = 0$$

otherwise.

Proposition 7.2.1 Peierls' lemma. *The expectation, $E_\beta(I_\gamma)$ of the random variable I_γ satisfies the inequality*

$$E_\beta(I_\gamma) \le e^{-2K|\gamma|}. \tag{7.2.6}$$

Proof. Let $S = S(\gamma)$ denote the set of σ for which γ is a component of $C(\sigma)$. For each $\sigma \in S(\gamma)$ let σ^* denote the state obtained from σ by changing the signs at all the vertices of the original lattice which are inside γ. This has the effect of eliminating γ as a component, and $|\sigma| = b + |\sigma^*|$ where $b = |\gamma|$. Thus

$$E_\beta(I_\gamma) = \frac{\sum_{\sigma \in S} e^{-2K|\sigma|}}{\sum_{\text{all } \sigma} e^{-2K|\sigma|}} = e^{-2Kb} \frac{\sum_{\sigma \in S} e^{-2K|\sigma^*|}}{\sum_{\text{all} \sigma} e^{-2K|\sigma|}}.$$

But this last quotient is ≤ 1 since every term in the sum in the numerator occurs in the sum in the denominator. QED.

Now to say that $\sigma(0) = -1$ means that there is some component, γ, of $C(\sigma)$ which contains the origin in its interior. For any such γ the probability of γ occuring as a component of is at most e^{-2Kb} by Peierls' lemma, where $b = |\gamma|$. Thus the left hand side of (7.2.2) is at most

$$\sum_{b \ge 4} e^{-2Kb} B_b.$$

where B_b is the number of closed polygons of length b containing 0 in the interior.

Now the number B_b is difficult to compute exactly, but we can estimate it as follows: Any closed path γ enclosing the origin must cross the positive

$x-$axis somewhere. If $|\gamma| = b$, then there at most $b/2$ possibilities for the crossing point nearest to the origin. Taking the upper vertex of the vertical crossing nearest the origin as starting point, there are at most three choices of edge at each vertex. Hence we have the estimate

$$B_b \leq \frac{b}{2} 3^b.$$

Hence the probability on the left of (7.2.2) is estimated by

$$\sum_{b \geq 4} \frac{b}{2} e^{(\ln 3 - 2K)b}.$$

So if K is sufficiently large, the above sum is small. In particular the left hand side of (7.2.2) can be made as small as we like, independent of N, if we choose K large enough.

The problem the arises: what is the critical temperature? That is, find the value of K_c such that the boundary effects disappear for $K < K_c$ as $N \to \infty$ but persist for $K > K_c$. In the standard texts, cf. for example [105], it is shown that this definition of critical point is equivalent to the following: For which values of K does the limit of the Massieu function per unit site,

$$\lim_{N \to \infty} \frac{1}{N} \ln Z_N$$

develop a singularity.

In 1940 Kramers and Wannier [714] evaluated this critical value as

$$K = \frac{1}{2} \ln(2^{1/2} + 1) \tag{7.2.7}$$

under the additional assumption that there is exactly one critical point. Their argument was an ingenious extension of the Peierls argument presented in this section. Then, in a celebrated paper of 1944, Onsager proved that there is exactly one critical point, justifying the Kramers Wannier computation, cf. [962]. This paper was a mathematical tour de force leading to the solution of the square lattice Ising model. Onsager then extended the computation to the case of the triangular lattice via the introduction of the star triangle relation. It is safe to say that these two papers of Onsager represent the starting point of our whole subject.

7.3 The Kramers - Wannier computation.

We go back to the general expression for Z_N with free boundary values. Recall the notations: N denotes the number of lattice sites and M the number of horizontal edges = the number of vertical edges $\sim N$. Any σ determines a polygon $C(\sigma)$, but now there are two $\sigma's$ which correspond to any $C(\sigma)$ since reversing the signs at all the sites does not change $C(\sigma)$.

Let

$$r = r(\sigma)$$

denote the number of horizontal bonds in $C(\sigma)$ and

$$s = s(\sigma)$$

the number of vertical bonds of $C(\sigma)$. So

$$Z_N = 2\exp[M(K+L)]\sum_C \exp(-2Lr - 2Ks) \tag{7.3.1}$$

where the sum is over all closed (possibly disconnected) polygons in the dual lattice inside our square. (The L multiplies the r in (7.3.1) since a horizontal bond, α appears in $C(\sigma)$ if σ takes opposite values at the ends of the vertical bond of the original lattice bisected by α. Similarly K multiplies the number of vertical bonds in $C(\sigma)$. Formula (7.3.1) is exact for all temperatures, but becomes useful at low temperatures as we have seen above. It is known as the *low temperature representation.*

We are interested in the limit of the Massieu function per unit site as N gets large, i.e. in

$$\psi(K, L) = \lim_{N\to\infty} (1/N)\ln Z_N.$$

From the low temperature representation we get

$$\psi(K, L) = K + L + \phi(e^{-2L}, e^{-2K}) \tag{7.3.2}$$

where

$$\phi(v, w) = \lim_{N\to\infty} N^{-1}\ln\sum_C v^r w^s. \tag{7.3.3}$$

Of course, at the moment, we have no idea how to compute the function ϕ or even be convinced that the limit defining it exists. But there is another representation for Z_N deriving from the fact that

$$e^x = \cosh x + \sinh x.$$

We can write

$$Z_N = (\cosh K \cosh L)^M \sum_{\sigma} \prod (1 + v\sigma_i\sigma_j) \prod (1 + w\sigma_i\sigma_j) \qquad (7.3.4)$$

where

$$v = \tanh K \text{ and } w = \tanh L.$$

The first product is over all horizontal edges and the second is over all vertical edges in our original graph, Γ. We can give a graphical description of each term in the expansion of the double product by marking the horizontal edge $e = (ij)$ if we select the term $v\sigma_i\sigma_j$ (as opposed to the term 1) in the first product expansion and similarly marking the vertical edge (ij) if we choose the term $w\sigma_i\sigma_j$ in the second. In this way there is a one to one correspondence between the 2^{2M} terms in expanding out the products in (7.3.4) and the set of all subsets of edges. Given a set S of edges, with r horizontal and s vertical edges, the corresponding term in the expansion will be

$$v^r w^s \sigma_1^{n_1} \sigma_2^{n_2} \sigma_3^{n_3} \dots$$

where n_i denotes the number of edges in S having site i as an end point.

We must sum this over all S and then over all σ. Interchanging the order of summation, the contribution of any S where any n_i is odd will be zero, since both $+1$ and -1 occur at the same position, multiplying a common factor. So the sum is only over those S for which all the n_i are even - the S corresponding to closed paths.

For a fixed such S all σ give the same value, $v^r w^s$. Hence for fixed S the contribution to Z_N is $2^N v^r w^s$ and so

$$Z_N = 2^N (\cosh K \cosh L)^M \sum_{C} v^r w^s \qquad (7.3.5)$$

where now the sum is over all closed curves in the original lattice. This equation is known as the *high temperature expansion*. So

$$\psi(K, L) = \ln[2 \cosh K \cosh L] + \phi(\tanh K, \tanh L), \qquad (7.3.6)$$

again assuming that we have proved that the limit defining ϕ exists. But now consider two different pairs of values (K, L) and (K^*, L^*) related by

$$\tanh K^* = e^{-2L} \text{ and } \tanh L^* = e^{-2K}. \qquad (7.3.7)$$

Substituting (7.3.7) into (7.3.6) we get

$$\psi(K^*, L^*) = \ln[2\cosh K^* \cosh L^*] + \phi(e^{-2L}, e^{-2K})$$

so by (7.3.2)

$$\psi(K^*, L^*) = \psi(K, L) + \ln[2\cosh K^* \cosh L^*] - (K + L). \tag{7.3.8}$$

This last equation does not explicitly involve the unknown ϕ! It gives a relation between the Massieu functions at different temperatures. In order to understand it, we must examine the "duality transformation" relating the starred and unstarred variables. Let us set

$$b = e^{-2L} \text{and } a = e^{-2K^*}.$$

Then the relation between L and K^* is given by

$$b = \frac{-a+1}{1+a}$$

which is the fractional linear transformation corresponding to the matrix

$$T = \begin{pmatrix} -1 & 1 \\ 1 & 1 \end{pmatrix}.$$

But $T^2 = 2I$ as a matrix, hence the fractional linear transformation corresponding to T has square equal to the identity. In other words the relation between L and K^* is reciprocal. Indeed we have

$$b^{-1} = \frac{a+1}{1-a}$$

so

$$b^{-1} - b = \frac{4a}{1-a^2} = \frac{4}{a^{-1} - a}$$

and thus

$$\sinh 2K^* \sinh 2L = 1. \tag{7.3.9}$$

Now suppose that $K = L$ and hence $K^* = L^*$. Then, setting

$$\Psi(K) = \psi(K, K)$$

we have

$$\Psi(K^*) = \Psi(K) + \ln[2(\cosh K^*)^2] - 2K. \tag{7.3.10}$$

If K_c is a value of K at which the function Ψ fails to be analytic, the Ψ must also fail to be analytic at the corresponding value, K_c^*. Hence if we *assume* that the limits exist and that there is only one critical point, it must be located where $K = K^*$ i.e. at the value

$$\sinh 2K = 1$$

or

$$K = \frac{1}{2}\log(2^{1/2} + 1) = 0.44068679\ldots.$$

This is the Kramers-Wannier computation, later justified by Onsager. Onsager extended the Kramers-Wannier computation to determine the critical temperature for the triangular and hexagonal lattices. For this he was led to introduce the star triangle relation, as we shall now explain.

7.4 Honeycomb-triangular duality.

We would like to find the critical temperature of the honeycomb or hexagonal lattice by the Kramers-Wannier method. We will have a low temprature representation, an expression which is the analogue of (7.3.1), and involves a sum over the dual lattice. We will also have a high temperature representation, the analogue of (7.3.5) which involves a sum over the original lattice. But these two lattices are not the same, as in the square lattice case: The dual of the honeycomb lattice is the triangular lattice whose vertices are at the centers of the hexagons of the honeycomb. So we can not immediately apply the Kramers-Wannier method which involves an involution relating the high and low temperature expansions on the same lattice. The star triangle relation will allow us to overcome this difficulty.

The honeycomb has three parallel classes of edges which we call types 1, 2 and 3. Thus, to fix the ideas, let us say that the type 1 edges are at 30° to the positive x−axis, the type 2 edges are at 150° and the type 3 edges are at 270°. Each vertex in the honeycomb has three nearest neighbors, connected to it each of the three types of edges. If we start with a vertex which we color white, and call of type A, we then color its nearest neighbors black and call them of type B. We then color the neighbors of the black vertices white, and call them of type A. Proceeding in this way we color alternatively each vertex of the honeycomb black, or type A, and white or type B, so black vertices

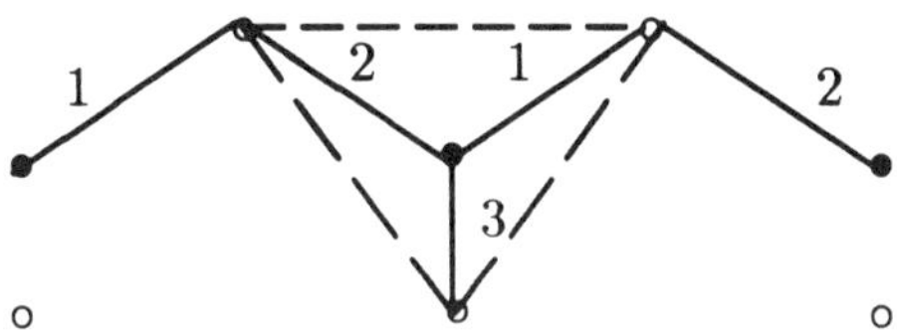

Figure 7.2:

form a triangular lattice as do the white vertices. The star triangle relation will allow us to eliminate the type A vertices from the expression for the high temperature representation and get a partition function for the triangular lattice of the type B vertices. Relating the high and low representations for the triangular lattice allows Onsager to compute the critical temperature for the honeycomb:

Let $\mathbf{L} = (L_1 L_2, L_3)$ where the L_i are positive and define the hexagon partition function with $2N$ sites, $Z_{H,N}$ (with H now standing for "hexagon") as

$$Z_{H,N}(\mathbf{L}) = \sum_{\sigma} \exp(L_1 \Sigma^1 \sigma_i \sigma_j + L_2 \Sigma^2 \sigma_i \sigma_j + L_3 \Sigma^3 \sigma_i \sigma_j) \tag{7.4.1}$$

where Σ^i denotes sum over edges of type i. The number, N_i, of edges of type i is not exactly N, but the error involved in replacing N_i by N will disappear when we pass to the thermodynamic limit which involves taking log, dividing by N and letting $N \to \infty$. So we have the low temperature representation

$$Z_{H,N}(\mathbf{L}) = \exp N(L_1 + L_2 + L_3) \sum_{P} \exp(-2L_1 r_1 - 2L_2 r_2 - 2L_3 r_3), \tag{7.4.2}$$

where the sum is over all closed polygons P of the dual triangular lattice and $r_j = r_j(P)$ is the number of edges of type j in P. If we set

$$\Psi_N(v_1, v_2, v_3) = \sum_{P} v_1^{r_1} v_2^{r_2} v_3^{r_3}$$

Then we have

$$Z_{H,N}(\mathbf{L}) = \exp\,(N(L_1 + L_2 + L_3)\Psi_N(e^{-2L_1}, e^{-2L_2}, e^{-2L_3}). \qquad (7.4.3)$$

We now turn to the high temperature expansion: Each type A vertex is connected to exactly three type B vertices and no others. Hence we can write each summand, in (7.4.1) as

$$\prod \exp\ \sigma_m(L_1\sigma_{1(m)} + L_2\sigma_{2(m)} + L_3\sigma_{3(m)}) \qquad (7.4.4)$$

where the product is over all type A vertices m, and where 1(m) is the type B vertex connected to m by an edge of type 1, etc. Each type A vertex occurs exactly once in this product expansion. To compute the sum in (7.4.1) we first sum over all the values of σ_i corresponding to type A vertices. For each such vertex, m, the sum at m over the two values, ± 1, of σ_m becomes

$$\exp[L_1\sigma_{1(m)} + L_2\sigma_{2(m)} + L_3\sigma_{3(m)}] + \exp[-(L_1\sigma_{1(m)} + L_2\sigma_{2(m)} + L_3\sigma_{3(m)})]$$

which equals

$$2\cosh\ (L_1\sigma_{1(m)} + L_2\sigma_{2(m)} + L_3\sigma_{3(m)}).$$

Let us write $i = 1(m)$, $j = 2(m)$, $k = 3(m)$. Then

$$Z_{H,N} = \sum_{\sigma} \prod 2\cosh\ (L_1\sigma_i + L_2\sigma_j + L_3\sigma_k) \qquad (7.4.5)$$

where the product is over all the triangles centered at the type A vertices and the sum is over all the σ restricted to the type B vertices.

Now cosh is a positive even function. So even though there are eight possibilities for the values of σ at the vertices of the triangle, the function

$$2\cosh(L_1\sigma_i + L_2\sigma_j + L_3\sigma_k)$$

takes on only four values. Hence, given L, we should be able to find four numbers R, K_1, K_2, K_3 such that

$$2\cosh(L_1\sigma_i + L_2\sigma_j + L_3\sigma_k) = R\exp(K_1\sigma_j\sigma_k + K_2\sigma_i\sigma_k + K_3\sigma_i\sigma_j) \quad (7.4.6)$$

for all eight possible values of the σ, allowing us to express $Z_{H,N}$ in terms of a partition function, $Z_{T,N}$ of the triangular lattice of type B vertices.

7.5 The star triangle relation.

Equation (7.4.6) is known as the star triangle relation. It relates the contribution of the "star" centered at an A type vertex of the hexagonal lattice to a contribution to the partition function of the associated triangle surrounding this vertex made up of B type vertices.

A graphical version of the star triangle relation is given in figure 7.3. In this figure, the variables at the vertices take on the values ± 1. An edge corresponds to an exponential function of its end points with indicated weights. The top edge of the triangle, for example, corresponds to the function $\exp K_3ab$. We multiply the functions associated to all the edges, so the triangle corresponds to the function $\exp(K_1bc+K_2ac+K_3ab)$ of the variables attached to its vertices. In the star, associate a variable f to the center circle, so that the bottom edge, for example, represents the function $\exp L_3fc$. The fact that the circle is black indicates summation over the associated variable. So the function associated to the star is $\cosh(L_1a + L_2b + L_3c)$. Equation (7.4.6) (with $(\sigma_i, \sigma_j, \sigma_k)$ replaced by (a, b, c) asserts that the two functions are proportional. The product occuring in the sum on the right of (7.4.5) is over all black vertices, i.e. over all stars. Similarly, the partition function for the triangular lattice can be written as a sum of products, where the products are over all triangles.

Explicitly, equation(7.4.6) reduces to the four equations

$$\begin{aligned} 2\cosh\,(L_1 + L_2 + L_3) &= R\exp(K_1 + K_2 + K_3), \\ 2\cosh\,(-L_1 + L_2 + L_3) &= R\exp(K_1 - K_2 - K_3), \\ 2\cosh(L_1 - L_2 + L_3) &= R\exp(-K_1 + K_2 - K_3), \\ 2\cosh(L_1 + L_2 - L_3) &= R\exp(-K_1 - K_2 + K_3) \end{aligned} \tag{7.5.1}$$

Multiplying the first two equations and dividing by the product of the last two equations gives $\exp[4K_1]$ on the right hand side in terms of an expression involving the L's, and similarly for each of the remaining K's.

Having solved for **K**, we can then solve for R. We will do the details of this computation in a moment. The upshot will be that

$$\sinh 2K_i \sinh 2L_i = k^{-1} \quad i = 1, 2, 3 \tag{7.5.2}$$

where

$$k^{-1} = k(\mathbf{L})^{-1} = \prod \sinh 2L_i/2(cc_1c_2c_3)^{1/2} \quad i = 1, 2, 3 \tag{7.5.3}$$

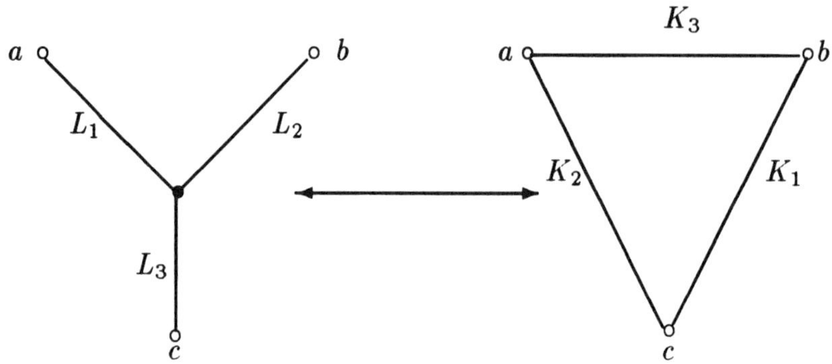

Figure 7.3: The star triangle relation. Each of the variables at the open circles can take on the values ± 1. Each edge represents an exponential function of its end points with indicated weights. We take the product over all edges. A black circle indicates summation of the corresponding variable.

where

$$\begin{aligned} c &= \cosh(L_1 + L_2 + L_3) \\ c_1 &= \cosh(-L_1 + L_2 + L_3) \\ c_2 &= \cosh(L_1 - L_2 + L_3)\text{and} \\ c_3 &= \cosh(L_1 + L_2 - L_3). \end{aligned}$$

Given a solution of (7.5.3) we can then multiply all four equations relating the K's to the L's to get

$$R^4 = 2^4 c c_1 c_2 c_3$$

so

$$R^2 = 2 \cdot 2(c c_1 c_2 c_3)^{1/2} = 2k s_1 s_2 s_3 = 2/(\sinh 2K_1 \sinh 2K_2 \sinh 2K_3).$$

Any given set of solutions of (7.5.3) determines the R by the preceding equation and hence a solution of our original four equations, (7.5.1), relating the exponentials of the L's to the cosh of the K's.

The point is that we can solve for R, k and $\mathbf{K}$ as a function of $\mathbf{L}$ or for k, R and $\mathbf{L}$ as a function of $\mathbf{K}$.

In any event we get

$$\begin{aligned} Z_{H,N}(\mathbf{L}) &= R^N Z_{T,N}(\mathbf{K}(\mathbf{L})) \\ &= \exp\left(N(L_1+L_2+L_3)\Psi_N(e^{-2L1}, e^{-2L2}, e^{-2L3}\right). \end{aligned}$$

For a triangular lattice we can apply the high temperature expansion, the triangular analogue of (7.3.5), to get, for $\mathbf{J} = (J_1, J_2, J_3)$ the formula

$$\begin{aligned} Z_{T,N}(\mathbf{J}) &= (2\cosh J_1 \cosh J_2 \cosh J_3)^N \Psi_N(v_1, v_2, v_3) \\ v_i &= \tanh J_i, \quad i = 1, 2, 3. \end{aligned}$$

If we choose the $\mathbf{J}$ so that $\tanh J_i = \exp(-2L_i)$, $i = 1, 2, 3$ we see from the preceding two formulas that we will get a relation between $Z_{T,N}(\mathbf{J})$ and $Z_{T,N}(\mathbf{K})$. It turns out that the map $\mathbf{K} \rightsquigarrow \mathbf{J}(\mathbf{L}(\mathbf{K}))$ is an involution, and this determines the critical temperature of the isotropic triangular lattice (and the isotropic hexagonal lattice) under the assumption that there is exactly one critical point. This was Onsager's argument. The details for the isotropic case yielding

$$\begin{aligned} \sinh 2K_c &= 3^{-\frac{1}{2}} \\ \sinh 2L_c &= 3^{\frac{1}{2}} \end{aligned}$$

can be found on pp86-87 of [105].

For the proof of (7.5.2) and (7.5.3) solve for K_1 as indicated above so that

$$e^{4K_1} = \frac{cc_1}{c_2c_3},$$

hence

$$e^{4K_1} - 1 = \frac{cc_1}{c_2c_3} - 1 = \frac{cc_1 - c_2c_3}{c_2c_3} = \frac{s_2s_3}{c_2c_3}$$

where

$$s_i = \sinh 2L_i.$$

Multiplying by

$$e^{4K_1} = \left(\frac{cc_1}{c_2c_3}\right)^{-1/2},$$

gives

$$e^{2K_1} - e^{-2K_1} = \frac{s_2s_3}{(cc_1c_2c_3)^{1/2}}$$

and multiplying by sinh $2L_1$ gives (7.5.3) fcr $i = 1$. But the equations are symmetric in 1, 2,and 3 . Hence the equation holds for all i.

For a better understanding of the star triangle equations, (7.5.1), let us introduce the matrices

$$P(A) = \begin{pmatrix} e^A & 0 & 0 & 0 \\ 0 & e^{-A} & 0 & 0 \\ 0 & 0 & e^{-A} & 0 \\ 0 & 0 & 0 & e^A \end{pmatrix}, \quad Q(A) = \begin{pmatrix} e^A & 0 & e^{-A} & 0 \\ 0 & e^A & 0 & e^{-A} \\ e^{-A} & 0 & e^A & 0 \\ 0 & e^{-A} & 0 & e^A \end{pmatrix}$$

Then our equations can be written as

$$RP(K_1)Q(K_2)P(K_3) = Q(L_3)P(L_2)Q(L_1). \tag{7.5.4}$$

This equation should be compared with the "braid relation"

$$\sigma_i\sigma_{i+1}\sigma_i = \sigma_{i+1}\sigma_i\sigma_{i+1}.$$

Indeed, if we replace Q by RQ, then the star triangle relation becomes a "parametrized version" of the braid relations. So the computation due to Onsager of the critical temperature for the hexagonal lattice Ising model is the starting point in statistical mechanics for the "Yang -Baxter" equations.

The star triangle relations have a lot of symmetry. For example, if we start with a given solution set $\mathbf{K}, \mathbf{L}$, and R then the substitution

$$K_2 \rightsquigarrow L_2,\ L_2 \rightsquigarrow K_2\,,\ R \rightsquigarrow R\,\text{sinh}2K_2/\text{sinh}2L_2$$

gives another solution.

We have illustrated the use of the star triangle relations in the computation of the critical temperature of the honeycomb and triangular lattice. But the main use is in proving the commutativity of various matrices that arise in the square lattice Ising model, thereby leading to a complete solution of this model. This is thoroughly explained in Chapter 7 of [105]. This technique was then generalized to the method using the Yang Baxter equation, and we will take up this stage in our story in Section 7.7.

7.6 Star triangle and commutation.

In 1971 Onsager observed that one can use the star triangle relation to establish the commutativity of diagonal to diagonal transfer matrices of the

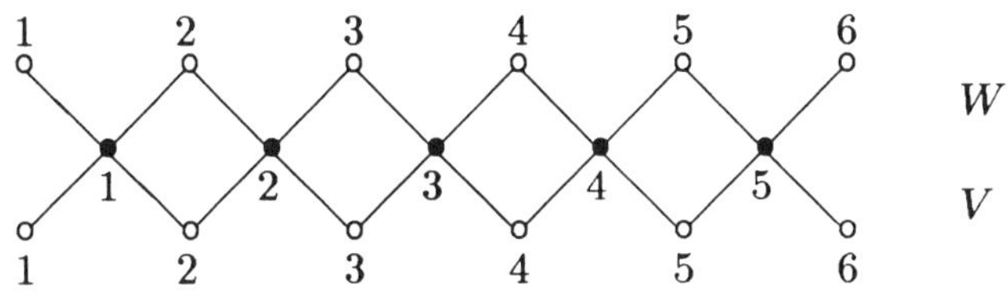

Figure 7.4: Alternating rows of the square lattice drawn diagonally, and the corresponding transfer matrices.

square lattice Ising models with appropriate boundary conditions, for differing values of K and L so long as they have the same value of $\sinh 2k \sinh 2L$. We will explain the key step in this argument following Baxter [105] sections 7.2 and 7.3 almost verbatim.

The first step is to observe that the square lattice (indeed the cubic lattice in any dimension) is a bipartite graph. We may divide its vertices into type A and type B, along parallel diagonals. So if we rotate the square lattice through 45 degrees we may draw it as in Figure . We impose periodic boundary conditions on this rotated lattice. Consider two successive rows, and let $\alpha = (\alpha_1, \ldots, \alpha_n)$ denote the spins on the lower row, and $\alpha' = (\alpha'_1, \ldots, \alpha'_n)$ denote the spins on the upper row.

Define

$$\begin{aligned} V_{\alpha,\alpha'} &= \exp[\textstyle\sum_1^n (K\alpha_{j+1}\alpha'_j + L\alpha_j\alpha'_j)] \\ W_{\alpha,\alpha'} &= \exp[\textstyle\sum_1^n (K\alpha_j\alpha'_j + L\alpha_j\alpha'_{j+1})] \end{aligned} \tag{7.6.1}$$

Here $n+1 = 1$ in the sum expressing the periodicity in the "horizontal" direction. Consider the α as parametrizing a basis in a vector space of dimension 2^n. Then, using the periodicity in the "vertical" direction (so that the number of "rows" is even), we see that the partition function can be written as

$$Z = \operatorname{tr}(VW)^m$$

where $2m$ is the number of rows. Thus

$$Z_N = \lambda_1^m + \cdots + \lambda_n^m$$

where the λ_i are the eigenvalues of Z; and hence it becomes important to determine these eigenvalues (as functions of K and L). An important tool

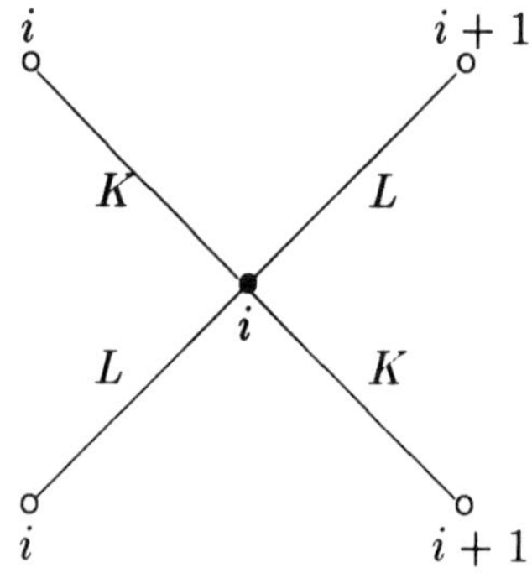

Figure 7.5:

in finding the eigenvalues of matrix is to embed it in a family of commuting matrices. In the case at hand, we want to know when transfer matrices corresponding to different values of K and L commute. An elementary argument given in Baxter's book shows that this boils down to the following question: when does the equation

$$V(K,L)W(K',L') = V(K',L')W(K,L) \tag{7.6.2}$$

hold?

We will prove

Proposition 7.6.1 *If*

$$\sinh 2K \sinh 2L = \sinh 2K' \sinh 2L' \tag{7.6.3}$$

then (7.6.2) holds.

Proof. We can write the α, α' component of the matrix product on the left of (7.6.2) as

$$\sum_{\beta_1}\sum_{\beta_2}\cdots\sum_{\beta_n}\prod_{j=1}^{n} \exp[\beta_j(K\alpha_{j+1} + L\alpha_j + K'\alpha'_j + L'\alpha'_{j+1}].$$

The sum is over all black circles wedged between a lower and upper row of white circles. But each black circle occurs in only one factor in the product, so the sum simplifies to

$$\prod_{j=1}^{n} X(\sigma_j, \sigma_{j+1}; \sigma'_j, \sigma'_{j+1}) \tag{7.6.4}$$

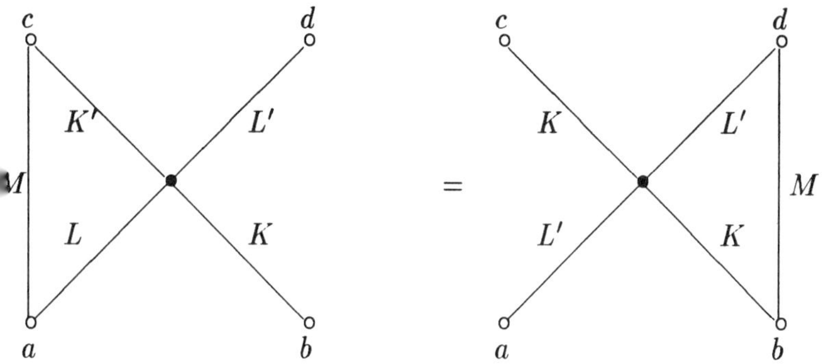

Figure 7.6: The equation $e^{Mac}X(a,b;c,d) = X'(a,b;c,d)e^{Mbd}$ in diagrammatic form.

where

$$X(a,b;c,d) = 2\cosh(La + Kb + K'c + L'd).$$

The (a,b) refer to the lower row and the (c,d) refer to the upper row. The entire product in (7.6.4) is over a horizontal row of black circles wedged between pairs of white circles. It really is a product over a ring on account of the horizontal periodicity. Clearly this product is unchanged if we replace $X(a,b;c,d)$ by

$$e^{Mac}X(a,b;c,d)e^{-Mbd}$$

since the exponential factors will cancel in the product. So we will have a solution of (7.6.2) if there exists a number M such that

$$e^{Mac}X(a,b,;c,d) = X'(a,b;c,d)e^{Mbd} \tag{7.6.5}$$

where X' is obtained from X by interchanging K with K' and L with L'. This equation can be written graphically as in Figure 7.?.

Consider the triangle on the left of the figure, but with the center black circle replaced by an open circle. We may replace this triangle by a star provided that the star triangle conditions hold: If we define

$$K_1 = L, \quad K_2 = K', \quad K_3 = M$$

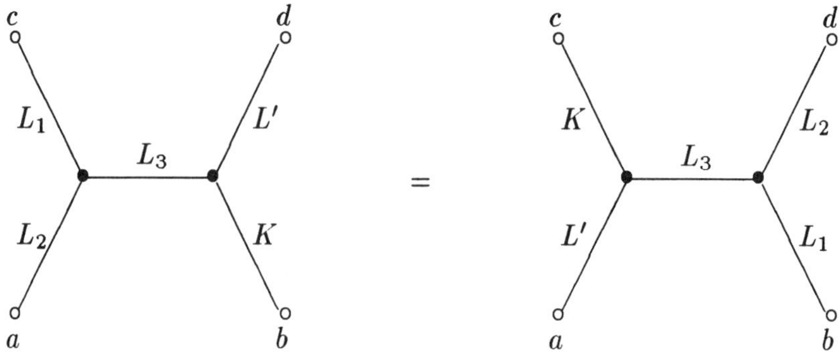

Figure 7.7: The equation $e^{M_{ac}}X(a,b;c,d) = X'(a,b;c,d)e^{M_{bd}}$ after performing the triangle to star transformation on both sides.

then we may replace the triangle by the star where

$$\sinh 2K_i \sinh 2L_i = k^{-1} \quad i = 1,2,3.$$

If we then sum over the right hand vertex of the star, we see that the left hand side of (7.6.5) is the expression symbolized by the four legged graph on the left of Figure 7.7, up to a factor of R. If we perform the same operation on the triangle on the right in Figure 7.7, we get the expression symbolized by the four legged graph to the right in Figure 7.7. Clearly these two expressions will be equal if

$$L_1 = K, \quad L_2 = L'$$

which implies that (7.6.3) holds. Conversely, if (7.6.3) holds, we can choose R, L_3, and M so that the star triangle relation is satisfied, and so replace the four legged graphs by those of Figure 7.? and conclude that (7.6.2) holds.QED

The reader should compare this argument with Kennelly's solution of the Wheatstone bridge presented at the end of this chapter.

7.7 Vertex models.

In the late 1960's work by Lieb [789] on the "six vertex model" and Yang on the scattering theory for δ function potentials was a major step forward in understanding the phenomena of integrable lattice models, and was greatly advanced by Baxter who isolated a particular algebraic equation in his famous solution of the "eight vertex" model, see equation (7.7.13) below. The constraints on the matrix R occurring in this equation give a condition identical to one introduced by Yang in his generalization the Bethe ansatz in his analysis of the δ function potential. This constraint has become known as the Yang Baxter equation, cf. (7.7.15). It is also kind of "parametrized braid relation" like the star triangle equations of the preceding section. The importance of the equations (7.7.13) and (7.7.15) was emphasized by the Fadeev school, and it was eventually realized that there was a close connection between these equations and Mac Lane's hexagon constraints. In this section and the next we outline the key ideas, but don't go into the details, which can get quite complicated, and can be found in the classical book by Baxter [105] and the fundamental paper by Fadeev and Takhtadjan [355].

In the statistical models considered, the states are associated to the edges of a graph rather than to the vertices: To each edge emanating from a vertex in a square lattice attach either a + sign or a - sign. Since there are four edges emanating from each vertex, there are 16 possibilities altogether. Attach an "energy" ϵ_j to each of the sixteen possible types, so j runs from 1 to 16.

A choice of $\pm$ at each edge (called a state or a configuration) determines one of the sixteen possibilities at every vertex. (We consider a toroidal, i.e doubly periodic, lattice of size M horizontally and N vertically.) Let

$$E = \sum_{1}^{16} N_j \epsilon_j, \tag{7.7.1}$$

where N_j denotes the number of sites of type j. Define the partition function to be

$$Z = \sum \exp(-\beta E), \tag{7.7.2}$$

where the sum is over all configurations. We are interested in computing the limiting Massieu function per unit site which is defined as

$$\lim_{M,N,\ \to\infty} \frac{1}{MN} \log Z. \tag{7.7.3}$$

Introduce the following definition: At a given vertex, let $\alpha = \pm$ denote the choice for the edge emanating downward from the vertex and α' the choice for the edge emanating upward. Similarly, let γ correspond to the left pointing horizontal edge, and γ' to the right pointing edge. So specifying the values of α, α', γ, γ' specifies the type. Let

$$L_{\alpha}^{\alpha'}(\gamma,\gamma') = \exp(-\beta\epsilon_j) \tag{7.7.4}$$

where j on the right is the type determined by the choice on the left.

A specific configuration makes the contribution

$$\Pi L_{\alpha}^{\alpha'}(\gamma,\gamma')$$

to the partition function, where the product is taken over all sites. We must sum over all configurations. Organize this sum as a double sum, the inner sum being over all choices in the rows for a definite choice on the vertical edges, and then sum over the vertical choices. So let

$$\alpha = (\alpha_1, \ldots, \alpha_N)$$

denote the choice of vertical edges emanating downward from a definite row, and similarly α' the choice for the upward edges. The contribution of all choices on a given row with the specified values of α and α' will then be the sum

$$T_{\alpha}^{\alpha'} = \sum_{\gamma_1} \cdots \sum_{\gamma_N} \prod_{n=1}^{N} L_{\alpha_n}^{\alpha'_n}(\gamma_n, \gamma_{n+1}), \tag{7.7.5}$$

where it is understood that $\gamma_{N+1} = \gamma_1$.

We can think of the α as parametrizing a basis in a space of dimension 2^N, namely the N-fold tensor product of the standard two dimensional vector space, $\mathbf{C}^2$. Then $T_{\alpha}^{\alpha'}$ can be thought of as the matrix of an operator T_N on this space, and

$$Z = \operatorname{tr}(T_N)^M. \tag{7.7.6}$$

For fixed α and α' regard $L_{\alpha}^{\alpha'}$ as defining a two by two matrix in $\mathbf{C}^2$, where the γ and γ' parametrize the rows and columns. Then we can rewrite (7.7.5) as

$$T_{\alpha}^{\alpha'} = \operatorname{tr} L_{\alpha_1}^{\alpha'_1} \cdots L_{\alpha_N}^{\alpha'_N}. \tag{7.7.7}$$

Let $\Lambda_{\max}(N)$ denote the eigenvalue of maximum absolute value of $T = T_N$. (Assume, for the sake of discussion, that it is unique for each N) If the limit of the Massieu function per unit site exists and does not depend on the order of the limit operations over M and N, then clearly it is given by

$$\lim_{N\to\infty} \frac{1}{N} \log|\Lambda_{\max}(N)|. \tag{7.7.8}$$

We now give a slightly different interpretation to the preceding discussion. Let us allow the γ's to range over a possibly different index set than the α's. So we let W be an auxiliary vector space, not necessarily $\mathbf{C}^2$. We can think of L as the matrix of an element of $\mathrm{End}(\mathbf{C}^2 \otimes W)$ relative to a preferred basis. Consider the space

$$U = \mathbf{C}^2 \otimes \cdots \mathbf{C}^2 \otimes W \quad N \quad \mathbf{C}^2 \ \text{ factors.}$$

An $L \in \mathrm{End}(\mathbf{C}^2 \otimes W)$ determines an element, $L_i \in \mathrm{End}(U)$ by considering U as the tensor product of $N-1$ copies of $\mathbf{C}^2$ with $\mathbf{C}^2 \otimes W$ where the $\mathbf{C}^2$ factor is placed in the i–th position, and then tensoring with the identity in all other factors. With this notation, $T_\alpha^{\alpha'}$ is the matrix of the operator

$$T_N = \mathrm{tr}_W\, L_N \cdots L_1, \tag{7.7.9}$$

where

$$\mathrm{tr}_W : \mathrm{End}(X \otimes W) \to \mathrm{End}(X)$$

is defined, for example, be regarding

$$\mathrm{End}(X \otimes W) = X \otimes W \otimes X^* \otimes W^*$$

and perform the trace (contraction) on $W \otimes W^*$.

In terms of a basis of W we can regard $\mathrm{End}(X \otimes W)$ as the set of square matrices of size $\dim W$ with entries from the ring $\mathrm{End}(X)$. From this point of view, tr_W is obtained by summing the diagonal elements. If we set

$$S_N = L_n \cdots L_1$$

then

$$T_N = \mathrm{tr}_W\, S_N,$$

and our problem is to find the eigenvalues of T_N, in particular the largest eigenvalue as a function of N. This was the problem that was solved by Lieb in the six vertex model (where only certain six of the sixteen possible vertex configurations are allowed) and by Baxter in the eight vertex model. The key to Baxter's method was to find a whole family of values of the parameters entering into the definition of the L so that the corresponding matrices T_N commute. Having a large commuting family of matrices aids in the diagonalization process for each matrix in the family.

Baxter produced families of commuting T_N by the following argument: Let $n = \dim W$ and consider n by n matrices over a ring A. (In the case of interest $A = \mathrm{End}(X)$.) If P and Q are two such matrices, define their tensor product, $P \otimes Q$ to be the n^2 by n^2 matrix given by

$$P \otimes Q = \begin{pmatrix} P_{11}Q & \cdots & P_{1n}Q \\ \vdots & \vdots & \vdots \\ P_{n1}Q & \cdots & P_{nn}Q \end{pmatrix}.$$

Here $P_{ij} \in A$ are the matrix entries of P. Notice that

$$\mathrm{tr}_{W\otimes W}\, P \otimes Q = \mathrm{tr}_W\, P \cdot \mathrm{tr}_W\, Q. \tag{7.7.10}$$

Suppose that R is a matrix whose entries lie in the center of A and X is any matrix with entries in A. The we still have the usual formula $\mathrm{tr}\, RXR^{-1} = \mathrm{tr}\, X$. Indeed,

$$\begin{aligned} \mathrm{tr}\, RXR^{-1} &= \sum_{i,j,k} R_{ij} X_{jk} (R^{-1})_{ki} \\ &= \sum_{i,j,k} (R^{-1})_{ki} R_{ij} X_{jk} \\ &= \sum_{j,k} \delta_{kj} X_{jk} \\ &= \mathrm{tr}\, X. \end{aligned}$$

In particular, if we can find an invertible matrix, R, of size n^2 and with scalar entries, such that

$$R(P \otimes Q)R^{-1} = Q \otimes P \tag{7.7.11}$$

then

$$tr_W P \cdot tr_W Q = tr_W Q \cdot tr_W P. \tag{7.7.12}$$

We want to apply (7.7.12) to $P = S_N$, $Q = S'_N$ where the $'$ refers to a different choice of the parameters that enter into the definition of L. But S_N and S'_N are products of the individual L_i and L'_i, so we can apply the following observation: Let P, Q, P' and Q' be four matrices with entries in A, and suppose that all the matrix entries, P'_{ij} of P' commute with all the matrix entries of Q. Then matrix multiplication shows that

$$(P \otimes Q)(P' \otimes Q') = PP' \otimes QQ'.$$

So if

$$R(P \otimes Q)R^{-1} = Q \otimes P \text{ and } R(P' \otimes Q')R^{-1} = Q' \otimes P'$$

then

$$R(PP' \otimes QQ')R^{-1} = QQ' \otimes PP'.$$

In our case, the "matrix entries" of L_i commute with the matrix entries of L'_j if $i \neq j$, since they operate on differing $\mathbf{C}^2$ factors in the tensor product. So if we can find an invertible (four by four) matrix R such that

$$R(L_i \otimes L'_i) = (L'_i \otimes L_i)R \tag{7.7.13}$$

for all i, then the corresponding transfer matrices, T_N and T'_N commute.

So a considerable portion of the problem is devoted to finding solutions of (7.7.13). For the actual solution, involving Jacobi elliptic functions in the eight vertex model or trigonometric functions in the six vertex model, we refer the reader to [105] of [355]. We will content ourselves with comments about the structure of the equations. In (7.7.13) all the matrices depend on some auxiliary parameters (the "spectral parameter") but we will suppress this from the notation. That is, we will use the generic symbol, L, to denote members of a whole family of matrices, and rewrite (7.7.13) as

$$RL_1L_2 = L_2L_1R. \tag{7.7.14}$$

Here L_1 denotes $L \otimes id$ and L_2 denotes $id \otimes L$. In other words the subscript determines the tensor position.

A first step is to observe that we can use (7.7.14) to get an equation for R: For $L_1L_2L_3$ acting on the triple tensor product we have

$$R_{12}L_1L_2L_3 = L_2L_1L_3R_{12}$$

where the subscript 12 denotes the tensor factors on which R operates. Then we can move from $L_1L_2L_3$ to $L_3L_2L_1$ in two different ways:

$$R_{23}R_{13}R_{12}L_1L_2L_3 = L_3L_2L_1R_{23}R_{13}R_{12}$$

and

$$R_{12}R_{13}R_{23}L_1L_2L_3 = L_3L_2L_1R_{12}R_{13}R_{23}.$$

In other words, the element

$$\mathbf{YB}(R) = (R_{12}R_{13}R_{23})^{-1}R_{23}R_{13}R_{12}$$

commutes with all triple products $L_1L'_2L_3''$. We do not want this commutation condition to impose new constraints on the L. One way of guaranteeing this is to require that the expression $\mathbf{YB}(R)$ be a multiple of the identity, $\mathbf{YB}(R) = aI$. Taking the determinant, we see that a is a root of unity, and by a continuity argument we can assume that it must be 1. Thus

$$R_{12}R_{13}R_{23} = R_{23}R_{13}R_{12}. \tag{7.7.15}$$

This is the celebrated Yang Baxter equation. Like the star triangle relation, it is a parametrized version of the braid relations.

7.8 Yang-Baxter and the Mac Lane hexagons.

Let A be an associative algebra. If M and N are A modules, then $M \otimes N$ is an $A \otimes A$ module where

$$(A \otimes A) \otimes (M \otimes N) \to M \otimes N$$

is given by the composite map

$$A \otimes A \otimes M \otimes N \xrightarrow{s_{23}} A \otimes M \otimes A \otimes N \to M \otimes N,$$

where the second map is the tensor product of the two maps

$$A \otimes M \to M, \quad A \otimes N \to N$$

given by the module structures. (In the above formulas, and throughout this section, all tensor products are over the ground field.)

Suppose that we would like to make $M \otimes N$ into an A module. For this we need an algebra morphism

$$\Delta : A \to A \otimes A.$$

This suggests that to endow the category, $\mathbf{Mod}(A)$ with a tensor product, A should have the structure of a bialgebra. Let Q be a third A module. By definition, the action of A on $(M \otimes N) \otimes Q$ is given by the map

$$(\Delta \otimes id) \circ \Delta : A \to A \otimes A \otimes A$$

while the action of A on $M \otimes (N \otimes Q)$ is given by the map

$$(id \otimes \Delta) \circ \Delta : A \to A \otimes A \otimes A.$$

So if the comultiplication, Δ is coassociative, the standard vector space identification of $(M \otimes N) \otimes Q$ with $M \otimes (N \otimes Q)$ is an A module morphism and so $\mathbf{Mod}(A)$ is a monoidal category.

Now let us consider the linear isomorphism:

$$s : X \otimes Y \to Y \otimes X$$

given by the symmetry which transposes the factors. Is this a morphism of A modules? Since the action of A on a tensor product is given by $\Delta : A \to A \otimes A$, if we want s to be an A morphism for all modules, we must require that

$$s \circ \Delta = \Delta,$$

in other words that the comultiplication be cocommutative. Suppose that we allow the symmetry $s_{X,Y}$ to have a more general form requiring only that there be some element $R \in A \otimes A$ such that

$$s_{X,Y} = s \circ R : X \otimes Y \to Y \otimes X$$

is an A morphism. Then this weakens the cocommutativity requirement to

$$R \Delta R^{-1} = s \circ \Delta. \tag{7.8.1}$$

But now we must check whether this new symmetry operator satisfies the Mac Lane hexagon constraints. One of these, equation (6.21), is that the two maps

$$s_{X \otimes Y, Z} : (X \otimes Y) \otimes Z \to Z \otimes (X \otimes Y)$$

and

$$a \circ (s_{X,Z} \otimes id) \circ a \circ (id \otimes s_{Y,Z}) \circ a : \quad (X \otimes Y) \otimes Z \to Z \otimes (X \otimes Y)$$

be equal. Here the a's denote various associativities and the second map corresponds to the five step trip

$$(X \otimes Y) \otimes Z \to X \otimes (Y \otimes Z) \to X \otimes (Z \otimes Y) \to (X \otimes Z) \otimes Y \to (Z \otimes X) \otimes Y \to Z \otimes (X \otimes Y)$$

around the hexagon. Now the map $s_{X \otimes Y, Z}$, when applied to the element $(x \otimes y) \otimes z$ is

$$[(id \otimes \Delta) R_{21}](z \otimes x \otimes y).$$

(This is because, before applying the transposition, the first factor in R acts by the A module structure on the tensor product $X \otimes Y$ which uses Δ.)

In the five step trip we can ignore the assocativities, since they are all the standard vector space identifications. the crucial two steps are the symmetry operators, and the five step trip sends $x \otimes y \otimes z$ into

$$R_{12}R_{23}(z \otimes x \otimes y).$$

With a similar argument for the second hexagon, we see that the two hexagon constraints yield the conditions

$$(\Delta \otimes id)R = R_{13}R_{23} \tag{7.8.2}$$
$$(id \otimes \Delta)R = R_{13}R_{12} \tag{7.8.3}$$

We now will show that (7.8.2) and (7.8.1) give the Yang-Baxter equations. Apply $s \otimes id$ to (7.8.2) and right multiply by R_{12}. Then use (7.8.1).

$$R_{23}R_{13}R_{12} = [(s \circ \Delta) \otimes id](R)R_{12} \tag{7.8.4}$$
$$= R_{12}(\Delta \otimes id)(R) \tag{7.8.5}$$
$$= R_{12}R_{13}R_{23}. \tag{7.8.6}$$
$$\tag{7.8.7}$$

Obviously we could just as easily have used the second hexagon equation (7.8.3) to derive (7.8.4). For a strictly monoidal category, one for which the associativity operator is the identity, the Yang-Baxter equation expresses the braid relation

$$s_1s_2s_1 = s_2s_1s_2$$

for the generalized symmetries of the triple tensor product, s_1 acting on the first two factors and s_2 acting on the second and third factors. This equation asserts the identity of the two morphisms $X \otimes Y \otimes Z \to Z \otimes Y \otimes X$ (dropping the parentheses and the associativities)

$$X \otimes Y \otimes Z \to X \otimes Z \otimes Y \to Y \otimes Z \otimes X \to Z \otimes Y \otimes X$$

and

$$X \otimes Y \otimes Z \to Y \otimes X \otimes Z \to Z \otimes X \otimes Y \to Z \otimes Y \otimes X.$$

The diagram expressing this identity is a composite of two copies of the MacLane hexagon in figure (6.1) and the diagram

$$\begin{array}{ccc} (X \otimes Y) \otimes Z & \to & Z \otimes (X \otimes Y) \\ \downarrow & & \downarrow \\ (Y \otimes X) \otimes Z & \to & Z \otimes (Y \otimes X) \end{array}$$

In the case that the category is not strictly monoidal, and a nontrivial associativity operator enters the hexagon identities, there is a generalized Yang Baxter equation which expresses the braid relations for operators constructed from both the symmetries and the associativities. We will come back to this point in the next chapter on Drinfeld algebras and in Chapter 12.

This leads to the strategy of studying those solutions to the Yang-Baxter equations which arise from the hexagon equations, (7.8.2,7.8.3), and thus the theory of monoidal categories and bialgebras links up with statistical mechanics.

Let us add one more equation to guarantee that the transposition is compatible with the natural identifications of $k \otimes X$ and $X \otimes k$ with X,

$$(\epsilon \otimes id)R = (id \otimes \epsilon)R = 1 \otimes 1.$$

A bialgebra satisfying this equation together with (7.8.1) and the hexagon equations is what Drinfeld calls a quasi-triangular bialgebra. As remarked above when we relax the strict monoidal condition in the category we have to include an associativity operator in the hexagon equations and hence in the Yang Baxter equations. The appropriate algebraic structure in this case is a quasi-triangular Drinfeld algebra, which we will describe in the next chapter.

7.9 Electrical star triangle.

A paper by A.E. Kenelly appeared in the September 16, 1899 issue of *Electrical World and Engineer* entitled " The Equivalence of Triangles and Three-Pointed Stars in Conducting Networks." It opening paragraphs are;

> Computations relating to conducting networks are often capable of being simplified by the use of the following proposition, which appears to be new:
>
> *To every triangle there corresponds a Y which is its external equivalent, so that one may always be replaced by the other, without disturbing the electrical distribution external to it.*

The situation envisaged by Kennelly is as in Figure 7.3:

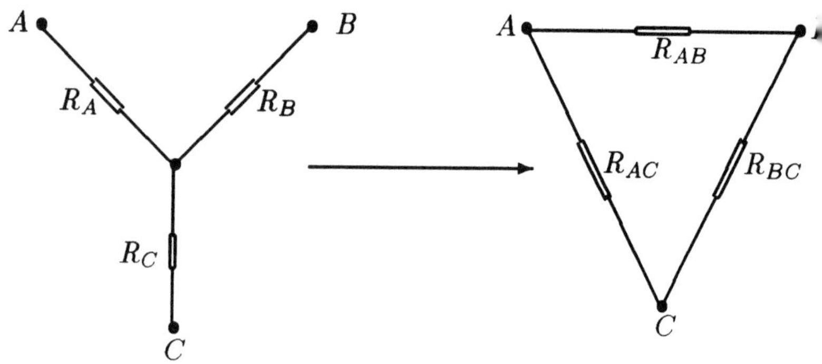

Electrical star triangle transformation

Figure 7.8:

For these two networks to be equivalent, the resistance must be the same between any pair of terminals. In the star, the resistances R_A and R_B are in series, hence they add to give $R_A + R_B$. In the triangle, the resistance R_{AB} is parallel to the resistances R_{AC} and R_{BC} which are in series. The reciprocals of parallel resistances add, so if R is the resistance in the triangle between A and B we have

$$R^{-1} = \frac{1}{R_{AC} + R_{BC}} + \frac{1}{R_{AB}} = \frac{R_{AB} + R_{AC} + R_{BC}}{R_{AB}(R_{BC} + R_{BC})}.$$

Setting$R = R_A + R_B$ and doing the same computation for the two other pairs of terminals gives

$$\begin{aligned} R_A + R_B &= \frac{R_{AB}(R_{BC} + R_{AC})}{R_{AB} + R_{AC} + R_{BC}} \\ R_B + R_C &= \frac{R_{BC}(R_{AC} + R_{AB})}{R_{AB} + R_{AC} + R_{BC}} \\ R_A + R_C &= \frac{R_{AC}(R_{AB} + R_{BC})}{R_{AB} + R_{AC} + R_{BC}}. \end{aligned}$$

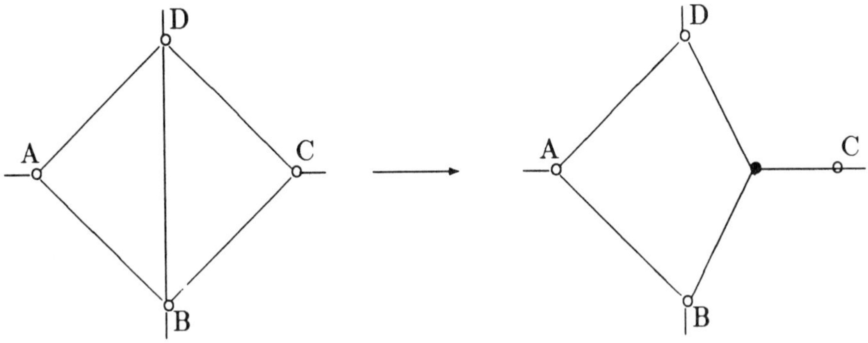

Figure 7.9: The Wheatstone bridge solved by applying the triangle to star transformation to the triangle BCD. The network on the right is two sets of parallel resistors in series and so can be immediately solved. We have not drawn in the resistances.

Subtracting the second relation from the first and adding the last, gives, after a little algebra

$$R_A = \frac{R_{AB}R_{AC}}{R_{AB} + R_{BC} + R_{CA}} \tag{7.9.1}$$

with cyclic permutations giving the equations for R_B and R_C. A bit of further computation shows that the inverse relations are

$$R_{AB} = \frac{R_A R_B + R_A R_C + R_B R_C}{R_C} \tag{7.9.2}$$

etc.

Kennelly goes on to give some applications of this method, for example to the Wheatstone bridge. This technique of reducing networks to simpler form became part of the standard courses in electricity as taught in the high schools.

Chapter 8

Drinfeld algebras and Majid algebras

8.1 Drinfeld algebras.

Let A be an associative algebra. If M and N are A modules, then $M \otimes N$ is an $A \otimes A$ module where

$$(A \otimes A) \otimes (M \otimes N) \to M \otimes N$$

is given by the composite map

$$A \otimes A \otimes M \otimes N \xrightarrow{s_{23}} A \otimes M \otimes A \otimes N \to M \otimes N,$$

where the second map is the tensor product of the two maps

$$A \otimes M \to M, \quad A \otimes N \to N$$

given by the module structures. We saw in Section 7.8, that we need a bialgebra structure to endow the category **Mod**(A) with a tensor product, and if we want the usual associativity for vector spaces to be an A module morphism, the comultiplication, Δ should be coassociative.

Drinfeld has shown that for some very important applications it is necessary to relax the condition of coassociativity, but retain enough to keep the monoidal structure on **Mod**(A). The associativity map

$$\phi = a^{-1} : (M \otimes N) \otimes Q \to M \otimes (N \otimes Q)$$

will no longer be the standard vector space identification. Rather, he assumes that the associativity operator be the standard identification followed by the action of an element $\Phi \in A \otimes A \otimes A$. In other words, that the associativity operator ϕ be given by

$$x \otimes y \otimes z \rightsquigarrow \Phi(x \otimes y \otimes z), \quad \text{where } \Phi \in A \otimes A \otimes A. \tag{8.1.1}$$

(Notice that this is analogous to how we relaxed cocommutativity in Section 7.8 and derived the Yang Baxter equations.) In order that ϕ be an A module morphism for all of **Mod**(A) it is clearly necessary and sufficient that

$$\Phi((\Delta \otimes id) \circ \Delta)\Phi^{-1} = (id \otimes \Delta) \circ \Delta \tag{8.1.2}$$

hold. This condition is called coassociativity up to conjugation and Φ is also called a coassociativity operator. The MacLane pentagon condition then imposes the following constraint on Φ, $A \otimes A \otimes A \otimes A$:

$$(1 \otimes \Phi)([id \otimes \Delta \otimes id](\Phi))(\Phi \otimes 1) = ([id \otimes id \otimes \Delta](\Phi))([\Delta \otimes id \otimes id](\Phi)). \tag{8.1.3}$$

where the three multiplications on the left and the two multiplications on the right all take place in $A \otimes A \otimes A \otimes A$. A bialgebra with associative multiplication and whose comultiplication satisfies the two conditions (8.1.2), (8.1.3) and the third condition

$$(id \otimes \epsilon \otimes id)\Phi = 1 \tag{8.1.4}$$

which guarantees the commutativity of the diagram

$$\begin{array}{ccc} (X \otimes I) \otimes Y & \longrightarrow & X \otimes (I \otimes Y) \\ r_x \otimes id \searrow & & \swarrow id \otimes l_y \\ & X \otimes Y & \end{array}$$

be called a *Drinfeld algebra*. Drinfeld algebras constitute the main object of study of the remaining chapters of this book.

In Chapter 2 we saw that a coalgebra can be defined purely categorically, built out of a functor $F : \mathcal{C} \to$ **Vec**, where the coalgebra, C, is the representing object for the associated functor

$$V \rightsquigarrow \mathcal{N}\mathbf{at}(F, {}_V F), \quad \mathbf{Vec} \to \mathbf{Set}.$$

The comultiplication on C is coassociative. This ultimately derives from the associativity of the tensor product on **Vec**. As we will see in the next section, if the category $\mathcal{C}$ is monoidal, then the coalgebra C has a multiplication. This multiplication need not be associative, but it satisfies a condition dual to (8.1.2),(8.1.3), and (8.1.4) cf. equations (8.2.8) and (8.2.10) below. Algebras of this sort will be called *Majid algebras*. The idea of getting the algebra and coalgebra structures by reconstruction from categories and functors originated in the work of Grothendieck and Saavedro Rivano [1068], see also [286]. The results of the next sections are due to Ulbrich [1173] and Majid [857], [848]. We follow the treatment of Majid.

8.2 Monoidal reconstruction

Recall that a monoidal category, **C**, is a category with a tensor product, i.e. a functor sending (X,Y) to $X \otimes Y$. It is assumed that there is an associativity ϕ which is a natural transformation between the two functors $(X,Y,Z) \rightsquigarrow (X \otimes Y) \otimes Z$ and $(X,Y,Z) \rightsquigarrow X \otimes (Y \otimes Z)$. So for each triple (X,Y,Z) we have a

$$\phi_{X,Y,Z} \in \mathrm{Hom}((X \otimes Y) \otimes Z, X \otimes (Y \otimes Z)).$$

This ϕ is subject to the MacLane pentagon identity.

Let $\mathcal{C}$ and $\mathcal{D}$ be two monoidal categories. A functor $F : \mathcal{C} \to \mathcal{D}$ is called *multiplicative*, as opposed to monoidal, if the objects $F(I_{\mathcal{C}})$ and $I_{\mathcal{D}}$ are isomorphic by an isomorphism $f \in \mathrm{Hom}_{\mathcal{D}}(I_{\mathcal{C}}, I_{\mathcal{D}})$ making the bottom two diagrams in Figure 6.2 commute, and the two functors $(X,Y) \rightsquigarrow F(X \otimes_{\mathcal{C}} Y)$ and $(X,Y) \to F(X) \otimes_{\mathcal{D}} F(X)$ are naturally equivalent. We do not require that the hexagon diagram in Figure 6.2 commute. That is there are functorial isomorphisms

$$b_{X,Y} \in \mathrm{Hom}_{\ \mathcal{D}}(F(X) \otimes F(Y), F(X \otimes Y))$$

for every pair of objects X, Y in $\mathcal{C}$ subject to the usual consistency condition for functors.

Suppose that $\mathcal{C}$ is a monoidal category and $F : \mathcal{C} \to$ **Vec** is a multiplicative functor such that the functor **Vec** $\to$ **Set** , $V \rightsquigarrow \mathcal{N}\mathbf{at}(F, {}_V F)$ is representable. Let C be the representing object. So C is a coassociative coalgebra by the results of Section 2.5.

For each positive integer n let F^n denote the functor $\mathcal{C}^n \to$ **Vec**

$$F^n(X,Y,\ldots,Z) = F(X) \otimes F(Y) \ldots \otimes F(Z).$$

We get the corresponding functor

$$V \rightsquigarrow \mathcal{N}\mathbf{at}(F^n, V \otimes F^n).$$

This functor is representable as follows: The coalgebra structure on each $F(X)$ gives us maps

$$F(X) \otimes F(Y) \ldots \otimes F(Z) \to C \otimes F(X) \otimes C \otimes F(Y) \ldots \otimes C \otimes F(Z)$$

and applying a succession of twists gives us a map

$$F(X) \otimes F(Y) \ldots \otimes F(Z) \quad \to (C \otimes C \otimes \ldots C) \otimes F(X) \otimes F(Y) \ldots \otimes F(Z).$$

Composing this with $f \otimes id$ where $f \in \mathrm{Hom}(C^{\otimes n}, V)$ defines a map

$$\rho_V^n : \mathrm{Hom}\ (C^{\otimes n}, V) \to \mathcal{N}\mathbf{at}(F^n, V \otimes F^n).$$

Using a recursion on the relations

$$\begin{aligned} \mathcal{N}\mathbf{at}(F(X) \otimes F(Y), V \otimes F(X) \otimes F(Y)) \; &\simeq \\ &\simeq \; \mathcal{N}\mathbf{at}(F(Y), \mathcal{N}\mathbf{at}(F(X), V \otimes F(X)) \otimes F(Y) \\ &\simeq \; \mathcal{N}\mathbf{at}(F(Y), \mathbf{Lin}(C, V) \otimes F(Y) \\ &\simeq \; \mathbf{Lin}(C, \mathbf{Lin}(C, V)) \\ &\simeq \; \mathbf{Lin}(C \otimes C, V) \end{aligned}$$

one proves, as in Section 2.5, that this is an isomorphism. Thus $C^{\otimes n}$ is a representing object for the functor $V \rightsquigarrow \mathcal{N}\mathbf{at}(F^n, V \otimes F^n)$.

In particular, $C \otimes C$ is a representing object for $V \rightsquigarrow \mathcal{N}\mathbf{at}(F^2, V \otimes F^2)$. But now consider the composition

$$F(X) \otimes F(Y) \xrightarrow{b_{X,Y}} F(X \otimes Y) \xrightarrow{c_{X \otimes Y}} C \otimes F(X \otimes Y) \xrightarrow{id \otimes b^{-1}_{X,Y}} C \otimes F(X) \otimes F(Y)$$

for every pair of objects X and Y in $\mathcal{C}$. (The middle map, $c_{X \otimes Y}$, is the comodule structure defined on $F(X \otimes Y)$ as described in Sections 2.5 and 2.7.) The composite map, which we shall denote by $m_{X,Y}$ is an element of $\mathcal{N}\mathbf{at}(F^2, C \otimes F^2)$. Hence it must be represented by an element, $m \in Lin(C \otimes C, C)$. This map $m : C \otimes C \to C$ will be thought of as a multiplication, and we will write

$$m(a \otimes b) = a \cdot b.$$

To repeat, the map $m_{X,Y}$ is defined so as to make the diagram

$$\begin{array}{ccc} F(X)\otimes F(Y) & \stackrel{m_{X,Y}}{\longrightarrow} & C\otimes F(X)\otimes F(Y) \\ \big\downarrow b_{X,Y} & & \big\downarrow id\otimes b_{X,Y} \\ F(X\otimes Y) & \stackrel{c_{X\otimes Y}}{\longrightarrow} & C\otimes F(X\otimes Y) \end{array} \tag{8.2.1}$$

commute, and *the multiplication* m *is the representing map of* $m_{X,Y}$.

The unit, $u : k \to C$ is defined as follows: Let I denote the unit element in the category $\mathcal{C}$. So we are given functorial isomorphism $I\otimes X \sim X \sim X\otimes I$ for every object X of $\mathcal{C}$. We then have

$$\underbrace{k \stackrel{f}{\longrightarrow} F(I) \stackrel{c_k}{\longrightarrow} C\otimes F(I) \stackrel{\sim}{\longrightarrow} C}_{u\rightarrow}$$

and u is defined as the composite. Let us show that this is indeed a unit for the multiplication. Let $\tilde{m}_{I,X} = (id\otimes f\otimes id)\circ m_{I,X}\circ f\otimes id$. Consider the diagram,

$$\begin{array}{ccccccc} F(X) & \longrightarrow & k\otimes F(X) & \stackrel{\tilde{m}_{I,X}}{\longrightarrow} & C\otimes k\otimes F(Y) & & \\ \big\downarrow & & b_{I,X}\big\downarrow & & \big\downarrow id\otimes b_{I,X} & & \\ F(X) & \longrightarrow & F(I\otimes X) & \longrightarrow & C\otimes F(I\otimes X) & \longrightarrow & C\otimes F(X) \end{array}$$

Naturality implies that the composition along the bottom row is just c_X. So if we let v_X denote the isomorphism $v_X : F(X) \rightsquigarrow k\otimes F(X)$ and $c_k = id\otimes f^{-1}\circ c_1\circ f$ then the commutativity of the above diagram shows that the following diagram commutes

$$\begin{array}{ccccc} F(X) & & & & \\ v_X\big\downarrow & & & & \\ k\otimes F(X) & \stackrel{c_k\otimes c_X}{\longrightarrow} & C\otimes k\otimes C\otimes F(X) & \stackrel{(\tilde{m}\circ s_{23}\otimes id}{\longrightarrow} & C\otimes k\otimes F(X) \\ & u\otimes c_X\searrow & & & \big\downarrow id\otimes v_X^{-1} \\ & & C\otimes C\otimes F(X) & \stackrel{m\otimes id}{\longrightarrow} & C\otimes F(X) \end{array}$$

where the composition of the last two arrows on the top row is $\tilde{m}_{I,X}$ and the map from $F(X)$ to $C\otimes F(X)$ along the bottom route is c_X. Hence

$$\begin{aligned} c_X &= id\otimes v_X^{-1}\circ m_{1,X}\circ v_X \\ &= m\circ(u\otimes c_X)\circ v_X \\ &= m\circ(u\otimes id)\circ(id\otimes c_X)\circ v_X. \end{aligned}$$

Let $v : C \rightsquigarrow k \otimes C$, then $v \circ c_X = (id \otimes c_X) \circ v_X : F(X) \rightsquigarrow k \otimes C \otimes F(X)$, so

$$c_X = m \circ (u \otimes id) \circ v \circ c_X.$$

Hence $m(u \otimes id) = id$ as maps of $k \otimes C \sim C \to C$. This proves that u is a left unit, and similarly we prove that it is a right unit.

Let us show that this multiplication makes C into a bialgebra (with possibly non-associative multiplication). We first must show that $\Delta(a \cdot b) = \Delta(a) \cdot \Delta(b)$, i.e. that.

$$\Delta \circ m = (m \otimes m)s(\Delta \otimes \Delta) \tag{8.2.2}$$

where s is a twist map. We will prove this in gory detail. By the definition of multiplication

$$c_{X \otimes Y} \circ b_{X,Y} = (id \otimes b_{X,Y}) \circ (m \otimes id \otimes id) \circ [c_X \otimes c_Y] \tag{8.2.3}$$

where the notation $[c_X \otimes c_Y]$ means that we have included a twist map moving all the C factors to the left, i.e. $[c_X \otimes c_Y] = (id \otimes s \otimes id) \circ (c_X \otimes c_Y)$.

So

$$(\Delta \otimes id) \circ c_{X \otimes Y} \circ b_{X,Y} = (id \otimes id \otimes b_{X,Y}) \circ (\Delta \circ m \otimes id \otimes id) \circ [c_{X \otimes} c_Y].$$

Recall that $(\Delta \otimes id) \circ c_Z = (id \otimes c_Z) \circ c_Z$ for any $Z \in \mathcal{C}$. Substituting this into the left hand side we get

$$(id \otimes id \otimes b_{X,Y}) \circ (\Delta \circ m \otimes id \otimes id) \circ [c_{X \otimes} c_Y] = (id \otimes c_{X \otimes Y}) \circ c_{X \otimes Y} \circ b_{X,Y}$$

$$= (id \otimes c_{X \otimes Y}) \circ (id \otimes b_{X,Y}) \circ (m \otimes id \otimes id) \circ [c_{X \otimes} c_Y] \qquad \text{by} 8.2.3$$

$$= (id \otimes id \otimes b_{X,Y}) \circ (id \otimes m \otimes id \otimes id) \circ [id \otimes c_X \otimes c_Y] \\ \circ (m \otimes id \otimes id) \circ [c_X \otimes c_Y] \quad \text{by } (m)$$

again. So after applying the appropriate twists, this last expression becomes

$$(id \otimes id \otimes b_{X,Y}) \circ (m \otimes id) \circ (id \otimes m)(id \otimes [c_{X \otimes} c_Y] \circ [c_{X \otimes} c_Y])$$

which is the natural transformation corresponding to $(m \otimes m) \circ s \circ \Delta_{C \otimes C}$. This proves (8.2.2). A similar argument proves that $\Delta \circ u = u \otimes u$. Thus Δ is a homomorphism of algebras.

Now let us turn to the question of the associativity of multiplication. We wish to compare $m(\cdot, m(\cdot, \cdot))$ with $(\cdot, \cdot), \cdot)$ as maps of $C \otimes C \otimes C \to C$. First we prove two lemmas.

Lemma 8.2.1 *If $\nu \in \mathcal{N}\mathbf{at}(F, {}_VF)$ is a natural transformation represented by $N \in Lin(C,V)$ then the natural transformation*

$$F(X)\otimes F(Y) \overset{b_{X,Y}}{\longrightarrow} F(X\otimes Y) \overset{\nu_{X\otimes Y}}{\longrightarrow} V\otimes F(X\otimes Y) \overset{id\otimes b_{X,Y}^{-1}}{\longrightarrow} V\otimes F(X)\otimes F(Y)$$

is represented by

$$N \circ m \in \mathrm{Lin}(C\otimes C, V).$$

Proof. In the diagram

$$\begin{array}{ccccccc} F(\cdot)\otimes F(\cdot) & \overset{b}{\rightarrow} & F(\cdot\otimes\cdot) & \overset{c}{\rightarrow} & C\otimes F(\cdot\otimes\cdot) & \overset{id\otimes b^{-1}}{\rightarrow} & C\otimes F(\cdot)\otimes F(\cdot) \\ id\downarrow & & id\downarrow & & \downarrow N\otimes id & & \downarrow N\otimes id\otimes id \\ F(\cdot)\otimes F(\cdot) & \overset{b}{\rightarrow} & F(\cdot\otimes\cdot) & \overset{\nu}{\rightarrow} & V\otimes F(\cdot\otimes\cdot) & \overset{id\otimes b^{-1}}{\rightarrow} & V\otimes F(\cdot)\otimes F(\cdot) \end{array}$$

The middle square commutes by the definition of N. By equation 8.2.3 the composition of the top row is $m\otimes id\otimes id[c_X\otimes c_Y]$ and

$$(N\circ m)\otimes id\otimes id\circ[c_X\otimes c_Y] = (id\otimes b_{X,Y}^{-1})\circ \nu_{X\otimes Y}\circ b_{X,Y}$$

as asserted. QED

Lemma 8.2.2 *Let the natural transformation $F(X) \overset{\nu_X}{\longrightarrow} V\otimes F(X)$ be represented by N, and let the natural transformation $F(X) \overset{\omega_X}{\longrightarrow} W\otimes F(X)$ be represented by M. Then the composite natural tranformation*

$$(id\otimes\omega_X)\circ\nu_X : F(X) \longrightarrow V\otimes W\otimes F(X)$$

is represented by the "convolution"

$$N * M = (N\otimes M)\circ\Delta, \tag{8.2.4}$$

where $\Delta : C \longrightarrow C\otimes C$ is the comultiplication.

Proof. In the commutative diagram

$$\begin{array}{ccccc} F(X) & \overset{c_x}{\longrightarrow} & C\otimes F(X) & \overset{id\otimes c_X}{\longrightarrow} & C\otimes C\otimes F(X) \\ id\downarrow & & \downarrow N\otimes id & & \downarrow N\otimes M\otimes id \\ F(X) & \overset{\nu_X}{\longrightarrow} & V\otimes F(X) & \overset{id\otimes\omega_X}{\longrightarrow} & V\otimes W\otimes F(X) \end{array}$$

the composition along the top row is represented by Δ and the squares commute by the defintions of N and M. QED

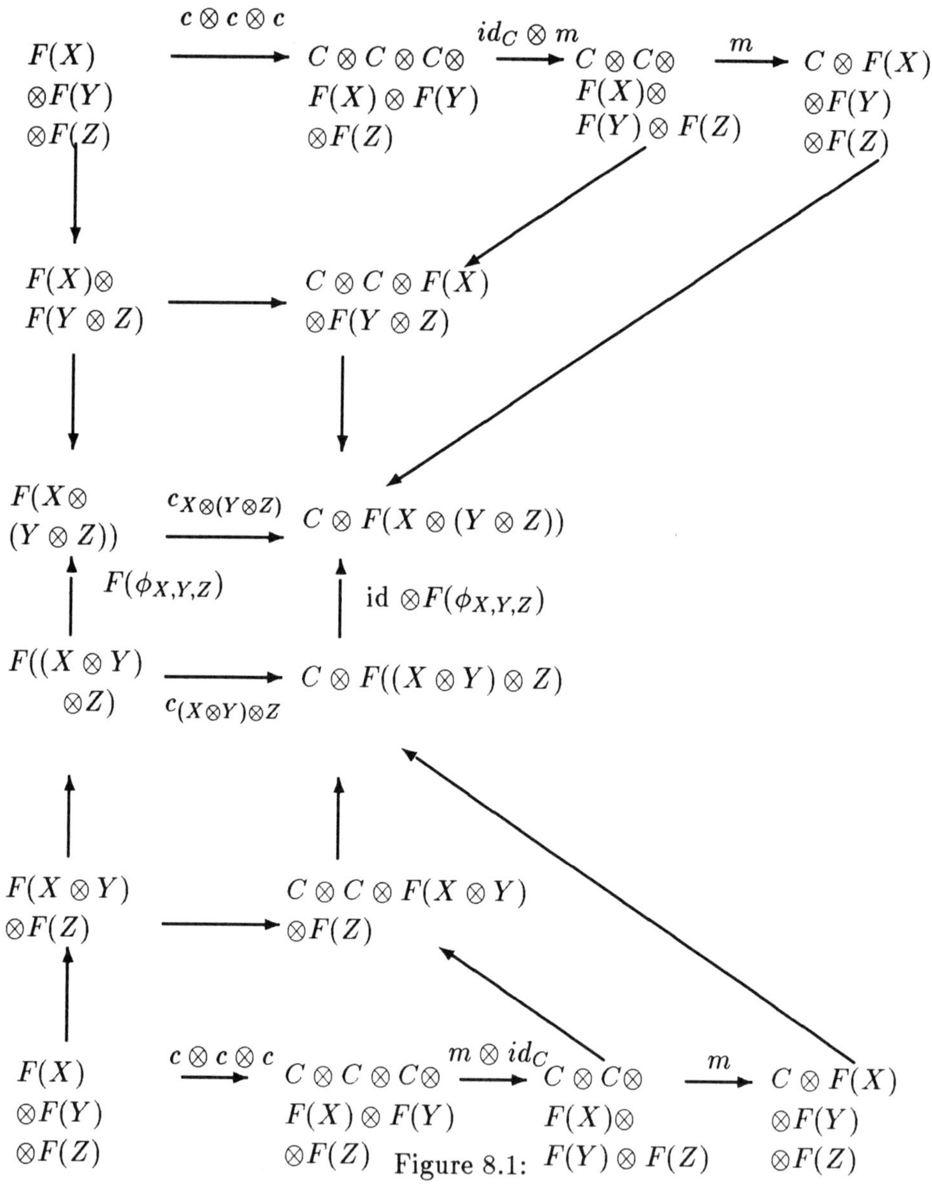
$F(X) \otimes F(Y) \otimes F(Z)$
$c \otimes c \otimes c$
$C \otimes C \otimes C \otimes F(X) \otimes F(Y) \otimes F(Z)$
$id_C \otimes m$
$C \otimes C \otimes F(X) \otimes F(Y) \otimes F(Z)$
m
$C \otimes F(X) \otimes F(Y) \otimes F(Z)$
$F(X) \otimes F(Y \otimes Z)$
$C \otimes C \otimes F(X) \otimes F(Y \otimes Z)$
$F(X \otimes (Y \otimes Z))$
$c_{X \otimes (Y \otimes Z)}$
$C \otimes F(X \otimes (Y \otimes Z))$
$F(\phi_{X,Y,Z})$
$\mathrm{id} \otimes F(\phi_{X,Y,Z})$
$F((X \otimes Y) \otimes Z)$
$c_{(X \otimes Y) \otimes Z}$
$C \otimes F((X \otimes Y) \otimes Z)$
$F(X \otimes Y) \otimes F(Z)$
$C \otimes C \otimes F(X \otimes Y) \otimes F(Z)$
$F(X) \otimes F(Y) \otimes F(Z)$
$c \otimes c \otimes c$
$C \otimes C \otimes C \otimes F(X) \otimes F(Y) \otimes F(Z)$
$m \otimes id_C$
$C \otimes C \otimes F(X) \otimes F(Y) \otimes F(Z)$
m
$C \otimes F(X) \otimes F(Y) \otimes F(Z)$

Figure 8.1:

Consider now the diagram of Figure 8.1. The middle square commutes because c is a natural transformation. By lemma 8.2.1 the morphism corresponding to going around the left, bottom and right sides of the upper trapezoid,

$$(id \otimes id \otimes b_{Y,Z})^{-1} \circ (id \otimes b_{X,Y\otimes Z})^{-1} \circ c_{X\otimes(Y\otimes Z)} \quad \circ b_{X,Y\otimes Z} \circ (id \otimes b_{Y,Z}),$$

is represented by $(m \otimes id) \circ m$. Therefore the morphism which corresponds to going straight along the top is

$$\rho_C^3(m(\cdot, m(\cdot,\cdot))) \in \mathcal{N}\mathbf{at}(F^3, C \otimes F^3),$$

Going around the outer rim of the large trapezoid on the bottom gives the corresponding expression for $\rho_C^3(m(m(\cdot,\cdot),\cdot)))$. Consider the element φ of $\mathcal{N}\mathbf{at}(F^3, C \otimes F^3)$ given by going down the left hand column. So

$$\varphi_{X,Y,Z} = (id \otimes b_{Y,Z}^{-1})) \circ b_{X,Y\otimes Z}^{-1} \circ F(\phi_{X,Y,Z}) \circ b_{X\otimes Y,Z} \circ (b_{X,Y} \otimes id). \quad (8.2.5)$$

Putting this all together we get

$$(id \otimes \varphi) \circ \rho_C^3(m(m(\cdot,\cdot),\cdot))) = \rho_C^3(m(\cdot, m(\cdot,\cdot))) \circ \varphi \quad (8.2.6)$$

as elements of $\mathcal{N}\mathbf{at}(F^3, C \otimes F^3)$.

We must translate this equation into an assertion relating $m(m(\cdot,\cdot),\cdot))$ to $m(\cdot, m(\cdot,\cdot))$ in $\mathrm{Lin}(C \otimes C \otimes C, C)$.

Let $[\Delta^{\otimes 3}]$ be the comultiplication on $C^{\otimes 3} = C\otimes C\otimes C$ given by composing $\Delta \otimes \Delta \otimes \Delta$ with the appropriate permutation and let $\Phi \in \mathrm{Hom}(C\otimes C\otimes C, k)$ be the element corresponding to φ under ρ_C^3. Then, applying Lemma 8.2.2 to the representing object $C^{\otimes 3}$, (8.2.6) translates into the Majid associativity condition

$$(m(\cdot, m(\cdot,\cdot)) \otimes \Phi) \circ [\Delta^{\otimes 3}] = (\Phi \otimes m(m(\cdot,\cdot),\cdot) \circ [\Delta^{\otimes 3}]. \quad (8.2.7)$$

Or in terms of convolution product

$$m(\cdot, m(\cdot,\cdot)) * \Phi = \Phi * m(m(\cdot,\cdot),\cdot).$$

Expressing the last equation in terms of Φ^{-1}, we have associativity up to conjugation:

$$\Phi * m(m(\cdot,\cdot),\cdot) * \Phi^{-1} = m(\cdot, m(\cdot,\cdot)). \quad (8.2.8)$$

(This reduces to the usual associative law when $\Phi = \epsilon \otimes \epsilon \otimes \epsilon$).

In terms of Sweedler notation the associativity condition (8.2.7) can be written as

$$\sum a_{(1)}\cdot(b_{(1)}\cdot c_{(1)})\Phi(a_{(2)}\otimes b_{(2)}\otimes c_{(2)}) = \sum \Phi(a_{(1)}\otimes b_{(1)}\otimes c_{(1)})(a_{(2)}\cdot b_{(2)})\cdot c_{(2)}. \tag{8.2.9}$$

In view of the isomorphisms $1\otimes X \sim X \sim X\otimes 1$ in $\mathcal{C}$, we must have

$$\Phi(1\otimes b\otimes c) = \epsilon(b)\epsilon(c)\,,\ \Phi(a\otimes 1\otimes c) = \epsilon(a)\epsilon(c),\ \Phi(a\otimes b\otimes 1) = \epsilon(a)\epsilon(b).$$

Now ϕ satisfies the Mac Lane pentagon condition which translates into a condition on Φ. We state the result of this translation, which can be verified directly from the definitions, (we will also give another derivation shortly):

$$\sum \Phi(a_{(1)}, b_{(1)}, c_{(1)}\cdot d_{(1)})\Phi(a_{(2)}\cdot b_{(2)}, c_{(2)}, d_{(2)}) = \tag{8.2.10}$$

$$\sum \Phi(b_{(1)}, c_{(1)}, d_{(1)})\Phi(a_{(1)}, b_{(2)}\cdot c_{(2)}, d_{(2)})\Phi(a_{(2)}, b_{(2)}, c_{(2)}).$$

Let us summarize what we have established so far in this section. We define a Majid bialgebra to be a bialgebra C which is coassociative as a coalgebra, and for which there exists a $\Phi\in \mathrm{Lin}(C\otimes C\otimes C, k)$, invertible in the convolution algebra and satisfying (8.2.10) such that the associativity condition (8.2.8) holds for multiplication. We have proved the following Tanaka-Krein-Ulbrich-Majid reconstruction theorem:

Theorem 8.2.1 *Let $\mathcal{C}$ be a monoidal category equivalent to a small category. Let $F:\mathcal{C}\to$ **Vec** be a multiplicative functor with b the associated natural transformation. Then there exists a Majid algebra C , unique up to isomorphism, and universal with respect to the property that F factorizes as a multiplicative functor through $\mathcal{C}\to$ **Com**(C) followed by the forgetful functor.*

The comultiplication in a Majid bialgebra is coassociative, and it is the multiplication which satisfies a complicated associativity condition. This means that a comodule satisfies the usual definition, as introduced in section 2.1, but that the definition of a module requires a complicated definition involving an identity similar to (8.2.7). For Drinfeld algebras, the modules are the familiar objects, but the comodules satisfy a complicated coassociativity condition. It is clear that the Majid conditions are dual to the Drinfeld conditions. So C is a finite dimensional Majid algebra if and only if $A=C^*$ is a finite dimensional Drinfeld algebra. Under the identification of $\mathbf{Lin}(C\otimes C\otimes C,k)$ with $A\otimes A\otimes A$ we can think of Φ either as an element of

$A \otimes A \otimes A$ or as an element of $\mathbf{Lin}(C \otimes C \otimes C, k)$ and the Drinfeld pentagon condition, (8.1.3) dualizes to the Majid pentagon condition (8.2.10). This provides an alternative proof of (8.2.10) for the finite dimensional case.

When we have a braided monoidal category, the braiding $s_{X,Y}$ defines a natural transformation $F(X) \otimes F(Y) \to F(X) \otimes F(Y)$ given by the composition

$$\begin{aligned} F(X) \otimes F(Y) \overset{b_{X,Y}}{\to} F(X \otimes Y) \overset{F(s_{X,Y})}{\to} F(Y \otimes X) & \\ \overset{b^{-1}_{X,Y}}{\to} F(Y) \otimes F(X) & \overset{s_{12}}{\to} F(X) \otimes F(Y). \end{aligned} \tag{8.2.11}$$

which is represented by $R \in \mathrm{Lin}(C \otimes C, k)$.

Proposition 8.2.1 *When the category $\mathcal{C}$ is a braided monoidal category multiplication in the representing Majid algebra C is "commutative up to conjugation" in the sense that there exists an element of $R \in \mathrm{Lin}(C \otimes C, k)$ such that*

$$R * (m \circ s) = m * R. \tag{8.2.12}$$

Proof. The following diagram proves (8.2.12).

$$\begin{array}{ccccc}
F(X)\otimes F(Y) & \xrightarrow{c\otimes c} & C\otimes F(X)\otimes C\otimes F(Y) & \xrightarrow{s} & C\otimes C\otimes F(X)\otimes F(Y) \\
{\scriptstyle b_{X,Y}}\downarrow & & & & \downarrow{\scriptstyle m\otimes id\otimes id} \\
F(X\otimes Y) & \xrightarrow{c} & C\otimes F(X\otimes Y) & \xrightarrow{id\otimes b^{-1}} & C\otimes F(X)\otimes F(Y) \\
{\scriptstyle F(s_{X,Y})}\downarrow & & \downarrow{\scriptstyle id\otimes F(s_{X,Y})} & & \downarrow{\scriptstyle id\otimes\rho(R)} \\
F(Y\otimes X) & \xrightarrow{c} & C\otimes F(Y\otimes X) & \xrightarrow{s\circ(id\otimes b^{-1})} & C\otimes F(X)\otimes F(Y) \\
{\scriptstyle b^{-1}_{Y,X}}\downarrow & & & & \uparrow{\scriptstyle m\otimes id\otimes id} \\
F(Y)\otimes F(X) & \xrightarrow{c\otimes c} & C\otimes F(Y)\otimes C\otimes F(X) & \xrightarrow{s\circ s} & C\otimes C\otimes F(X)\otimes F(Y) \\
{\scriptstyle s_{12}}\downarrow & & & & \uparrow{\scriptstyle s_{12}} \\
F(X)\otimes F(Y) & \xrightarrow{c\otimes c} & C\otimes F(X)\otimes C\otimes F(Y) & \xrightarrow{s} & C\otimes C\otimes F(X)\otimes F(Y)
\end{array}$$

The diagram consisting of the top two rows commutes because of the definition of multiplication. The square on the right in rows two and three commutes because of the definition of R, and the square to the left of it commutes by the naturality assumption. The commutativity of the diagram consisting of the third and fourth rows follows from the definition of multiplication if we keep track of the permutations involved. The diagram consisting of the bottom two rows commutes because of a simple identity between the permutations involved.

We have the following correspondence between paths and maps

1. Down left column $\rightsquigarrow \rho_C^2(R)$,
2. Along bottom row, up right column to third row $\rightsquigarrow \rho_C^2(m \circ s_{12})$,
3. Along top row, down right column to second row $\rightsquigarrow \rho_C^2(m)$,
4. Right column, second row, to right column, third row $\rightsquigarrow id \otimes \rho_C^2(R)$.

Since the diagram commutes, the map, $R * (m \circ s)$, representing the composition of the first two paths, is the same as the map, $m * R$, representing the composition of the third and fourth paths. QED

From the two hexagon identities we derive two other equations involving R. These are the Majid algebra duals of equations (11.6.5) and (11.6.6) which appear in the definition of quasi-triangular Drinfeld algebras in Chapter 11. This means that when the category C is braided, the associated Majid algebra has an additional structure which Majid calls *dual quasi-triangular* [857].

8.3 Twisting.

We saw in Section 2.7 that the functor F in the reconstruction theorem determines the coalgebra C without any reference to the monoidal structure. The diagram in Figure 8.1 shows how the multiplication is derived from the tensor product and the natural transformation, b. Following Majid [848] we will show that changing b defines a transformation on the Majid algebra which is dual to the operation called twisting on Drinfeld algebras.

Suppose that b and b' are two natural transformations from $F \otimes F$ to $F \circ \otimes$. Then

$$t \stackrel{\text{def}}{=} b'^{-1} \circ b : F \otimes F \to F \otimes F$$

is a natural transformation from $F \otimes F$ to itself and hence is represented by

$$T \in \text{Lin}_k(C \otimes C, k) = (C \otimes C)^*.$$

The left hand column in Figure 8.1 defines a natural transformation

$$\varphi : F^{\otimes 3} \longrightarrow F^{\otimes 3}$$

associated to $F(\phi)$ which is represented by

$$\Phi \in \text{Lin}_k(C \otimes C \otimes C, k).$$

By definition, φ is given by (8.2.5) with a similar expression for the φ' associated to b'. The relation between φ' and φ is given by

$$\varphi'_{X,Y,Z} =$$

$$[(id\otimes t_{Y,Z})(id\otimes b^{-1})t_{X,Y\otimes Z}(id\otimes b)]\varphi_{X,Y,Z}[b^{-1}\otimes id)t^{-1}_{X\otimes Y,Z}(b\otimes id)(t^{-1}_{X,Y}\otimes id)]. \tag{8.3.1}$$

The natural transformations φ', φ and t are represented by Φ', Φ and T respectively. Applying the Lemmas 8.2.1 and 8.2.2 to (8.3.1) gives

$$\Phi' = T_{23} * (id \otimes \Delta)T * \Phi * (\Delta \otimes id)T^{-1} * T_{12}^{-1}. \tag{8.3.2}$$

This is Majid's formula for the "twisting" effect of the change of b on Φ. To see the effect of the change in b on the multiplication, consider the diagram

$$\begin{array}{ccccccc}
F(\cdot)\otimes F(\cdot) & \xrightarrow{b} & F(\cdot\otimes\cdot) & \xrightarrow{c} & C\otimes F(\cdot\otimes\cdot) & \xrightarrow{id\otimes b^{-1}} & C\otimes F(\cdot)\otimes F(\cdot) \\
t\downarrow & & id\downarrow & & \downarrow N\otimes id & & \downarrow id\otimes t \\
F(\cdot)\otimes F(\cdot) & \xrightarrow{b'} & F(\cdot\otimes\cdot) & \xrightarrow{c} & C\otimes F(\cdot\otimes\cdot) & \xrightarrow{id\otimes b'^{-1}} & C\otimes F(\cdot)\otimes F(\cdot)
\end{array}$$

The natural transformation representing composition along the top is m, and along the bottom is m'. Going down the left hand column and across the bottom row gives a natural transformation represented by $m' * T$ (where the $*$ represents convolution relative to $C \otimes C$). Going across the top and then down the right hand arrow gives $T * m$. Therefore we get

$$T * m = m' * T. \tag{8.3.3}$$

Dualizing gives the form of the corresponding equations in the Drinfeld algebra setting. Fore example, this last equation dualizes to a conjugation:

$$T\Delta = \Delta' T. \tag{8.3.4}$$

8.4 Modules over a monoidal category

The material in the next two sections describes some aspects of Majid's theory of representations of one monoidal category in another. We prefer the terminology "modules over a monoidal category relative to a monoidal functor."

What is a (left) module for a bialgebra, B? If V is a left B module and $F(X)$ is a (left) B comodule then we can define non-trivial map,

$$\lambda_{X,V} : F(X) \otimes V \to V \otimes F(X) \tag{8.4.1}$$

as follows: use the comodule structure on $F(X)$ to map

$$F(X) \otimes V \longrightarrow B \otimes F(X) \otimes V,$$

then apply the trivial switch to to get to $B \otimes V \otimes F(X)$, then the left module map to get to $V \otimes F(X)$. So

$$\lambda_{V,X} = (\mu \otimes id) \circ s_{23} \circ c_X, \tag{8.4.2}$$

where $\mu : B \otimes V \to V$ is the left action of B on V.

It is reasonable to expect that if you know $\lambda_{V,X}$ for enough X, then you know the module structure on V. So we can try to *define* a left module to be an object, $V \in Ob(\mathbf{Vec})$ together with a natural transformation of functors (8.4.1).

So let $\mathcal{C}$ be a monoidal category, and (F, b, f) be a monoidal functor from $\mathcal{C}$ to **Vec** with representing bialgebra, B. How does the existence of the natural transformation (8.4.1) determine a left B-module structure on V? Define a transformation

$$\tilde{\lambda}_{X,V} : F(X) \to \mathbf{Lin}(V, V \otimes F(X)), \quad \tilde{\lambda}_{X,V}(x)(v) = \lambda_{X,V}(x \otimes v) \tag{8.4.3}$$

where $x \in X$, $v \in V$. When $F(X)$ is finite dimensional, we have the canonical isomorphism

$$\mathbf{Lin}(V, V \otimes X) \cong \mathrm{End}(V) \otimes F(X).$$

So we can consider $\tilde{\lambda}_{X,V}$ as a natural transformation

$$\tilde{\lambda}_{X,V} : F(X) \longrightarrow \mathrm{End}(V) \otimes F(X).$$

Since B is a representing object for F, there is a linear map

$$\tilde{\rho}_V : B \to \mathrm{End}(V)$$

such that $\tilde{\lambda}_{X,V}$ is given as the composition

$$F(X) \longrightarrow B \otimes F(X) \stackrel{\tilde{\rho}_V \otimes id}{\longrightarrow} \mathrm{End}(V) \otimes F(X).$$

We can then define

$$\rho_V : B \otimes V \longrightarrow V$$

by

$$\rho_V(h \otimes v) = \tilde{\rho}_V(h)(v), \quad h \in B.$$

In order to have the associativity law which is required in order that this be a left module action, we need to place an additional condition on the natural transformation $\lambda_{X,V}$. We require that the next diagram commute:

$$\begin{array}{ccccc} F(X) \otimes F(Y) \otimes V & \stackrel{id \otimes \lambda_{Y,V}}{\longrightarrow} & F(X) \otimes V \otimes F(Y) & \stackrel{\lambda_{X,V} \otimes id}{\longrightarrow} & V \otimes F(X) \otimes F(Y) \\ {\scriptstyle b_{X,Y}} \downarrow & & & & \downarrow {\scriptstyle id \otimes b_{X,Y}} \\ F(X \otimes Y) \otimes V & & \stackrel{\lambda_{X \otimes Y, V}}{\longrightarrow} & & V \otimes F(X \otimes Y). \end{array} \tag{8.4.4}$$

Proposition 8.4.1 *When (8.4.4) commutes the $\tilde{\rho}_V$ is a homomorphism, defining a left module structure.*

Proof. Lemma 8.2.1 implies that the element of $\mathbf{Lin}(B \otimes B, \mathrm{End}(V))$ representing the composite

$$F(X) \otimes F(Y) \to F(X \otimes Y) \stackrel{\tilde{\lambda}_{X \otimes Y,V}}{\longrightarrow} \mathrm{End}(V) \otimes F(X \otimes Y) \to \mathrm{End}(V) \otimes F(X) \otimes F(Y)$$

is

$$a \otimes b \mapsto \tilde{\rho}_V(ab).$$

The top row of (8.4.4) is the map

$$x \otimes y \otimes v \mapsto \tilde{\lambda}_{X,V}(x)\left(\tilde{\lambda}_{Y,V}(y)(v)\right)$$

which is represented by

$$a \otimes b \mapsto \tilde{\rho}(a) \circ \tilde{\rho}(b).$$

Commutativity of (8.4.4) implies that $\tilde{\rho}$ is a homomorphism. As a corollary we conclude

Proposition 8.4.2 *V is a left B module if and only if there exists a natural transformation (8.4.1) such that (8.4.4) commutes.*

This way of describing B–modules has a natural generalization to the case of any monoidal functor (F, b, f), $F : \mathcal{C} \to \mathcal{V}$ between two monoidal categories: We consider pairs $(V, \lambda_{\cdot,V})$ where $V \in \mathrm{Ob}(\mathcal{V})$ and $\lambda_{\cdot,V}$ is a natural transformation of type (8.4.1) for which 8.4.4 commutes. These are the objects in a third category, $\mathcal{C}_{\mathcal{V},F}$, *the category of $\mathcal{C}$ modules in $\mathcal{V}$ relative to F*. The morphisms in this new category are the linear maps

$$\alpha : V_1 \to V_2$$

such that the diagram

$$\begin{array}{ccc} F(X)\otimes V_1 & \stackrel{\lambda_{X,V_1}}{\longrightarrow} & V_1\otimes F(X) \\ {\scriptstyle id\otimes\alpha}\downarrow & & \downarrow{\scriptstyle \alpha\otimes id} \\ F(X)\otimes V_2 & \stackrel{\lambda_{X,V_2}}{\longrightarrow} & V_2\otimes F(X) \end{array} \tag{8.4.5}$$

commutes for all $X \in \mathrm{Obj}\mathcal{C}$.

The tensor product for the monoidal structure on this new category is given by

$$(V_1, \lambda_{\cdot,V_1}) \otimes (V_2, \lambda_{\cdot,V_2}) = (V_1 \otimes V_2, (id_{V_1} \otimes \lambda_{\cdot,V_2}) \circ (\lambda_{\cdot,V_1} \otimes id_{V_2}).$$

and the identity object is

$$(\mathbf{1}_{\mathcal{V}}, l_{\mathcal{V}}).$$

We leave to the reader the verification that the condition for a monoidal structure are satisfied, or refer to the paper by Majid, [854][858].

Suppose that we have, in addition to the multiplicative functor between the monoidal categories $\mathcal{C}$ and $\mathcal{V}$, a multiplicative functor from $\mathcal{V}$ to **Vec** giving a commutative diagram

$$\begin{array}{ccccc} \mathcal{C} & & \stackrel{G}{\longrightarrow} & & \mathcal{V} \\ F\circ G & \searrow & & \swarrow & F \\ & & \mathbf{Vec} & & . \end{array}$$

Then the category $\mathcal{C}_{\mathcal{V},G}$ also has a monoidal functor $\tilde{F}$ to **Vec**, $\tilde{F}(V, \lambda_V) = F(V)$, forgetting about λ.

If we start with two bialgebras and a monoidal functor between their categories of modules, $\mathcal{C}$ and $\mathcal{V}$, then we generate a third bialgebra as the representing bialgebra of $\mathcal{C}_{\mathcal{V},G}$.

Consider the special case $\mathcal{C} = \mathcal{V}$ and $G = id$, which arises whenever we have a monoidal functor. In particular, given a bialgebra B and its category of left comodules $\mathbf{Com}(B)$, consider $\mathbf{Com}(B)_{\mathbf{Com}(B),id}$. The objects will be modules over B as well as comodules over B. If B is finite dimensional, then we can identify modules over B with comodules over B^*, so in this case the objects in $\mathbf{Com}(B)_{\mathbf{Com}(B),id}$ will be comodules over both B and B^*. It is reasonable then to expect that the representing object will be a bialgebra, $\tilde{B}$, which contains both B and B^* as subbialgebras.

Let us see how all of this works for the category of comodules over $B = k[G]$, the group algebra of a finite group, G or equivalently the category of modules over $k(G)$, the Hopf algebra dual of $k[G]$. (David Kazhdan described this example to us.) The Hopf algebra $k(G)$ consists of functions on G with the usual pointwise multiplication of functions,

$$f_1 f_2(a) = f_1(a) f_2(a) \quad a \in G.$$

Under the identification of $k(G \times G)$ with $k(G) \otimes k(G)$ the comultiplication is is given by

$$\Delta f(a,b) = f(ab).$$

If we take $k(G)$ as the starting point and consider $k[G]$ as its dual then $k[G]$ consists of measures on G. Multiplication on $k[G]$ is convolution so

$$m(\mu_1, \mu_2) = \mu_1 * \mu_2$$

where

$$\mu_1 * \mu_2(z) = \sum_{xy=z} \mu_1(x)\mu_2(y).$$

If $M : G \times G \to G$ denotes the multiplication map, and M_* denotes the corresponding push forward of measures, then we can write

$$\mu_1 * \mu_2 = M_*(\mu_1 \times \mu_2)$$

where $\mu_1 \times \mu_2$ denotes the product measure on $G \times G$. Similarly, the comultiplication is defined as the push forward under the diagonal map,

$$\Delta\mu = (\text{diag})_*\mu, \ \text{diag} : G \to G \times G, (\text{diag})(a) = (a, a).$$

In algebraic language, $k[G]$ is the usual group algebra with comultiplication determined by $\Delta(g) = g \otimes g$. Any measure is a linear combination of point measures μ_x.

$$\mu_x * \mu_y = \mu_{xy} \qquad \Delta\mu_x = \mu_x \otimes \mu_x.$$

1. What is the category of $k[G]$ comodules? Since $k[G]$ is finite dimensional, a $k[G]$ comodule is the same as an $k(G)$ module, and since G is a

finite set, a module N over $k(G)$ is the same as a "vector bundle" over G, where the "fiber" N_x over a point x is given as

$$N_x = N/m_xN, \quad m_x = \text{ functions vanishing at } x,$$

and N_x is identified as the subspace of N given by

$$N_x = \delta_x N$$

where δ_x is the δ function based at x.

The category of vector bundles over G is the same as the category of $k[G]$ comodules and thus $k[G]$ is its representing bialgebra.

2. What is the monoidal structure? We must describe the tensor product. If E and F are two vector bundles over G, then we can form their "external tensor product", $E \bigcirc F$, which is the vector bundle over $G \times G$ whose fiber over the point (x, y) is given by

$$(E \bigcirc F)_{(x,y)} = E_x \otimes F_y.$$

The space of sections of this vector bundle is a module over $k(G \times G) = k(G) \otimes k(G)$ and we must consider it as a module over $k(G)$ via the comultiplication on $k(G)$. But

$$\Delta\delta_z = \sum_{xy=z} \delta_x \otimes \delta_y$$

so we see that the corresponding fiber over z is given by

$$(E \otimes F)_z = \bigoplus_{xy=z} (E_x \otimes F_y).$$

This *defines* the tensor product structure on the space of vector bundles over G.

3. What is the functor $\lambda_{V,X}$ when we take the identity functor? In other words, what is the data associated to natural transformations from $E \otimes F$ to $F \otimes E$? Any vector bundle can be written as a direct sum of the bundles $L(x)$ as x varies over G, where

$$L(x)_x = \mathbf{C}, \quad \text{and} \quad L(x)_y = \{0\} \quad \text{for} \quad y \neq x.$$

So we must be given a family of isomorphisms from $L(x) \otimes F$ to $F \otimes L(x)$. But from the definition of $L(x)$ and the definition of tensor product, we see that

$$(L(x) \otimes F)_z = F_{x^{-1}z} \quad \text{and} \quad (F \otimes L(x))_z = F_{zx^{-1}}.$$

Taking $z = xw$, we must be given an isomorphism of

$$F_w \text{ with } F_{xwx^{-1}}$$

for all x and w in G. This means that we are given an action of G as vector bundle automorphisms of F, covering the conjugation action of G on itself. Thus the category associated to the category of vector bundles on G is the category of G homogeneous vector bundles over G just defined. This is the new category.

4. Since we are dealing with finite dimensional bialgebras we can consider the dual of the usual representing bialgebra with respect to which our category is the category of modules instead of comodules. That is, we view the situation from the point of view of modules and ask for which bialgebra B is our category of vector bundles with a G action the category of B modules.

What is this dual representing bialgebra for our new category? We are considering modules E over $k(G)$ which are also modules over G, $\rho : G \to \mathrm{End}(E)$. The consistency condition between these two actions is given by

$$\rho(x)(f \cdot e) = (x \cdot f)(\rho(x)e)$$

for $x \in G$, $f \in k(G)$ and $e \in E$. If we think of e as a section of a vector bundle E, with G acting by

$$(\rho(x)e)(y) = x \cdot e(x^{-1}yx),$$

then the induced G action on $k(G)$ is the coadjoint action, dual to the conjugation action.

Recall the definition of semidirect product. If we are given a representation of ρ of G on a vector space A, this makes A into a left $k[G]$ module. If A is a commutative algebra, then we can form the semidirect product, $A \times_\rho k[G]$ which, as a vector space is $A \otimes k[G]$ and with multiplication given by

$$(a, x)(b, y) = \sum (a\rho(x_{(1)}(b), x_{(2)}y),$$

for $a, b \in A$ and $x, y \in k[G]$. When A is $k(G)$ the bialgebra of functions on G and ρ is the coadjoint representation, the semidirect product $k(G) \times_{ad^*} k[G]$ is the dual representing bialgebra just described.

8.5 The Dual Category and Drinfeld Double

Let us return to the case where $\mathcal{V} = \mathbf{Vec}$. The forgetful functor

$$Fgt : \mathcal{C}_{\mathbf{Vec},F} \to \mathbf{Vec} \quad (V, \lambda) \mapsto V$$

is clearly a strictly monoidal functor. Let B' be its representing object, and recall that B is the representing object for F.

Proposition 8.5.1 *There is a dual pairing between B and B'.*

Proof. The defining property for morphisms in $\mathcal{C}_{\mathbf{Vec},F}$, namely the commutativity of (8.4.5), implies that for fixed X,

$$\lambda^{-1}_{X,V} : V \otimes F(X) \to F(X) \otimes V$$

is a natural transformation in V.

Let $\tilde{\rho}_V \in \mathbf{Lin}(B, \mathrm{End}(V))$ be the linear map representing $\tilde{\lambda}_{X.V}$ as defined in (8.4.3). Then for any $b \in B$,

$$\tilde{\rho}_V(b) : V \to V$$

is a natural transformation.

Since B' is the representing object for $(\mathcal{C}_{\mathbf{Vec},F}, Fgt)$, $\tilde{\rho}_V(b)$ defines an element, $f(b)$ of $\mathbf{Lin}(B', k)$.

Define the pairing

$$B \times B' \to k$$

by

$$(b, b') \mapsto f(b)(b')$$

and check that this is non-degenerate.

Without the assumption of a multiplicative functor to $\mathbf{Vec}$ we cannot talk about a dual pairing of representing objects. In the general case the duality is expressed by the fact that

$$X \in \mathcal{C} \mapsto (F(X), \lambda^{-1}_{X,\cdot}) \in (\mathcal{C}^o)^o$$

defines an embedding of the original monoidal category into the double dual.

These last two facts lead Majid [869] to call the category of modules of $\mathcal{C}$ in $\mathcal{V}$ relative to F, $\mathcal{C}_{\mathcal{V},F}$, the *dual category* to $\mathcal{C}$ relative to $F : \mathcal{C} \to \mathcal{V}$, and to denote it simply by $\mathcal{C}^o$ when F and $\mathcal{V}$ are fixed.

Let us consider the example given in the previous section a little further. The representing object $k(G) \times_{ad^*} k[G]$ is not simply the semidirect product of algebras. It is a bialgebra (in fact Hopf algebra) with comultiplication which restricts to the usual comultiplication on $k[G]$ and and $k(G)$. There is an algebraic construction called the *Drinfeld double* which, starting with a Hopf algebra, H, generates a new Hopf algebra which contains H and $H^{*\mathrm{cop}}$ (H^* with opposite comultiplication) as sub-Hopf algebras. This construction has many applications in the theory of quantum groups, the most important of which is the definition of the quantization of the full enveloping algebra of a simple Lie algebra from the quantization of the universal enveloping algebra of a Borel subalgebra, see for example[1004].

There are two alternative definitions, the one adopted by Majid has $H^{*\mathrm{op}}$, H^* with the opposite multiplication, imbedded as a sub-Hopf algebra. The map $id \otimes T^*$, where T^* is the antipode of of H^* defines an isomorphism between them. We give Drinfeld's definition as formulated by Radford [1002]. In order to avoid some of the complications arising in the infinite dimensional case, we will assume that the Hopf algebra H is finite dimensional. Then the full linear dual H^* =Hom(H, k) is also a Hopf algebra. Let $H^{*\mathrm{cop}}$ be H^* with the opposite comultiplication $\Delta^{\mathrm{op}} = s \circ \Delta$, where s is transposition. In addition, because H is finite dimensional,the antipode T of H is necessarily invertible. The antipode of H^* is T^*.

Define the left coadjoint action of $h \in H$ on $p \in H^*$ by

$$ad^*(h) \cdot p = \sum \langle p_{(3)}, h_{(1)} \rangle \langle p_{(1)}, T^{-1}(h_{(2)}) \rangle p_{(2)}$$

and the right coadjoint action of p on h by

$$h \cdot ad^*(p) = \sum \langle T^{*-1}(p_{(1)}), h_{(3)} \rangle \langle p_{(2)}, h_{(1)} \rangle h_{(2)}.$$

Definition[317](paragraph 13)[842][1002] Let H be a finite dimensional Hopf algebra with antipode T. The *Drinfeld double* of H is a (quasi-triangular) Hopf algebra $D(H)$ which contains both H and $H^{*\mathrm{cop}}$ as sub-Hopf algebras. As a coalgebra it is isomorphic to $H^{*\mathrm{cop}} \otimes H$ and multiplication is given by

$$(p \otimes h) \diamond (p' \otimes h') = \sum p(ad^*(h_{(1)}) \cdot p'_{(2)}) \otimes (h_{(2)} \cdot ad^*(p'_{(1)}))h'. \qquad (8.5.1)$$

The antipode of $D(H)$ is defined by

$$\tilde{T}(p \otimes h) = (\epsilon \otimes T(h))(T^{*-1}(p) \otimes 1).$$

The quasi-triangular structure (see Chapter 11) is given by

$$R = \epsilon \otimes Id \otimes 1 \in (H^{*\mathrm{cop}} \otimes H) \otimes (H^{*\mathrm{cop}} \otimes H),$$

where $Id \in H \otimes H^{*\mathrm{cop}}$ represents the identity map on H.

The equation for multiplication is simplified using the formula [1002]

$$\sum ad^*(h_{(1)}) \cdot p_{(2)} \otimes h_{(2)} \cdot ad^*(p_{(1)}) = \sum \langle p_{(3)}, h_{(1)} \rangle \langle p_{(1)}, T^{-1}(h_{(3)}) \rangle p_{(2)} \otimes h_{(2)}.$$

When H is the group algebra $k[G]$ and $h \in G$, we have $T(h) = h^{-1}$ and

$$\sum h_{(1)} \otimes h_{(2)} \otimes h_{(3)} = (\Delta \otimes id)\Delta(h) = h \otimes h \otimes h.$$

Then

$$\sum ad^*(h_{(1)}) \cdot p_{(2)} \otimes h_{(2)} \cdot ad^*(p_{(1)}) = \sum \langle p_{(3)}, h \rangle \langle p_{(1)}, h^{-1} \rangle p_{(2)} \otimes h.$$

and

$$\sum \langle p_{(3)}, h \rangle \langle p_{(1)}, h^{-1} \rangle \langle p_{(2)}, g \rangle = p(h^{-1}gh).$$

The map $T^* \otimes id$ defines an isomorphism between the Radford-Drinfeld double, which uses $H^{*\mathrm{cop}}$, and the Majid-Drinfeld double (which we will not describe explicitly here), which uses $H^{*\mathrm{op}}$ see [848]. Since the antipode for $k(G)$ is an involution, $T^{*-1} = T^*$, we can compute the form of the multiplication in the Majid-Drinfeld double as follows

$$\begin{aligned} T^* \otimes id[(\epsilon \otimes h) \diamond (T^*(p) \otimes 1)] &= T^* \otimes id[\sum \langle (T^*(p_))(3), h \rangle \langle (T^*(p))_{(1)}, h^{-1} \rangle \\ &\quad \times (T^*(p))_{(2)} \otimes h] \\ &= \sum \langle T^*(p_{(1)}), h \rangle \langle T^*(p_{(3)}), h^{-1} \rangle p_{(2)} \otimes h \\ &= \sum \langle p_{(1)}, h^{-1} \rangle \langle p_{(3)}, h \rangle p_{(2)} \otimes h \end{aligned}$$

for $h \in G$. This agrees with the product rule in the semidirect product. Moreover $k(G)$ is commutative, so $k[G]^{*\mathrm{op}} = k(G)^{\mathrm{op}} = k(G)$ the Majid-Drinfeld double is precisely the semidirect product $k(G) \times_{ad^*} k[G]$ with the usual comultiplications on $k(G)$ and $k[G]$.

Proposition 8.5.2 *[848] Let $k[G]$ be the group algebra of a finite group and $k(G)$ the dual bialgebra of functions on G. Then the Drinfeld double is $k(G) \times_{ad^*} k[G]$, the semidirect product of algebras induced by the coadjoint action of the group algebra on the algebra of functions with $k(G)$ and $k[G]$ imbedded as sub-Hopf algebras.*

See also [241]

The example demonstrates that *If $\mathcal{M}$ is the category of comodules over the Hopf algebra $k[G]$ (which is the same as the category of modules over the dual function algebra $k(G)$), then $\mathcal{M}_{\mathcal{M},id}$ is the category of modules over the Drinfeld double $D(k[G])$.*

More generally, Majid and Drinfeld [854][860] have shown that for any monoidal category $\mathcal{C}$ the dual category $\mathcal{C}_{\mathcal{C},id}$ of modules in $\mathcal{C}$ relative to the identity is braided with braiding given by

$$s_{(V,\lambda),(V',\lambda')}(V\otimes V',\lambda_{V\otimes V'}) = \lambda_{V,V'}(V\otimes V'),(\lambda_{V,V'}\otimes id)\circ\lambda_{V\otimes V'}\circ(id\otimes\lambda^{-1}_{V,V'})).$$

Therefore, the representing bialgebra will be quasi-triangular. In the context of Hopf algebras, their general theorem states

Theorem 8.5.1 *If H is a (finite dimensional) Hopf algebra, which is quasi-triangular (see Chapter 11 section 6) and $\mathcal{M}$ is the category of H comodules them $\mathcal{M}_{\mathcal{M},id}$ is the category of modules for the Drinfeld double $D(H)$.*

In the situation of the example, $k[G]$ is a cocommutative Hopf algebra and hence, trivially, a quasi-triangular one with $R = 1$. Thus the example above is a particular case of the general theorem.

Chapter 9

Gerstenhaber's Deformation theory

The classical deformation theory of associative algebras, due to Gerstenhaber, [446],[447],[449], says the *derivative* or *infinitesimal* of an algebra deformation is an element of $H^2(A, A)$ (the second Hochschild cohomology, [542] of the algebra A with values in A). We will recall its definition later on. It also says that step by step, the obstruction to extending a finite Taylor series deformation to one higher order lies in $H^3(A, A)$. By simply dualizing, one can get a theory of deformations of coalgebras. It was realized by Gerstenhaber and Schack,[462] that putting the two together gives a double complex, whose total cohomology measures the deformation of the bialgebra in the same way. (For the convenience of the reader, will review some of the formalism of double complexes in the next chapter.)

The total H^2 of the double complex describes the potential "tangent space" to the deformation of a bialgebra and the the obstruction to extending the infinitesimal deformation step by step lies in the total H^3. It was then realized by Shnider [1094] that if we deform a bialgebra in the framework of Drinfeld algebras, thus allowing for coassociativity only up to conjugation (with the pentagon identity) the computation of the total cohomology simplifies in the important cases, and one gets a direct proof of Drinfeld's theorem on the rigidity of deformations of $U(g)$ where g is a simple Lie algebra.

This introductory section of has been expanded in order to give some perspective, historical and conceptual, on the material in this chapter.

Gerstenhaber's theory of deformations of associative algebras [444] grew out of the analytic theory of deformations of complex structure developed by Fröhlicher, Kodaira, Nijenhuis and Spencer [416][677]. In the first of Gerstenhaber's fundamental series of papers in the Annals of Mathematics,[445], "On the deformation of ring and algebras," he listed the basic requirements of a deformation theory. **1.** A definition of the class of objects within which the deformation takes place and the identification of the equivalence classes of infinitesimal deformations of a given object with the elements of a suitable cohomology group.

2. A theory of obstructions to the integration of an infinitesimal deformation.

3. A parametrization of the set of objects obtainable by deformation from a given one.

4. A determination of the natural automorphisms of the parameter space (the modular group of the theory) and determination of the rigid objects.

For the Kodaira-Spencer theory, the relevant cohomology theory is cohomology with values in the holomorphic tangent sheaf. For Gerstenhaber's theory of deformation of associative algebras the relevant theory is Hochschild cohomology. In order to identify the appropriate cohomology theory, one considers a parametrized family of structures, which for heuristic purposes we assume depends smoothly on the parameter t, and differentiates the basic structure equations of the theory. In the theory of deformations of complex manifolds, the basic structure equation can be considered either as the "cocycle condition" on the transition functions $g_{ij}\ g_{jk} = g_{ik}$ which leads to Cech cohomology with values in the sheaf of holomorphic tangent vectors, $T^{1,0}(M)$, or the integrability condition for the almost complex structure which leads to the Dolbeault complex for the cohomology of the same sheaf. In the theory of the deformation of associative algebras, the basic structure equation is the associativity of multiplication. Differentiating the equation

$$m(t) \circ (m(t) \otimes id) - m(t) \circ (id \otimes m(t)) = 0$$

and evaluating at $t = 0$ gives the equation $dm'(0) = 0$ where d is the Hochschild coboundary. Later on we will do this calculation together with the dual calculation for coalgebras.

The basic observation here is that the coboundary operator in a particular dimension (one for complex manifolds, two for associative algebras) is the first derivative of a structure equation. The equivalence relation defined by conjugacy, when differentiated, should give the coboundary operator in one dimension less. A cohomology theory which has these two properties satisfies the first requirement in Gerstenhaber's list.

These conditions relate just to the first derivative of the deformation, the so-called "infinitesimal deformation." The second requirement relates to higher derivatives. In order to translate the problem to a cohomological one, one views the deformation as a Taylor series. (This, in fact, is completely consistent with a rigorous approach based on a formal algebraic definition.) We want to determine, given a Taylor series defined up to order n, which satisfies the structure equations to order u, if there is an extension to order $n+1$ which satisfies the structure equations to order $n+1$.

One converts this into a question in cohomology by considering the $n+1$st derivative of the structure equation and separating it into two parts, one involving the $n+1$st derivative of the basic operation (multiplication in the case of associative algebras), and the other part, involving all derivatives up to order $n+1$. The latter term is an expression in the Taylor coefficients which have already been defined. Since the first term is the same formal expression as appeared in considering the infinitesimal $m'(0)$, we get an equation $d\ m^{(n+1)}(0) =$ combination of $(m(0), m'(0), \ldots, m^{(n)}(0))$.

The expression on the right defines a cochain which must be a coboundary in order for an $n+1$ st order Taylor series expansion to exist. Requirement 2 is met once we show that this cochain is a cocycle.

The condition for the existence of an extension is that the cohomology class of the obstruction cocycle is 0.

In this chapter we show how the cohomology theory defined by Gerstenhaber and Schack [462] meets requirements 1 and 2 for deformations of bialgebras, and, in fact, for deformations of bialgebras to Drinfeld algebras. In chapter 11 we show how this theory can be applied to requirements 3 and 4 in the case of $U(\mathcal{G})$, the universal enveloping algebra of a Lie algebra. $\mathcal{G}$. In section 1 we consider the deformations of coalgebras from the point of view just discussed and define the appropriate cohomology theory originally defined by Cartier [213]. In section 2 we define the cobar complex and describe the relation to the results of section 1. We need the cobar complex and the (dual) bar complex for the calculations in Chapter 10. In section 3 we define the Hochschild cohomology and review quickly the theory of deformations of algebras. In section 4 we show how these theories combine to give the double complex of Gerstenhaber - Schack cohomology. In section 5 we verify that requirements 1 and 2 are satisfied. In section 6, we give a formal definition of deformations of a bialgebra and state fully all the propositions relating to requirements 1 and 2. In section 7 we show how an extension of the double complex by an extra row provides a a cohomology suited to the study of deformations of bialgebras to Drinfeld algebras, and in the final section we discuss the Gerstenhaber algebra structure on Hochschild cohomology.

In the sections on deformations of coalgebras and algebras we will use t as deformation of parameter in accordance with the usual conventions. In the sections on bialgebras we will use h as deformation parameter as Drinfeld has done in order suggest Planck's constant, an appropriate convention for the theory of quantum groups.

9.1 Deformation of coalgebras.

Starting with a coalgebra C, we are given a comultiplication $\Delta : C \to C \otimes C$. Consider the problem of deforming this map Δ by replacing it with

$$\Delta(t) = \Delta + t\Delta^{(1)} + t^2\Delta^{(2)} + \cdots \tag{9.1.1}$$

where the $\cdots$ indicate terms which are higher order in t. Look at the equation expressing the coassociativity condition for $\Delta(t)$:

$$(\Delta(t) \otimes id) \circ \Delta(t) - (id \otimes \Delta(t)) \circ \Delta(t) = 0$$

which asserts coassociativity of the comultiplication. Then the coefficient of t on the left side of this equation is

$$(\Delta \otimes id) \circ \Delta^{(1)} - (id \otimes \Delta^{(1)}) \circ \Delta + (\Delta^{(1)} \otimes id) \circ \Delta - (id \otimes \Delta) \circ \Delta^{(1)} \stackrel{def}{=} \delta\Delta^{(1)}. \tag{9.1.2}$$

We can think of the operator δ defined by (9.1.2) as being one component of a coboundary operator. The general complex can be defined as follows: Suppose that M is a bicomodule for the coalgebra C. That is, suppose that M is a both a right C comodule under $\rho : M \to M \otimes C$ and a left comodule under $\lambda : M \to C \otimes M$ and that these two operations commute in the obvious sense. Define

$$C^n(M) = \mathrm{Hom}_k(M, C^{\otimes n}) \tag{9.1.3}$$

and the differential is

$$\delta = \delta^n : C^n(M) \to C^{n+1}(M)$$

by

$$\delta\alpha = (id \otimes \alpha) \circ \lambda + \Sigma(-1)^i \Delta_i \circ \alpha + (-1)^{n+1}(\alpha \otimes id) \circ \rho \tag{9.1.4}$$

where Δ_i means, as usual, to apply the comultiplication Δ to the i-th factor in the tensor product.

With this definition, if we take $M = C$ with $\lambda = \rho = \Delta$ and think of $\Delta^{(1)}$ as an element of $C^2(C)$ then (9.1.4) becomes (9.1.2). So the infinitesimal coassociativity condition on $\Delta^{(1)}$ is precisely that

$$\delta\Delta^{(1)} = 0.$$

We must still check that the $C^n(M)$, δ_n do form a chain complex. We will do this in the next section.

Let us call two deformations, $\Delta(t)$ and $\Delta'(t)$ *equivalent*, if there is a linear map $L_t = id + t\chi + \dots$ with

$$\Delta'(t) = (L_t \otimes L_t) \circ \Delta(t) \circ L_t^{-1}. \tag{9.1.5}$$

Then $\Delta'(t) = \Delta + t\Delta^{(1)\prime} + \dots$ where

$$\Delta^{(1)\prime} = \Delta^{(1)} + (\chi \otimes id) \circ \Delta - \Delta \circ \chi + (id \otimes \chi) \circ \Delta.$$

In other words

$$\Delta^{(1)\prime} = \Delta^{(1)} + \delta\chi. \tag{9.1.6}$$

So the cocycles of equivalent deformations differ by a coboundary, and H^2 measures infinitesimal deformations up to equivalence.

The next step in the Gerstenhaber theory is to examine the conditions on $\Delta^{(1)}$ which allow us to find a term $\Delta^{(2)}$ so that $\Delta + t\Delta^{(1)} + t^2\Delta^{(2)}$ satisfies the coassociativity law up to terms of order t^2. If we compute the coefficient of t^2 in the coassociativity equation we get

$$\delta\Delta^{(2)} + (\Delta^{(1)} \otimes id) \circ \Delta^{(1)} - (id \otimes \Delta^{(1)}) \circ \Delta^{(1)}. \tag{9.1.7}$$

So a necessary and sufficient condition to be able to extend $\Delta + t\Delta^{(1)}$ to a multiplication which is coassociative up to terms of order is that

$$(\Delta^{(1)} \otimes id) \circ \Delta^{(1)} - (id \otimes \Delta^{(1)}) \circ \Delta^{(1)} \tag{9.1.8}$$

be a coboundary. A straightforward but messy computation that we leave to the reader shows that the above expression is a cocycle (as an element of $\mathrm{C}^3(C)$ if $\Delta^{(1)}$) is a cocycle (as an element of $\mathrm{C}^2(C)$). We shall do the analogous computation for the case of multiplication in Section 9.3. The principle behind both computations is the same, as we will explain in Section 9.3. Thus $H^3(C)$ measures the obstruction to extending an infinitesimal deformation one higher order in t. This is the gist of the Gerstenhaber theory.

9.2 The cobar complex.

We must still show that $\mathrm{C}(M)$ is a complex. For this it is convenient to give the $C^n(M)$ a slightly different interpretation. If $\mathcal{B}$ denotes the category of C bicomodules, we claim that the map

$$\beta \mapsto \tilde{\beta} = (\epsilon \otimes id^{\otimes n} \otimes \epsilon) \circ \beta : \quad \mathrm{Hom}_{\mathcal{B}}(M, C^{\otimes(n+2)}) \to \mathrm{Hom}_k(M, C^{\otimes^n}) \tag{9.2.1}$$

is an isomorphism of vector spaces. Indeed, define the inverse map by

$$\alpha \mapsto \hat{\alpha} = (id \otimes \alpha \otimes id) \circ (\lambda \otimes id) \circ \rho : \mathrm{Hom}_k(M, C^{\otimes^n}) \to \mathrm{Hom}_{\mathcal{B}}(M, C^{\otimes^(n+2)}). \tag{9.2.2}$$

The various coalgebra and bicomodule axioms show that these two maps are indeed mutually inverse.

Define the (unnormalized) *cobar resolution*, $\mathrm{Cob}(C) = \bigoplus \mathrm{Cob}^n(C)$, as follows: the $n-$th term in the cobar resolution is given as

$$\mathrm{Cob}^n(C) = C^{\otimes(n+2)}. \tag{9.2.3}$$

The differential

$$\delta_c = \delta_c^n : \mathrm{Cob}^n(C) \to \mathrm{Cob}^{n+1}(C)$$

is defined as

$$\begin{aligned} \delta_c &= \Delta \otimes id \otimes \ldots \otimes id - id \otimes \Delta \otimes id \otimes \ldots \\ &\quad \pm id \otimes \ldots \otimes id \otimes \Delta \\ &= \sum (-1)^{i+1} \Delta_i . \end{aligned} \tag{9.2.4}$$

Let us check that $\delta^2 = 0$. For example,

$$\delta^0 = \Delta \otimes id - id \otimes \Delta$$

and

$$\delta^1 = \Delta \otimes id \otimes id - id \otimes \Delta \otimes id + id \otimes id \otimes \Delta$$

so

$$\begin{aligned} \delta^1 \circ \delta^0 &= [(\Delta \otimes id - id \otimes \Delta) \circ \Delta] \otimes id + (id \otimes id \otimes \Delta) \circ (\Delta \otimes id) \\ &\quad -(\Delta \otimes id \otimes id) \circ (id \otimes \Delta) + id \otimes [(\Delta \otimes id - id \otimes \Delta) \circ \Delta]. \end{aligned}$$

the second and third terms on the right are just $\Delta \otimes \Delta$ with oppositie signs and so cancel. The first and last terms on the right cancel by coassociativity. In general, if we write

$$\Delta_{i,p} = \Delta_i : C^{\otimes(p+2)} \to C^{\otimes(p+3)}$$

to keep track of the space where Δ_i is operating, so that

$$\delta_c^p = \sum_i (-1)^{i-1} \Delta_{i,p}$$

then

$$\delta_c^{p+1} \circ \delta_c^p = \Sigma_1 + \Sigma_2 + \Sigma_3$$

where

$$\begin{aligned} \Sigma_1 &= \sum_{i<j} a(-1)^{i+j} \Delta_{i,p+1} \circ \Delta_{j,p} \\ \Sigma_2 &= \sum_i [\Delta_{i,p+1} \circ \Delta_{i,p} - \Delta_{i+1,p+1} \circ \Delta_{i,p}] \\ \Sigma_3 &= \sum_{i<j} (-1)^{i+j+1} \Delta_{j+1,p+1} \circ \Delta_{i,p}. \end{aligned}$$

There is term by term cancellation between the first sum and the third, and every summand in the second sum vanishes because of coassociativity. So we have proved that

$$\delta \circ \delta = 0. \tag{9.2.5}$$

Notice also that we have the "homotopy operator"

$$h = \epsilon_{1,p} = \epsilon_1 : \mathrm{Cob}^p(C) \to \mathrm{Cob}^{p-1}(C)$$

and it follows immediately from $(\epsilon \otimes id) \circ \Delta = id$ that

$$\delta h + h\delta = id. \tag{9.2.6}$$

Equations (9.2.5) and (9.2.6) together prove

Proposition 9.2.1 Cob(C) *is an acyclic complex.*

The $C(M)$ of the preceding section is obtained by applying Hom$(M, \cdot)$ to the complex Cob(C). It was first defined by Cartier [213]. The coboundary operator δ defined in (9.1.4) should be called the *Cartier coboundary or*

Cartier differential and the associated cohomology should be called *Cartier cohomology.*

To show that the coboundary δ defined in (9.1.4) has square zero, we note that using (9.2.1) and (9.2.2) we have

$$(\tilde{\delta_c \hat{\alpha}}) = \delta(\alpha). \tag{9.2.7}$$

Indeed, for $\alpha \in \mathrm{Hom}_k(M, C^{\otimes n})$ we have

$$\begin{aligned} \delta_c(\hat{\alpha}) &= \Delta_1 \circ (id \otimes \alpha \otimes id) \circ (\lambda \otimes id) \circ \rho \\ & \sum_{i=2}^{n+1} (-1)^{i+1} (id \otimes \Delta_{i-1} \circ \alpha \otimes id) \circ (\lambda \otimes id) \circ \rho \\ & \quad (-1)^{n+3} \Delta_{n+2} \circ (id \otimes \alpha \otimes id) \circ (\lambda \otimes id) \circ \rho. \end{aligned}$$

Applying $\epsilon \otimes id^{\otimes(n+1)} \otimes \epsilon$ gives

$$\begin{aligned} (\tilde{\delta_c \hat{\alpha}}) &= (id \otimes \alpha) \circ \lambda + \sum_{i=1}^{n} (-1)^i \Delta_i \circ \alpha + (-1)^{n+1} (\alpha \otimes id) \circ \rho \\ &= \delta \alpha, \end{aligned}$$

proving(9.2.7). So $\delta^2 = 0$ since $\delta_c^2 = 0$.

9.3 Deformation of algebras.

We have considered infinitesimal deformations of the comultiplication of a coalgebra. Equally well, we can consider infinitesimal deformations of the multiplication of an algebra. (This is the classical setting of the Gerstenhaber deformation theory.) So A is an algebra with multiplication, $m : A \otimes A \to A$. We deform the multiplication into

$$m(t) = m + tm^{(1)} + t^2 m^{(2)} + \cdots$$

and temporarily write μ for $m^{(1)}$ so as not to have too many superscripts. The coefficient of t in the associativity condition gives

$$m(\mu(a, b), c) + \mu(m(a, b), c) = m(a, \mu(b, c)) + \mu(a, m(bc))$$

or

$$d\mu = 0 \tag{9.3.1}$$

where

$$d\mu(a,b,c) = a\mu(b,c) - \mu(ab,c) + \mu(a,bc) - \mu(a,b)c. \tag{9.3.2}$$

This brings us to the general definition of the Hochschild cohomology . Let M be an A bimodule and $C_p(A,M) = \mathrm{Hom}_k(A^{\otimes p}, M)$. Define the Hochschild differential or Hochschild coboundary operator $d : C_p \to C_{p+1}$ by

$$\begin{aligned} d\mu(a_1,\ldots,a_{p+1}) &= a_1\mu(a_2,\ldots,a_{p+1}) - \mu(a_1a_2,a_3,\ldots,a_{p+1}) \\ &\quad +\mu(a_1,a_2a_3,\ldots,a_{p+1})\ldots(-1)^{p+1}\mu(a_1,\ldots,a_p)a_{p+1}. \end{aligned} \tag{9.3.3}$$

As before, this can be described in terms of an (unnormalized) bar resolution of A. We discuss the definitions more precisely in Chapter 10 and the first section of Chapter 11.

We define equivalent deformations as before, and see that the $\mu's$ of two equivalent deformations differ by a cocycle.

Suppose that μ is a cocycle, and we want to check the coefficient of t^2 in the associativity condition, the so called first "obstruction cocycle". It is given by β where

$$\beta(a,b,c) = \mu(\mu(a,b),c) - \mu(a,\mu(b,c)). \tag{9.3.4}$$

We claim that

$$d\beta = 0, \quad \text{i.e.} \quad \beta \text{ is a three cocycle} \tag{9.3.5}$$

and if

$$\beta = d\nu$$

then the multiplication

$$m + t\mu + t^2\nu$$

satisfies associativity up to t^2.

Proof of $d\beta = 0$: In the formula

$$\begin{aligned} d\beta(x,y,z,w) &= x\beta(y,z,w) - \beta(xy,z,w) + \beta(x,yz,w) \\ &\quad -\beta(x,y,zw) + \beta(x,y,z)w \end{aligned}$$

we can replace the $\mu's$ that occur in the definition of β by the "multiplication" $m(t) = m + t\mu$. Since the lower order terms, i.e. the constant and linear terms, in the expression for the associativity vanish, we see that

$$\beta = \text{coeff of } t^2 \text{ in } m(t) \circ (m(t) \otimes id) - m(t) \circ (id \otimes m(t)).$$

In fact this was our original definition of β.

For the same reason, cancellation of lower order terms, we can now replace every multiplication in the the expression for $d\beta$ by $m(t)$ and collect the coefficient of t^2.

To simplify notation define

$$a \diamond b = ab + t\mu(a, b).$$

Now β is the coefficient of t^2 in

$$(a \diamond b) \diamond c - a \diamond (b \diamond c),$$

and $d\beta(x, y, z, w)$ is the coefficient of t^2 in

$$\begin{aligned}
&x \diamond ((y \diamond z) \diamond w) - x \diamond (y \diamond (z \diamond w)) \\
&\quad -((x \diamond y) \diamond z) \diamond w + (x \diamond y) \diamond (z \diamond w) \\
&\quad\quad +(x \diamond (y \diamond z)) \diamond w - x \diamond ((y \diamond z) \diamond w) \\
&\quad\quad\quad -(x \diamond y) \diamond (z \diamond w) + x \diamond (y \diamond (z \diamond w)) \\
&\quad\quad\quad\quad +((x \diamond y) \diamond z) \diamond w - (x \diamond (y \diamond z)) \diamond w \\
&\quad\quad\quad\quad = 0.
\end{aligned}$$

In this expression we get term by term cancellation proving that $d\beta = 0$, i.e. that β is a three cocycle.

Suppose that add a term νt^2 to the definite of m_t, so that now

$$m(t) = m + \mu t + \nu t^2.$$

Then the expression for $m(t)(m(t) \otimes id) - m(t)(id \otimes m(t))$ is modified by

$$\beta \mapsto \beta - d\nu$$

according to (9.3.2). This shows that if $\beta = d\nu$ then $m(t) = m + \mu t + \nu t^2$ is associative up to order t^2. Thus the cohomology class of β in H^3 gives the obstruction to extending the multiplication to make it associative up to order t^2.

In general, the same argument shows that if we have the Taylor series expansion

$$m_n(t) = m + tm^{(1)} + \ldots + t^n m^{(n)}$$

which is associative through order t^n, then the obstruction class is defined by

$$\beta_{n+1} t^{n+1} = m_{(n)}(t) \circ (m_{(n)}(t) \otimes id - id \otimes m_{(n)}(t)) \mod t^{n+2}.$$

To show that β_{n+1} is a cocycle, we replace the multiplication, m in the coboundary operator by $m_{(n)}(t)$ and calculate modulo t^{n+2}. Exactly the same calculation as above shows that $d\beta_{n+1} = 0$, and that if $\beta_{n+1} = d\nu$ then

$$m_{(n+1)}(t) = m_{(n)}(t) + t^{n+}\nu$$

is associative modulo t^{n+2}.

A completely parallel argument applies in the coalgebra deformation theory. To prove that the obstruction class is a cocycle we use a general principle that we mentioned in section 1: Wherever the initial multiplication or comultiplication enters into the definition of the coboundary operator, replace it with the expansion up to order n. In the resulting expression there will be a term by term cancellation, not dependent on any further assumptions of associativity or coassociativity. The same principle will reappear again in section 9.5 in the theory of bialgebra deformations.

9.4 Deformation of bialgebras.

Let B be a bialgebra. Define

$$C^{p,q} = \mathrm{Hom}_k(B^{\otimes q}, B^{\otimes p}).$$

Now Δ^{p-1} is an unambiguous homomorphism of of B to $B^{\otimes p}$ in view of the coassociativity of Δ. We make $B^{\otimes p}$ into a right or left module by right or left multiplication by $\Delta^{p-1}(a)$.

To illustrate the comodule structure, for $p = 3$, use the Sweedler notation, $\Delta a = \Sigma a_{(1)} \otimes a_{(2)}$. Then

$$\lambda : B^{\otimes 3} \to B \otimes B^{\otimes 3},\ \lambda(a \otimes b \otimes c) = \Sigma a_{(1)} b_{(1)} c_{(1)} \otimes a_{(2)} \otimes b_{(2)} \otimes c_{(2)}$$

and

$$\rho : B^{\otimes^3} \to B^{\otimes 3} \otimes B, \rho(a \otimes b \otimes c) = \Sigma a_{(1)} \otimes b_{(1)} \otimes c_{(1)} \otimes a_{(2)} b_{(2)} c_{(2)}.$$

The generalization to $B^{\otimes p}$ is straightforward: To define

$$\lambda(a_1 \otimes \cdots \otimes a_p)$$

we use the comodule structure on each factor, rearrange terms, and multiply all the $a_{i(1)}$. Similarly for the right comodule structure, ρ.

We have defined the differentials δ in (9.1.4) using the coalgebra structure and the differential d in (9.3.3 using the algebra structure where

$$\begin{aligned} d : C^{p,q} &\rightarrow C^{p.q+1}, \\ \delta : C^{p,q} &\rightarrow C^{p+1,q}. \end{aligned}$$

The complex can be pictured as a square lattice in the plane as in the accompanying diagram. The entry in position in (p,q), column p, row q is $C^{p,q}$. The Cartier differential is represented by a horizontal arrow pointing to the right and the Hochschild differential is represented by a vertical arrow pointing up.

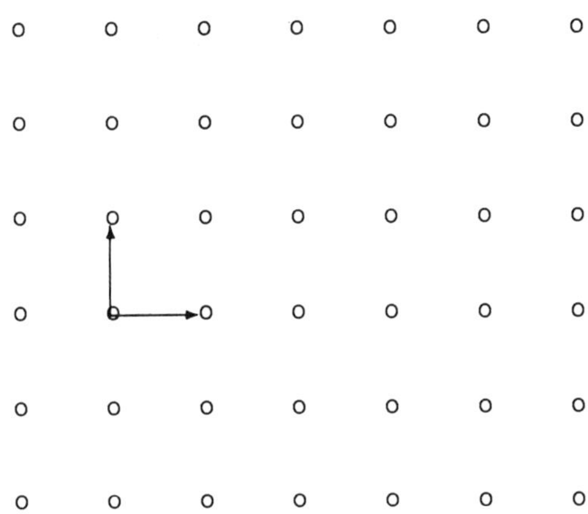

The bialgebra axiom

$$\Delta \circ m = (m \otimes m) \circ s_{23} \circ (\Delta \otimes \Delta)$$

asserting the compatibility of multiplication and comultiplication implies that the comodule maps λ and ρ are bimodule maps, and that the bimodule structure maps are bicomodule morphisms. (These conditions are in fact equivalent. See the discussion of B-4 modules in Chapter 10.) This, in turn, implies that d and δ commute. We define the operators $D^{p,q}$ on $C^{p,q}$ by

$$D^{p,q} = d + (-1)^q \delta. \tag{9.4.1}$$

and

$$D = \bigoplus D^{p,q} \text{ on } \bigoplus C^{p,q}.$$

Then

$$D^2 = 0$$

follows immediately from $d\delta = \delta d$.

Now let us simultaneously deform the multiplication and comultiplication, so $m(t) = m + t\mu$ and $\Delta(t) = \Delta + t\Delta^{(1)}$. Since $m^{(1)} \in C^{1,2}$ and $\Delta^{(1)} \in C^{2,1}$,

$$D(m^{(1)}, \Delta^{(1)}) = (dm^{(1)}, \delta m^{(1)}, +d\Delta^{(1)}, -\delta\Delta^{(1)}).$$

We have already seen that the associativity of the deformed multiplication (through order t) implies that $dm^{(1)} = 0$ and that the associativity of the deformed comultiplication implies that $\delta\Delta^{(1)} = 0$. So the first and last terms on the right hand side of above expression vanish. We will now show that the consistency of multiplication and comultiplication implies that the sum of the remaining two terms vanishes. For this let us examine the coefficient of t in the bialgebra consistency condition

$$(m(t) \otimes m(t)) \circ s_{23} \circ (\Delta(t) \otimes \Delta(t)) = \Delta(t)\mu(t).$$

Let us collect the coefficient of t in

$$\begin{aligned}&((m + tm^{(1)}) \otimes (m + tm^{(1)}))s_{23}((\Delta + t\Delta^{(1)}) \otimes (\Delta + t\Delta^{(1)}))(a \otimes b)\\ &-(\Delta + t\Delta^{(1)})(m + tm^{(1)})(a \otimes b).\end{aligned}$$

In the first expression there will be four terms corresponding to one choice of a multiple of t in each factor, and in the second expressions there will be two terms occurring with a minus sign. We list the terms (remembering the module and comodule structures on $A \otimes A$):

$$\begin{aligned}
mm\Delta\Delta^{(1)} &: \quad (a \otimes b) \rightarrow \quad a\Delta^{(1)}(b)\\
mm\Delta^{(1)}\Delta &: \quad (a \otimes b) \rightarrow \quad (\Delta^{(1)}(a))b\\
mm^{(1)}\Delta\Delta &: \quad (a \otimes b) \rightarrow \quad (id \otimes m^{(1)}) \circ \lambda(a \otimes b)\\
m^{(1)}m\Delta\Delta &: \quad (a \otimes b) \rightarrow \quad (m^{(1)} \otimes id) \circ \rho(a \otimes b)\\
-\Delta m^{(1)} &: \quad (a \otimes b) \rightarrow \quad -\Delta m^{(1)}(a \otimes b)\\
-\Delta^{(1)}m &: \quad (a \otimes b) \rightarrow \quad -\Delta^{(1)}(ab).
\end{aligned}$$

Combining terms 1, 2 and 6 gives $d\Delta^{(1)}$ and combining terms 3, 4 and 5 gives $\delta m^{(1)}$. Thus an infinitesimal bialgebra deformations satisfies the three component condition

$$D(m^{(1)}, \Delta^{(1)}) = 0, \qquad (9.4.2)$$

and so is a cocycle for D, the Gerstenhaber-Schack differential.

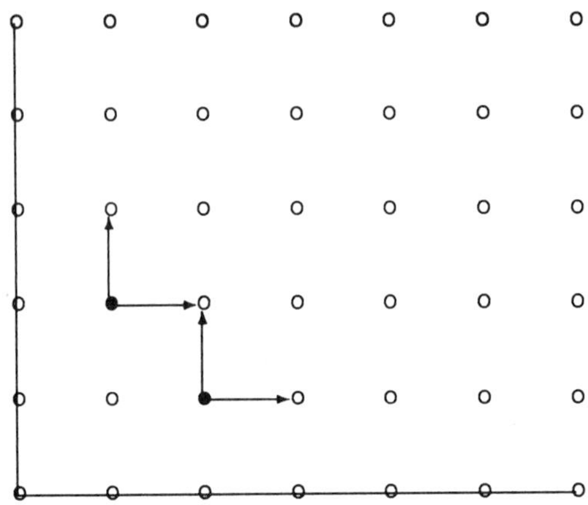

Here we have considered the cohomology in dimension 3. In the next section we consider the cohomology in dimensions 4 and 2.

9.5 Deformation cohomology of bialgebras

In the last section we defined the spaces $C^{p,q}$ and the operators $D^{p,q}$ acting on $C^{p,1}$. We defined the total differential $D = \oplus D^{p,q}$, satisfying $D^2 = 0$ and showed that the infinitesimal of a bialgebra deformation $(m^{(1)}, \Delta^{(1)}) \in C^{1,2} \oplus C^{2,1}$ satisfies the D cocycle condition

$$D^{1,2}\, m^{(1)} + D^{2,1}\, \Delta^{(1)} = 0.$$

Continuing Gerstenhaber's program, we must show that equivalence is a coboundary condition and that the obstruction class is a cocycle and deal with the situations described in the following diagrams.

Equivalence is described by the diagram.

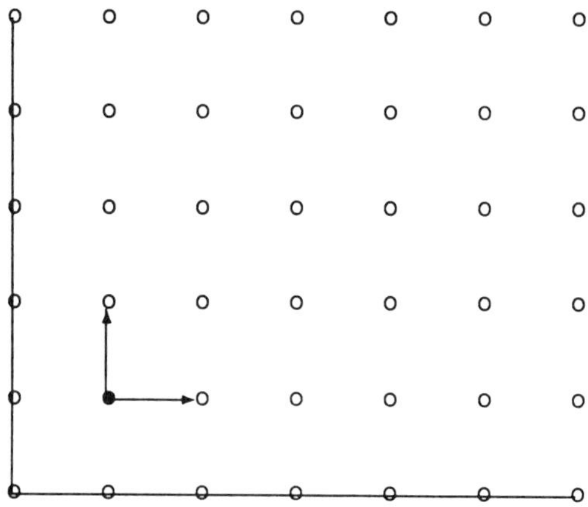

The cocycle condition on the obstruction class is described in the diagram:

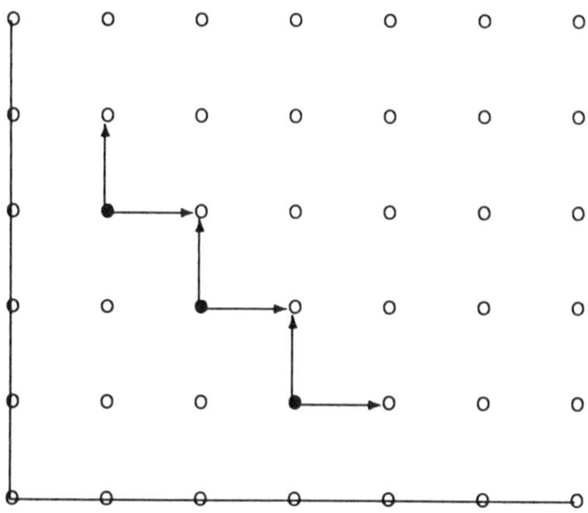

Two infinitesimal deformations $(m^{(1)}, \Delta^{(1)})$ and $(m'^{(1)}, \Delta'^{(1)})$ represent equivalent first order deformations if there exists a $\chi \in C^{1,1}$ such that, defining $L_{(1)} = id + t\chi$, we have

$$L_{(1)} \circ m_{(1)} \circ \left(L_{(1)}^{-1} \otimes L_{(1)}^{-1}\right) = m'_{(1)} \bmod t^2 \qquad (9.5.1)$$

$$(L_{(1)} \otimes L_{(1)}) \circ \Delta_{(1)} \circ L_{(1)}^{-1} = \Delta'_{(1)} \bmod t^2 \qquad (9.5.2)$$

We saw in section 1, equation (9.1.6) that (9.5.2) is equivalent to

$$\Delta'^{(1)} = \Delta^{(1)} + \delta\, \chi.$$

Equation (9.5.1) becomes

$$\begin{aligned} m'^{(1)}(a\otimes b) &= m^{(1)}(a\otimes b)+\chi(m(a\otimes b)) \\ &\quad -m(a\otimes \chi(b))-m(\chi(a)\otimes b) \\ &= m^{(1)}(a\otimes b)-(a\chi(b))-\chi(ab)+\chi(a)b) \\ &= m^{(1)}(a\otimes b)-d\ \chi\ (a\otimes b) \end{aligned} \tag{9.5.3}$$

Recall that

$$D^{1,1}\ \chi = d\chi - \delta\chi.$$

Then putting the two equations (9.1.6) and (9.5.3) gives

$$(m'^{(1)},\ \Delta'^{(1)}) = (m^{(1)},\ \Delta^{(1)}) - \ D^{1,1}\ \chi \tag{9.5.4}$$

This proves the only if part of the following:

Proposition 9.5.1 *Two infinitesimal bialgebra deformations* $(m^{(1)},\ \Delta^{(1)})$ *and* $(m'^{(1)},\ \Delta'^{(1)})$ *define equivalent first order deformations if and only if they differ by a coboundary in the Gerstenhaber - Schack complex.*

The if part of the proposition is proved by the same calculation read backwards.

Next we define the obstruction class and prove that it is a D cocycle. There are 3 components $(\beta^{1,3}\ ,\ \beta^{2,2}\ ,\ \beta^{3,1})$ corresponding to associativity, compatibility and coassociativity respectively, defined mod t^3 by

$$t^2\beta^{1,3} = m_{(1)}\circ(id\otimes m_{(1)} - m_{(1)}\otimes id) \tag{9.5.5}$$

$$t^2\beta^{2,2} = m_{(1)}\otimes m_{(1)}\circ s_{2,3}\circ\Delta_{(1)}\otimes\Delta_{(1)} - \Delta_{(1)}\circ m_{(1)} \tag{9.5.6}$$

$$t^2\beta^{3,1} = (\Delta_{(1)}\otimes id - id\otimes\Delta_{(1)})\circ\Delta_{(1)} \tag{9.5.7}$$

The terms $\beta^{1,3}$ was already defined in equation (9.3.4) and $\beta^{3,1}$ in equation (9.1.8). We showed that $d\beta^{1,3} = 0$ in (9.3.5) and explained how an analogous calculation proves that $\delta\beta^{3,1} = 0$. Since

$$D(\beta^{,1}\ ,\ \beta^{2,2}\ ,\ \beta^{3,}) = (d\ \beta^{1,3}, -\delta\ \beta^{1,3} + d\ \beta^{2,2}\ ,\ \delta\beta^{2,2} + d\ \beta^{3,1}\ ,\ \delta\beta^{3,1})$$

we need to check two more equations, the second and third components of the coboundary in $C^{2,3}$ and $C^{3,2}$ respectively. For the $C^{2,3}$ component we must show

$$\delta\beta^{1,3} = d\ \beta^{2,2}$$

$$\delta\beta^{1,3}(a\otimes b\otimes c) = id\otimes\beta^{1,3}(\lambda(a\otimes b\otimes c)) - \Delta\circ\beta^{1,3}(a\otimes b\otimes c) + \beta^{1,3}\otimes id(\rho(a\otimes b\otimes c))$$

is minus the t^2 component of

$$\begin{aligned}
&[\tilde{m}^3 \otimes (m_{(1)} \circ (m_{(1)} \otimes id - id \otimes m_{(1)}))] \circ \tilde{\Delta}^3 \\
&\quad -\Delta \circ (m_{(1)} \circ (m_{(1)} \otimes id - id \otimes m_{(1)})) \\
&+[(m_{(1)} \circ (m_{(1)} \otimes id - id \otimes m_{(1)})) \otimes \tilde{m}^3] \circ \tilde{\Delta}^3
\end{aligned} \tag{9.5.8}$$

where

$$\begin{aligned}
\tilde{\Delta}^3(a \otimes b \otimes c) &= \sum a_{(1)} \otimes b_{(1)} \otimes c_{(1)} \quad \otimes a_{(2)} \otimes b_{(2)} \otimes c_{(2)} \\
\tilde{m}^3(a \otimes b \otimes c) &= a\ b\ c. \\
\text{Whereas} & \\
d\beta^{2,2}(a \otimes b \otimes c) &= \Delta(a) \cdot \beta^{2,2}(b \otimes c) - \beta^{2,2}(ab \otimes c) \\
&+ \beta^{2,2}(a \otimes bc) - \beta^{2,2}(a \otimes b) \cdot \Delta(c)
\end{aligned}$$

is the t^2 component of

$$\begin{aligned}
&m \otimes m \circ \{(\Delta(a)\overline{\otimes}[m_{(1)} \otimes m_{(1)} \circ (\Delta_{(1)}(b)\overline{\otimes}\Delta_{(1)}(c)) - \Delta_{(1)} \circ m_{(1)}(b \otimes c)]\} \\
&\quad -[m_{(1)} \otimes m_{(1)} \circ (\Delta_{(1)}\overline{\otimes}\Delta_{(1)}) - \Delta_{(1)} \circ m_{(1)}] \circ (m \otimes id)(a \otimes b \otimes c) \\
&\quad +[m_{(1)} \otimes m_{(1)} \circ (\Delta_{(1)}\overline{\otimes}\Delta_{(1)}) - \Delta_{(1)} \circ m_{(1)}] \circ (id \otimes m)(a \otimes b \otimes c) \\
&-m \otimes m \circ \{[m_{(1)} \otimes m_{(1)} \circ (\Delta_{(1)}(a)\overline{\otimes}\Delta_{(1)}(b)) - \Delta_{(1)} \circ m_{(1)}(a \otimes b)]\overline{\otimes}\Delta(c)\}
\end{aligned} \tag{9.5.9}$$

where

$$(a_1 \otimes a_2)\overline{\otimes}(b_1 \otimes b_2) = (a_{(1)} \otimes b_{(1)}) \otimes (a_{(2)} \otimes b_{(2)})$$

Using the same idea as in the proof that $d\ \beta^{1,3} = 0$, expression (9.5.8) can be written as the t^2 component of

$$\begin{aligned}
&-[(m_{(1)} \circ (m_{(1)} \otimes id)) \otimes m_{(1)} \circ (m_{(1)} \otimes id - id \otimes m_{(1)})]\ \Delta_{(1)}(a)\overline{\otimes}\Delta_{(1)}(b)\overline{\otimes}\Delta_{(1)}(c) \\
&-[\Delta_{(1)} \circ m_{(1)} \circ (m_{(1)} \otimes id - id \otimes m_{(1)})](a \otimes b \otimes c) \\
&-[m_{(1)} \circ (m_{(1)} \otimes id - id \otimes m_{(1)}) \otimes (m_{(1)} \circ id \otimes m_{(1)})]\Delta_{(1)}(a)\overline{\otimes}\Delta_{(1)}(b)\overline{\otimes}\Delta_{(1)}(c)
\end{aligned}$$

and expression (9.5.9) can be written as the t^2 component of

$$\begin{aligned}
&[(m_{(1)} \circ id \otimes m_{(1)}) \otimes (m_{(1)} \circ id \otimes m_{(1)})]\Delta_{(1)}(a)\overline{\otimes}\Delta_{(1)}(b)\overline{\otimes}\Delta_{(1)}(c) \\
-\ &[m_{(1)} \otimes m_{(1)}] \circ [\Delta_{(1)}(a)\overline{\otimes}\Delta_{(1)} \circ m_{(1)}(b \otimes c)]
\end{aligned}$$

$$
\begin{aligned}
- & \quad [(m_{(1)} \otimes m_{(1)}] \circ [\Delta_{(1)} \circ m_{(1)}(a \otimes b) \overline{\otimes} \Delta_{(1)}(c) - \Delta_{(1)}(a) \overline{\otimes} \Delta_{(1)} \circ m_{(1)}(b \otimes c)] \\
+ & \quad \Delta_{(1)} \circ m_{(1)} \circ (m_{(1)} \otimes id)(a \otimes b \otimes c) - \Delta_{(1)} \circ m_{(1)} \circ (id \otimes m_{(1)})(a \otimes b \otimes c) \\
- & \quad [m_{(1)} \circ (m_{(1)} \otimes id) \otimes m_{(1)}(\circ m_{(1)} \otimes id)] \; \Delta(a) \overline{\otimes} \Delta(b) \overline{\otimes} \Delta(c) \\
+ & \quad [m_{(1)} \otimes m_{(1)}] \circ [\Delta_{(1)} \circ m_{(1)}(a \otimes b) \overline{\otimes} \Delta_{(1)}(c)].
\end{aligned}
$$

After internal cancellations of the terms with $[m_{(1)} \otimes m_{(1)}]$ in the second expression and the terms involving $m_{(1)} \circ (m_{(1)} \otimes id) \otimes m_{(1)} \circ (id \otimes m_{(1)})$ in the first expression, we see that the two expressions agree. The dual calculation, reversing roles of multiplications and comultiplications, which we spare the tired reader, proves the other relation

$$d\, \beta^{3,1} + \delta\, \beta^{2,2} = 0$$

The change of sign in the differential is taken care of by the fact that $\beta^{1,3}$ and $\beta^{3,1}$ are dual only up to a change in sign.

9.6 Formal bialgebra deformations

The proof that the obstruction class of the infinitesimal deformation is a cocycle leads us to the second item in Gerstenhaber's list, the theory of obstructions, which describes the recursive process for finding the Taylor series of a deformation. At this point we really have to deal more seriously with the question of the dependence of the deformation on the parameter t. What exactly is the the meaning of the calculations that we did in the previous sections? The algebraic operations on expressions $a + bt$ computed modulo t^2 can be thought of as referring to first order Taylor series, expressing relations between the initial value and the first derivative of the deformation. However, no differentiability assumptions are necessary in order to fully justify the calculations. Instead of thinking of the basic operations, $m(t)$ and $\Delta(t)$ as being a functions of t, differentiable or otherwise, we can consider them as having an expansion in powers of t with the operations of composition, tensor product and so forth behaving just as if we were dealing with Taylor series of differentiable functions. The appropriate structure here is the algebra of formal power series over a field.

Definition. The algebra of formal power series over the field k, denoted $k[[t]]$, is the k vector space of infinite formal sums $\sum_{0 \leq n} a_n t^n$ with the rules for the sum and the product being formally the same as for the convergent power series over **R** or **C**.

$$\sum_n a_n t^n + \sum_m b_m t^m = \sum_n (a_n + b_n) t^n$$

$$(\sum a_n t^n)(\sum b_m t^m) = \sum c_n t^n, \quad \text{where} \quad c_n = \sum_{i+j=n} a_i b_j.$$

There is a unique maximal ideal $\mathcal{M} \subset k[[t]]$ given by the power series with zero constant term, $a_0 = 0$. The powers $\mathcal{M}^n$ define a neighborhood basis at 0 for what is called the Krull topology, or formal power series topology. A sequence $\sum a_{m,n} t^n$ converges to $\sum a_n t^n$ in this topology if the sequence (in m) $a_{m,n}$ of nth coefficients for a fixed n is eventually constant and equal to a_n. Relative to this topology the polynomial algebra $k[t]$ is a dense subalgebra. If V is a k vector space, the tensor product $V \otimes_k k[[t]]$ consists of formal sums $\sum v_n t^n$ where, for any particular sum, the coefficients v_n span a finite dimensional subspace. The completion of $V \otimes_k k[[t]]$ relative to the topology generated by the subspaces $V \otimes_k \mathcal{M}^n$, called the completed tensor product, can be identified with the formal sums $\sum v_n t^n$ with no restriction on the coefficients. We shall denote this completed tensor product by $V[[t]]$.

Definition Let B be a bialgebra defined over the field k with multiplication m and comultiplication Δ, unit u and counit (augmentation) ϵ.

By a *formal deformation* of B we mean a bialgebra, which we denote B_t, such that

1. There is a $k[[t]]$ module isomorphism $B_t \cong B[[t]]$.
2. The operations, which we denote $m(t)$ and $\Delta(t)$ are $k[[t]]$ linear and satisfy the usual axioms, associativity, compatibility, coassociativity, unit and counit.
3. The isomorphism $B_t/tB_t \cong B$ induced by the isomorphism in 1. is an isomorphism of bialgebras.

The isomorphism described in 1. allows us to identify B_t with $B[[t]]$ and consider the operators $m(t)$ and $\Delta(t)$ as defined on $B[[t]]$. It is important to note that the identification is not canonical and two different isomorphisms will lead to different operators $m(t)$ and $\Delta(t)$ on $B[[t]]$. Although different, they will be related by an automorphism of $B[[t]]$ of the form $id + \sum \chi^{(n)} t^n$ with $\chi \in \mathrm{End}_k(B)$. Henceforth we will identify B_t with $B[[t]]$.

It follows immediately from the definition that $m(t)$ is determined by its restriction to $B \otimes B$ and Δ is determined by its restriction to B and that there is an expansion

$$m(t) = m + \sum_{1 \leq n} t^n m^{(n)} \qquad \Delta(t) = \Delta + \sum_{1 \leq n} t^n \Delta^{(n)}.$$

Where $m = m^{(0)}$ and $\Delta = \Delta^{(0)}$ are the multiplication and comultiplication on B and

$$m^{(n)} \in \mathrm{Hom}_k(B \otimes B, B) = C^{1,2}$$

and

$$\Delta^{(n)} \in \mathrm{Hom}_k(B, B \otimes B) = C^{2,1}.$$

In writing a power series expansion of an operator, $S(t)$, we will write the operator coefficients on the right $S(t) = \sum t^n S^{(n)}$ as though the t^n was a scalar multiplying the operator $S^{(n)}$.

Throughout our discussion of deformations we have not discussed the unit and counit. Although they are an essential part of the bialgebra structure, they play no essential role in the deformation theory. We will show in the next chapter that for any deformation there is an equivalent one satisfying

$$\begin{aligned} m^{(n)}(1 \otimes a) = m^{(n)}(a \otimes 1) &= 0 \\ (\epsilon \otimes id)\Delta^{(n)} = (id \otimes \epsilon)\Delta^{(n)} &= 0 = \Delta^{(n)}(1) \end{aligned} \tag{9.6.1}$$

for $a \in B \quad n \geq 1$,

From here on by bialgebra deformation we will mean formal bialgebra deformation.

Definition By an nth *order bialgebra deformation* of B we mean a bialgebra $B_{(n)}$ satisfying the conditions of the previous definition with $k[[t]]$ replaced by $k[[t]]/t^{n+1}k[[t]]$.

In the discussion of infinitesimals we have been dealing with first order deformations.

The theory of obstructions described in Gerstenhaber's list deals with the problem: Given an nth order deformation $B_{(n)}$, does there exists an $n+1$ st order deformation $B_{(n+1)}$ such that $B_{(n+1)}/t^{n+1}B_{(n)} \cong B_{(n)}$.

Just as we identified any deformation B_t with $B[[t]]$ we can identify an n th order deformation with $B \otimes_k k_{(n)}$ where $k_{(n)} = k[[t]]/t^{n+1}k[[t]] \cong k[t]/t^{n+1}k[t]$ is the truncated power series algebra, which is, in turn, isomorphic to the truncated polynomial algebra.

Suppose we have an n th order bialgebra deformation $B_{(n)}$, where we have identified $B_{(n)}$ with $B \otimes k_{(n)}$. As in the case of a deformation B_t the multiplication $m_{(n)}$ and comultiplication $\Delta_{(n)}$ are determined by their restrictions to B and we have expansions

$$m_{(n)} = m + tm^{(1)} + \cdots + t^n m^{(n)} \quad \Delta_{(n)} = \Delta + t\Delta^{(1)} + \cdots + t^n\Delta^{(n)}.$$

where $m^{(p)} \in \mathrm{Hom}_k(B \otimes B, B) = C^{1,2}(B)$ and $\Delta^{(p)} \in \mathrm{Hom}_k(B, B \otimes B) = C^{2,1}(B)$, for $0 \le p \le n$.

If we extend $m_{(n)}$ and $\Delta_{(n)}$ to be $k_{(n+1)}$ linear on $B_{(n+1)} = B \otimes_k k_{(n+1)}$ they define a multiplication and comultiplication which satisfy the axioms of a bialgebra modulo t^{n+1}, that is, on the quotient $B_{(n+1)}/t^{n+1}B_{(n+1)}$. The error term in the structure equations defines the obstruction class in exactly as in the same way as in the last section.

Definition The n th obstruction class $\beta_{(n)}$ has 3 components

$$\beta_{(n)} = (\beta^{1,3}_{(n)}\ ,\ \beta^{2,2}_{(n)}\ ,\beta^{3,1}_{(n)})$$

corresponding to associativity, compatibility and coassociativity respectively, defined, mod t^{n+2} by

$$t^{n+1}\beta^{1,3}_{(n)} = m_{(n)} \circ (id \otimes m_{(n)} - m_{(n)} \otimes id) \tag{9.6.2}$$

$$t^{n+1}\beta^{2,2}_{(n)} = m_{(n)} \otimes m_{(n)} \circ s_{2,3} \circ \Delta_{(n)} \otimes \Delta_{(n)} - \Delta_{(n)} \circ m_{(n)} \tag{9.6.3}$$

$$t^{n+1}\beta^{3,1}_{(n)} = (\Delta_{(n)} \otimes id - id \otimes \Delta_{(n)}) \circ \Delta_{(n)} \tag{9.6.4}$$

The basic proposition states

Proposition 9.6.1 *The obstruction class $\beta_{(n)}$ of an nth order deformation is a 4 cocycle in the Gerstenhaber-Schack complex. If it is a coboundary $\beta_{(n)} + D(\mu,\nu) = 0$ then the multiplication*

$$m_{(n)} + t^{n+1}\mu = m + tm^{(1)} + \cdots + t^n m^{(n)} + t^{n+1}\mu,$$

and comultiplication

$$\Delta_{(n)} + t^{n+1}\nu = \Delta + t\Delta^{(1)} + \cdots + t^n\Delta^{(n)} + t^{n+1}\nu$$

on $B \otimes_k k_{(n+1)}$ *define an* $n+1$*st order deformation. If we change* (μ, ν) *by a coboundary* $D\chi$ *then the* $n+1$*st order deformations defined by* $(m_{(n)} + t^{n+1}\mu, \Delta_{(n)} + t^{n+1}\nu)$ *and* $(m_{(n)} + t^{n+1}\mu + d\chi, \Delta_{(n)} + t^{n+1}\nu - \delta\chi)$ *are equivalent under conjugation by* $id - t^{(n+1)}\chi$.

Proof. The same calculations by which we proved that the obstruction class of a first order deformation is a cocycle prove the general result. We simply substitute $m_{(n)}$ and $\Delta_{(n)}$ for $m_{(1)}$ and $\Delta_{(1)}$ and calculate modulo t^{n+2} instead of modulo t^3. The second part of the statement comes from the fact that if we define $m_{(n+1)}$ and $\Delta_{(n+1)}$ as in the statement of the proposition, then, modulo t^{n+2}

$$t^{n+1}(d\mu + \beta^{1,3}_{(n)}) = m_{(n+1)} \circ (id \otimes m_{(n+1)} - m_{(n+1)} \otimes id) \tag{9.6.5}$$

$$t^{n+1}(\delta\mu + d\nu + \beta^{2,2}_{(n+1)}) = m_{(n+1)} \otimes m_{(n+1)} \circ s_{2,3} \circ \Delta_{(n+1)} \otimes \Delta_{(n+1)} - \Delta_{(n+1)} \circ m_{(n+1)} \tag{9.6.6}$$

$$t^{n+1}(-\delta\nu + \beta^{3,1}_{(n+1)}) = (\Delta_{(n+1)} \otimes id - id \otimes \Delta_{(n+1)}) \circ \Delta_{(n+1)} \tag{9.6.7}$$

Define $L_{(n)} = id - t^{n+1}\chi$. (Note the change in sign of χ from the calculation in section 5.) The last part follows from calculating the coefficient of $t^{(n+1)}$ in the expressions

$$L_{(n)} \circ m_{(n)} \circ \left(L^{-1}_{(n)} \otimes L^{-1}_{(n)}\right) = m'_{(n)} \bmod t^{n+2} \tag{9.6.8}$$

$$(L_{(n)} \otimes L_{(n)}) \circ \Delta_{(n)} \circ L^{-1}_{(n)} = \Delta'_{(n)} \bmod t^{n+2} \tag{9.6.9}$$

just as we did for 9.5.1 and 9.5.2 in the last section. The result is $(\mu + d\chi, \nu - \delta\chi) = (\mu, \nu) + D^{1,1}\chi$.

We can restate the proposition

The obstructions to extending an nth order bialgebra deformation lie in the fourth Gerstenhaber-Schack cohomology group. If the obstruction cocycle defines the zero cohomology class an extension exists. The equivalence classes of extensions are parametrized by the third cohomology group.

Thus the Gerstenhaber-Schack cohomology satisfies the second requirement of the Gerstenhaber program.

9.7 Bialgebra to Drinfeld algebra deformations.

In a Drinfeld algebra the coassociativity condition is relaxed to allow coassociativity only up to conjugacy by an element $\Phi \in B \otimes B \otimes B$:

$$\Phi[(\Delta \otimes id) \circ \Delta]\Phi^{-1} = (id \otimes \Delta) \circ \Delta. \tag{9.7.1}$$

Since the differential (coboundary operator) of the deformation cohomology is determined by the structure equations, we would expect the new structure equation to lead to a new differential. Indeed this is what happens as we will show in chapter 13. However if we assume that the original structure is a bialgebra with strict coassociativity, i.e. coassociativity operator $1 \otimes 1 \otimes 1$, and that we allow the deformation to satisfy the Drinfeld condition of coassociativity up to conjugation, then the coboundary operator does not change. (Recall that by definition of a deformation the expansion of $\Phi(t)$ as a formal power series is required to begin with the undeformed coassociativity operator $1 \otimes 1 \otimes 1$.) The only change required in order to have an appropriate deformation cohomology is to extend the double complex by adding an extra row at the bottom. So we make the replacement

$$\bigoplus_{1 \leq p,q} C^{p,q} \rightsquigarrow \bigoplus_{1 \leq p, 0 \leq q} C^{p,q}$$

For a further explanation of any comments made here about double complexes, see the first section of the next chapter.

The deformation cohomology complex for bialgebra deformations was the subcomplex of the full Gerstenhaber-Schack complex given by deleting the bottom row and the leftmost column. In this section we will show that the deformation cohomology complex for deformations from a bialgebra to a Drinfeld algebra is given by simply deleting the leftmost column $p = 0$ but using the same differential $D^{p,q} = d + (-1)^q \delta$ or equivalently $D^{p,q} = (-1)^p d + \delta$. We will denote this extended complex by $\tilde{C}(B)$.

As before, we identify our deformation with $B \otimes k[[t]]$ but now with a three part structure:

$$\begin{aligned} m(t) &= m + \sum_{1 \leq n} t^n m^{(n)} \\ \Delta(t) &= \Delta + \sum_{1 \leq n} t^n \Delta^{(n)}, \\ \Phi(t) &= 1 \otimes 1 \otimes 1 + \sum_{1 \leq n} \Phi_n t^n \end{aligned}$$

The structure equations are the associativity of $m(t)$, the compatibility of $m(t)$ and $\Delta(t)$, which are the same as for bialgebras, the conjugate coassociativity of $\Delta(t)$ relative to $\Phi(t)$, which we will write:

$$\Phi(t)[(\Delta(t)\otimes id)\circ\Delta(t)]-[(id\otimes\Delta(t))\circ\Delta(t)]\Phi(t)=0. \tag{9.7.2}$$

and the pentagon identity for $\Phi(t)$:

$$1\otimes\Phi(t)[(id\otimes\Delta(t)\otimes id)\Phi(t)]\Phi(t)\otimes 1-[(\Delta(t)\otimes id\otimes id)\Phi(t)][(id\otimes id\otimes\Delta(t))\Phi(t)]= \tag{9.7.3}$$

The infinitesimal deformation has the form $(m^{(1)},\Delta^{(1)},-\Phi_1)\in C^{1,2}\oplus C^{2,1}\oplus C^{3,0}$. Note the negative sign of the third component.

The first step in the Gerstenhaber program is to show that the infinitesimals are cocycles. The relevant equations are

$$dm^{(1)}=0,\quad \delta m^{(1)}+d\Delta^{(1)}=0\quad \delta\Delta^{(1)}+d\Phi_1=0\quad \delta\Phi_1=0.$$

The third equation, which replace $\delta\Delta^{(1)}=0$ in the theory of bialgebra deformations, involves the Hochschild differential of Φ_1,

$$d\Phi_1(a)=\Delta^{[3]}(a)\cdot\Phi_1-\Phi_1\Delta^{[3]}(a).$$

Here $\Delta^{[3]}(a)=(\Delta\otimes id)\Delta(a)=(id\otimes\Delta)\Delta(a)$, since the original comultiplication is coassociative. Using the expression $(id\otimes\Delta)\Delta(a)$ in the first summand and the expression $(\Delta\otimes id)\Delta(a)$ in the second and recalling that

$$\begin{aligned}\delta(\Delta^{(1)}) &= (id\otimes\Delta^{(1)})\circ\Delta-(\Delta\otimes id)\Delta^{(1)}\\ &+ (id\otimes\Delta)\Delta^{(1)}-(\Delta^{(1)}\otimes id)\Delta.\end{aligned}$$

we see that the third equation is (-1) times the t term of (9.7.2). For a first order deformation (9.7.2) is satisfied to order t^2 so this component of the cocycle condition is satisfied.

The last component will follow from the t term in the pentagon identity (9.7.3). We consider Φ_1 as the value at $1\in k$ of an element $\tilde{\Phi}_1\in C^{3,0}=\mathrm{Hom}(k,B^{\otimes 3})$. To compute δ we use the B bicomodule structure on k

$$\lambda_k : k\to k\otimes k\overset{u\otimes id}{\to} B\otimes k$$

$$\rho_k : k\to k\otimes k\overset{id\otimes u}{\to} k\otimes B.$$

For any $\alpha\in B^{\otimes p}$ consider δ acting on $\tilde{\alpha}\in C^{p,0}$

$$\delta(\tilde{\alpha})=(id\otimes\tilde{\alpha})\circ\lambda_k+\sum(-1)^i\Delta_i\circ\tilde{\alpha}+(-1)^{n+1}(\tilde{\alpha}\otimes id)\circ\rho_k.$$

Evaluating a $1 \in k$ gives the formula for the differential on the $\sum_{1 \leq p} B^{\otimes p}$ which we identify with the bottom row of the extended complex:

$$\delta(\alpha) = 1 \otimes \alpha + \sum (-1)^i \Delta_i(\alpha) + (-1)^{n+1} \alpha \otimes 1. \tag{9.7.4}$$

Because of the conditions (9.6.1) and the fact that the constant term in $\Phi(t)$ is $1 \otimes 1 \otimes 1$ the t component of the pentagon identity of a first order deformation does not involve either $m^{(1)}$ or $\Delta^{(1)}$ but gives exactly

$$\delta(\Phi_1) = 1 \otimes \Phi_1 - (\Delta \otimes id \otimes id)\Phi_1 + (id \otimes \Delta \otimes id)\Phi_1 - (id \otimes id \otimes \Delta)\Phi_1 + \Phi_1 \otimes 1.$$

This establishes the last component of the cocycle condition for the infinitesimal.

The next step is to show that equivalent infinitesimals are cohomologous. The equivalence relation now involves the operation called twisting which we introduced in the last chapter. Twisting by an element $T(t) \in B \otimes B[[t]]$ does not affect the multiplication, changes $\Delta(t)$ to $\Delta(t)'$

$$\Delta(t)' = T(t)\Delta(t)T(t)^{-1} \tag{9.7.5}$$

and changes $\Phi(t)$ to $\Phi(t)'$

$$\Phi(t)' = 1 \otimes T(t)(id \otimes \Delta)T(t)\Phi(t)(\Delta \otimes id)T(t)^{-1}T(t)^{-1} \otimes 1. \tag{9.7.6}$$

The t component of the equation (9.7.6) is

$$\Phi_1' = \Phi_1 + \delta T^{(1)}. \tag{9.7.7}$$

The t component of the equation (9.7.5) is

$$\Delta^{(1)'} = \Delta^{(1)} - dT^{(1)}. \tag{9.7.8}$$

Combining equations (9.7.8) and (9.7.7) shows that the change in the infinitesimals under twisting by $1 \otimes 1 + tT^{(1)}$ is

$$(m^{(1)}, \Delta^{(1)}, -\Phi^{(1)} - (m^{(1)}, \Delta^{(1)'}, -\Phi'^{(1)}) = (0, dF_1, \delta T^{(1)} = (d+\delta)T^{(1)} = D^{2,0}T(1).$$

Conversely, reading the same calculation backwards shows that if two infinitesimals differ by a coboundary of the form $D^{1,1}T^{(1)}$ then the corresponding first order deformations are equivalent under twisting by $1 \otimes 1 + tT^{(1)}$.

Proposition 9.7.1 *An infinitesimal Drinfeld algebra deformation of a bialgebra B defines a cocycle in the complex $\tilde{C}(B)$. Two infinitesimal deformations define equivalent first order deformations if and only if they are cohomologous.*

Proof. We have shown that the infinitesimals are cocycles and verified two infinitesimals are equivalent by twisting if and only if they are cohomologous. We saw earlier that equivalence under conjugation by $id + t\chi$ is the same as differing by a coboundary $D^{1,1}\chi$. Since the "first derivative" linearizes composition, the full equivalence relation allowing the composition of conjugation and twisting by $1 + t\tau$ is the same as differing by a sum of terms $D^{1,1}\chi + D^{2,0}\tau = D(\chi, \tau)$, completing the proof.

Next we consider the obstruction class which now has four components

$$\tilde{\beta}_{(n)} = (\beta^{1,3}_{(n)}, \beta^{2,2}_{(n)}, \beta^{3,1}_{(n)}, \beta^{4,0}_{(n)}).$$

The first component comes from the coefficient of t^{n+1} of the associativity equation for the trivial extension of $m_{(n)}$ to $B \otimes k_{(n+1)}$ as in (9.6.2). The second component comes from the coefficient of t^{n+1} in the compatibility equation, according to 9.6.3. The third component comes from the same procedure applied to the conjugate coassociativity of $\Delta_{(n)}$ relative to $\Phi_{(n)}$, and the fourth from the pentagon identity. Calculating modulo t^{n+2}

$$t^{n+1}\beta^{3,1}_{(n)} = \Phi_{(n)}[(\Delta_{(n)} \otimes id) \circ \Delta_{(n)}] - [(id \otimes \Delta_{(n)}) \circ \Delta_{(n)}]\Phi_{(n)} \tag{9.7.9}$$

$$\begin{aligned} t^{n+1}\beta^{4,0}_{(n)} &= [(\Delta_{(n)} \otimes id \otimes id)\Phi_{(n)}][(id \otimes id \otimes \Delta_{(n)})\Phi_{(n)}] \\ &\quad -1 \otimes \Phi_{(n)}[(id \otimes \Delta_{(n)} \otimes id)\Phi_{(n)}]\Phi_{(n)} \otimes 1 \end{aligned} \tag{9.7.10}$$

There are five components of the $D\tilde{\beta}_{(n)}$ which must vanish. Two of them are the same as before. The equation $\delta\beta^{4,0} = 0$, which is actually the most complicated, is proved in greater generality in Chapter 13. The proof of the remaining two is a tedious calculation using the same principle as in the earlier proofs. Without going into the details we state the final proposition.

Proposition 9.7.2 *The obstruction class $\tilde{\beta}_{(n)}$ is a cocycle in $\tilde{C}(B)$. If it cobounds an element (μ, χ, ξ) then the triple $(m_{(n)} - t^{n+1}\mu, \Delta_{(n)} - t^{n+1}\chi, \Phi_{(n)} + t^{n+1}\xi)$ extended $k_{(n+1)}$ linearly to $B \otimes k_{(n+1)}$ defines an $n+1$ st order deformation. Two extensions define equivalent deformations under the composition of conjugacy by $id + t^{n+1}\sigma$ and twisting by $1 \otimes 1 - t^{n+1}\tau$ if and only if $(\mu - \mu', \chi - \chi', \xi - \xi') = D(\sigma, \tau)$.*

9.8 Poisson algebras and Gerstenhaber algebras.

The expression for the obstruction to extending an infinitesimal deformation, μ, one step further is a quadratic expression in the cohomology class, $[\mu]$.

This is the tip of an iceberg: there is an algebraic structure discovered by Gerstenhaber carried by the Hochschild cohomology of an associative algebra, [444]. This structure is beginning to play a steadily increasing role in various areas of mathematics and physics. In this section we give an introductory exposition to the subject.

We first backtrack one step to recall the notion of Poisson algebra. Let A be an associative and commutative algebra. A bilinear map

$$A \otimes A \to A, \quad f \otimes g \mapsto \{f, g\}$$

is called a *Poisson bracket* if
(**i**) The bracket $\{\ ,\ \}$ satisfies the axioms for a Lie bracket, so that A becomes a Lie algebra under $\{\ ,\ \}$ and
(**ii**)Left bracket by a fixed element is a derivation of the commutative multiplication of A.

In formulas, these conditions amount to the conditions

$$\{f, g\} = -\{g, f\}, \tag{9.8.1}$$
$$\{f, \{g, h\}\} = \{\{f, g\}, h\} + \{g, \{f, h\}\} \tag{9.8.2}$$
$$\{f, gh\} = \{f, g\}h + g\{f, h\}. \tag{9.8.3}$$

The pair $(A, \{\ ,\ \})$ consisting of an associative commutative algebra together with a Poisson bracket is called a *Poisson algebra*. . We discuss a refinement of this structure at the end of Chapter 11. Examples are:

1) the algebra of smooth functions on a symplectic manifold. This is the original example in classical mechanics cf. for example [501] or [501].

2) the algebra of smooth functions or the algebra of polynomials on the dual space, $\mathcal{G}^*$ of a Lie algebra, $\mathcal{G}$. The space $\mathcal{G}$ itself consists of the linear functions, and the restriction of the Poisson bracket to $\mathcal{G}$ coincides with the Lie bracket. The derivation property then determines it on all polynomials. This Poisson structure is known as the Kirillov-Kostant- Souriau bracket.

The theory of algebraic quantization as developed by Flato et. al. (see a long series of references in the bibliography) seeks to deform the (commutative) multiplication on A into an associative but not necessarily commutative multiplication

$$m_t(f, g) = fg + t\mu(f, g) + \cdots$$

in such a way that the commutator

$$m_t(f, g) - m_t(g, f) = t(\mu(f, g) - \mu(g, f)) + \cdots$$

has linear term equal to $\{\ ,\ \}$. In other words that

$$\mu(f,g) - \mu(g,f) = \{f,g\}.$$

If $\{\ ,\ \}$ is given as such a first order approximation to a commutator, then the conditions (9.8.1), (9.8.2) and (9.8.3) follow.

Suppose that we wish to extend the definition of Poisson algebra to the setting where A is a commutative superalgebra:

$$A = A_0 \oplus A_1.$$

One possibility is to simply follow the usual procedure and put in the expected minus sign whenever you have to move an odd element past another odd element. So, for example, in the superalgebra setting we would replace (9.8.3) by

$$\{f,gh\} = \{f,g\}h + (-1)^{deg(f)deg(g)}g\{f,h\},$$

and replace the Lie algebra axioms by the axioms for a Lie superalgebra [249] [1076]. This gives what we might call a Poisson superalgebra, and plays a key role the theory of reduction of mechanical systems and their quantization associated to the BRST transformation [704] [501].

However there is an alternative choice available. We let $\hat{A}$ denote the $\mathbf{Z}_2$ graded vector space A with its gradation reversed:

$$(\hat{A})_0 = A_1, \quad (\hat{A})_1 = A_0.$$

We can then demand that $\{\ ,\}$ make $\hat{A}$ into a Lie superalgebra which acts (with its altered gradation) as superderivations of A with its original structure. The axioms, expressed in terms of the *original* gradation of A are:

$$\begin{aligned}
\{f,g\} &= -(-1)^{[deg f-1][deg\ g-1]}\{g,f\}, && (9.8.4)\\
\{f,\{g,h\}\} &= \{\{f,g\},h\} + (-1)^{[deg f-1][deg\ g-1]}\{g,\{f,h\}\} && (9.8.5)\\
\{f,gh\} &= \{f,g\}h + (-1)^{[deg f-1]deg\ g}g\{f,h\}. && (9.8.6)
\end{aligned}$$

An object satisfying these axioms is called a *Gerstenhaber algebra.*

A basic example of a Gerstenhaber algebra is the algebra of multivector-fields on a manifold, M:

$$\begin{aligned}
A_0 &= C^\infty(\wedge^0(TM)) \oplus C^\infty(\wedge^2(TM)) \oplus C^\infty(\wedge^4(TM)) \oplus \cdots && (9.8.7)\\
A_1 &= C^\infty(\wedge^1(TM)) \oplus C^\infty(\wedge^3(TM)) \oplus \cdots && (9.8.8)
\end{aligned}$$

where $C^\infty(E)$ means the space of C^∞ sections of the vector bundle E. The supercommutative multiplication is just exterior product. So, in this example,

$$\mathcal{F} = \mathcal{F}(M) = C^\infty(\wedge^0(M))$$

is a commutative subalgebra of A_0 and

$$\mathrm{Vect}(M) = C^\infty(\wedge^1(M)),$$

the Lie algebra of vector fields on M is an ordinary Lie subalgebra, $\mathcal{L}$, of A_1. In this example, the functions and the vector fields "generate" (up to completion of the tensor product) the entire algebra under multiplication.

Notice that $\mathcal{L}$ is a module over $\mathcal{F}$ and acts as derivations of F. The axioms for a Gerstenhaber algebra guarantee that the module action of $\mathcal{F}$ on $\mathcal{L}$, the Lie bracket of $\mathcal{L}$ and the derivation action of $\mathcal{L}$ on $\mathcal{F}$ are all consistent. Conversely, starting with a pair consisting of a commutative algebra $\mathcal{F}$ and an ordinary Lie algebra, $\mathcal{L}$ with the consistent actions just described, on can endow the exterior algebra of $\mathcal{L}$ over $\mathcal{F}$ with the structure of a Gerstenhaber algebra. Such pairs $(\mathcal{F}, \mathcal{L})$ have arisen in various areas of mathematics and physics and their properties are studied extensively in a forthcoming treatise by Mackenzie[823].

Chapter 10

Homological tools

In the preceding chapter we saw how the problem of constructing and classifying deformations, step by step in powers of t could be translated into questions of cohomology. Homological algebra provides tools which allow us to answer many of these questions. We will actually need to use only two very basic tools of homological algebra: the long exact sequence of (co)homology and the theory of projective and injective resolutions.

The Gerstenhaber-Schack complex is a double complex. In the next chapter we use a technique for normalizing a deformation which is really a particular case of a general argument in the theory of double complexes. Also, one can establish the long exact sequence of cohomology from Weil's theorem about double complexes. For these reasons we include, as a first section, an introductory discussion of double complexes.

The theory of projective and injective resolutions will allow us to replace the Gerstenhaber- Schack complex by a homologically equivalent one built on two Koszul resolutions (see below for their precise definitions). This new complex is much more tractable for computations. Its use will allow us to prove that if $\mathcal{G}$ is a Lie algebra over a field of characteristic zero then

$$H^{p,q}_{GS}(U(\mathcal{G})) = H^q(\mathcal{G}, \wedge^p\mathcal{G}). \tag{10.0.1}$$

This result and related results for the deformation cohomology of function algebras have been announced by Lazarev and Movshev [748].

We have already introduced bimodules over an algebra and bicomodules over a coalgebra. To use the theory of projective and injective resolutions in the study of bialgebra cohomology we need to introduce a four-fold structure combining the bimodule and bicomodule structures with appropriate consistency conditions). Such a structure will be called a B-4 structure and an object with this structure will be called a B-4 module or tetramodule. This concept appears in the original paper of Gerstenhaber and Schack [462] where it is call a $(B - B)$ birepresentation. In fact, they define a more general concept, an (A, B) birepresentation. Joseph Bernstein emphasized to us the importance of tetramodules in the theory of quantum groups. In section 10.2 we study B-4 modules. In section 10.3 we study the general theory of resolutions. In the remaining sections we introduce, study, and compute with the Koszul resolutions.

10.1 Double complexes.

André Weil's celebrated proof of the de Rham theorem, [1198] gave impetus to the use of double complexes as a computational tool in algebraic topology. In this chapter we collect various facts about them that we will need for the study of deformation theory. All the material in this section is completely standard. We follow the exposition in the beautiful book by Bott and Tu [181] very closely, and refer the reader there for a full treatment. See also [904].

Let X and Y be manifolds. If we consider differential forms on $X \times Y$ then we can take their "partial exterior derivatives", d_X and d_Y. Then

$$d_X d_Y = -d_Y d_X$$

and the total exterior derivative is given by

$$d = d_X + d_Y.$$

We can consider the degree of a form in the X and Y variables separately, and so speak of a form of bidegree p, q. Then d_X increases p by one while d_Y increases q by one.

More generally, we can consider a bigraded module

$$K = \bigoplus_{p,q} K^{p,q}$$

with "horizontal " differentials

$$D' : K^{p,q} \to K^{p+1,q}$$

and vertical differentials

$$D'' : K^{p,q} \to K^{p,q+1}$$

which satisfy,

$$D'^2 = 0, \quad D''^2 = 0, \quad D'D'' + D''D' = 0.$$

Such an object is called a double complex. We let

$$T = \bigoplus_n T^n$$

where T^n is the sum of the terms along the diagonal $p + q = n$:

$$T^n = \bigoplus_{p+q=n} K^{p,q}.$$

Then

$$D = D' + D'' : \quad T^n \to T^{n+1}$$

and

$$D^2 = 0.$$

The complex (T, D) is called the *total complex* of the double complex. Sometimes, as in the preceding chapter, we are given horizontal operators $\delta : K^{p,q} \to K^{p+1,q}$ satisfying $\delta^2 = 0$ and vertical operators $d : K^{p,q} \to K^{p,q+1}$ with $d^2 = 0$ which commute:

$$d\delta = \delta d.$$

Then just modify one or the other of d or δ by a sign change setting

$$D' = \delta, \quad D'' = (-1)^p d$$

or

$$D' = (-1)^q \delta, \quad D'' = d$$

on $K^{p,q}$ to get a double complex. So we will use either description, the (D', D'') or the (δ, d) description when talking about double complexes.

Unless otherwise specified, all of our double complexes will be "first quadrant"- i.e. satisfy

$$K^{p,q} = (0) \text{ if } p < 0 \text{ or } q < 0.$$

Let (A, d) be the complex

$$A = \bigoplus_q A^q, \quad A^q = \ker D' : \ K^{0,q} \to K^{1,q}$$

where

$$d : A^q \to A^{q+1}$$

is just the restriction of $D'' : K^{0,q} \to K^{0,q+1}$ to A^q. Let

$$r : A^q \to K^{0,q}$$

denote the natural embedding, and we think of r as a map of A into T. Then

$$\begin{aligned} D \circ r &= (D' + D'') \circ r \\ &= D'' \circ r \\ &= r \circ d \end{aligned}$$

since r is inclusion into the kernel of δ. Thus

$$(A, d) \xrightarrow{r} (T, D)$$

is a morphism of complexes.

Theorem 10.1.1 Weil. *If the rows of K are exact under D' then r induces an isomorphism on homology.*

We let r_* denote the induced map on homology. We must show that r_* is surjective and injective.

Proof of surjectivity. If

$$k = k_{0,n} + k_{1,n-1} + \ldots + k_{n,0}$$

satisfies

$$Dk = 0,$$

then

$$D'k_{n,0} = 0.$$

By row exactness we can find a $c_{n-1,0} \in K^{n-1,0}$ with

$$D'c_{n-1,0} = k_{n,0}.$$

Hence

$$k' = k - Dc$$

is in the same D cohomology class as k with zero entry in the n-th column. Hence

$$D'k'_{n-1,1} = 0$$

and we can find c' with $k' - Dc'$ zero in the n-th and $(n-1)$st column. Proceeding in this way we see that k is cohomologous under D to a $z = z_{0,n}$

with non-zero entries only in the zero'th column. but then $Dz = 0$ implies that $D'z = 0$ and $dz = 0$. So $[k] = [z]$ is in the image of r_*.

Proof of injectivity. We must show that if $z = z_{0,n}$ is a D coboundary, then it is d coboundary. So suppose that

$$z = Dc.$$

Then $D''c_{0,n-1} = z_{0,n}$ and $D'c_{0,n-1} = -D''c_{1,n-2}$ etc. At the bottom of the staircase we have $D'c_{n-1,0} = 0$. By row exactness, $c_{0,n-1} = D'f_{n-2,0}$. so we can replace c by $c' = c - Df_{n-2,0}$ and iterate until we get a c whose only non-zero entry is in the zero'th column. But this then implies that $z_{0,n}$ is a d coboundary. Q.E.D.

Applications.

The long exact sequence of homology. Let

$$0 \to A \to B \to C \to 0$$

be an exact sequence of complexes and think of $B \to C \to 0$ as the δ of a double complex with $T^n = K^{0,n} \oplus K^{1,n-1} = B^n \oplus C^{n-1}$, that is all columns are zero except for the columns 0 and 1. A cocycle of the total complex is an element of the form $b_n + c_{n-1}$ with $db_n = 0$ and $dc_{n-1} = \delta b_n$. Consider the subspace consisting of those cycles with $b_n = 0$. Such an element will be a coboundary if there is a b_{n-1} with

$$db_{n-1} = 0$$

and

$$\delta b_{n-1} = c_{n-1} + dc_{n-2}.$$

Thus we have the exact sequence

$$H_d^{n-1}(B) \to H_d^{n-1}(C) \to H_D^n(T).$$

The map

$$(b_n + c_{n-1}) \mapsto b_n$$

maps

$$H_D^n(T) \to H_d(B)$$

with kernel $H_d^{n-1}(C)$. Thus

$$H_d^{n-1}(B) \to H_d^{n-1}(C) \to H_D^n(T) \to H_d^n(B)$$

is exact. An element of $H_d^n(B)$ is in the image of $H_D^n(T)$ if it is represented by a class b_n such that there is a c_{n-1} with $\delta b_n = dc_{n-1}$. But this means that

$$\to H_d^{n-1}(B) \to H_d^{n-1}(C) \to H_D^n(T) \to H_d^n(B) \to H_d^n(C) \to$$

is exact. Weil's theorem allows us to replace $H_D(T)$ by $H_d(A)$ giving the long exact homology sequence

$$\to H^{n-1}(B) \to H^{n-1}(C) \to H^n(A) \to H^n(B) \to H^n(C) \to \quad .$$

De Rham's theorem. Although we will not need it in what follows, for the sake of completeness we include Weil's elegant proof of De Rham's theorem. Suppose that $\mathcal{S}$ is a fine sheaf with partitions of unity $\{\rho_\alpha\}$ subordinate to any cover $\{U_\alpha\}$. For example $\mathcal{S}$ could be the sheaf of forms of degree q on a manifold. Let $U_{\alpha\beta\gamma}$ denote $U_\alpha \cap U_\beta \cap U_\gamma$ etc. Let ∂_0 denote the inclusion obtained by dropping the first index, ∂_1 by dropping the second etc. So

$$\partial_0 : \coprod_\alpha U_{\alpha\beta\gamma} \to U_{\beta\gamma}, \quad \partial_1 : \coprod_\beta U_{\alpha\beta\gamma} \to U_{\alpha\gamma}, \quad \partial_2 : \coprod_\gamma U_{\alpha\beta\gamma} \to U_{\alpha\beta}.$$

Let

$$C^p(\mathcal{U}, \mathcal{S}) = \prod \mathcal{S}(U_{\alpha_0\alpha_1\ldots\alpha_p}).$$

Define $\delta_0 : C^p(\mathcal{U}, \mathcal{S}) \to C^{p+1}(\mathcal{U}, \mathcal{S})$ as composition with the restriction map corresponding to ∂_0 etc.

Finally, define

$$\delta = \delta_0 - \delta_1 + \delta_2 - \ldots .$$

So, for example, if $\xi \in C^1(\mathcal{U}, \mathcal{S})$ then

$$(\delta\xi)_{\alpha\beta\gamma} = \xi_{\beta\gamma} - \xi_{\alpha\gamma} + \xi_{\alpha\beta}.$$

Notice that

$$\begin{aligned}
(\delta(\delta\xi))_{\alpha\beta\gamma\delta} &= (\delta\xi)_{\beta\gamma\delta} - (\delta\xi)_{\alpha\gamma\delta} + (\delta\xi)_{\alpha\beta\delta} - (\delta\xi)_{\alpha\beta\gamma} \\
&= \xi_{\gamma\delta} - \xi_{\beta\delta} + \xi_{\beta\gamma} \\
&\quad -\xi_{\gamma\delta} + \xi_{\alpha\delta} - \xi_{\alpha\gamma} \\
&\quad\quad +\xi_{\beta\delta} - \xi_{\alpha\delta} + \xi_{\alpha\beta} \\
&\quad\quad\quad -\xi_{\beta\gamma} + \xi_{\alpha\gamma} - \xi_{\alpha\beta} = 0 .
\end{aligned}$$

More generally, $\delta^2 = 0$ because

$$\begin{aligned}&(\delta^2\xi)_{\alpha_0\alpha_1\ldots\alpha_{p+2}}\\&=\sum(-1)^i(\delta\xi)_{\alpha_0\alpha_1\ldots\hat{\alpha}_i\ldots\alpha_{p+2}}\\&=\sum_{j<i}(-1)^i(-1)^j\xi_{\alpha_0\alpha_1\ldots\hat{\alpha}_j\ldots\hat{\alpha}_i\ldots\alpha_{p+2}}\\&\ +\sum_{i<j}(-1)^i(-1)^{j-1}\xi_{\alpha_0\alpha_1\ldots\hat{\alpha}_i\ldots\hat{\alpha}_j\ldots\alpha_{p+2}}\\&=0.\end{aligned}$$

We define the operator κ by $(\kappa\xi)_{\alpha_0\alpha_1\ldots\alpha_{p-1}} = \sum_\alpha \rho_\alpha\xi_{\alpha\alpha_0\alpha_1\ldots\alpha_{p-1}}$. Then a direct check shows that $\kappa\delta + \delta\kappa = 1$. So the δ sequence has zero cohomology in positive degrees.

In particular, if we take the sheaves of $q-$ forms, we get a double complex with exact rows. An element of the first column in the kernel of δ is just a globally defined form. The corresponding d cohomology on globally defined forms (the cohomology of A) is just the De Rham cohomology. Hence the De Rham cohomology of M is isomorphic to the cohomology of the total complex, T, of the double complex, K.

If all the intersections in the cover are contractible, then the columns of the double complex have trivial cohomology by the Poincaré lemma. By interchanging the roles of the rows and columns in Weil's theorem, we conclude that the Cech cohomology with real coefficients (relative to the given cover), which corresponds to the δ complex on the kernel of d at the bottom row, is isomorphic also to the cohomology of the total complex. Hence De Rham's theorem, that the De Rham cohomology is isomorphic to the Cech cohomology (and proving that the Cech cohomology is independent of the contractible cover). This was Weil's proof.

Mayer-Vietoris.

Even if the intersections in the cover are not contractible, the theorem above applies to give the de Rham cohomology as the cohomology of the double complex. In the extreme special case where there are only two elements in the cover, U and V, only the zero-th and first column are non-vanishing. So we are in the situation we described above for the long exact sequence: the zero'th column has $H_d^*(B) = H^*(U) \oplus H^*(V)$ and first column $H_d^*(C) = H^*(U \cap V)$. Thus we get the Mayer-Vietoris sequence

$$\begin{aligned}&\to H^{q-1}(U)\oplus H^{q-1}(V)\to H^{q-1}(U\cap V)\to H^q(M)\\&\to H^q(U)\oplus H^q(V)\to H^q(U\cap V)\to\end{aligned}$$

Attaching a cell. For example, let e^n denote the unit disk and S^{n-1} its boundary. Given a space X, and a map $f: S^{n-1} \to X$, we define the "space obtained from X by attaching the n-cell, e^n, along the map f, denoted by X_f to be the space

$$X_f = X \cup_f e^n,$$

the disjoint union of X and e^n with the points u and $f(u)$ identified, for all $u \in S^{n-1}$.

Let $U = X_f - \{p\}$ where p is the origin of e^n and let V be the disk of radius $1/2$ in e^n. Thus U is homotopic to X, V is contractible and $U \cap V$ is homotopic to S^{n-1}. Let us assume for the moment that we know that S^{n-1} has non-zero cohomology only in degrees 0 and $n-1$, where it is one dimensional. For positive $q \neq n, n-1$, the above sequence becomes

$$0 \to H^q(X_f) \to H^q(X) \oplus \{0\} \to 0.$$

So the cohomology of X_f is the same as the cohomology of X in all positive dimensions except n and $n-1$. We also have

$$\begin{aligned} H^{n-2}(S^{n-1}) &\to H^{n-1}(X_f) \to H^{n-1}(X) \to H^{n-1}(S^{n-1}) \\ &\to H^n(X_f) \to H^n(X) \to 0. \end{aligned}$$

If $n-2 > 0$, the leftmost term is 0. If $n = 2$ then we have, to the left of this term,

$$H^0(X_f) \to H^0(X) \oplus H^0(e^n) \to H^{n-2}(S^{n-1})$$

which is

$$\mathbf{R} \to \mathbf{R} \oplus \mathbf{R} \to \mathbf{R}$$

if X is connected. So we can replace the $H^{n-2}(S^{n-1})$ by 0. Thus in all cases we have

$$0 \to H^{n-1}(X_f) \to H^{n-1}(X) \to H^{n-1}(S^{n-1}) \to H^n(X_f) \to H^n(X) \to 0.$$

For example, suppose that X has nonvanishing cohomology only in even degrees and we add a cell of even dimension $n = 2m$, so $H^{n-1}(X) = 0$. We then see that $H^{n-1}(X_f) = 0$ and $H^n(X_f) = H^n(X) \oplus \mathbf{R}$. Thus if a space is built up by adding cells of even dimension, its odd cohomology vanishes and its cohomology in degree $2m$ is $\mathbf{R}^{\#m}$ where $\#m$ denotes the number of m cells attached.

The cohomology of spheres. S^0 consists of two points so $H^0(S^0) = \mathbf{R} \oplus \mathbf{R}$ and all higher cohomology vanishes.

S^1: cover S^1 by two arcs, the top arc, U, a bit bigger than the upper semicircle and the bottom arc, V, a bit bigger than the bottom semicircle. So $U \cap V$ consists of two arcs. Then Mayer-Vietoris reads

$$0 \to H^0(S^1) \to \mathbf{R} \oplus \mathbf{R} \to \mathbf{R} \oplus \mathbf{R} \to H^1(S^1) \to 0 \to 0.$$

The image of $H^0(S^1) = \mathbf{R}$ is one dimensional, hence the image of the second map is one dimensional. Hence $H^1(S^1) = \mathbf{R}$.

S^2 : Cover S^2 by U, a little more that the upper hemisphere and V, a little more than the bottom hemisphere. So $U \cap V$ is homotopic to S^1. So M-V becomes

$$0 \to H^0(S^2) \to \mathbf{R} \oplus \mathbf{R} \to \mathbf{R} \to H^1(S^2) \to 0 \oplus 0 \to H^1(S^1) \to H^2(S^2) \to 0 \oplus 0$$

Since the image of $H^0(S^2) = \mathbf{R}$ is one dimensional, we conclude that the image of $\mathbf{R} \oplus \mathbf{R} \to \mathbf{R}$ is all of $\mathbf{R}$, so $H^1(S^2) = 0$ and from the end that $H^2(S^2) = \mathbf{R}$.

S^n : Suppose we know that $H^k(S^{n-1}) = 0$ for $k \neq 0, n-1$ and is $\mathbf{R}$ in these two cases, $n > 2$. Then using the U and V as above, Mayer-Vietoris implies that $H^1(S^n) = 0$ as above, and for $1 < k \neq n-1$ Mayer-Vietoris says

$$0 \to H^k(S^n) \to 0 \oplus 0,$$

so $H^k(S^n) = 0$ for $k \neq 0, n$ while at the end Mayer-Vietoris says

$$0 \oplus 0 \to H^{n-1}(S^{n-1}) \to H^n(S^n) \to 0.$$

So $H^n(S^n) = \mathbf{R}$.

10.2 B-4 modules

Definition. A left-left *dimodule* M over a bialgebra B is a left B-module and left B-comodule for which the comodule structure map is a map of modules (using the comultiplication of B to define the tensor product of modules), or, equivalently, the module structure map is a map of comodules (using multiplication of B to define the tensor product of comodules). That is, the diagram in the upper left corner in Figure 10.1 commutes.

A B-4 *module* (*tetramodule, bidimodule*) M is a B-bimodule and a B-bicomodule such that the both comodule structure maps are maps of left

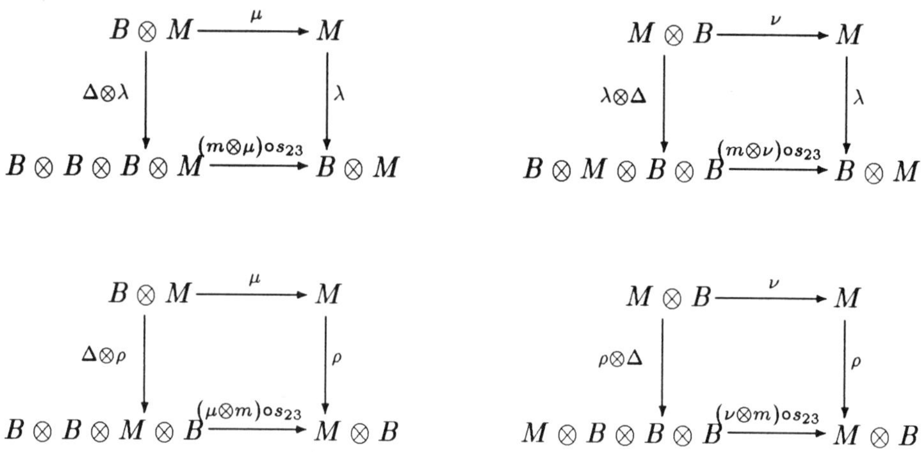

Figure 10.1: The four consistency diagrams relating module to comodule structures.

and right modules, or equivalently both module structure maps are maps of left and right comodules. Thus, instead of one commutative diagram as in the case of dimodules, there are four. The second one in the first row is for compatibility of right module and left comodule structures. The second row has the diagrams for left module-right comodule and right module-right comodule.

When B is a Hopf algebra, a dimodule M is called a Hopf module. In this case the existence of an antipode allows one to prove that as a comodule M is isomorphic to the tensor product of B with a trivial comodule. For the sake of completeness, we will prove this structure theorem at the end of the chapter.

For any fixed k-algebra A define the functor

$$\begin{array}{rcl} \beta : \mathcal{V}\mathbf{ec}_k & \rightarrow & \mathcal{B}\mathbf{imod}_A \\ \beta(N) & = & A \otimes N \otimes A \end{array}$$

with the bimodule structure given by multiplication on the left and right.

$$a(b \otimes n \otimes b')a' = ab \otimes n \otimes b'a'$$

Let F_A be the forgetful functor

$$F_A : \mathcal{B}\mathbf{imod}_A \to \mathcal{V}\mathbf{ec}_k.$$

Then, for any A-bimodule M and k vector space N, there is a bijection

$$\mathrm{Hom}_k(N, F_A(M)) \leftrightarrow \mathrm{Hom}_{A\text{-bimod}}(\beta(N), M) \tag{10.2.1}$$

with the correspondence given by

$$\phi \mapsto \beta(\phi)$$

where

$$\beta(\phi)(a \otimes n \otimes b) = a\phi(n)b \tag{10.2.2}$$

and

$$\psi \mapsto \beta^{-1}(\psi)$$

where

$$\beta^{-1}(\psi)(n) = \psi(1 \otimes n \otimes 1). \tag{10.2.3}$$

Similarly for a coalgebra C define

$$\begin{aligned} \gamma : \mathcal{V}\mathbf{ec}_k &\to \mathcal{B}\mathbf{icomod}_C \\ \gamma(N) &= C \otimes N \otimes C. \end{aligned}$$

The bicomodule structure is given by

$$\lambda(a \otimes n \otimes b) = \sum a_{(1)} \otimes a_{(2)} \otimes n \otimes b,$$

$$\rho(a \otimes n \otimes b) = \sum a \otimes n \otimes b_{(1)} \otimes b_{(2)}.$$

If F_C is the forgetful functor

$$F_C : \mathcal{B}\mathbf{icomod}_C \to \mathcal{V}\mathbf{ec}_k$$

then there is a bijection

$$\mathrm{Hom}_k(F_C(M), N) \leftrightarrow \mathrm{Hom}_{C\text{-bicomod}}(M, \gamma(N)) \tag{10.2.4}$$

with the correspondence given by

$$\phi \mapsto \gamma(\phi)$$

where

$$\gamma(\phi)(m) = (id \otimes \phi \otimes id)((\lambda_M \otimes id)\rho_M(m)) \tag{10.2.5}$$

and

$$\psi \mapsto \gamma^{-1}(\psi)$$

where

$$\gamma^{-1}(\psi)(m) = (\epsilon \otimes id \otimes \epsilon)\psi(m). \tag{10.2.6}$$

Suppose that N is a B-bicomodule, then we can endow the B-bicomodule $\beta(N) = A \otimes N \otimes A$ with the B-bimodule structure coming from the tensor product of comodules. For example, if the left comodule structure $\lambda : N \to B \otimes N$ is given in Sweedler notation

$$\lambda(n) = \sum \lambda_{(1)}(n) \otimes \lambda_{(2)}(n)$$

then the left comodule structure on $\beta(N)$ will be

$$\lambda(a \otimes n \otimes b) = \sum a_{(1)}\lambda_{(1)}(n)b_{(1)} \otimes a_{(2)} \otimes \lambda_{(2)}(n) \otimes b_{(2)}.$$

The B-4 condition of compatibility with the left module structure on $\beta(N)$ is

$$\begin{aligned} \lambda(a'a \otimes n \otimes b) &= \sum (a'a)_{(1)}\lambda_{(1)}(n)b_{(1)} \otimes (a'a)_{(2)} \otimes \lambda_{(2)}(n) \otimes b_{(2)} \\ &= \sum a'_{(1)}a_{(1)}\lambda_{(1)}(n)b_{(1)} \otimes a'_{(2)}a_{(2)} \otimes \lambda_{(2)}(n) \otimes b_{(2)} \\ &= a' \sum \cdot (a_{(1)}\lambda_{(1)}(n)b_{(1)} \otimes a_{(2)} \otimes \lambda_{(2)}(n) \otimes b_{(2)}) \\ &= a' \cdot \lambda(a \otimes n \otimes b). \end{aligned}$$

The verification of the other 3 conditions is equally straightforward.

Similarly, if N is a B-bimodule, then we can endow the B bicomodule $\gamma(N)$ with the B-bimodule structure coming from the tensor product of B-bimodules. Once again we get a B-4 module structure.

Proposition 10.2.1 *The functors β and γ satisfy the following bijective correspondences describing adjoint functor relations, where M is a B-4 module and F is the appropriate forgetful functor:*

$$\textit{For } N \textit{ a } B\textit{-bicomodule}: \operatorname{Hom}_{B\text{-}4}(\beta(N), M) \leftrightarrow \operatorname{Hom}_{B\text{-}bicomod}(N, F(M))$$

$$\textit{For } N \textit{ a } B\textit{-bimodule}: \operatorname{Hom}_{B\text{-}4}(M, \gamma(N)) \leftrightarrow \operatorname{Hom}_{B\text{-}bimod}(F(M), N)$$

Proof. The only condition that remains to be verified is that $\beta(\phi)$ is a B-4 module map for any B-bicomodule map ϕ and $\gamma(\psi)$ is a B-4 module map for any B-bimodule map ψ. Since $\beta(\phi)$ is a B bimodule map by construction, we need only check that it is a B-bicomodule map for the tensor product structure.

$$\begin{aligned}
\lambda_M\beta(\phi)(a\otimes n\otimes b) &= \lambda_M(a\phi(n)b)\\
&= \sum \lambda_{M(1)}(a\phi(n)b)\otimes\lambda_{M(2)}(a\phi(n)b)\\
&= \sum a_{(1)}\lambda_{M(1)}(\phi(n))b_{(1)}\otimes a_{(2)}\lambda_{M(2)}(\phi(n))b_{(2)}\\
&= \sum a_{(1)}\lambda_{N(1)}(n)b_{(1)}\otimes a_{(2)}\phi(\lambda_{N(2)}(n))b_{(2)}\\
&= \sum a_{(1)}\lambda_{N(1)}(n)b_{(1)}\otimes \beta(\phi)(a_{(2)}\otimes\lambda_{N(2)}(n)\otimes b_{(2)})\\
&= (id\otimes\beta(\phi))(\lambda_N(a\otimes n\otimes b))
\end{aligned}$$

where the passage from the second to the third equality comes from the fact that M is a $B-4$ module and the passage from the third to the fourth follows from the fact that ϕ is a left B-comodule map. The verification that $\beta(\phi)$ is a right B-comodule map is completely analogous.

Let $a\mapsto\sum a_{(1)}\otimes a_{(2)}\otimes a_{(3)}$ denote the composition of comultiplications, $(\Delta\otimes id)\circ\Delta=(id\otimes\Delta)\circ\Delta$, and $m\mapsto\sum m_{(1)}\otimes m_{(2)}\otimes m_{(3)}$ the composition of right and left comodule structures $(\lambda\otimes id)\circ\rho=(id\otimes\rho)\circ\lambda$. Then we have

$$\begin{aligned}
\gamma(\phi)(am) &= \sum (am)_{(1)}\otimes\phi((am)_{(2)})\otimes(am)_{(3)}\\
&= \sum a_{(1)}m_{(1)}\otimes\phi(a_{(2)}m_{(2)})\otimes a_{(3)}m_{(3)}\\
&= \sum a_{(1)}m_{(1)}\otimes a_{(2)}\phi(m_{(2)})\otimes a_{(3)}m_{(3)}\\
&= \sum a\cdot(m_{(1)}\otimes\phi(m_{(2)})\otimes m_{(3)})\\
&= a\cdot\gamma(\phi)(m),
\end{aligned}$$

showing that $\gamma(\phi)$ is a left module map. A similar verification of right module condition completes the proof of the proposition.

As a corollary we obtain

Proposition 10.2.2 *For a B-bicomodule M and a B-bimodule N there is a well-defined bijection*

$$\mathrm{Hom}_{B\text{-}4}(\beta(M),\gamma(N))\leftrightarrow\mathrm{Hom}_k(M,N)$$

given by

$$\phi\mapsto\gamma(\beta(\phi))=\beta(\gamma(\phi)).$$

Proof. We must prove the equality in the proposition. With the notation as above,

$$\begin{aligned}\beta(\gamma(\phi))(a\otimes m\otimes b) &= a\gamma(\phi)(m)b\\ &= \sum a(m_{(1)}\otimes\phi(m_{(2)})\otimes m_{(3)})b\\ &= \sum a_{(1)}m_{(1)}b_{(1)}\otimes a_{(2)}\phi(m_{(2)})b_{(2)}\otimes a_{(3)}m_{(3)}b_{(3)}\end{aligned}$$

$$\begin{aligned}&\gamma(\beta(\phi))(a\otimes m\otimes b)\\ &= \sum a_{(1)}m_{(1)}b_{(1)}\otimes\beta(\phi)(a_{2)}\otimes m_{(2)}\otimes b_{(2)})\otimes a_{(3)}m_{(3)}b_{(3)}\\ &= \sum a_{(1)}m_{(1)}b_{(1)}\otimes a_{(2)}\phi(m_{(2)})b_{2)}\otimes a_{(3)}m_{(3)}b_{(3)}.\end{aligned}$$

Thus $\gamma(\beta(\phi))=\beta(\gamma(\phi))$.

10.3 B-4 complexes.

In this section we define the complexes of $B-4$ modules which appear in Gerstenhaber-Schack cohomology. The bar complex $\mathrm{Bar}(A)$ of an associative algebra A is the direct sum of of q-th degree components

$$\mathrm{Bar}_q(A)=A^{\otimes(q+2)}\quad q\geq -1.$$

For $q\geq 0$ this is $\beta(A^{\otimes q})$.

The differential is

$$\partial_q(a_0\otimes a_1\otimes\ldots\otimes a_{q+1})=\sum_{i=0}^{q}(-1)^i a_0\otimes\ldots\otimes a_ia_{i+1}\otimes\ldots\otimes a_{q+1}\quad q>-1 \tag{10.3.1}$$

$$\partial_{-1}(a)=0.$$

One verifies immediately that associativity of multiplication implies $\partial_{q-1}\partial_q=0$ for all $q\geq 0$ so $\mathrm{Im}\partial_q\subset\mathrm{Ker}\partial_{q-1}$. To show the reverse inclusion define the chain homotopy

$$h_q(a_0\otimes a_1\otimes\ldots\otimes a_{q+1})=1\otimes a_0\otimes\ldots\otimes a_{q+1}\quad q\geq -1 \tag{10.3.2}$$

which satisfies the equation

$$\partial_{q+1}h_q + h_{q-1}\partial_q = id \quad q \geq -1. \tag{10.3.3}$$

This proves the inclusion $\mathrm{Ker}\partial_{q-1} \subset \mathrm{Im}\partial_q$ so $(\mathrm{Bar}(A), \partial)$ is acyclic.

The A bimodule structure on $\mathrm{Bar}_q(A)$ for $q \geq 0$ is the one defined by the functor β, for $q = -1$ it is given by left and right multiplication. More explicitly this A-bimodule structure or equivalently a left $A \otimes A^{\mathrm{op}}$ module structure on $\mathrm{Bar}(A)$ is given by

$$(a \otimes b)(a_0 \otimes a_1 \otimes \ldots \otimes a_q) = aa_0 \otimes a_1 \otimes \ldots \otimes a_q b. \tag{10.3.4}$$

The differential ∂ is an A-bimodule map.

Any bialgebra B is a bicomodule over itself, and for $q \geq 0$ $B^{\otimes q}$ has a bicomodule structure coming from the tensor product of comodules, which, in turn, induces on $\beta(B^{\otimes q})$ a bicomodule structure as described in section 2. More explicitly the B-bicomodule structure on $\mathrm{Bar}(B)$ is given by

$$\lambda(a_0 \otimes a_1 \otimes \ldots \otimes a_{q+1})$$

$$= \sum a_{0(1)}a_{1(1)} \ldots a_{(q+1)(1)} \otimes a_{0(2)} \otimes a_{1(2)} \otimes \ldots \otimes a_{q+1(2)} \tag{10.3.5}$$

and

$$\rho(a_0 \otimes a_1 \otimes \ldots \otimes a_{q+1}) =$$

$$\sum a_{0(1)} \otimes a_{1(1)} \otimes \ldots \otimes a_{q+1(1)} \otimes a_{0(2)}a_{1(2)} \ldots a_{(q+1)(2)}. \tag{10.3.6}$$

Both the right and left module structures are compatible with both of the comodule structures, giving $\mathrm{Bar}(B)$ the structure of a B-4 module, and in fact a symmetric $B-4$ module structure.

Note that the homotopy is a right A-module map but not a left A-module map. A more important fact that plays a crucial role in the proof of the main theorem is that it is a bicomodule map.

$$\begin{aligned} & \lambda h(a_0 \otimes a_1 \otimes \cdots \otimes a_{q+1}) = \\ = \; & \lambda(1 \otimes a_0 \otimes a_1 \otimes \cdots \otimes a_{q+1}) \\ = \; & \sum a_{0(1)}a_{1(1)} \cdots a_{q+1(1)} \otimes 1 \otimes a_{0(1)} \otimes a_{1(1)} \otimes \cdots \otimes a_{q+1(1)} \\ = \; & id \otimes h\lambda(a_0 \otimes a_1 \otimes \cdots \otimes a_{q+1}). \end{aligned}$$

Similarly for the right comodule structure.

Definition A sequence

$$\cdots \to Y^{i+1} \overset{\alpha^{i+1}}{\to} Y^i \overset{\alpha^i}{\to} Y^{i-1} \to \cdots$$

of B bimodules (bicomodules) is called *split exact* at Y^i if it is exact at Y^i and Y^i is a direct sum of sub-bimodules (sub-bicomodules) $Y^{i\prime}$ and $Y^{i\prime\prime}$ such that $Y^{i\prime} = \mathrm{Ker}\alpha^i = \mathrm{Im}\alpha^{i+1}$.

(A similar definition makes sense in any abelian category.)

Proposition 10.3.1 *The bar complex of a bialgebra B is split exact as a sequence of bicomodules.*

Proof. The basic equation (1.10) implies that Bar^q is the direct sum of $\mathrm{Im}h^{q-1}$ and $\mathrm{Im}\partial^{q+1} = \mathrm{Ker}\partial^q$, and the fact that both h and ∂ are bicomodule maps implies that the summands are sub-bicomodules.

The cobar complex of a bialgebra has analogous properties. For $q \geq 0$ $\mathrm{Cob}^q(C) = \gamma(C^{\otimes q})$ and $\mathrm{Cob}^{-1} = C$. The differential is

$$\delta(a_0 \otimes \cdots \otimes a_{q+1}) = \sum (-1)^i a_0 \otimes \cdots \otimes \Delta(a_i) \otimes \cdots \otimes a_{q+1}.$$

Coassociativity of Δ implies that $\delta^2 = 0$. The existence of a contracting homotopy

$$h(a_0 \otimes \cdots \otimes a_{q+1}) = \epsilon(a_0) a_1 \otimes \cdots \otimes a_{q+1}$$

implies that $\mathrm{Cob}(C)$ is acyclic.

The left comodule structure associated to the functor γ is

$$\lambda = \Delta_1 = \Delta \otimes id_{q+1} \tag{10.3.7}$$

and right comodule structure

$$\rho = \Delta_{q+2} = id_{q+1} \otimes \Delta. \tag{10.3.8}$$

Relative to these structures δ is a bicomodule map.

For any bialgebra B, $B^{\otimes q}$ has a bimodule structure as a tensor product of bimodules. This induces on $\gamma(B^{\otimes q})$ a $B-4$ module structure and δ is a $B-4$ module map.

More explicitly the left B-module structure on $\mathrm{Cob}^q(B)$ is given by

$$\begin{aligned} & a \cdot (b_0 \otimes b_1 \otimes \ldots \otimes b_q) = \\ = \; & \Delta^{[q+1]}(a) \cdot (b_0 \otimes b_1 \otimes \ldots \otimes b_{q+1}) \\ = \; & \sum a_{(1)} b_0 \otimes a_{(2)} b_1 \otimes \ldots \otimes a_{(q+2)} b_{q+1} \end{aligned} \tag{10.3.9}$$

where the $q+1$-st iterate of Δ, $\Delta^{[q+1]}$ is written in Sweedler notation

$$\Delta^{[q+1]}(a) = \sum a_{(1)} \otimes a_{(2)} \otimes \ldots \otimes a_{(q+2)},$$

and right module structure is given by the composition of $\Delta^{[q+1]}$ and right multiplication. This defines a B-4 module structure on $\mathrm{Cob}(B)$ and δ is compatible with all the structures.

Here also the homotopy operator respects three of the four structures, the right comodule structure and both right and left module structures. The compatibility with the modules structures follows from the identity,

$$h(\Delta^{[q+1]}(a) = (m \otimes id_q)(\epsilon \otimes id_{q+1})\Delta^{[q+1]} = \Delta^{[q]}(a),$$

and the fact that h is an algebra map.

$$\begin{aligned} h(\Delta^{[q+1]}(a)a_0 \otimes a_1 \otimes \cdots \otimes a_{q+1}) &= h(\Delta^{[q+1]}(a))h(a_0 \otimes a_1 \otimes \cdots a_{q+1}) \\ &= h(\Delta^{[q]}(a))h(a_0 \otimes a_1 \otimes \cdots \otimes a_{q+1}) \end{aligned}$$

Proposition 10.3.2 *The cobar complex of a bialgebra B is split exact as a complex of of B bimodules.*

Definition. Given an acyclic complex defined in degrees $q \geq -1$, we call the complex in degrees $q \geq 0$ a *resolution* of the component in degree -1. The cohomology (or homology) of the resolution vanishes in degree > 0, and in degree 0 equals the degree -1 component of the original acyclic complex.

For example, the bar resolution is given

$$X^q = \mathrm{Bar}_q(A), \quad q \geq 0$$

with ∂_q given by (10.3.1) for $q > 0$, with a modified $\partial_0 = 0$ and with

$$\epsilon : Bar_0(A) \to A. \tag{10.3.10}$$

defined by

$$\epsilon(a_0 \otimes a_1) = a_0 a_1,$$

and

$$H_0(\mathrm{Bar}(A)) = A. \tag{10.3.11}$$

We make a similar (dual) definition of a resolution of a coalgebra, C. The subcomplex of $\mathrm{Cob}(C)$ formed by components of non-negative degree is coaugmented over C by $\Delta_{-1} = \Delta : C \to C \otimes C = \mathrm{Cob}^0(C)$, thus forming a resolution, the cobar resolution of C.

With a slight abuse of notation, we will use the same symbol for the resolution and the associated acyclic complex, since the context will make clear whether the -1 degree component is present or not.

Proposition 10.3.3 Gerstenhaber and Schack.*[462] The Gerstenhaber-Schack complex can be described as*

$$C^{p,q}_{G.S.} = \mathrm{Hom}_{B\text{-}4}(\mathrm{Bar}^q(B), \mathrm{Cob}^p(B))$$

with differential

$$D^{p,q}\alpha = (-1)^q \delta^p \circ \alpha + \alpha \circ \partial^q. \tag{10.3.12}$$

Proof. Combining the adjoint relations (10.2.1) and (10.2.4) gives the following isomorphism of vector spaces

$$\mathrm{Hom}_k(B^{\otimes q}, B^{\otimes p}) \leftrightarrow \mathrm{Hom}_{B\text{-}4}(\mathrm{Bar}^q(B), \mathrm{Cob}^p(B))$$

$$\phi \mapsto \hat{\phi} \quad \phi \in \mathrm{Hom}_k(B^{\otimes q}, B^{\otimes p})$$

where

$$\hat{\phi}(a_0 \otimes \ldots \otimes a_{q+1}) =$$

$$\sum (a_{0(1)} \cdots a_{(q+1)(1)}) \otimes a_{0(2)}\phi(a_{1(2)} \otimes \ldots \otimes a_{q(2)})a_{q+1(2)} \otimes (a_{0(3)} \cdots a_{(q+1)(3)}) \tag{10.3.13}$$

and

$$\psi \mapsto \tilde{\psi} \quad \psi \in \mathrm{Hom}_{B\text{-}4}(\mathrm{Bar}^q(B), \mathrm{Cob}^p(B))$$

where

$$\tilde{\psi}(a_1 \otimes \cdots \otimes a_q) = (\epsilon \otimes id \otimes \epsilon) \circ \psi(1 \otimes a_1 \otimes \ldots \otimes a_q \otimes 1) \tag{10.3.14}$$

Proposition 2 implies that this correspondence is in fact an isomorphism of complexes:

$$\widetilde{\delta\hat{\phi}} = d_C \phi \tag{10.3.15}$$

and

$$\widetilde{\hat{\phi} \circ \partial} = d_H \phi \tag{10.3.16}$$

To verify (10.3.15), let $\phi_{(i)} = \phi(a_{1(i)} \otimes \ldots \otimes a_{q(1)})$ and $\overline{a} = a_0 \ldots a_{q+1}$, then

$$\hat{\phi}(a_0 \otimes \ldots \otimes a_{q+1}) = \sum \overline{a}_{(1)} \otimes a_{0(2)} \phi_{(2)} a_{q+1(2)} \otimes \overline{a}_{(3)}$$

so

$$\begin{aligned} \delta \circ \hat{\phi} &= \sum \overline{a}_{(1)} \otimes \overline{a}_{(2)} \otimes a_{0(3)} \phi_{(3)} a_{q+1(3)} \otimes \overline{a}_{(4)} \\ &\quad + \sum (-1)^i \sum \overline{a}_{(1)} \otimes \Delta_i (a_{0(2)} \phi_{(2)} a_{q+1(2)}) \otimes \overline{a}_{(3)} \\ &\quad +(-1)^{p+1} \sum \overline{a}_{(1)} \otimes a_{0(2)} \phi_{(2)} a_{q+1(2)} \otimes \overline{a}_{(3)} \otimes \overline{a}_{(4)} \end{aligned}$$

and therefore

$$\begin{aligned} \widetilde{\delta \circ \hat{\phi}} &= \sum a_{1(1)} \ldots a_{q(1)} \otimes \phi(a_{1(2)} \otimes \ldots \otimes a_{q(2)}) \\ &\quad + \sum (-1)^i \Delta_i \circ \phi(a_1 \otimes \ldots \otimes a_q) \\ &\quad +(-1)^{p+1} \sum \phi(a_{1(1)} \otimes \ldots \otimes a_{q(2)}) \otimes a_{1(2)} \ldots a_{q(2)} \end{aligned}$$

which agrees with (9.1.4). In the same way one finds that $\widetilde{\hat{\phi} \circ \partial}$ is the Hochschild differential, proving Proposition 5.

10.4 Projective and injective resolutions.

The adjoint relations have another important consequence which allows one to use different resolutions.

Definition. Let A be a k algebra. An A-bimodule P is called *projective* if the functor $\mathrm{Hom}_{A\text{-bimod}}(P, \cdot)$ is exact, i.e., takes short exact sequences of A-bimodules into short exact sequences of k vector spaces.

The condition in the definition is equivalent to requiring that for any surjection $\pi : M \to N$ and $\tau : P \to N$ there is a $\tau' : P \to M$ completing the diagram

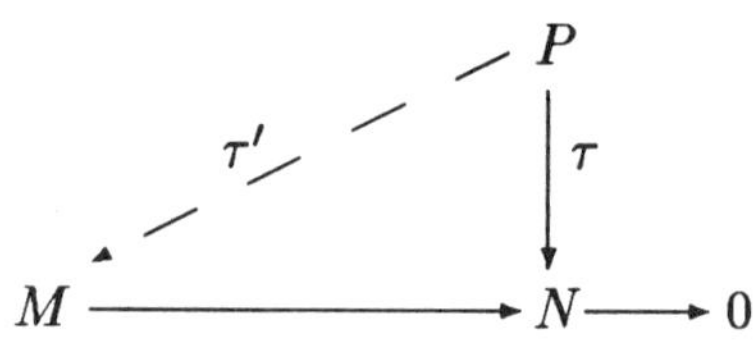

Definition. Let C be a coalgebra. A C-bicomodule Q is called *injective* if the functor $\mathrm{Hom}_{C\text{-bicomod}}(\cdot, Q)$ is exact.

This is equivalent to requiring that for any injection $i : M \to N$ and $\tau : M \to Q$ there is an extension to $\tau' : N \to Q$ completing the diagram

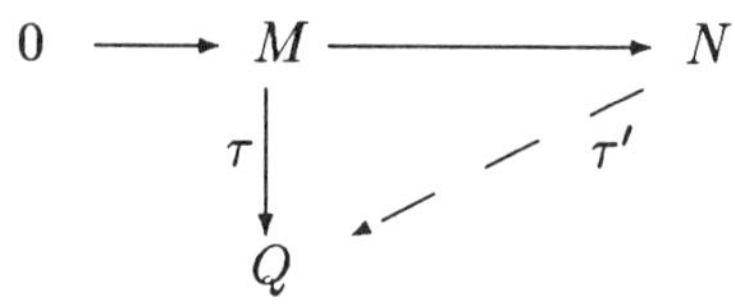

(Obviously, one can define left-projective, right-projective, $B - 4$ projective and similarly for injectives by considering the appropriate Hom functor.)

Proposition 10.4.1 *Let V be a vector space over k, A an algebra over k and C a coalgebra over k. Then $\beta(V) = A \otimes V \otimes A$ is a projective A-bimodule and $\gamma(V) = C \otimes V \otimes C$ is an injective C-bicomodule.*

Proof. This follows immediately from the adjoint relations for β and γ and the fact that $\mathrm{Hom}_k(V, \cdot)$ and $\mathrm{Hom}_k(\cdot, V)$ are exact for any vector spaces V.

In particular this implies that $\mathrm{Bar}^q(B)$ is a projective bimodule and $\mathrm{Cob}^p(B)$ is an injective bicomodule for all non-negative p, q.

The following proposition will allow us to change resolutions.

Proposition 10.4.2 *Let A be a k-algebra. Given $X \to A$ and $Y \to A$, two complexes of A-bimodules augmented over A, such that X^q is projective for all $q \geq 0$ and $Y \to A$ is acyclic. Then there exists a map of complexes of A-bimodules ϕ such that the triangle*

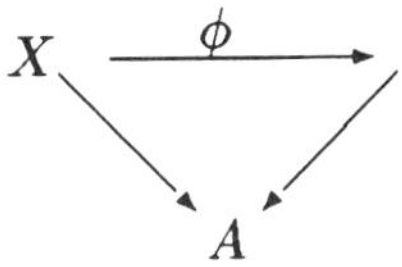

commutes, and ϕ is unique up to homotopy which is also an A bimodule map.

Proof. To begin we have the diagram

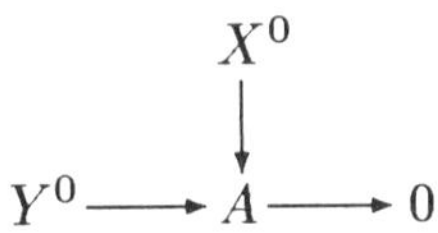

Since X^0 is projective there exists a $\phi^0 : X^0 \to Y^0$ such that

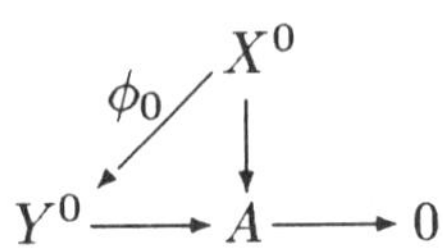

commutes. Next consider the diagram

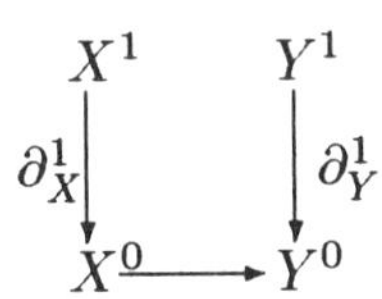

Let $\psi^0 = \phi^0 \circ \partial_X^1 : X^1 \to Y^0$. The image of ∂_X^1 is contained in the kernel of $X^0 \to A$ thus the image of ψ^0 is contained in the kernel of $Y^0 \to A$, but this equals the image of ∂_Y^1 by acyclicity of Y, so again we have a diagram

$$\begin{array}{ccccc} & & X^1 & & \\ & & \downarrow & & \\ Y^1 & \longrightarrow & \mathrm{Im}\partial_Y^1 & \longrightarrow & 0 \end{array}$$

Since X^1 is projective, there exists a $\phi^1 : X^1 \to Y^1$ filling in the diagram. Then

$$\partial_Y^1 \circ \phi^1 = \psi = \phi^0 \circ \partial_X^1.$$

The same argument proves the induction step from the existence of ϕ^n to the existence of ϕ^{n+1}.

Next suppose $\tilde{\phi} : X^0 \to Y^0$ is another map satisfying the conditions of the theorem, then we have $\mathrm{Im}(\phi^0 - \tilde{\phi}^0) \subset \mathrm{Ker}(Y^0 \to A) = \mathrm{Im}\partial_Y^1$ and so the diagram

$$\begin{array}{ccccc} & & X^0 & & \\ & & \downarrow & & \\ Y^1 & \longrightarrow & \mathrm{Im}\partial_Y^1 & \longrightarrow & 0 \end{array}$$

Thus there exists an $h^0 : X^0 \to Y^1$ such that $\phi^0 - \tilde{\phi}^0 = \partial_Y^1 \circ h^0$. Suppose we have defined $h^i : X^i \to Y^{i+1}$ for $i < n$ satisfying

$$\phi^i - \tilde{\phi}^i = h^{i-1}\partial_X^i + \partial_Y^{i+1}h^i.$$

Then consider $\psi = \phi^n - \tilde{\phi}^n - h^{n-1}\partial_X^n : X^n \to Y^n$. Its image is in the kernel of ∂_Y^n since

$$\begin{aligned} \partial_Y^n(\phi^n - \tilde{\phi}^n - h^{n-1}\partial_X^n) &= \phi^{n-1}\partial_X^n - \tilde{\phi}^{n-1}\partial_X^n - \partial_Y^n h^{n-1}\partial_X^{n-1} \\ &= (\phi^{n-1} - \phi^{n-1} - \partial_Y^n h^{n-1} - h^{n-2}\partial_X^{n-1})\partial_X^{n-1} \\ &= 0. \end{aligned}$$

As before we have h^n filling in the diagram

$$\begin{array}{ccccc} & & X^n & & \\ & \swarrow^{h^n} & \downarrow \psi & & \\ Y^{n+1} & \longrightarrow & \mathrm{Im}\partial_Y^{n+1} & \longrightarrow & 0 \end{array}$$

giving

$$\partial_Y^{n+1} h^n = \phi^n - \tilde{\phi}^n - h^{n+1}\partial_X^n,$$

and completing the proof of the proposition.

A complex $X^{\cdot}$ augmented over A is called a *projective resolution* of A if each X^q is projective and $X^{\cdot} \to A$ is acyclic. Since a projective resolution satisfies the requirements for both X and Y of the preceding proposition, we get, as a corollary,

Proposition 10.4.3 *Given two projective resolutions of A, X and Y, there exist maps of A-bimodule complexes ϕ, ψ :*

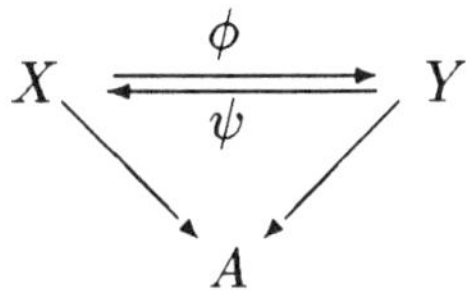

and homotopies $h_X^q : X^q \to X^{q+1}$, $h_Y^q : Y^q \to Y^{q+1}$ satisfying

$$id_X - \psi^q \phi^q = \partial_X^{q+1} h_X^q + h_X^{q-1} \partial_X^q$$

$$id_Y - \phi^q \psi^q = \partial_Y^{q+1} h_Y^q + h_Y^{q-1} \partial_Y^q.$$

The maps ϕ and ψ are called *homotopy equivalences.*

A similar result holds for coalgebras:

Proposition 10.4.4 *Let C be a k-coalgebra. Given $C \to X$ and $C \to Y$, two complexes of C bicomodules coaugmented over C such that X^q is injective for all $a \geq 0$ and $C \to Y$ is acyclic then there exists a map ϕ of C-bicomodule complexes such that*

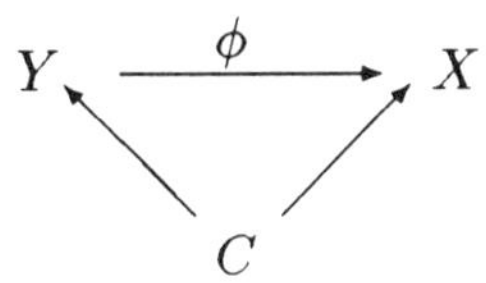

commutes and ϕ is unique up to an A bicomodule homotopy.

Given two injective resolutions of C, $C \to X$, $C \to Y$, then there exist homotopy equivalences ϕ, ψ

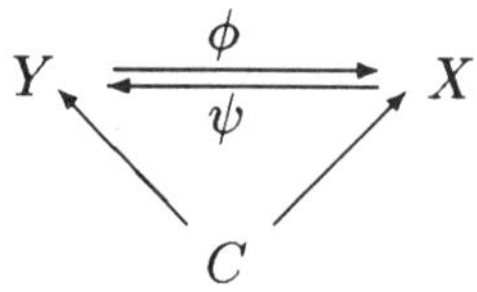

Proof. In order to prove this theorem, we simply make the following changes in the proof of the previous theorem: replace A with C, reverse all the arrows, replace $\mathrm{Im}\partial_Y$ by $Y/\mathrm{Ker}\partial_Y$.

At this point it is worthwhile summing up what we have know so far and describing what we plan to do in the rest of the chapter. In section 3 we showed that the Gerstenhaber-Schack complex is a double complex constructed from the Hom functor between two resolutions. In section 4 we saw that one of these resolutions, $\mathrm{Bar}(B)$ is in fact a B bimodule projective resolution of B and the other, $\mathrm{Cob}(B)$ is an injective B bicomodule resolution of B. We also proved that any two resolutions of B of the same type are homotopy equivalent. The strategy for computing the G-S cohomology, and also for giving an explicit representations of the cohomology classes which can be used in the uniqueness and existence proofs in the next chapter is to replace the bar and cobar resolutions with the Koszul resolutions K_b and K_c (defined in section 6). In general, if X, Y, and Z are three complexes, the following basic lemma explains how to derive a homotopy equivalence between the double complexes $\mathrm{Hom}(X,Y)$ and $\mathrm{Hom}(X,Z)$ or between $\mathrm{Hom}(Y,X)$ and $\mathrm{Hom}(Z,X)$ if we are given a homotopy equivalence between Y and Z.

Lemma 10.4.1 *Suppose that we have three complexes X, Y, and Z of either bimodules, bicomodules or tetramodules such that there exist homotopy equivalences of the appropriate type*

$$\phi : Y \to Z \quad \psi : Z \to Y.$$

Define double complexes

$$C_1 = \mathrm{Hom}(X,Y) \quad \textit{and} \quad C_2 = \mathrm{Hom}(X,Z)$$

$$C_3 = \mathrm{Hom}(Y, X) \quad \textit{and} \quad C_4 = \mathrm{Hom}(Z, X)$$

Where Hom *is the appropriate homomorphism functor, with differentials*

$$D_1\alpha_1 = (-1)^q d_Y \circ \alpha_1 + \alpha_1 \circ d_X \quad \alpha_1 \in C_1^{p,q}$$

$$D_2\alpha_2 = (-1)^q d_Z \circ \alpha_2 + \alpha_2 \circ d_X \quad \alpha_2 \in C_2^{p,q}.$$

$$D_3\alpha_3 = (-1)^q d_X \circ \alpha_3 + \alpha_3 \circ d_Y \quad \alpha_3 \in C_3^{p,q}$$

$$D_4\alpha_4 = (-1)^q d_X \circ \alpha_4 + \alpha_4 \circ d_Z \quad \alpha_4 \in C_4^{p,q}.$$

In this situation, we define

$$\overline{\phi} : C_1 \to C_2$$

$$\overline{\psi} : C_2 \to C_1$$

$$\tilde{\psi} : C_3 \to C_4$$

$$\tilde{\phi} : C_4 \to C_3$$

by $\overline{\phi}(\alpha_1) = \phi \circ \alpha_1$, $\overline{\psi}(\alpha_2) = \psi \circ \alpha_2$, $\tilde{\psi}(\alpha_3) = \alpha_3 \circ \psi$, *and* $\tilde{\phi}(\alpha_4) = \alpha_4 \circ \phi$. *Then* $\overline{\phi}$, $\overline{\psi}$, $\tilde{\psi}$ *and* $\tilde{\psi}$ *are homotopy equivalences.*

Proof. It is clear that $\overline{\phi}$, $\overline{\psi}$ $\tilde{\psi}$ and $\tilde{\psi}$ are chain maps. Define $\overline{h}_1(\alpha) = (-1)^q h_Y \circ \alpha$ for $\alpha \in C_1^{p,q}$ and $\overline{h}_2(\beta) = (-1)^q h_Z \circ \beta$ for $\beta \in C_2$, where h_Y and h_Z are the homotopies satisfying

$$\begin{aligned} (d_Z h_Z + h_Z d_Z) &= (id - \phi \circ \psi) \\ (d_Y h_Y + h_Y d_Y) &= (id - \psi \circ \phi) \end{aligned}$$

We will prove one of the homotopy equivalence relations.

$$\begin{aligned} D_1\overline{h}_1(\alpha_1) + \overline{h}_1 D_1\alpha_1 &= d_Y \circ (h_Y \circ \alpha_1) + (-1)^q (h_Y \circ \alpha_1) \circ d_X \\ &\quad + h \circ (d_Y \circ \alpha_1) + (-1)^{q+1} h \circ (\alpha_1 \circ d_X) \\ &= (d_Y h_Y + h_Y d_Y) \circ \alpha_1 \\ &= (id - \psi \circ \phi) \circ \alpha_1 \\ &= \alpha_1 - \overline{\psi}(\overline{\phi}(\alpha_1)). \end{aligned}$$

Similarly for the remaining relations.

We want to apply this lemma twice, first to the case when the second variable changes, to get a homotopy equivalence

$$\mathrm{Hom}_{U(\mathcal{G})-4}(\mathrm{Bar}(U(\mathcal{G}), \mathrm{Cob}(U(\mathcal{G}))) \simeq \mathrm{Hom}_{U(\mathcal{G})-4}(\mathrm{Bar}(U(\mathcal{G}), K_c)$$

then to the case in which the first variable is changing to show

$$\mathrm{Hom}_{U(\mathcal{G})-4}(\mathrm{Bar}(U(\mathcal{G}), K_c) \simeq \mathrm{Hom}_{U(\mathcal{G})-4}(K_b, K_c).$$

However to use the lemma the initial homotopy equivalences have to be be $U(\mathcal{G}) - 4$ maps. Propositions 10.4.3 and 10.4.4 give homotopy equivalences of a more restricted type. The complex Cob $= \gamma(\otimes\mathcal{G})$ is a B-bicomodule injective resolution by Proposition 10.4.1. We shall see that $K_c = \gamma(\wedge\mathcal{G})$ and thus it is also a $U(\mathcal{G})$-bicomodule injective resolution. Proposition 10.4.4 proves that they are homotopy equivalent as $U(\mathcal{G})$ bicomodules *not* as $U(\mathcal{G})-4$ modules. Similarly Bar $= \beta(\otimes\mathcal{G})$ and $K_b = \beta(\wedge\mathcal{G})$, so the analogous argument gives an $U(\mathcal{G})_4$ bimodule homotopy equivalence, weaker than what is required.

A closer look at the proof of Proposition 10.4.3 reveals a way out of this predicament. The existence of the required map from X to Y does not require that $\mathrm{Hom}(X_i, \cdot)$ preserve the exactness of arbitrary sequences, only the sequences

$$\to Y_{j+1} \to Y_j \to Y_{j-1}$$

and

$$\to X_{j+1} \to X_j \to X_{j-1}$$

making up the complexes X and Y. Thus it is enough that X and Y have a projectivity property relative to a particular class of sequences. (This crucial observation goes back to Hochschild [543] and is the basis of the theory of relative homological algebra developed by Eilenberg and Moore [337].)

10.5 Relative projectives and injectives

In this short section we define more precisely the relative structures which we discussed at the end of the previous section.

The following elementary lemma is stated explicitly in order to clarify the definition of relative projectives and injectives which follows.

Lemma 10.5.1 *If*

$$\cdots \to Y_{i+1} \xrightarrow{\alpha_{i+1}} Y_i \xrightarrow{\alpha_i} Y_{i-1} \to \cdots$$

is a sequence of B bimodules (bicomodules) which is split exact at Y_i and X is any B bimodule (bicomodule), then the following sequences are also split exact

$$\cdots \to \mathrm{Hom}(X, Y_{i+1}) \xrightarrow{\alpha_{i+1*}} \mathrm{Hom}(X, Y_i) \xrightarrow{\alpha_{i*}} \mathrm{Hom}(X, Y_{i-1}) \to \cdots$$

$$\cdots \leftarrow \mathrm{Hom}(Y_{i+1}, X) \xleftarrow{\alpha_{i+1*}} \mathrm{Hom}(Y_i, X) \xleftarrow{\alpha_{i*}} \mathrm{Hom}(Y_{i-1}, X) \leftarrow \cdots$$

where the functor Hom *is either* $\mathrm{Hom}_{B-bimod}$ *or* $\mathrm{Hom}\, B_{bicomod}$ *according to the situation under consideration.*

The complexes we are interested in are either projective B-bimodule resolutions or injective B-bicomodule resolutions. In order to extend their exactness properties to the category of $B-4$ modules we need to deal with the B-bicomodule structure for the B-bimodule projectives and the B-bimodule structure for the B-bicomodule injectives. As explained above we do this by considering a particular class of exact sequences.

Definitions A $B - 4$ module X is called a *relative* $B - 4$ *projective* if $\mathrm{Hom}_{B-4}(X, \cdot)$ takes exact sequences of $B - 4$ modules which are split exact as sequences of B-bicomodules into exact sequences.

A $B - 4$ module Y is called a *relative* $B - 4$ *injective* if $\mathrm{Hom}_{B-4}(\cdot, Y)$ takes exact sequences of $B - 4$ modules which are split exact as sequences of B-bimodules into exact sequences.

A *relative* $B - 4$ *injective resolution* Y of B is a resolution in which all the modules are relative $B - 4$ injectives and the sequences are split exact as sequences of B-bimodules. A *relative* $B - 4$ *projective resolution* Y of B is a resolution in which all the modules are relative $B - 4$ projectives and the sequences are split exact as sequences of B-bicomodules.

Now we can state the basic propositions

Proposition 10.5.1 *If M is a B bicomodule then any $B - 4$ module of the form $\beta(M)$ is a relative $B - 4$ projective.*

If N is a B bimodule then any $B - 4$ module of the form $\gamma(N)$ is a relative $B - 4$ injective.

Proof. This follows immediately from the lemma on split exact sequences and the adjoint functor relations for β and γ described in Proposition 10.2.1.

Proposition 10.5.2 *Let B be a bialgebra. The bar complex* Bar(B) *is a relative $B-4$ projective resolution. The cobar complex* Cob(B) *is a relative $B-4$ injective resolution.*

Proof. The relative projectivity and injectivity respectively follow from the preceeding proposition. The split exactness was proven in section 4.

Proposition 10.5.3 *Any two relative injective(projective) resolutions of the bialgebra B are homotopy equivalent as $B-4$ complexes.*

Proof. Formally the same as for Proposition 10.4.2.

At this point we are ready to proceed with the construction of the Koszul complexes.

10.6 The Koszul complexes.

The *Koszul complexes*, K_b and K_c, are defined as follows.

The first complex K_b is $\beta(\bigwedge \mathcal{G})$ where $\bigwedge \mathcal{G}$ is considered as a trivial $U(\mathcal{G})$-bicomodule with $\lambda(u) = 1 \otimes u$, $\rho(u) = u \otimes 1$.

The second complex K_c is $\gamma(\bigwedge \mathcal{G})$ where $\bigwedge \mathcal{G}$ has a $U(\mathcal{G})$ bimodule structure to be described below.

The complex K_b has the form $U(\mathcal{G}) \otimes L_b$ and the complex K_c has the form $U(\mathcal{G}) \otimes L_c$ where both L_b and L_c are complexes which are isomorphic to $U(\mathcal{G}) \otimes \wedge(\mathcal{G})$ as vector spaces. For purposes of a later calculation, it is useful have the inclusion map of $U(\mathcal{G}) \otimes \wedge\mathcal{G} \otimes U(\mathcal{G})$ into Bar$(U(\mathcal{G}))$ to be a chain map of K_b into Bar$(U(\mathcal{G}))$. Similarly we want the inclusion of K_c into Cob$(U(\mathcal{G}))$ to be a chain map. For this to work, we can not simply identify L_b or L_c with $1 \otimes \wedge\mathcal{G} \otimes U(\mathcal{G})$ or with $U(\mathcal{G}) \otimes \wedge\mathcal{G} \otimes 1$. Instead, we will express the $U(\mathcal{G})$ bimodules $\beta(\wedge\mathcal{G})$ and $\gamma(\wedge\mathcal{G})$ as left $U(\mathcal{G}) \otimes U(\mathcal{G})^{\mathrm{op}}$ left modules. We then decompose $U(\mathcal{G}) \otimes U(\mathcal{G})^{\mathrm{op}}$ into a product of subalgebras corresponding to a decomposition of $\mathcal{G} \oplus \mathcal{G}^{\mathrm{op}}$ into a sum of Lie subalgebras. Here are the details:

We will identify the underlying vector spaces of $\beta(\bigwedge \mathcal{G})$ and $\gamma(\bigwedge \mathcal{G})$, both of which are copies of $U(\mathcal{G}) \otimes \bigwedge \mathcal{G} \otimes U(\mathcal{G})$, with

$$U(\mathcal{G}) \otimes U(\mathcal{G}) \otimes \bigwedge \mathcal{G}.$$

So an element of either $\beta(\bigwedge \mathcal{G})$ or $\gamma(\bigwedge \mathcal{G})$ can be written as a sum of terms

$$w \otimes X_1 \wedge \cdots \wedge X_p, \quad w \in U(\mathcal{G}) \otimes U(\mathcal{G}), \quad X_i \in \mathcal{G}.$$

The $U(\mathcal{G})$ bimodule structure on K_b comes from the functor β so it is just an extension of the natural $U(\mathcal{G})$ bimodule structure on $U(\mathcal{G}) \otimes U(\mathcal{G})$ given by

$$a(u \otimes v)b = au \otimes vb.$$

We regard the bimodule structure as a left $U(\mathcal{G}) \otimes U(\mathcal{G})^{\mathrm{op}}$ module structure where $U(\mathcal{G})^{\mathrm{op}}$ has the opposite multiplication

$$(u \cdot v)^{\mathrm{op}} = v \cdot u.$$

$U(\mathcal{G})^{\mathrm{op}}$ is isomorphic to the enveloping algebra of $\mathcal{G}^{\mathrm{op}}$ where $\mathcal{G}^{\mathrm{op}}$ is $\mathcal{G}$ as a vector space but with the bracket $[X,Y]^{\mathrm{op}} = -[X,Y]$. Then

$$U(\mathcal{G}) \otimes U(\mathcal{G})^{\mathrm{op}} \simeq U(\mathcal{G}) \otimes U(\mathcal{G}^{\mathrm{op}}) \simeq U(\mathcal{G} \oplus \mathcal{G}^{\mathrm{op}}).$$

Define the following subspaces of $\mathcal{G} \oplus \mathcal{G}^{\mathrm{op}}$

$$\mathcal{G}' = \{X' = (X, -X) \mid X \in \mathcal{G}\}$$

$$\mathcal{G}'' = \{X'' = (X, 0) \mid X \in \mathcal{G}\},$$

then $[X', Y'] = [X,Y]'$; $[X', Y''] = [X,Y]''$, and $[X'', Y''] = [X,Y]''$ so $\mathcal{G}'$ and $\mathcal{G}''$ are subalgebras isomorphic to $\mathcal{G}$ and $\mathcal{G}''$ is a $\mathcal{G}'$ module isomorphic to the adjoint representation.

For any choice of ordered basis $\{Y_1, \ldots Y_m\}$ we can define the corresponding ordered basis of $\mathcal{G} \oplus \mathcal{G}^{\mathrm{op}}$, $\{Y_1'', \ldots Y_m'', Y_1' \ldots Y_m'\}$ and the Poincaré-Birkhoff-Witt (PBW) theorem gives an isomorphism of left $U(\mathcal{G}'')$, right $U(\mathcal{G}')$ modules

$$U(\mathcal{G} \oplus \mathcal{G}^{\mathrm{op}}) \simeq U(\mathcal{G}'')U(\mathcal{G}').$$

This isomorphism, is, in fact, independent of the basis.

The left $U(\mathcal{G})$ comodule structure on $\beta(\bigwedge \mathcal{G})$ is induced by the left comodule structure on $U(\mathcal{G}) \otimes U(\mathcal{G})$ as the tensor product of comodules

$$\lambda(u \otimes v \otimes X_1 \wedge \cdots \wedge X_p) = \sum u_{(1)}v_{(1)} \otimes u_{(2)} \otimes v_{(2)} \otimes X_1 \wedge \cdots \wedge X_p$$

since we are giving $\bigwedge \mathcal{G}$ the trivial bicomodule structure.

We now turn to the definition of the boundary operator.The differential ∂_b on the complex K_b is defined to be

$$\begin{aligned}\partial_b(w \otimes X_1 \wedge \ldots \wedge X_p) &= \sum(-1)^{i-1} w X_i' \otimes X_1 \wedge \ldots \wedge \hat{X}_i \wedge \ldots \wedge X_p \\ &\quad + \sum(-1)^{i+j} w \otimes [X_i, X_j] \wedge \ldots \hat{X}_i \ldots \hat{X}_j \wedge X_p\end{aligned}$$

which is clearly a left $U(\mathcal{G}'')U(\mathcal{G}')$ module map.

A straightforward calculation shows that the natural inclusion of the Koszul complex in Bar($U(\mathcal{G})$ is a chain map. Similarly, we want to define a $U(\mathcal{G})$-4 module structure on $\gamma(\wedge\mathcal{G})$ so that its inclusion in Cob($U(\mathcal{G})$) is a tetramodule chain map. The somewhat contorted definition given below achieves that purpose.

We first define a $U(\mathcal{G} \oplus \mathcal{G}^{\mathrm{op}})$ module structure on $\bigwedge\mathcal{G}$ by representing $U(\mathcal{G}'')$ via the augmentation and $U(\mathcal{G}')$ by the adjoint action,

$$\begin{aligned}\mu(X')(X_1 \wedge \ldots \wedge X_p) &= \mathrm{ad}X'(X_1 \wedge \ldots \wedge X_p) \\ &= \sum_i X_1 \wedge \ldots \wedge [X, X_i] \wedge \ldots \wedge X_p\end{aligned}$$

$$\begin{aligned}\mu(X'')(X_1 \wedge \ldots \wedge X_p) &= \epsilon(X'')(X_1 \wedge \ldots \wedge X_p) \\ &= 0.\end{aligned}$$

Then

$$\begin{aligned}\mu([X', Y'']) &= \mu([X, Y]'') = 0 \\ &= \mu(X')\mu(Y'') - \mu(Y'')\mu(X')\end{aligned}$$

so μ is well defined on $U(\mathcal{G} \oplus \mathcal{G}^{\mathrm{op}})$. We let $\gamma(\bigwedge\mathcal{G})$ have the B-bimodule structure given by the tensor product of modules as discussed in Section 10.2.

By definition, bimodule action of $U(\mathcal{G})$ on a tensor product is via the application of Δ. So the bimodule action of X' on $U(\mathcal{G}) \otimes U(\mathcal{G})$ is the difference between right and left multiplication by $\Delta(X)$. The action of X' on $w = a \otimes b \in U(\mathcal{G}) \otimes U(\mathcal{G})$ is given by

$$\begin{aligned}X'(a \otimes b) &= (X \otimes 1 + 1 \otimes X)(a \otimes b) - (a \otimes b)(X \otimes 1 + 1 \otimes X) \\ &= (Xa - aX) \otimes b + a \otimes (Xb - bX) \\ &= \mathrm{ad}X(a) \otimes b + a \otimes \mathrm{ad}X(b) \\ &= \mathrm{ad}X(a \otimes b)\end{aligned}$$

In other words, $\mathcal{G}'$ acts via the (tensor product of two copies of the) adjoint representation of $\mathcal{G}$, and this then extends to $U(\mathcal{G}')$ acting on $U(\mathcal{G})\otimes U(\mathcal{G})$. The action of X'' is via left multiplication by $\Delta(X)$. Hence the module action of $u''v' \in U(\mathcal{G}'')U(\mathcal{G}')$ on $w \in U(\mathcal{G})\otimes U(\mathcal{G})$ is given by

$$(u''v')w = \Delta(u)\mathrm{ad}(v)w. \tag{10.6.1}$$

The bimodule action on $\gamma(\wedge\mathcal{G})$ is obtained from this formula by applying Δ once more:

$$\mu(u''v')(w\otimes X_1\wedge\ldots X_p) = \sum \Delta(u_{(1)})\mathrm{ad}v_{(1)}(w)\otimes\epsilon(u_{(2)})\mathrm{ad}v_{(2)}(X_1\wedge\ldots\wedge X_p).$$

But

$$\sum u_{(1)}\otimes\epsilon(u_{(2)}) = (id\otimes\epsilon)\Delta(u) = u$$

so

$$\mu(u''v')(w\otimes X_1\wedge\ldots X_p) = \sum \Delta(u)\mathrm{ad}v_{(1)}(w)\otimes\mathrm{ad}v_{(2)}(X_1\wedge\ldots\wedge X_p) \tag{10.6.2}$$

The bicomodule structure is

$$\begin{aligned}\lambda &= \Delta\otimes id\otimes id\\ \rho &= s_{34}(id\otimes\Delta\otimes id)\end{aligned}$$

acting on $U(\mathcal{G})\otimes U(\mathcal{G})\otimes\wedge\mathcal{G}$.

To express the differential δ_c on K_c we use the PBW basis for $U(\mathcal{G})\otimes U(\mathcal{G})$ corresponding to the basis $\overline{Y}_i = (Y_i, Y_i)$ and Y_i'. In multi-index notation $a = (a_1,\ldots a_n)$, $\overline{Y}^a = \overline{Y}_1^{a_1}\cdots\overline{Y}_n^{a_n}$, $Y'^a = Y_1'^{a_1}\ldots Y_n'^{a_n}$, $\delta_i = (0,\ldots,1,\ldots 0)$ with 1 in the i-th position.

Let $\omega\in\wedge\mathcal{G}$ then

$$\delta_c(\overline{Y}^a Y'^b\otimes\omega) = \sum b_i\overline{Y}^a Y'^{b-\delta_i}\otimes Y_i\wedge\omega. \tag{10.6.3}$$

The $\overline{Y}_i$ do not span a subalgebra of $\mathcal{G}\oplus\mathcal{G}^{\mathrm{op}}$ so we can not simply write $U(\mathcal{G})\otimes U(\mathcal{G}^{\mathrm{op}} = U(\overline{\mathcal{G}})U(\mathcal{G}')$ as an algebra decomposition. But as a complex, K_c does not really depend on the Lie algebra structure. In fact, δ_c is defined relative to the underlying vector space on the tensor product of the symmetric algebra of $\mathcal{G}$ with the exterior algebra of $\mathcal{G}$ when we identify the monomials Y'^a with the corresponding monomial basis of $S(\mathcal{G})$. We do have the vector space decomposition

$$\mathcal{G}\oplus\mathcal{G}^{\mathrm{op}} = \overline{\mathcal{G}}\oplus\mathcal{G}'$$

which gives the decomposition

$$\mathrm{Sym}(\mathcal{G} \oplus \mathcal{G}^{\mathrm{op}}) = \mathrm{Sym}(\overline{\mathcal{G}}) \otimes \mathrm{Sym}(\mathcal{G}')$$

of the symmetric algebra.

In fact, for an vector space V, Koszul has defined

$$\delta : S^q(V) \otimes \wedge^q V \to S^{p-1}(V) \otimes \wedge^{q+1}(V)$$

independent of the choice of basis for V, the choice of basis for $\mathcal{G}$ merely allows us to identify $U(\mathcal{G})$ and $S(\mathcal{G})$. Also, once we have chosen a basis of the vector space, V, we may identify V with V^* and hence $S^q(V)$ with the homogeneous polynomials of degree q. Then the formula for δ_c in (10.6.3) (when we ignore the $\overline{Y}$) is just the usual exterior differential.

It is immediate that $\delta_c^2 = 0$ and is standard, or an easy verification to check that $\partial_b^2 = 0$. We must show that these complexes are acyclic and that ∂_b and δ_c are $U(\mathcal{G})$-4 morphisms.

10.7 B-4 properties of ∂_b and δ_c.

We now consider the $U(\mathcal{G})$-4 module properties of the differentials. Since ∂_b is a left $U(\mathcal{G}) \otimes U(\mathcal{G})^{\mathrm{op}}$ module extension of a map defined on $\wedge\mathcal{G}$, it is a bimodule map. Furthermore, since $\beta(\wedge\mathcal{G})$ is a free $U(\mathcal{G}) \otimes U(\mathcal{G})^{\mathrm{op}}$ module generated by $\bigwedge \mathcal{G}$ and both ∂_b and the left and right comodule structures are bimodule maps, it is enough to check the bicomodule map conditions on $\wedge\mathcal{G}$. That is, we must check whether the equation

$$\lambda\partial_b(1 \otimes 1 \otimes X_1 \wedge \cdots \wedge X_p) = (id \otimes \partial_b)\lambda(1 \otimes 1 \otimes X_1 \wedge \cdots \wedge X_p)$$

holds. We have

$$\lambda(1 \otimes 1 \otimes X_1 \wedge \cdots \wedge X_p) = 1 \otimes 1 \otimes 1 \otimes X_1 \wedge \cdots \wedge X_p$$

and hence

$$(id \otimes \partial_b)\lambda(1 \otimes 1 \otimes X_1 \wedge \cdots \wedge X_p) = 1 \otimes (\sum (-1)^{i-1} X_i' \otimes X_1 \wedge \cdots \wedge \hat{X}_i \wedge \cdots \wedge X_p$$

$$+ \sum (-1)^{i+j} w \otimes [X_i, X_j] \wedge \ldots \wedge \ldots \hat{X}_i \ldots \hat{X}_j \wedge X_p)$$

The only nontrivial part of computing $\lambda\partial_b(1 \otimes 1 \otimes X_1 \wedge \cdots \wedge X_p)$ involves computing the $\lambda X_i'$ which occur in the expansion of $\partial_b(1\otimes 1\otimes X_1 \wedge\cdots\wedge X_p)$. A simple calculation (see the last section) gives

$$\lambda(X') = 1 \otimes X'.$$

Thus applying λ to the X' terms in $\partial_b(1 \otimes 1 \otimes X_1 \wedge \ldots \wedge X_p)$ we find

$$\begin{aligned}\lambda\partial_b(1 \otimes 1 \otimes X_1 \wedge \ldots \wedge X_)p &= 1 \otimes \partial_b(1 \otimes 1 \otimes X_1 \wedge \ldots \wedge X_p) \\ &= (id \otimes \partial_b)\lambda(1 \otimes 1 \otimes X_1 \wedge \ldots \wedge X_p)\end{aligned}$$

and ∂_b is a left comodule map. Similarly for the right comodule structure.

Let us now show that δ_c is a bimodule map. According to (10.6.2), an element $Y'' \in \mathcal{G}$ acts by left multiplication by $\Delta(Y)$. Under the identification of $U(\mathcal{G}) \otimes U(\mathcal{G})^{\mathrm{op}}$ with $U(\mathcal{G} \oplus \mathcal{G}^{\mathrm{op}})$, $\Delta(Y)$ corresponds to $\overline{Y}$. By(10.6.3)

$$\delta_c(Y''^a \cdot (Y'^b \otimes \omega)) = \delta_c(\overline{Y}^a Y'^b \otimes \omega) = \overline{Y}^a \delta(Y'^b \otimes \omega) Y''^a \cdot \delta_c(Y'^b \otimes \omega).$$

So the action of $\mathcal{G}''$ commutes with δ_c. By (10.6.2), $\mathcal{G}'$ acts via the adjoint representation which commutes with δ_c. Hence δ_c is a bimodule map.

We now turn to the comodule conditions. The left and right comodule structures on $\gamma(\wedge\mathcal{G})$ are given by

$$\begin{aligned}\lambda_c(u \otimes v \otimes X_1 \wedge \cdots \wedge X_p) &= \sum u_{(1)} \otimes u_{(2)} \otimes v \otimes X_1 \wedge \cdots \wedge X_p \\ \rho_c(u \otimes v \otimes X_1 \wedge \cdots \wedge X_p) &= \sum u \otimes v_{(1)} \otimes X_1 \wedge \cdots \wedge X_p \otimes v_{(2)}.\end{aligned}$$

Also

$$\lambda_c(X') = \lambda_c(X\otimes 1 - 1\otimes X) = X\otimes 1\otimes 1 + 1\otimes(X\otimes 1 - 1\otimes X) = X\otimes 1\otimes 1 + 1\otimes X'.$$

Therefore

$$\lambda_c(Y'^a \otimes \omega) = \sum_b \begin{pmatrix} a \\ b \end{pmatrix} Y^{a-b} \otimes Y'^b \otimes \omega$$

where

$$\begin{pmatrix} a \\ b \end{pmatrix} = \begin{pmatrix} a_1 \\ b_1 \end{pmatrix} \cdots \begin{pmatrix} a_n \\ b_n \end{pmatrix}.$$

Then we have

$$(id \otimes \delta_c)\lambda(Y'^a \otimes \omega) = \sum_b \begin{pmatrix} a \\ b \end{pmatrix} Y^{a-b} \otimes \delta_c(Y'^b \otimes \omega)$$

$$= \sum \binom{a}{b} b_i Y^{a-b} \otimes Y'^{b-\delta_i} \otimes Y_i \wedge \omega$$

$$\lambda\delta_c(Y'^a \otimes \omega) = \lambda \sum a_i Y_i'^{a-\delta_i} \otimes Y \wedge \omega$$

$$= \sum a_i \binom{a-\delta_i}{b} Y^{a-b-\delta_i} \otimes Y'^b \otimes Y_i \wedge \omega$$

The relation

$$a_i \binom{a_i - 1}{b_i} = (b_i + 1)\binom{a_i}{b_i + 1}$$

then proves

$$(id \otimes \delta_c)\lambda = \lambda\delta_c$$

in this case. Since δ_c, λ_c and ρ_c are all $U(\overline{\mathcal{G}})$ module maps, the formula is proved in general. Similarly for the right comodule structure. Thus both differentials are $U(\mathcal{G})$ 4-module maps.

There is a natural imbedding of $\wedge^p(\mathcal{G})$ in $\otimes^p U(\mathcal{G})$

$$X_1 \wedge X_2 \wedge \cdots \wedge X_p \mapsto \sum (\text{sign}\sigma) X_{\sigma(1)} \otimes X_{\sigma(2)} \otimes \cdots \otimes X_{\sigma(p)}.$$

Applying the functor β gives an imbedding of K_b^q in Bar^q and similarly, applying the functor γ defines an imbedding of K_c in Cob. One verifies by a straightforward but tedious calculation that these maps are chain maps.

10.8 Acyclicity of the Koszul complexes.

In order to prove that the Koszul complexes are in fact resolutions of $U(\mathcal{G})$, (i.e., acyclic, we consider the subcomplexes $\tilde{K}_b \subset K_b$ and $\tilde{K}_c \subset K_c$ both of which have the same underlying vector space $U(\mathcal{G}') \otimes \bigwedge \mathcal{G}$ but differentials ∂_b and δ_c respectively. We shall show that these are acyclic complexes augmented over the base field k . The full complexes are freely generated as left $U(\mathcal{G})$ modules by tensoring these subcomplexes with $U(\mathcal{G})$. This will give acyclic resolutions of $U(\mathcal{G})$.

First consider the case when $\mathcal{G}$ is abelian, then $U(\mathcal{G}')$ can be identified with the symmetric algebra $S(\mathcal{G})$ and as such is graded by homogeneous degree. Let $\{X_1, \ldots X_n\}$ be a basis for $\mathcal{G}$ then the operators ∂_b and δ_c on $S(\mathcal{G}) \otimes \bigwedge \mathcal{G}$ are given by

$$\partial_b(u \otimes X_{i_1} \wedge \ldots \wedge X_{i_p}) = \sum (-1)^{j-1} u X_{i_j} \otimes X_{i_1} \wedge \ldots \wedge \hat{X}_{i_j} \wedge \ldots \wedge X_{i_p} \quad (10.8.1)$$

and

$$\delta_c(u \otimes X_{i_1} \wedge \ldots \wedge X_{i_p}) = \sum \frac{\partial u}{\partial X_i} \otimes X_i \wedge X_{i_1} \wedge \ldots \wedge X_{i_p} \qquad (10.8.2)$$

For u homogeneous of degree q a simple calculation proves

$$(\partial_b \delta_c + \delta_c \partial_b)(u \otimes X_{i_1} \wedge \ldots \wedge X_{i_p}) = (p+q)(u \otimes X_{i_1} \wedge \ldots \wedge X_{i_p}). \quad (10.8.3)$$

This proves the acyclicity of $\tilde{K}_b$ and $\tilde{K}_c$ since δ_c suitably normalized provides a homotopy operator for ∂_b and conversely. The complex $\tilde{K}_c$ is independent of the Lie bracket on $\mathcal{G}$ and formula (29) holds when the PBW basis for $U(\mathcal{G}')$ is identified with the corresponding monomial basis for $S(\mathcal{G})$. Thus $\tilde{K}_c$ is acyclic in general.

To prove that $\tilde{K}_b$ is acyclic when $\mathcal{G}$ is not necessarily abelian we argue inductively on the filtration degree on $U(\mathcal{G})$. Let $U^k(\mathcal{G})$ be the subspace of $U(\mathcal{G})$ generated by the elements which products of at most k elements of $\mathcal{G}$. Define a subcomplex of $\tilde{K}_b$

$$(\tilde{K}_b)_k = \bigoplus_{0 \leq \ell \leq k} U^{k-\ell}(\mathcal{G}) \otimes \wedge^\ell \mathcal{G}$$

Note that

$$(\tilde{K}_b)_k / (\tilde{K}_b)_{k-1} \simeq \bigoplus S^{k-\ell}(\mathcal{G}) \otimes \wedge^\ell \mathcal{G}$$

since

$$\bigoplus U^k(\mathcal{G})/U^{k-1}(\mathcal{G}) \simeq \bigoplus S^k(\mathcal{G})$$

as graded algebras. Now $(\tilde{K}_b)_0$ as a complex augmented over k is just $0 \to k \xrightarrow{id} k \to 0$ which is clearly acyclic. Assume $(\tilde{K}^b)_j$ is acyclic for $j = k-1$ then consider the short exact sequence of complexes

$$0 \to (\tilde{K}_b)_{k-1} \to (\tilde{K}_b)_k \to (\tilde{K}_b)_k/(\tilde{K}_b)_{k-1} \to 0.$$

The two complexes at the ends are acyclic, the left one by the induction assumption and the right one by the result for abelian $\mathcal{G}$, thus the long exact sequence for homology proves that the middle one is acyclic, proving the induction.

Putting together all the results that have been established regarding K_b and K_c gives

Proposition 10.8.1 *The Koszul complex K_b is a projective $U(\mathcal{G})$-bimodule resolution of $U(\mathcal{G})$. In addition it is a complex of a $U(\mathcal{G})_4$ modules.*

The Koszul complex K_c is an injective $U(\mathcal{G})$-bicomodule resolution of $U(\mathcal{G})$. In addition it is a complex of $U(\mathcal{G}-)$ 4 modules.

We have just shown that the complex $\tilde{K}_b = U(\mathcal{G}') \otimes \wedge\mathcal{G}$ is acyclic, therefore, asa complex of vector spaces it is split exact. In the last section we will show that as a B-bicomodule K_b is isomorphic to $U(\mathcal{G}) \otimes \tilde{K}_c$ where the bicomodule structure on $\tilde{K}_c$ is trivial. (This is a general property of dimodules over Hopf algebras.) But for a trivial bicomodule structure, any decomposition as a direct sum of vector spaces is a direct sum decomposition as bicomodule. Since the tensor product with $U(\mathcal{G})$ preserves direct sums the complex K_b is also split exact as a sequence of bicomodules.

Regarding K_c we have seen that the contracting homotopy is essentially the operator $\partial_{b,\mathcal{G}-abelian}$, the coboundary operator of K_b for $\mathcal{G}$ abelianized. This operator is clearly invariant with respect to the adjoint representation of $\mathcal{G}$ with its usual structure. Thus the contracting homotopy for K_c is a $U(\mathcal{G})$ bimodule map, and K_c is split exact as a complex of $U(\mathcal{G})$ bimodules. This proves

Proposition 10.8.2 *The complex K_b is a relative $U(\mathcal{G}) - 4$ projective resolution. The complex K_c is a relative $U(\mathcal{G}) - 4$ injective resolution.*

As an immediate consequence Proposition 10.5.3 we have

Proposition 10.8.3 *Let $\mathcal{G}$ be a Lie algebra over a field of characteristic 0. The complexes* Bar$(U(\mathcal{G})$ *and K_b are homotopy equivalent as $U(\mathcal{G}) - 4$ complexes. The complexes* Cob$(U(\mathcal{G})$ *and K_c are homotopy equivalent as $U(\mathcal{G}) - 4$ complexes.*

10.9 The main reduction.

With this background we can state the main theorem of this chapter.

Theorem 10.9.1 *For any Lie algebra, $\mathcal{G}$, the Gerstenhaber-Schack complex of $U(\mathcal{G})$ is homotopy equivalent to the Koszul double complex*

$$C_K = \mathrm{Hom}_{U(\mathcal{G})\text{-}4}(K_b, K_c)$$

with coboundary operator, for $\alpha \in C^{p,q}$

$$d\alpha = (-1)^q \delta_c \circ \alpha + \alpha \circ \partial_b. \tag{10.9.1}$$

Furthermore, there is an isomorphism of complexes

$$\mathrm{Hom}_{U(\mathcal{G})\text{-}4}(K_b, K_c) \cong \mathrm{Hom}_k(\bigwedge \mathcal{G}, \bigwedge \mathcal{G})$$

where the complex on the right is the Chevalley-Eilenberg complex for the cohomology of the Lie algebra $\mathcal{G}$ with values in the adjoint representation extended to the exterior algebra. Thus the Gerstenhaber-Schack cohomology in bidegree p, q is

$$H^q(\mathcal{G}, \wedge^p \mathcal{G})$$

Proof. As a consequence of the last proposition and Lemma 10.4.1 we have

$$\mathrm{Hom}_{U(\mathcal{G})-4}(\mathrm{Bar}, \mathrm{Cob}) \simeq \mathrm{Hom}_{U(\mathcal{G})-4}(\mathrm{Bar}, K_c) \simeq \mathrm{Hom}_{U(\mathcal{G})-4}(K_b, K_c)$$

The following calculation shows that using the adjoint functor relation

$$\mathrm{Hom}_{U(\mathcal{G})-4}(\mathrm{Bar}, K_c) = \mathrm{Hom}_{U(\mathcal{G})-4}(\mathrm{Bar}, \gamma(\bigwedge \mathcal{G})) = \mathrm{Hom}_{U(\mathcal{G})\text{-bimod}}(\mathrm{Bar}, \bigwedge \mathcal{G})$$

to reduce the second complex the differential δ_c becomes 0, and the double complex in fact reduces to a simple complex.

Let $\alpha \in \mathrm{Hom}_{U(\mathcal{G})-4}(\mathrm{Bar}, K_c) = \mathrm{Hom}_{U(\mathcal{G})-4}(\mathrm{Bar}, \gamma(\wedge \mathcal{G}))$. Then

$$\begin{aligned}
&(\delta_c \circ \hat{\alpha})^{\sim}(u_0 \otimes u_1 \otimes \cdots \otimes u_{q+1}) = \\
&= (\epsilon \otimes \epsilon \otimes id)\delta_c(u_{0(1)} \cdots u_{q+1(1)} \otimes u_{0(2)} \cdots u_{q+1(2)} \otimes \alpha(u_{0(3)} \otimes \cdots \otimes u_{q+1(3)}) \\
&= (\epsilon \otimes \epsilon \otimes id)\delta_c(\Delta(u_{0(1)} \cdots u_{q+1(1)})(1 \otimes 1 \otimes \alpha(u_{0(2)} \otimes \cdots \otimes u_{q+1(2)})) \\
&= (\epsilon \otimes \epsilon \otimes id)(\Delta(u_{0(1)} \cdots u_{q+1(1)})\delta_c(1 \otimes 1 \otimes \alpha(u_{0(2)} \otimes \cdots \otimes u_{q+1(2)})) \\
&= 0.
\end{aligned}$$

Thus at this stage we have already reduced to a direct sum of simple complexes, the sum over p of the Hochschild complexes for the cohomology of $U(\mathcal{G})$ with coefficients in $\wedge^p \mathcal{G}$. The cohomology in bidegree p, q is $H^q(\mathcal{G}, \wedge^p \mathcal{G})$ as asserted in the statement of the theorem.

Regarding the identification of the complex $\mathrm{Hom}_{U(\mathcal{G})-4}(K_b, K_c)$ with the Chevalley-Eilenberg cohomology, we need to calculate the operators defined using the adjoint relations connecting $\mathrm{Hom}_{U(\mathcal{G})\text{-}4}(K_b, K_c)$ and $\mathrm{Hom}_k(\bigwedge \mathcal{G}, \bigwedge \mathcal{G})$ (10.3.13) and (10.3.14).

$$\hat{\alpha}(u''v' \otimes X_1 \wedge \ldots \wedge X_q) = \Delta(u) \otimes \mathrm{ad}v(\alpha(X_1 \wedge \ldots \wedge X_q))$$

and

$$\tilde{\beta}(X_1 \wedge \ldots \wedge Y_q) = (\epsilon \otimes \epsilon \otimes id)\beta(1 \otimes 1 \otimes X_1 \wedge \ldots \wedge X_q).$$

Then

$$(\hat{\alpha} \circ \partial_b)^{\sim}(X_1 \wedge \ldots \wedge X_{q+1}) =$$

$$\begin{aligned}
&= (\epsilon \otimes \epsilon \otimes id)\hat{\alpha} \circ \partial_b(1 \otimes 1 \otimes X_1 \wedge \ldots \wedge X_{q+1}) \\
&= (\epsilon \otimes \epsilon \otimes id)\hat{\alpha}(\sum (-1)^i X_i' \otimes X_1 \wedge \ldots \wedge \hat{X}_i \wedge \ldots \wedge X_{q+1} \\
&\quad + \sum (-1)^{i+j} 1 \otimes 1 \otimes [X_i, X_j] \wedge \ldots \wedge \hat{X}_i \ldots \hat{X}_j \wedge \ldots \wedge X_{q+1}) \\
&= \sum (-1)^i \mathrm{ad} X_i \alpha(X_1 \wedge \ldots \wedge \hat{X}_i \wedge \ldots \wedge X_{q+1}) \\
&\quad + \sum (-1)^{i+j} \alpha([X_i, X_j] \wedge \ldots \wedge \hat{X}_i \wedge \ldots \wedge \hat{X}_j \ldots X_{q+1}) \\
&= d_{C.E.}\alpha(X_1 \wedge \ldots \wedge X_{q+1}).
\end{aligned}$$

This completes the proof of the theorem.

For the bialgebra deformation cohomology we use the subcomplex

$$\hat{C} = \bigoplus_{p>0, q>0} C^{p,q}.$$

The quotient complex is the sum of two subcomplexes, the left column and the bottom row.

$$C_1 = \bigoplus_{q>0} C_{GS}^{0,q} = \bigoplus_{q>0} \mathrm{Hom}_{U(\mathcal{G})\text{-bimod}}(\mathrm{Bar}_q, k)$$

and

$$C_2 = \bigoplus_{p \geq 0} C_{GS}^{p,0} = \bigoplus_{p \geq 0} \mathrm{Hom}_{U(\mathcal{G})\text{-bicomod}}(k, Cob^p).$$

The cohomology of C_1 is the Hochschild cohomology of $U(\mathcal{G})$ with values in k. The homotopy equivalence of Bar and K_b proves that this is isomorphic to the Chevalley-Eilenberg cohomology of $\mathcal{G}$ with values in k. The homotopy equivalence between Cob and K_c proves that the cohomology of C_2 is the cohomology of $\mathrm{Hom}_k(k, \bigwedge \mathcal{G})$ but we have just seen that the differential on this complex is 0, so the cohomology of C_2 is $\bigwedge \mathcal{G}$. With these preliminaries, the proof of the following proposition is almost immediate.

Proposition 10.9.1 *The bialgebra deformation cohomology of $U(\mathcal{G})$ in bidegree p,q is*

$$\begin{aligned}
\hat{H}^{p,q} &= H^q(\mathcal{G}, \wedge^p \mathcal{G}) \quad \textit{for} \quad q > 1 \\
\hat{H}^{p,1} &= Z^1(\mathcal{G}, \wedge^p \mathcal{G})
\end{aligned}$$

Proof. The short exact sequence of complexes

$$0 \to \hat{C} \to C_{GS} \to C_1 \oplus C_2 \to 0$$

gives the long exact sequence in cohomology

$$H^{n-1}(C_1 \oplus C_2) \to \hat{H}^n \to H^n_{GS} \to H^n(C_1 \oplus C_2)$$

that is

$$\wedge^{n-1}\mathcal{G} \oplus H^{n-1}(\mathcal{G}, k) \to \bigoplus_{\substack{n>p>0 \\ p+q=n}} \hat{H}^{p,q} \to \bigoplus_{p+q=n} H^q(\mathcal{G}, \wedge^p\mathcal{G}) \to \wedge^n\mathcal{G} \oplus H^n(\mathcal{G}, k)$$

The map on the right is the identity on $H^n(\mathcal{G}, \wedge^0\mathcal{G}) = H^n(\mathcal{G}, k)$ and is the inclusion of $H^0(\mathcal{G}, \wedge^n\mathcal{G}) = (\wedge^n\mathcal{G})^{\mathcal{G}}$ into $\wedge^n\mathcal{G}$, and zero on the remaining components. Thus we have the short exact sequence

$$0 \to \wedge^{n-1}\mathcal{G}/(\wedge^{n-1}\mathcal{G})^{\mathcal{G}} \to \bigoplus_{n>p>0, p+q=n} \hat{H}^{p,q} \to \bigoplus_{n>p>0, p+q=n} H^q(\mathcal{G}, \wedge^p\mathcal{G}) \to 0.$$

The map on the left is induced by the Hochschild differential. For $q > 1$ this gives the isomorphism in the statement of the proposition, and for $q = 1$ it gives the sequence

$$0 \to \wedge^{n-1}\mathcal{G}/(\wedge^{n-1}\mathcal{G})^{\mathcal{G}} \to \hat{H}^{n-1,1} \to H^1(\mathcal{G}, \wedge^{n-1}\mathcal{G}) \to 0.$$

Comparison of this sequence with the sequence

$$0 \to \wedge^{n-1}\mathcal{G}/(\wedge^{n-1}\mathcal{G})^{\mathcal{G}} \to Z^1(\mathcal{G}, \wedge^{n-1}\mathcal{G}) \to H^1(\mathcal{G}, \wedge^{n-1}\mathcal{G}) \to 0$$

proves the last statement of the proposition.

For Drinfeld algebras deformations of $U(\mathcal{G})$ the appropriate complex is

$$\tilde{C} = \bigoplus_{p>0} C^{p,q}_{GS}$$

and the quotient complex is C_1. In this case the long exact sequence gives immediately

Proposition 10.9.2 *The Drinfeld deformation cohomology for $U(\mathcal{G})$ in bidegree p, q is*

$$\tilde{H}^{p,q} = H^q(\mathcal{G}, \wedge^p\mathcal{G})$$

10.10 Hopf modules

Recall that a Hopf algebra B is an associative and coassociative bialgebra with an antipode $S : B \to B$ satisfying

$$m \circ (S \otimes id) \circ \Delta = \epsilon \tag{10.10.1}$$
$$m \circ (id \otimes S) \circ \Delta = \epsilon, \tag{10.10.2}$$

where m is multiplication in B. In this section we describe the structure of Hopf modules, as defined in section 2. First we need another definition

Definition A *coinvariant* of a comodule M over a coalgebra C is an element $n \in M$ such that

$$\lambda(n) = 1 \otimes n. \tag{10.10.3}$$

The space of coinvariants of a comodule M will be denoted by $M^{\#}$.

Proposition 10.10.1 *Let B be a Hopf algebra and M a left dimodule over B. Then*

$$M \cong B \otimes_k M^{\#} \tag{10.10.4}$$

where $B \otimes M^{\#}$ has the left dimodule structure

$$\begin{aligned} b(a \otimes n) &= ba \otimes n \\ \lambda'(a \otimes n) &= \Delta(a) \otimes n \\ &= \sum a_{(1)} \otimes a_{(2)} \otimes n. \end{aligned}$$

Proof. Define

$$\chi : B \otimes M^{\#} \to M$$

by

$$\chi(a \otimes n) = an.$$

It is clearly a left module morphism. We claim that it is also a left comodule morphism. Indeed,

$$\begin{aligned} \lambda(\chi(a \otimes n)) &= \lambda(an) \\ &= \Delta(a)\lambda(n) \\ &= (\sum a_{(1)} \otimes a_{(2)}) \cdot (1 \otimes n) \\ &= \sum a_{(1)} \otimes a_{(2)}n \\ &= (id \otimes \chi) \sum a_{(1)} \otimes a_{(2)} \otimes n \\ &= (id \otimes \chi)\lambda'(a \otimes n), \end{aligned}$$

proving that χ is a left comodule morphism. To prove the proposition, we wish to show that χ is an isomorphism by constructing its inverse. For this we begin by constructing a map from M to $M^{\#}$. Here is where we use the antipode:

Lemma 10.10.1 *For any* $m \in M$ *the element*

$$\mu(S \otimes id)(\lambda(n)) = \sum S(\lambda_{(1)}(n))\lambda_{(2)}(n)$$

is an element of $M^{\#}$.

We have

$$\begin{aligned} \lambda \circ (\mu \circ (S \otimes id) \circ \lambda) &= \text{by compatibility of } \lambda \text{ and } \mu \\ &= (m \otimes \mu) \circ s_{23} \circ (\Delta \otimes \lambda) \circ (S \otimes id) \circ \lambda \\ &= (m \otimes \mu) \circ s_{23} \circ (\Delta \circ S \otimes id \otimes id) \circ (id \otimes \lambda) \circ \lambda \end{aligned}$$

Now

$$\Delta \circ S = s_{12}(S \otimes S) \circ \Delta$$

is a general property of antipodes. Substituting this and $(\Delta \otimes id)\lambda = (id \otimes \lambda)\lambda$ into the last expression gives

$$\begin{aligned} &(m \otimes \mu) \circ s_{23} \circ s_{12} \circ (S \otimes S \otimes id \otimes id) \circ (\Delta \otimes id \otimes id) \circ (\Delta \otimes id) \circ \lambda \\ = \; &(m \otimes \mu) \circ s_{132} \circ (S \otimes S \otimes id \otimes id) \circ (id \otimes \Delta \otimes id) \circ (\Delta \otimes id) \circ \lambda \end{aligned}$$

$$= (id \otimes \mu) \circ [(m \otimes id \otimes id) \circ s_{132} \circ (S \otimes S \otimes id \otimes id) \circ (id \otimes \Delta \otimes id)] \circ (\Delta \otimes id) \circ \lambda.$$

Let us examine the portion in brackets in this last expression. It only acts on $B \otimes B$. That is, it is of the form

$$Q \otimes id : B \otimes B \otimes M \to B \otimes B \otimes M$$

where

$$Q = (m \otimes id) \circ s_{132} \circ (S \otimes S \otimes id) \circ (id \otimes \Delta) : \; B \otimes B \to B \otimes B.$$

Applying the constituent maps in Q successively to $a \otimes b \in B \otimes B$ we have

$$\begin{aligned} a \otimes b &\mapsto \sum a \otimes b_{(1)} \otimes b_{(2)} \\ &\mapsto \sum S(a) \otimes S(b_{(1)}) \otimes b_{(2)} \\ &\mapsto \sum S(b_{(1)})b_{(2)} \otimes S(a) \\ &\mapsto \epsilon(b) \otimes S(a), \end{aligned}$$

the last step coming from (10.10.2). If we write

$$(\Delta \otimes id)(\lambda(n)) = \sum \lambda_{((1))} \otimes \lambda_{((2))}(n) \otimes \lambda_{((3))}(n)$$

and substitute into the full expression for $\lambda \circ \mu \circ (S \otimes id) \circ \lambda$ we get

$$\begin{aligned}
[\lambda \circ \mu \circ (S \otimes id) \circ \lambda](n) &= \sum \epsilon(\lambda_{((2))}(n)) \otimes S(\lambda_{((1))}(n)) \otimes \lambda_{((3))}(n) \\
&= 1 \otimes \sum S(\epsilon(\lambda_{((2))}(n))\lambda_{((1))}(n))\lambda_{((3))}(n) \\
&= 1 \otimes \mu \circ (S \otimes id) \circ (\epsilon \otimes id \otimes id) \circ (\Delta \otimes id) \circ \lambda \\
&= 1 \otimes [\mu \circ (S \otimes id) \circ \lambda](n).
\end{aligned}$$

Thus the image of $\mu \circ (S \otimes id) \circ \lambda$ lies in $M^{\#}$ which is what we want to prove.

We now complete the proof of the proposition by showing that

$$(id \otimes \mu) \circ (id \otimes S \otimes id) \circ (\Delta \otimes id) \circ \lambda : M \to B \otimes M^{\#}$$

is a two sided inverse to χ.

Right inverse:

$$\begin{aligned}
&\chi \circ (id \otimes \mu) \circ (id \otimes S \otimes id) \circ (\Delta \otimes id) \circ \lambda \\
&= \mu \circ (id \otimes \mu) \circ ((id \otimes S) \circ \Delta \otimes id) \circ \lambda \\
&= \mu \circ (m \otimes id) \circ ((id \otimes S) \circ \Delta \otimes id) \circ \lambda \\
&= \mu \circ ((m \circ (id \otimes S) \circ \Delta) \otimes id) \circ \lambda \\
&= \mu \circ (\epsilon \otimes id) \circ \lambda \\
&= id.
\end{aligned}$$

left inverse:

$$\begin{aligned}
&(id \otimes \mu) \circ (id \otimes S \otimes id) \circ (\Delta \otimes id) \circ \lambda \circ \chi = \\
&= (id \otimes \mu) \circ (id \otimes S \otimes id) \circ (\Delta \otimes id) \circ (id \otimes \chi) \circ \lambda' \\
&= (id \otimes \mu) \circ (id \otimes S \otimes id) \circ (\Delta \otimes id) \circ (id \otimes \chi) \circ (\Delta \otimes id) \\
&= (id \otimes \mu) \circ (id \otimes id \otimes \mu) \circ (id \otimes S \otimes id \otimes id) \circ (\Delta \otimes id \otimes id) \circ (\Delta \otimes id) \\
&= (id \otimes \mu) \circ (id \otimes m \otimes id) \circ (id \otimes S \otimes id \otimes id) \circ (id \otimes \Delta \otimes id) \circ (\Delta \otimes id) \\
&= (id \otimes \mu) \circ (id \otimes (m \circ (S \otimes id) \circ \Delta) \otimes id) \circ (\Delta \otimes id) \\
&= (id \otimes \mu) \circ (id \otimes \epsilon \otimes id) \circ (\Delta \otimes id) \\
&= id
\end{aligned}$$

QED

With the proof of Proposition 10.10.1 we have completed all the details of the proof of the main theorem of this chapter. The remaining discussion in this section is not needed in the proof, and may be omitted. We consider the issue of getting a structure theorem for $B-4$ analogous to Proposition 10.10.1.

Suppose that M is a $B-4$ module over a Hopf algebra, then we can consider the right structures relative to the representation in Proposition 10.10.1. Computing χ^{-1} gives

$$\chi^{-1}(mb) = (id \otimes \mu) \circ (id \otimes S \otimes id) \circ (\Delta \otimes id) \circ \lambda(mb) = \sum b_{(1)} \otimes S(b_{(2)}) m b_{(3)}$$

for $m \in M^{\#}$.

Definition. Let M be a bimodule over a Hopf algebra B. The left and right adjoint actions are defined by

$$a * m = \sum a_{(1)} m S(a_{(2)}) \text{ and } m * a = \sum S(a_{(1)}) m a_{(2)}.$$

Relative to the original bimodule action, $M^{\#}$ is neither a right nor a left submodule, but it is a right module under the right adjoint action. In terms of the decomposition given in the preceding proposition, the right module action is given by

$$(a \otimes m)b = \sum ab_{(1)} \otimes m * b_{(2)}.$$

One checks easily that $M^{\#}$ is also a right B-comodule, but relative to the right adjoint action it is not a right B-Hopf module so we can't apply the proposition a second time to the right structures to get a further decomposition. However, when bicomodule structures are *symmetric* in the sense that

$$\tau_{12} \circ \lambda = \rho,$$

we can give a structure theorem in the sense of $B-4$ modules: Obviously when the comodule structure is symmetric the left coinvariants are also right coinvariants. In this case we have a full description of $B-4$ module structure of M.

Proposition 10.10.2 *Let B be a Hopf algebra and M a non zero $B-4$ module, which is symmetric as a bicomodule. Then B is cocommutative, and*

$$M \cong B \otimes M^{@}$$

where $M^{@} = \{m \in M | \lambda(m) = 1 \otimes m, \text{and } \rho(m) = m \otimes 1\} = M^{\#}$ *since* M *is symmetric as a bicomodule.*

$$b(a \otimes m) = ba \otimes m \tag{10.10.5}$$

$$(a \otimes m)b = \sum ab_{(1)} \otimes m * b_{(2)} \tag{10.10.6}$$

$$\lambda'(a \otimes m) = \sum a_{(1)} \otimes (a_{(2)} \otimes m) \tag{10.10.7}$$

$$\rho'(a \otimes m) = \sum (a_{(1)} \otimes m) \otimes a_{(2)}. \tag{10.10.8}$$

$$\tag{10.10.9}$$

Proof. The right comodule structure is given by

$$\begin{aligned} \rho'(a \otimes m) &= (\chi^{-1} \otimes id) \circ \rho \circ \chi(a \otimes m) \\ &= \sum (\chi^{-1} \otimes id)(a_{(1)}m \otimes a_{(2)}) \qquad \text{since} \quad \rho(m) = m \otimes 1 \\ &= \sum (a_{(1)} \otimes m) \otimes a_{(2)}. \end{aligned}$$

and the symmetry condition imply that for $0 \neq m \in M^{\#}$ and any $a \in B$

$$\tau \circ \lambda'(a \otimes m) = \sum (a_{(2)} \otimes m) \otimes a_{(1)} = \rho'(a \otimes m) = \sum (a_{(1)} \otimes m) \otimes a_{(2)},$$

so B is cocommutative.

Consider the space $\otimes^q U(\mathcal{G})$, as a tensor product of $U(\mathcal{G})$ bicomodules, it is a symmetric bicomodule, hence $\beta(\otimes^q U(\mathcal{G}))$ is a symmetric $B-4$ module. More generally $M = \beta(N)$ is a $B-4$ module for any N which is a symmetric B bicomodule. This structure arises in the case of the bar complex and Koszul complex.

In the left comodule structure of $U(\mathcal{G}) \otimes U(\mathcal{G})$ as a tensor product of comodules we have

$$\lambda(X \otimes 1) = X \otimes 1 \otimes 1 + 1 \otimes X \otimes 1$$

since

$$\Delta(X) = X \otimes 1 + 1 \otimes X \quad \text{and} \quad \Delta(1) = 1 \otimes 1.$$

Similarly

$$\lambda(1 \otimes X) = X \otimes 1 \otimes 1 + 1 \otimes 1 \otimes X.$$

Hence

$$\begin{aligned}
\lambda(X') &= \lambda(X \otimes 1 - 1 \otimes X) \\
&= X \otimes 1 \otimes 1 + 1 \otimes X \otimes 1 - X \otimes 1 \otimes 1 - 1 \otimes 1 \otimes X \\
&= 1 \otimes (X \otimes 1 - 1 \otimes X) \\
&= 1 \otimes X'.
\end{aligned}$$

Thus

$$(U(\mathcal{G}) \otimes U(\mathcal{G}))^{\#} \supset U(\mathcal{G}').$$

To prove the reverse inclusion, note that $U(\mathcal{G}) \otimes U(\mathcal{G})''U(\mathcal{G})'$ is a free $U(\mathcal{G}) \otimes U(\mathcal{G})''$ module generated by $1 \otimes U(\mathcal{G})'$. Let $\{v_i'\}$ be a basis, and suppose that

$$\lambda(\sum u_i'' v_i') = \sum u_{i(1)} \otimes u_{i(2)}'' v_i' = 1 \otimes (\sum u_i'' v_i').$$

Then because of the freeness condition we have the identity

$$\sum u_{i(1)} \otimes u_{i(2)} = 1 \otimes u_i$$

for all i. Applying $m \circ T \otimes id$, we conclude that $u_i \in k$ for all i and our coinvariant is in $U(\mathcal{G})'$. A similar argument shows that

$$(\beta(\wedge\mathcal{G}))^{\#} = U(\mathcal{G}') \otimes \wedge\mathcal{G}.$$

and the decomposition

$$\beta(\wedge\mathcal{G}) = U(\mathcal{G}'')\mathcal{U}(\mathcal{G}') \otimes \wedge\mathcal{G} \cong \mathcal{U}(\mathcal{G}'') \otimes (\mathcal{U}(\mathcal{G}') \otimes \bigwedge \mathcal{G})$$

is in fact the decomposition of Proposition 10.10.2.

Chapter 11

Deformations of $U(\mathcal{G})$

In this chapter we apply deformation cohomology to study the structure of deformations of universal enveloping algebras and to prove some existence and uniqueness theorems. In section 2 we prove, in general, that any bialgebra deformation is equivalent to one in which the unit and counit are undeformed. For this purpose we need the normalized bar and cobar resolutions which we define in section 1. In section 2 we also provethe existence of an antipode for any bialgebraa deformation of a Hopf algebra. After that we turn to the particular case of deformation theory of universal enveloping algebras, in particular, $U_h(\mathcal{G})$, Drinfeld's quantized universal enveloping algebras, which are really our main interest in this book. Let $\mathcal{G}$ be a simple Lie algebra. In section 4 we give a complete proof of Drinfeld's first uniqueness theorem, originally stated in his famous paper on quantum groups presented to the International Congress of Mathematicians in 1986. Roughly speaking, this theorem asserts that up to equivalence of deformation and change of deformation parameter there is a unique non-trivial bialgebra deformation of $U(\mathcal{G})$ which also deforms a Cartan involution. We give the precise formulation in section 4. In section 5 we use an extension of the Gerstenhaber-Schack deformation cohomology to prove a general uniqueness result on Drinfeld algebra (as opposed to bialgebra) deformations of $U(\mathcal{G})$ due to the first author [1094]. We prove that if $\mathcal{G}$ is semi-simple, any deformation of $U(\mathcal{G})$ within the class of Drinfeld algebras is equivalent (up to twisting) to one in which the multiplication and comultiplication are undeformed, so that the entire deformation occurs in the Φ. Then we show that if $\mathcal{G}$ is simple, there is a unique non-trivial deformation, i.e. a unique Φ, up to twisting and change of parameter. In the last two sections we give a survey of Drinfeld's theory of quasi-triangular structures. This is where the "universal R matrix" and the hexagon identities enter the theory of Drinfeld algebra deformations of universal enveloping algebras. A quasi-triangular deformation of $U(\mathcal{G})$ is one in which the deformed algebra is cocommutative up to conjugation by an R satisfying two hexagon identities, and coassociative up to conjugation by a Φ satisfying the pentagon identity. (Thus the category of its modules form a braided category.) Using only coalgebra cohomology and clever reductions of the R matrix, Drinfeld proved the striking result that for any Lie algebra $\mathcal{G}$ (not necessarily semisimple) any quasi-triangular deformation of $U(\mathcal{G})$ is equivalent under the usual equivalence and twisting to one which is cocommutative and coassociative. It follows that it is the universal enveloping algebra of a Lie algebra deformation of $\mathcal{G}$. The deformation is called triangular if R satisfies an identity, cf.(11.6.8) below, which makes this category symmetric. In section 6, as an introduction to the theory, we present Drinfeld's proof that any triangular deformation of $U(\mathcal{G})$ is equivalent to one which arises as the universal enveloping algebra of a Lie algebra deformation. In section 7 we complete the proof of Drinfeld's structure theorem for quasi-triangular Drinfeld quantum universal enveloping algebras. An existence theorem, also due to Drinfeld, based on the Knizhnik-Zamolodchikov equation is presented in the next chapter. In section 8 we describe the interpretation of the infinitesimals in terms of quasi-Lie bialgebras and Poisson-Lie groups. In section 9 we discuss quantum $sl(2)$ as an example.

11.1 The normalized bar and cobar resolutions

The normalized bar resolution, which we define in this section, will allow us to prove that for any algebra deformation there exists an equivalent deformation for which the unit is undeformed, that is, the unit 1 for m remains the unit for $m(t)$. Similarly the normalized cobar resolution will allow us to prove that for any coalgebra deformation there is an equivalent one for which the counit is undeformed. Then the general result for bialgebra deformations will follow by a double complex argument. Let A be an algebra defined over k and $u : k \to A$ the embedding given by the unit, $u(a) = a1$. Let $\tilde{A} = \text{Coker } u$. The normalized cobar complex is defined by

$$\begin{aligned} \text{Bar}_N^q(A) &= \beta(\otimes^q \tilde{A}) = A \otimes (\otimes^q \tilde{A}) \otimes A \quad \text{for } q \geq 0 && (11.1.1) \\ \text{Bar}_N^{-1}(A) &= A. && (11.1.2) \end{aligned}$$

Let

$$u_j = id \otimes \cdots \otimes u \otimes \cdots \otimes id : A^{\otimes j-1} \otimes k \otimes A^{\otimes q+2-j} \to A^{\otimes q+2}$$

where u occurs in the position j. For $q > 0$ let $\mathrm{Bar}^q_S(A)$ be the subspace of the (unnormalized) bar complex $\mathrm{Bar}^q(A)$ spanned by the image of all the u_j for $1 < j < q+2$, and $\mathrm{Bar}_S(A) = \bigoplus_{0<q} \mathrm{Bar}^q_S(A)$. Then we have the exact sequence of A bimodules

$$0 \to \mathrm{Bar}_S(A) \to \mathrm{Bar}(A) \to \mathrm{Bar}_N(A) \to 0. \qquad (11.1.3)$$

It is straightforward to check that $\mathrm{Bar}_S(A)$ is a subcomplex with respect to the bar differential, and that it is mapped into itself by the contracting homotopy (10.3.2). This fact together with the fact that $\mathrm{Bar}^1_S(A)$ is spanned by elements of the form $a \otimes 1 \otimes b$ (which are all cocycles) proves that $\mathrm{Bar}_S(A)$ is acyclic. In fact the existence of the homotopy implies it is split exact, as we saw in section 3 of the last chapter. One gives $\mathrm{Bar}_N(A)$ the structure of a quotient complex and the exact sequence above becomes a short exact sequence of complexes. The long exact sequence implies that $\mathrm{Bar}_N(A)$ is acyclic, since the other two complexes are. In fact the splittings of $\mathrm{Bar}_S(A)$ and $\mathrm{Bar}(A)$ are compatible (being given by the same homotopy) and so $\mathrm{Bar}_N(A)$ is split exact. Since $\beta(V)$ is always a projective bimodule, $\mathrm{Bar}_N(A)$ is projective resolution. In the case of a bialgebra B the maps

$$(u \otimes id) \circ \Delta : k \to k \otimes k \to B \otimes k$$

and

$$(id \otimes u) \circ \Delta : k \to k \otimes k \to k \otimes B$$

make k into an B bicomodule, and u becomes a bicomodule map. Thus the cokernel $\tilde{B}$ is a B bicomodule, and we can give $Bar^q_N(B) = \beta(\otimes \tilde{B})$ the structure of a $B - 4$ module. Since the contracting homotopy is a B bicomodule map, the complex is split exact as a sequence of B bicomodules. Combining these results we have

Proposition 11.1.1 *The normalized bar resolution* $\mathrm{Bar}_N(A)$ *is a projective A bimodule resolution. For a bialgebra B,* $\mathrm{Bar}_N(B)$ *is a relative projective $B - 4$ resolution.*

For a coalgebra C, let $\hat{C} = \mathrm{Ker}\ \epsilon : C \to k$ where ϵ is the counit (augmentation). We define the normalized cobar complex in degree q

$$\begin{aligned} \mathrm{Cob}^q_N(C) &= \gamma(\otimes^q \hat{C}) = C \otimes (\otimes^q \hat{C}) \otimes C \quad \text{for } q \geq 0 & (11.1.4) \\ \mathrm{Cob}^{-1}_N(C) &= C & (11.1.5) \end{aligned}$$

Let $\epsilon_j = id \otimes \cdots \otimes \epsilon \otimes \cdots \otimes id$ where ϵ occurs in the j th factor. Then

$$\mathrm{Cob}_N^q = \cap_{1<j<q+2}(\mathrm{Ker}\epsilon_j|_{\mathrm{Cob}^q(C)}).$$

The identity

$$(id \otimes \epsilon \otimes id)(\Delta(a) \otimes b - a \otimes \Delta(b)) = 0.$$

applied in any three adjacent factors gives

$$\epsilon_j \circ (\Delta_j - \Delta_{j+1}) = 0.$$

Together with the identities

$$\begin{aligned} \epsilon_j \circ \Delta_i &= \Delta_i \circ \epsilon_{j+1} \quad \text{for} \quad i+1<j \\ &= \Delta_i \circ \epsilon_j \quad\quad \text{for} \quad i \geq j+1 \end{aligned}$$

this implies that $\mathrm{Cob}_N(C)$ is a subcomplex relative to the cobar differential. There is an exact sequence

$$0 \to \mathrm{Cob}_N(C) \to \mathrm{Cob}(C) \to Cob_S(C) \to 0 \qquad (11.1.6)$$

where $\mathrm{Cob}_S(C)$ is defined as a quotient complex. Furthermore the contracting homotopy restricts to $\mathrm{Cob}_N(C)$ proving that it is acyclic. We have seen that for any vector space V, $\gamma(V)$ is an injective bicomodule. Since $\hat{B}$ is a bimodule over B, when we do the construction for a bialgebra, we can use the tensor product in the category of B-bimodules so that $\otimes^q \hat{B}$ is also a B bimodule. Then $\gamma(\otimes^q \hat{B})$ is a $B-4$ module. The contracting homotopy is a B bimodule map, therefore Cob_N is split exact as a sequence of B bimodules.

Proposition 11.1.2 *The normalized cobar complex,* $\mathrm{Cob}_N(C)$, *of a coalgebra* C *is an injective bicomodule resolution of* C. *For a bialgebra* B *it is a relative* $B-4$ *injective resolution.*

The splitting $B = k \oplus \hat{B}$ gives a splitting of the exact sequences (11.1.3) and (11.1.6) as complexes, so we can regard the normalized complexes as direct summands of the full complexes. Then $\mathrm{Hom}(\mathrm{Bar}_N(B), \mathrm{Cob}_N(B))$ can be considered as a direct summand of the full double complex. We will call elements of this subcomplex *doubly normalized cochains.* As a corollary of Lemma 10.4.1 we know

Proposition 11.1.3 *For any bialgebra* B *the double complexes*

$$\mathrm{Hom}(\mathrm{Bar}(B), \mathrm{Cob}(B))$$

and

$$\mathrm{Hom}(\mathrm{Bar}_N(B), \mathrm{Cob}_N(B))$$

are homotopy equivalent.

In particular any cocycle is cohomologous to a doubly normalized cocycle. However, in applying this result to deformation theory we have to be careful since the deformation cohomology complex is a subcomplex and the homotopies which prove the existence of a homotopy equivalence may not preserve this subcomplex.

11.2 Units, counits, antipodes under deformation

We begin this section with the proposition which motivated our introducing the normalized complexes.

Proposition 11.2.1 *Any bialgebra deformation is equivalent to one in which the unit and counit are undeformed.*

Proof. The unit is undeformed if and only if

$$m(h)(1 \otimes a) = m(h)(a \otimes 1) = a$$

for all $a \in B$, which is the same as the equations

$$m^{(n)}(1 \otimes a) = m^{(n)}(a \otimes 1) = 0 \quad \text{for} \quad n > 0. \tag{11.2.1}$$

Let

$$C^{p,q}_{*,N} = \mathrm{Hom}_{B-4}(\mathrm{Bar}^q_N, \mathrm{Cob}^p) \hookrightarrow \mathrm{Hom}_k(B^{\otimes q}, B^{\otimes p}),$$

be the image after applying the adjoint functor relations of the cochains normalized with respect to the bar complex. Equations (11.2.1) are equivalent to

$$m^{(n)} \in C^{1,2}_{*,N} \quad \text{for} \quad n > 0.$$

Let

$$C^{p,q}_{N,*} = Hom_{B-4}(\mathrm{Bar}^q, \mathrm{Cob}^p_N) \hookrightarrow \mathrm{Hom}_k(B^{\otimes q}, B^{\otimes p})$$

be the image after applying the adjoint functor relations of the cochains normalized with respect to the cobar complex. The counit is undeformed if and only if

$$m(h) \circ (\epsilon \otimes id) \circ \Delta(h)(a) = m(h) \circ (id \otimes \epsilon) \circ \Delta(h)(a) = a$$

for all $a \in B$, which is the same as

$$\Delta^{(n)} \in C^{2,1}_{N,*} \quad \text{for} \quad n > 0.$$

We will prove a stronger normalization.

Claim. Any deformation is equivalent to one with

$$\begin{aligned} m^{(n)} &\in C^{1,2}_{*,N} = \mathrm{Hom}_{B-4}(\mathrm{Bar}^2_N, \mathrm{Cob}^1_N) \\ \Delta^{(n)} &\in C^{2,1}_{N,N} = \mathrm{Hom}_{B-4}(\mathrm{Bar}^1_N, \mathrm{Cob}^2_N) \quad \text{for} \quad n > 0. \end{aligned} \tag{11.2.2}$$

Let π'' be the projection of C on $C_{*,N}$ and π' be the projection of $C_{*,N}$ on $C_{N,N}$. Then π' is a homotopy equivalence relative to δ and π'' a homotopy equivalence relative to d_H.

Suppose that we have normalized the deformation up to order n according to conditions (11.2.2). Let $(\beta^{1,3}, \beta^{2,2}, \beta^{3,1})$ be the obstruction cocycle at order n. Consider the complex $C_{*,N}$ as a simple complex relative to the coboundary d_H. Evaluating $\beta^{1,3}(a_1 \otimes a_2 \otimes a_3)$ involves only $m^{(k)}$ for $k < n$ and therefore vanishes when any $a_i = 1$. Therefore $\beta^{1,3} \in C^{1,3}_{*,N}$. In addition $d_H\beta^{1,3} = 0$ so $\beta^{1,3}$ is a cocycle in this complex. Let $m^{(n)''} = \pi''(m^{(n)})$. Applying π'' to the equation

$$d_H(m^{(n)}) = \beta^{1,3}$$

we get

$$d_H(m^{(n)''}) = \beta^{1,3}.$$

Then

$$d_H(m^{(n)} - m^{(n)''}) = 0.$$

By the chain homotopy equivalence condition on π'', $m^{(n)} - m^{(n)''} = d_H(\sigma)$ for some $\sigma \in C^{1,1}$. Conjugating the original deformation by $id + \sigma h^n$ gives an equivalent deformation in which the multiplication has been normalized through order n.

Turn now to the normalization of $\Delta^{(n)}$. By the induction assumption all $m^{(k)}$ and $\Delta^{(k)}$ are in the subcomplex $C_{*,N}$ for $k < n$; therefore, by the same argument as for $\beta^{1,3}$, we conclude that $\beta^{2,2} \in C^{2,2}_{*,N}$. Since we have already normalized $m^{(n)}$, and since δ preserves $C_{*,N}$ we have

$$d_H(\Delta^{(n)}) = \beta^{2,2} - \delta(m^{(n)}) \in C_{*,N}.$$

Evaluating on the element $1 \otimes 1$ we have

$$\Delta^{(n)}(1) = \Delta(1)\Delta^{(n)}(1) - \Delta^{(n)}(1) + \Delta^{(n)}(1)\Delta(1) = d_H(\Delta^{(n)})(1 \otimes 1) = 0.$$

Thus $\Delta^{(n)} \in C^{2,1}_{*,N}$. The normalization on $\Delta^{(k)}$ for $k < n$ implies that $\beta^{3,1} \in C^{3,1}_{N,N}$ and

$$\delta(\Delta^{(n)}) = \beta^{3,1} \in C^{3,1}_{N,N}.$$

Let $\Delta^{(n)'} = \pi'(\Delta^{(n)})$. Applying π' we have

$$\delta(\Delta^{(n)'}) - \Delta^{(n)}) = \pi'(\beta^{3,1}) - \beta^{3,1} = 0.$$

The homotopy equivalence condition on π' implies the existence of $\sigma' \in C^{1,1}_{*,N}$ such that

$$(\Delta^{(n)'}) - \Delta^{(n)} = \delta\sigma'.$$

Conjugating the deformation by $id + \sigma' h^n$ gives an equivalent deformation satisfying the normalization condition on $\Delta(h)$ through order n and preserving the normalization on $m(h)$.

Proposition 11.2.2 *Let $(B, m, \Delta, u, \epsilon, T)$ be a Hopf algebra and*

$$(B_h, m(h), \Delta(h), u, \epsilon)$$

a deformation of B as a bialgebra, where we will assume that we have normalized the deformation such that the unit, u, and counit, ϵ, are undeformed. Then there is a deformation $T(h)$ of the antipode such that

$$(B, m(h), \Delta(h), i, \epsilon, T(h))$$

is a Hopf algebra.

Proof. Recall the definition of the convolution product on $\mathrm{End}(B)$.

$$\alpha * \beta = m \circ (\alpha \otimes \beta) \circ \Delta.$$

The axiom for the counit ϵ says that it a two sided identity with respect to the convolution product. The axiom for the antipode T says that it is the inverse of id with respect to convolution. So we have

$$\epsilon * \alpha = \alpha * \epsilon = \alpha,$$

$$T * id = id * T = \epsilon.$$

We must show that relative to the deformed convolution product

$$\alpha *_h \beta = m(h) \circ (\alpha \otimes \beta) \circ \Delta(h)$$

the element $id \in \mathrm{End}(B_h)$ has a two sided inverse. It is enough to show that it has a right inverse and a left inverse, since associativity implies that they must agree. There is a general principle at work here.

Invertibility is preserved under deformation.

One can prove it explicitly in this case by expanding $T(h) = \sum T^{(k)}h^k$ and solving the equation

$$T(h) * id = \epsilon \tag{11.2.3}$$

term by term. Assume that $T^{(k)}$ has been determined for $k < n$, then, isolating the term $T^{(n)}$ on the left in (11.2.3) we have the equation

$$T^{(n)} * id = m \circ (T^{(n)} \otimes id) \circ \Delta = f_n(m^{(k)}, \Delta^{(k)}, T^{(j)})$$

f_n is some expression involving only $T^{(j)}, m^{(j)}$ and $\Delta^{(k)}$ for $n \geq k \geq 0, n > j \geq 0$). Using the fact that T is a right inverse for id relative to $*$ we can solve uniquely $T^{(n)} = f_n * T$.

11.3 Normalized forms of deformations of $U(\mathcal{G})$

Throughout this chapter we will assume that all our algebras are defined over k, an algebraically closed field of characteristic 0, and that all our deformations have been identified with $U(\mathcal{G})[[h]]$ as $k[[h]]$ modules.

$$B_h = U(\mathcal{G})[[h]].$$

We proved in the last chapter that for any Lie algebra $\mathcal{G}$ over k the bialgebra deformation cohomology in bidegree p, q is

$$\begin{aligned} \hat{H}^{p,q}(U(\mathcal{G})) &\cong H^q(\mathcal{G}, \wedge^p\mathcal{G}) \quad q > 1 \\ &\cong Z^1(\mathcal{G}, \wedge^p\mathcal{G}) \quad q = 1 \end{aligned}$$

The following two lemmas give a sharper result.

Lemma 11.3.1 Whitehead. *For $\mathcal{G}$ a semisimple Lie algebra over k, and V a $\mathcal{G}$ module, we have*

$$H^i(\mathcal{G}, V) \cong H^i(\mathcal{G}, k) \otimes V^{\mathcal{G}} \quad \textit{for all} \quad i.$$

and in particular

$$H^i(\mathcal{G}, V) = 0, \quad i = 1, 2$$

This is a basic elementary fact in the theory of Lie algebra cohomology. For a proof of the result for $i = 1, 2$ see, for example, [501]. The result for $i > 2$ is a little less standard so we sketch the proof. It is based on the Casimir element $\mathcal{C} = \sum X_a^2 \in U(\mathcal{G})$, where X_a is an orthonormal basis with respect to the Killing form. The element C acts as a non-zero scalar in any non-trivial irreducible representation of $\mathcal{G}$. For $X \in \mathcal{G}$ define the operator $i(X)$ mapping Hochschild p-cochains $\alpha \in C^p(\mathcal{G}, V)$ into $p-1$-cochains

$$i(X)\alpha(Y_1, \cdots, Y_{p-1}) = \alpha(X, Y_1, \cdots, Y_{p-1}).$$

If $X \cdot \alpha$ is the p-cochain given by having X act on the values of α, and d is the Hochschild coboundary, then

$$\begin{aligned} & [d(X \cdot i(X)\alpha) + X \cdot i(X)d\alpha](Y_1, \cdots, Y_p) \\ = & \sum (-1)^i([X, Y_i] \cdot \alpha(X, Y_1, \cdots, \hat{Y}_i, \cdots, Y_p) \\ + & X \cdot \alpha([X, Y_i], Y_1, \cdots, \hat{Y}_i, \cdots, Y_p)) \\ + & X^2 \cdot \alpha(Y_1, \cdots, Y_p), \end{aligned}$$

where ˆ means delete. Using this formula and the $ad_{\mathcal{G}}$ invariance of the element $\sum X_a \otimes X_a$ one proves that the operator

$$h(\alpha) = \sum_a X_a \cdot i(X_a)(\alpha)$$

satisfies

$$dh(\alpha) + hd(\alpha) = \sum_a X_a^2 \cdot \alpha = \mathcal{C} \cdot \alpha.$$

For any representation V of $\mathcal{G}$ we can decompose V into a direct sum of subrepresentations $V' \oplus V^{\mathcal{G}}$ where $\mathcal{C}$ is invertible on V'. This gives a decomposition of the cochain complex

$$C(\mathcal{G}, V) = C(\mathcal{G}, V') \oplus C(\mathcal{G}, V^{\mathcal{G}}).$$

The operator $\mathcal{C}^{-1} \cdot h$ is contracting homotopy on the subcomplex $C(\mathcal{G}, V')$ proving that it is acyclic. Therefore

$$H(\mathcal{G}, V) \cong H(\mathcal{G}, V^{\mathcal{G}}) \cong H(\mathcal{G}, k) \otimes V^{\mathcal{G}}.$$

We saw in the last chapter that the homotopy equivalence of the Koszul resolution and K_b proves that the Hochschild cohomology of $U(\mathcal{G})$ with values in a $U(\mathcal{G})$ bimodule M is the same as the Chevalley Eilenberg cohomology of the Lie algebra $\mathcal{G}$ with values in the M as a Lie module. Thus Whitehead's lemma also gives the calculation of Hochschild cohomology for $U(\mathcal{G})$:

$$\begin{array}{rcll} H^i(U(\mathcal{G}),M) &=& 0 & i=1,2 \\ H^i(U(\mathcal{G}),M) &=& H^i(\mathcal{G},k)\otimes(M)^{\mathcal{G}} & i\geq 0 \end{array}$$

Lemma 11.3.2 [Koszul] *If $\mathcal{G}$ is a simple Lie algebra over k, then $(\wedge^3\mathcal{G})^{\mathcal{G}}$ has dimension 1, and is generated by the element u defined by $\langle X\wedge Y\wedge Z,u\rangle = \langle [X,Y],Z\rangle$, where the inner product on $\wedge^3(\mathcal{G})$ is induced by the Killing form.*

For $\mathcal{G}$ simple, there are no invariants in $\wedge^2\mathcal{G}$, and by the Whitehead lemma, $H^1(\mathcal{G},\wedge\mathcal{G})=0$ and $H^2(\mathcal{G},\wedge\mathcal{G})=0$, so we have:

$$\begin{array}{rcll} \hat{H}^3(U(\mathcal{G})) &\cong& H^2(\mathcal{G},\mathcal{G})\oplus Z^1(\mathcal{G},\wedge^2\mathcal{G})\cong B^1(\mathcal{G},\wedge^2\mathcal{G}) & \\ &\cong& \wedge^2\mathcal{G} & (11.3.1) \\ \hat{H}^4(U(\mathcal{G})) &\cong& H^3(\mathcal{G},\mathcal{G})\oplus H^2(\mathcal{G},\wedge^2\mathcal{G})\oplus Z^1(\mathcal{G},\wedge^3\mathcal{G}) & \\ &\cong& B^1(\mathcal{G},\wedge^3\mathcal{G})\cong \wedge^3\mathcal{G}/(u) & (11.3.2) \end{array}$$

where (u) is the element defined in Koszul's lemma.

Proposition 11.3.1 *Let $\mathcal{G}$ be a semisimple Lie algebra over k. Any bialgebra deformation of $U(\mathcal{G})$ is equivalent to one in which the multiplication is undeformed.*

Proof. The proof is the same as the standard proof showing that if the deformation cohomology vanishes then any algebra deformation is trivial. Given a deformation $(U_h(\mathcal{G}),m(h),\Delta(h))$ assume that we have defined an equivalence $\sigma_{(n)}(h)$ such that the conjugated multiplication

$$m_{(n)}(h)=\sigma^{-1}_{(n)}(h)\circ m(h)\circ(\sigma_{(n)}(h)\otimes\sigma_{(n)}(h))$$

has $m_k=0$ for $0<k\leq n$. The same calculation as that showing that the infinitesimal $dm^{(1)}=0$ shows that if $m_{(n+1)}(h)=m+m^{(n+1)}h^{n+1}+O(h^{n+2})$ is associative modulo h^{n+2} then $dm^{(n+1)}=0$. Whitehead's lemma implies that the Hochschild 2-cohomology vanishes, therefore there exists an element $\sigma_{n+1}\in C^{1,1}=\mathrm{End}(U(\mathcal{G}))$ such that $d\sigma_{n+1}=m^{(n+1)}$. Conjugating $m_{(n+1)}(h)$ by $id+\sigma_{n+1}h^{n+1}$ is the same as conjugating $m(h)$ by $\sigma_{(n)}(h)\circ(id+\sigma_{n+1}h^{n+1})$ and gives a deformation $m'_{(n+1)}(h)$ equivalent to the original deformation, but now with $m^{(k)}=0$ for $0<k\leq n+1$. (According to the coboundary formula in the Gerstenhaber-Schack complex

$(m^{(n+1)\prime}, \Delta^{(n+1)\prime}) = (m^{(n+1)}, \Delta^{(n+1)}) + d\sigma_{n+1} = (0, \Delta^{(n+1)} - \delta\sigma_{n+.})$ and the effect of conjugating $\Delta(h)$ by $id + \sigma_{n+1}h^{n+1}$ is to subtract the term $\delta\sigma_{n+1}h^{n+1}$ but for the time being that does not interest us.) In this way we construct a sequence of equivalences in which the $(n+1)$st term, $\sigma_{(n+1)}(h)$ is the composition $\sigma_{(n)}(h) \circ (id + \sigma_{n+1}h^{n+1})$ and such that conjugation by $\sigma_{(n+1)}(h)$ gives an equivalent deformation $m_{(n+1)}(h)$, in which the multiplication equals m through order $n+1$. This sequence of equivalences converges relative to the formal power series topology (the Krull topology) to the desired equivalence, $\sigma(h)$ such that

$$\sigma(h)^{-1} \circ m(h) \circ (\sigma(h) \otimes \sigma(h)) = m.$$

Proposition 11.3.2 *Let $(U(\mathcal{G}), m, \Delta(h))$ be a deformation in the normalized form given by Proposition 11.3.1. We can further normalize so that the infinitesimal comultiplication deformation, $\Delta^{(1)}$ is given as*

$$\Delta^{(1)}(a) = d(r)(a) = [\Delta(a), r]$$

for $a \in U(\mathcal{G})$ where $r \in \wedge^2(\mathcal{G})$.

Note that this does not imply that the infinitesimal deformation is trivial since the bialgebra deformation complex does not include $\wedge^2\mathcal{G}$.

Proof: The cocycle condition for the infinitesimal, $(0, \Delta^{(1)})$ implies that $d\Delta^{(1)} = 0$. As $H^1(U(\mathcal{G}), U(\mathcal{G}) \otimes U(\mathcal{G})) = 0$, there exists a $u \in U(\mathcal{G}) \otimes U(\mathcal{G})$ such that $\Delta^{(1)}(a) = d(u)(a) = [\Delta(a), u]$. To prove the proposition we need to show that u can be chosen in $\wedge^2\mathcal{G}$. Since $\mathcal{G}$ is semisimple it is completely reductive and we can split the complex $C^{\cdot,0} = \oplus C^{q,0}_{G.S.} = \mathrm{Hom}_{B-4}(\mathrm{Bar}^0, \mathrm{Cob}) = \mathrm{Hom}(k, \otimes U(\mathcal{G}))$ into two subcomplexes, C_1, consisting of $\mathcal{G}$ invariants and C_2, a $\mathcal{G}$ invariant complement. Since $d = 0$ on the invariants, we can assume that u has been chosen in the complement C_2. Then the second half of the cocycle condition on $\Delta^{(1)}$ implies that

$$0 = \delta(\Delta^{(1)}) = \delta\ d(u) = d\delta(u)$$

but $\delta u \in C_2$ and $C_2 \cap \mathrm{Ker} d = 0$. Thus $\delta u = 0$. But $H^2(C^{\cdot,0}) = \wedge^2\mathcal{G}$ and the inclusion $\wedge\mathcal{G} \to \otimes U(\mathcal{G})$ (where the coboundary operator on $\wedge\mathcal{G}$ is 0) induces an isomorphism on cohomology. Therefore there exists a $v \in C^{1,0} = U(\mathcal{G})$ such that $u - \delta v = r \in \wedge^2\mathcal{G}$. Let $\sigma = d(v)$ then conjugating by $id + \sigma h$ gives a deformation with infinitesimal,

$$(0, \Delta^{(1)}) - D\sigma \quad = \quad (d\sigma, \Delta^{(1)} - \delta\sigma)$$

$$\begin{aligned} &= ((d)^2(v), \Delta^{(1)} - \delta\sigma) \\ &= (0, d(u) - \delta d(v)) \\ &= (0, d(u - \delta v)) \\ &= (0, d(r)), \end{aligned}$$

proving the proposition. This is clearly a double complex argument like those in Chapter 10, however we have gone outside the bialgebra deformation complex, using the bottom row of the Gerstenhaber-Schack complex to prove a result on deformations. The existence of this extra row simplifies the study of deformations significantly. Later on it will actually enter the Drinfeld algebra deformation cohomology.

Proposition 11.3.3 *For a deformation $(U_h(\mathcal{G}), m, \Delta(h))$ normalized as in Proposition 11.3.2, $(3,1)$ component of the obstruction cocycle $\beta^{3,1}(a) = (id \otimes \Delta^{(1)})\Delta^{(1)}(a) - (\Delta^{(1)} \otimes id)\Delta^{(1)}(a)$ is cohomologous to*

$$d([r^{12}, r^{13}+r^{23}]+[r^{13}, r^{23}]) = 2/3[\Delta^{[3]}(a), [r^{12}, r^{13}+r^{23}]+[r^{13}, r^{23}]], \quad (11.3.3)$$

where the expression in parentheses is the one appearing in the classical Yang-Baxter equation, and $\Delta^{[3]} = (id \otimes \Delta) \circ \Delta = (\Delta \otimes id) \circ \Delta$. The obstruction cocycle is cohomologous to zero if and only id $[r^{12}, r^{13} + r^{23}] + [r^{13}, r^{23}]$ is invariant.

Proof: Since $\Delta^{(1)}(a) = [\Delta(a), r]$, we have

$$\begin{aligned} (id \otimes \Delta^{(1)})\Delta^{(1)}(a) &= [(id \otimes \Delta)(\Delta^{(1)}(a)), r^{23}] \\ &= [(id \otimes \Delta)([\Delta(a), r]), r^{23}] \\ &= [[(id \otimes \Delta)(\Delta(a)), (id \otimes \Delta)(r)], r^{23}]] \\ &= [[\Delta^{[3]}(a), r^{12} + r^{13}], r^{23}] \end{aligned}$$

The summand $(\Delta^{(1)} \otimes id)\Delta^{(1)}(a)$ gives $[[\Delta^{[3]}(a), r^{13} + r^{23}], r^{12}]$. The cocycle condition implies that $d\beta = 0$. Since the first Hochschild cohomology group vanishes, just as in the proof of Proposition 11.3.2 there is a $\xi \in C_2^3 = C_2 \cap C_{GS}^{3,0}$ such that $\beta = d\xi$. Then $0 = \delta\beta = \delta d\xi = d\delta\xi$, and since $\delta\xi \in C_2$, $\delta\xi = 0$. Then there exists an $\eta \in C_2^2$ such that $\xi - \delta\eta \in \wedge^3\mathcal{G}$ since all the nontrivial cohomology of $C^{\cdot,0}$ is in this subcomplex, which is the image of the imbedding of the Koszul complex. Then

$$\beta - Dd\eta = \beta - \delta d\eta = d(\xi - \delta\eta) \in d(\wedge^3(\mathcal{G}).$$

Total antisymmetrization annihilates δ coboundaries and is the identity on $d(\wedge\mathcal{G})$, therefore, it defines the map from cocycles to cohomology classes. Thus to evaluate the obstruction on the cohomology level, we simply antisymmetrize. Antisymmetrizing any of the four summands and applying the Jacobi identity gives the right side of of the last equation in the proposition. Adding the four terms and dividing by 3! gives the constant.

From now on, in this chapter, we will assume that we have chosen a presentation of the deformation in the form $(U(\mathcal{G}), m, \Delta(h))$.

11.4 The first uniqueness theorem.

In this section we prove the uniqueness part of the following theorem, stated without proof in [317]

Theorem 11.4.1 [Drinfeld] *Let $\mathcal{G}$ be a simple Lie algebra over k. Modulo equivalence of deformation and change of the deformation parameter h, there is a unique quantum universal enveloping $U_h(\mathcal{G})$ satisfying the following additional conditions*

1. *$U_h(\mathcal{G})$ contains a cocommutative sub-Hopf algebra C such that $C/hC = U(\mathcal{H})$ for $\mathcal{H}$ a Cartan subalgebra of $\mathcal{G}$.*

2. *C is invariant under an algebra automorphism θ for which*

   2a. *θ is a coalgebra anti-automorphism*

   2b. *$\theta^2 = id$*

   2c. *θ induces the Cartan involution relative to the Cartan subalgebra, $\mathcal{H}$, on $\frac{U_h(\mathcal{G})}{hU_h(\mathcal{G})} = U(\mathcal{G})$.*

(Note: The involution acts by the standard formula on the basis of "root vectors" for the Drinfeld-Jimbo quantizations $U_h(\mathcal{G})$ of the classical finite dimensional simple Lie algebras.)

In the proof of the Theorem 1 it will be necessary for us to use the following description of u in terms of the Weyl basis. Let R be a system of roots for $\mathcal{G}$. The basis vectors $e_\alpha \in \mathcal{G}^\alpha$ satisfy

$$[e_\alpha, e_{-\alpha}] = \tau_\alpha$$

where τ_α is the element of the Cartan subalgebra $\mathcal{H}$ satisfying

$$\beta(\tau_\alpha) = \frac{2\langle\beta,\alpha\rangle}{\langle\alpha,\alpha\rangle} \qquad \beta \in \mathcal{H}^*.$$

The set of vectors e_α for all roots α and τ_α for α a simple positive root form a basis for $\mathcal{G}$. They satisfy the orthogonality relations:

$$\langle e_\alpha, \tau_\beta\rangle = 0$$

$$\langle \tau_\alpha, \tau_\alpha\rangle = \frac{4}{\langle\alpha,\alpha\rangle}$$

$$\langle e_\alpha, e_\beta\rangle = 0 \qquad \alpha+\beta \neq 0$$

$$\langle e_\alpha, e_{-\alpha}\rangle = \frac{2}{\langle\alpha,\alpha\rangle}.$$

Define $N_{\alpha,\beta}$ by

$$[e_\alpha, e_\beta] = N_{\alpha,\beta} e_{\alpha+\beta} \tag{11.4.1}$$

If we use the Weyl normalization the root vectors e_α

$$N_{-\alpha,-\beta} = -N_{\alpha,\beta}. \tag{11.4.2}$$

If $\alpha+\beta+\gamma=0$ for pairwise linearly independent roots, the complete anti-symmetry of the expression $\langle[e_\alpha,e_\beta],e_\gamma\rangle$ implies

$$\frac{N_{\alpha,\beta}}{\langle\gamma,\gamma\rangle} = \frac{N_{\beta,\gamma}}{\langle\alpha,\alpha\rangle} = \frac{N_{\gamma,\alpha}}{\langle\beta,\beta\rangle}. \tag{11.4.3}$$

Relative to this basis we can express a basis element u for the one dimensional space of invariants in $\wedge^3\mathcal{G}$. as

$$\sum_{\alpha+\beta+\gamma=0} \langle\alpha,\alpha\rangle\langle\beta,\beta\rangle N_{-\alpha,-\beta} e_\alpha \wedge e_\beta \wedge e_\gamma - 6\sum_{\alpha>0}\langle\alpha,\alpha\rangle^2 e_\alpha\wedge e_{-\alpha}\wedge\tau_\alpha =$$

$$6\Big(\sum_{\substack{\alpha>\beta>0 \text{ or} \\ \alpha<\beta<0}} \langle\alpha,\alpha\rangle\langle\beta,\beta\rangle N_{-\alpha,-\beta} e_\alpha\wedge e_\beta\wedge e_{-\alpha-\beta} - \sum_{\alpha>0}\langle\alpha,\alpha\rangle^2 e_\alpha\wedge e_{-\alpha}\wedge\tau_\alpha\Big)\Big) \tag{11.4.4}$$

For the purposes of Theorem 11.4.1, we choose a Weyl basis such that the involution $\theta(h) = \theta + \sum_{n>0}\theta^{(n)}h^n$ described in the theorem has leading term θ

$$\theta(e_\alpha) = -e_{-\alpha}, \quad \theta(\tau_\alpha) = -\tau_\alpha.$$

Proposition 11.4.1 *If B_h satisfies the conditions of Theorem 11.4.1 then, relative to the previous normalizations, the infinitesimal deformation has the special form*

$$r_1 = a \sum_{\alpha \in R_+} \langle \alpha, \alpha \rangle e_\alpha \wedge e_{-\alpha} \tag{11.4.5}$$

for an appropriate choice of positive roots.

Condition (2b) in Theorem 11.4.1 implies

$$(\theta(h) \otimes \theta(h)) \circ \Delta(h) = \tau \circ \Delta(h) \circ \theta(h), \tag{11.4.6}$$

where $\tau(a \otimes b) = b \otimes a$. Isolating the h term gives the equation

$$(\theta^{(1)} \otimes \theta) \circ \Delta + (\theta \otimes \theta^{(1)}) \circ \Delta + (\theta \otimes \theta) \circ \Delta^{(1)} = \tau \circ \Delta^{(1)} \circ \theta + \tau \circ \Delta \circ \theta^{(1)} \tag{11.4.7}$$

Apply θ to equation (11.4.7), recombine terms and use the fact that θ is an involution and coalgebra automorphism relative to Δ. (Since Δ is cocommutative there is no difference between automorphism and antiautomorphism). Then

$$\Delta^{(1)} - \tau \circ (\theta \otimes \theta) \circ \Delta^{(1)} \circ \theta = -\delta(\theta \circ \theta^{(1)}). \tag{11.4.8}$$

Subject to our normalizations, with $\Delta^{(1)} = d(r)$ the left side of equation (11.4.8) takes values in the skew symmetric tensor product and the right side takes values in the symmetric tensor product so both sides vanish. This implies that r is θ skew-invariant. The fact that the comultiplication on $U(\mathcal{H})$ is cocommutative implies that r is $\mathcal{H}$ invariant. Invariance implies that r is in the zero weight space of the adjoint action on $\wedge^2 \mathcal{G}$. Skew invariance relative to θ implies that there is no component in $\wedge^2 \mathcal{H}$. These two facts together imply that r can be written in the form

$$r_1 = \sum_{\alpha \in R} a_\alpha \langle \alpha, \alpha \rangle e_\alpha \wedge e_{-\alpha} \tag{11.4.9}$$

where the sum is over all roots and $a_{-\alpha} = -a_\alpha$. We show that the vanishing of the obstruction class in $\hat{H}^4(U(\mathcal{G}))$ forces (11.4.9) to take the form (11.4.5). By Proposition 11.3.3, the obstruction class is, up to a constant represented by

$$\begin{aligned} \sum a_\alpha a_\beta \langle \alpha, \alpha \rangle \langle \beta, \beta \rangle (&e_\alpha \wedge e_\beta \wedge [e_{-\alpha}, e_{-\beta}] - e_\alpha \wedge e_{-\beta} \wedge [e_{-\alpha}, e_\beta] \\ &+ e_{-\alpha} \wedge e_{-\beta} \wedge [e_\alpha, e_\beta] - e_{-\alpha} \wedge e_\beta \wedge [e_\alpha, e_{-\beta}]). \end{aligned} \tag{11.4.10}$$

If this vanishes in $\hat{H}^4(U(\mathcal{G}))$ then it is a multiple of (11.4.3) and we get equation

$$(a_\alpha\langle\alpha,\alpha\rangle)^2 = b\cdot\langle\alpha,\alpha\rangle^2 \tag{11.4.11}$$

and, collecting all multiples of the term $e_\alpha \wedge e_\beta \wedge e_\gamma$ we get

$$\begin{aligned} & a_\alpha a_\beta N_{-\alpha,-\beta}\langle\alpha,\alpha\rangle\langle\beta,\beta\rangle \\ + \ & a_\beta a_\gamma N_{-\beta,-\gamma}\langle\beta,\beta\rangle\langle\gamma,\gamma\rangle \\ + \ & a_\gamma a_\alpha N_{-\gamma,-\alpha}\langle\gamma,\gamma\rangle\langle\alpha,\alpha\rangle \ = \ -b\cdot\langle\alpha,\alpha\rangle\langle\beta,\beta\rangle N_{-\alpha,-\beta} \end{aligned} \tag{11.4.12}$$

Fix a choice of square root $\sqrt{b}\in k$ and define

$$\epsilon_\alpha = \frac{a_\alpha}{\sqrt{b}}$$

Equations (11.4.11) and (11.4.12) imply

$$\epsilon_\alpha^2 = 1 \tag{11.4.13}$$

and

$$\epsilon_\alpha\epsilon_\beta + \epsilon_\beta\epsilon_\gamma + \epsilon_\gamma\epsilon_\alpha = -1 \text{ for } \alpha+\beta+\gamma = 0. \tag{11.4.14}$$

Rewriting $\epsilon_\gamma = \epsilon_{-\alpha-\beta} = -\epsilon_{\alpha+\beta}$,

$$\epsilon_\alpha\epsilon_\beta - \epsilon_\beta\epsilon_{\alpha+\beta} - \epsilon_{\alpha+\beta}\epsilon_\alpha = -\epsilon_\alpha^2$$

$$(\epsilon_\alpha - \epsilon_{\alpha+\beta})(\epsilon_\alpha + \epsilon_\beta) = 0. \tag{11.4.15}$$

The set

$$S = \{\alpha\in R | \epsilon_\alpha = 1\}$$

satisfies the conditions in the following simple lemma (which we state without proof) and thus defines a system of positive roots for $\mathcal{G}$.

Lemma 11.4.1 *Let R be the set of roots of a simple Lie algebra $\mathcal{G}$ and $S\subset R$ a subset such that*

$$S\cup -S = R, \tag{a}$$

$$S\cap -S = \emptyset \tag{b}$$

and

$$\alpha,\beta\in S,\ \alpha+\beta\in R \Rightarrow \alpha+\beta\in S \tag{c}.$$

Then S is a system of positive roots for $\mathcal{G}$.

This lemma shows that r has the desired form and concludes the proof of the proposition. Suppose that we have two deformations satisfying the conditions of the theorem with data $(\Delta(h), \theta(h))$ and $(\Delta'(h), \theta'(h))$ respectively. Then for each one the infinitesimal has the form in Proposition 11.4.1.

$$\begin{aligned} r_1 &= a \sum_{\alpha \in S} \langle \alpha, \alpha \rangle e_\alpha \wedge e_{-\alpha} \\ r'_1 &= a' \sum_{\beta \in S'} \langle \beta, \beta \rangle e_\beta \wedge e_{-\beta}. \end{aligned}$$

Any two positive root systems are conjugate by the action of an element w of the Weyl group, $wS = S'$. We can conjugate the data $(\Delta(h), \theta(h))$ of the first deformation by w and get an equivalent deformation with infinitesimal proportional to that of the second deformation. This operation followed by a scaling in h makes the two infinitesimal deformations agree, therefore we will assume that the two deformations have the same infinitesimal deformation. Let k be the index for which

$$(\Delta^{(k)}, \theta^{(k)}) \neq (\Delta^{(k)'}, \theta^{(k)'}) \tag{11.4.16}$$

$$(\Delta^{(j)}, \theta^{(j)}) = (\Delta^{(j)'}, \theta^{(j)'}) \qquad j < k. \tag{11.4.17}$$

Since the deformations agree to order k, their obstruction cocycles at that order, $(0, 0, \beta_k)$, are the same. For both deformations the fact that they are defined through order k implies that $D(0, \Delta^{(k)}) = (0, 0, \beta_k) = D(0, \Delta^{(k)'})$ so the difference $(0, \Delta^{(k)} - \Delta^{(k)'})$ is a cocycle for the total complex and in particular $d(\Delta^{(k)} - \Delta^{(k)'}) = 0$. Just as in the case of the infinitesimal deformation we have can find equivalent deformations such that $\Delta^{(k)}(a) - \Delta^{(k)'}(a) = dr_k$, with $r_k \in \wedge^2 \mathcal{G}$. Equation (11.4.6) implies

$$\begin{aligned} ((\theta^{(k)} - \theta^{(k)'}) \otimes \theta) \circ \Delta + (\theta \otimes (\theta^{(k)} - \theta^{(k)'})) \circ \Delta + (\theta \otimes \theta) \circ (\Delta^{(k)} - \Delta^{(k)'}) \\ = \tau \circ (\Delta^{(k)} - \Delta^{(k)'}) \circ \theta + \tau \circ \Delta \circ (\theta^{(k)} - \theta^{(k)'}). \end{aligned} \tag{11.4.18}$$

After substituting in this equation, essentially the same argument as for the infinitesimal implies that r_k is θ invariant and $\mathcal{H}$ invariant and thus has the form (11.4.9). Next we prove that in fact it is a multiple of the infinitesimal.

Lemma 11.4.2 $r_k = ar_1$.

Proof. The difference between the obstruction classes at order $k+1$ is

$$\begin{aligned} & (id \otimes (\Delta^{(k)} - \Delta^{(k)'})) \circ \Delta^{(1)} + (id \otimes \Delta^{(1)}) \circ (\Delta^{(k)} - \Delta^{(k)'}) \\ - & ((\Delta^{(k)} - \Delta^{(k)'}) \otimes id) \circ \Delta^{(1)} - (\Delta^{(1)} \otimes id) \circ (\Delta^{(k)} - \Delta^{(k)'}). \end{aligned} \tag{11.4.19}$$

Since there is an extension to order $k+1$ for each of the deformation, each obstruction cobounds, therefore, so does the difference. Just as above we find that writing

$$r_1 = a \sum_{\alpha \in R} \epsilon_\alpha \langle \alpha, \alpha \rangle e_\alpha \wedge e_{-\alpha},$$

where $\epsilon_\alpha = +1$ for $\alpha \in R_+$, $\epsilon_\alpha = -1$ for $\alpha \in -R_+$ and

$$r_k = \sum_{\beta \in R} a_\beta \langle \beta, \beta \rangle e_\beta \wedge e_{-\beta}$$

the cohomology class of (11.4.19) is represented by

$$\sum a(\epsilon_\alpha a_\beta + \epsilon_\beta a_\alpha)\langle \alpha, \alpha \rangle \langle \beta, \beta \rangle \times \quad (e_\alpha \wedge e_\beta \wedge [e_{-\alpha}, e_{-\beta}] - e_\alpha \wedge e_{-\beta} \wedge [e_{-\alpha}, e_\beta] \\ + e_{-\alpha} \wedge e_{-\beta} \wedge [e_\alpha, e_\beta] - e_{-\alpha} \wedge e_\beta \wedge [e_\alpha, e_{-\beta}],$$

Analogously to (11.4.13) and (11.4.14) we get

$$2\epsilon_\alpha a_\alpha = b$$

$$a_\beta(\epsilon_\alpha + \epsilon_\gamma) + a_\gamma(\epsilon_\alpha + \epsilon_\beta) + a_\alpha(\epsilon_\beta + \epsilon_\gamma) = -b \text{ if } \alpha + \beta + \gamma = 0$$

Therefore

$$a_\alpha = \frac{b}{2}\epsilon_\alpha .$$

and r_k is a multiple of r_1, proving the lemma. This proves that $\Delta^{(k)} - \Delta^{(k)'} = a\Delta^{(1)}$ and a change of parameter $h \to h + ah^k$ in the primed deformation will change $\Delta^{(k)'}$ to $\Delta^{(k)'} + a\Delta^{(1)} = \Delta^{(k)}$ Now let's look at the conjugation at order k under the assumption that the comultiplications agree through order k. Equation (11.4.18) implies that $\xi = \theta \circ (\theta^{(k)} - \theta^{(k)'})$ is a δ cocycle, which says that

$$(id \otimes \xi) \circ \Delta - \Delta \circ \xi + (\xi \otimes id) \circ \Delta = 0.$$

Since we are assuming that the unit is undeformed and that $\theta(h)$ is an algebra automorphism, we have $\xi(1) = \theta \circ (\theta^{(k)}(1) - \theta^{(k)\prime}(1) = 0$. The cocycle equation implies that ξ maps Δ primitive elements into Δ primitive elements, i.e., maps $\mathcal{G}$ into $\mathcal{G}$. The algebra automorphism condition

$$m \circ (\theta(h) \otimes \theta(h)) = \theta(h) \circ m$$

implies that for all k

$$m \circ (\theta^{(k)} \otimes \theta + \theta \otimes \theta^{(k)}) = \theta^{(k)} \circ m.$$

Composing with θ gives $d(\theta \circ \theta^{(k)}) = 0$. This is true for $\theta^{(k)'}$ as well, so ξ also a d 1-cocycle , and thus a d coboundary, $d(Z)$. Since ξ maps $\mathcal{G}$ into itself, $Z \in \mathcal{G}$.

$$\theta \circ (\theta^{(k)} - \theta^{(k)'})(a) = [a, Z] = ad(Z)(a).$$

If $Y = \theta(Z)$,

$$(\theta^{(k)} - \theta^{(k)'})(a) = ad(Y)\theta(a).$$

The difference at order k of the two equations

$$\theta(h)^2 = id \qquad \theta'(h)^2 = id$$

implies

$$\theta \circ (\theta^{(k)} - \theta^{(k)'}) + (\theta^{(k)} - \theta^{(k)'}) \circ \theta = 0,$$

hence

$$Z = \theta(Y) = -Y.$$

Define the $k[[h]]$ module automorphism

$$\varphi(h) = id + \varphi^{(k)}\ h^k$$

where

$$\varphi^{(k)} = \frac{1}{2} ad(Y). \tag{11.4.20}$$

Conjugate $\theta'(h)$ by $\varphi(h)$ and compute modulo h^{k+1},

$$\begin{aligned}
\varphi(h)\theta'(h)(\varphi(h))^{-1} &= (id + \varphi^{(k)}h^k)\theta'(h)(id + \varphi^{(k)}h^k)^{-1} \\
&= (id + \varphi^{(k)}h^k)(\theta'(h) - \theta \circ \varphi^{(k)}h^k) \\
&= \theta'(h) - \theta \circ \varphi^{(k)}h^k + \varphi^{(k)} \circ \theta h^k \\
&= \theta'(h) + (2\varphi^{(k)} \circ \theta)h^k \\
&= \theta(h).
\end{aligned}$$

Modulo h^{k+1}

$$(\varphi(h) \otimes \varphi(h)) \circ \Delta'(h) \circ (\varphi(h)^{-1} = \Delta'(h)$$

since the only new terms that could arise come from

$$(\varphi(h) \otimes \varphi(h)) \circ \Delta \circ (\varphi(h))^{-1} - \Delta = 0$$

and $\varphi(h)$ is a coalgebra automorphism modulo h^{k+1}. Thus we have defined equivalences modulo h^{k+1} between the two deformations, including the involutions. As before iterating this procedure to all orders gives a well defined equivalence between the deformations, proving uniqueness. The corresponding existence theorem for this case is provided by the Drinfeld-Jimbo[317][576]. If we do not require a quantization of the Cartan involution it is possible to consider more general infinitesimals r. At present the existence problem for the quantization of an arbitrary solution to the classical Yang Baxter equation is still open. [452].

11.5 Uniqueness of Drinfeld algebra deformations.

We saw in section 9.7 that to study Drinfeld deformations of bialgebras, we consider a subcomplex $\tilde{C}(B)$, of the full Gerstenhaber-Schack complex which includes the terms in the bottom row,

$$C^{p,0} = \mathrm{Hom}_k(k, \otimes^p B) \quad p > 0.$$

Its cohomology, which will be denoted $\tilde{H}(B)$, classifies Drinfeld algebra deformations B_h of a bialgebra $B = B_0$, up to equivalence which includes twisting. The basic assumption about the deformation is that the undeformed coassociativity operator is the identity $\Phi = 1 \otimes 1 \otimes 1$. For any Lie algebra of $\mathcal{G}$ over k the Drinfeld algebra deformation cohomology of $U(\mathcal{G})$ is

$$\begin{aligned} \tilde{H}^{p,q}(U(\mathcal{G})) &\cong H^q(\mathcal{G}, \wedge^p \mathcal{G}) \quad q \geq 0,\ p > 0 \\ (&\cong (\wedge^p \mathcal{G})^{\mathcal{G}} \quad q = 0). \end{aligned}$$

For $\mathcal{G}$ a simple Lie algebra over k

$$\begin{aligned} \tilde{H}^3(U(\mathcal{G})) &\cong (\wedge^3 \mathcal{G})^{\mathcal{G}} = (u) \\ \tilde{H}^4(U(\mathcal{G})) &\cong (\wedge^4 \mathcal{G})^{\mathcal{G}} \end{aligned}$$

where (u) is the unique (up to scale) ad-invariant element of $\wedge^3 \mathcal{G}$.

Using this result we can prove the following stronger uniqueness result.

Theorem 11.5.1 *Let $\mathcal{G}$ be simple over k. Up to equivalence involving isomorphism of B_h as a $k[[h]]$ algebra, change of parameter h, and twisting there is at most one non-trivial Drinfeld algebra deformation of $U(\mathcal{G})$.*

Just as above, in order to the prove Theorem 11.5.1, we first prove that the deformations can be put in a normal form:

Proposition 11.5.1 *Any Drinfeld algebra deformation*

$$(U_h(\mathcal{G}), m(h), \Delta(h), \Phi(h))$$

for $\mathcal{G}$ a simple Lie algebra over a field of characteristic 0 is equivalent to one of the form

$$(U_h(\mathcal{G}), m, \Delta, \Phi(h)),$$

that is, one with the multiplication and comultiplication undeformed.

Proof. Assume that we have already normalized the deformation so that $m(h) = m$. If

$$\Delta(h) = \Delta \bmod h^n \tag{11.5.1}$$

then the obstruction cocycle has the form $\beta_n = (0, 0, \beta_n^{3,1}, \beta_n^{4,0})$, and therefore, the equation

$$D(0, \Delta^{(n)}, \Phi_n) = (0, d\Delta^{(n)}, -\delta(\Delta^{(n)}) + d(\Phi_n), \delta(\Phi_n)) = \beta_n$$

implies $d(\Delta^{(n)}) = 0$. As before, the vanishing of the first Hochschild cohomology of $U(\mathcal{G})$ implies that there exits $u \in U(\mathcal{G}) \otimes U(\mathcal{G})$ such that $d(u) = \Delta^{(n)}$. Twisting by $1 \otimes 1 + uh^n$ changes $\Delta^{(n)}$ to $\Delta^{(n)} - d(u)$ an gives an equivalent Drinfeld algebra deformation with $\Delta(h)$ vanishing to one higher order. The same induction we have already used several times, involving successive compositions completes the proof.

In order to facilitate the interpretation of infinitesimals as defining a quasi-Lie-bialgebra structure (see the last section of this chapter) Drinfeld assumes that the antisymmetric part of Φ_1 vanishes. This implies that after twisting we can asumme $\Phi_1 = 0$. That means that we have to deal with possibly nontrivial deformations in which the leading term is the coefficient of h^n for $n > 1$.

Proposition 11.5.2 *Let $\mathcal{G}$ be as in the preceding proposition. Given a deformation $(U(\mathcal{G}), m, \Delta, \Phi(h))$ with $\Phi(h) = 1 \otimes 1 \otimes 1 + \Phi_n h^n +$ higher order terms where $n > 1$ and $\Phi_n \neq 0$ spans $(\wedge^3 \mathcal{G})^{\mathcal{G}}$. After an appropriate equivalence we can assume that Φ_h contains only powers of h^n.*

Proof. Suppose that

$$\Phi(h) = 1 \otimes 1 \otimes 1 + \sum_{j=1\ldots m} \Phi_{jn} h^{jn} + \Phi_{mn+a} h^{mn+a} + \ldots$$

where $0 < a < n$. Since the obstruction cocycle at order $mn + a$ is 0, Φ_{mn+a} is a D cocycle. As above this means that it is a δ cocycle and is $\mathcal{G}$ invariant. Once again we use the fact that $C^{\cdot,0}$ splits into the subcomplex of $\mathcal{G}$ invariants and a $\mathcal{G}$ invariant complement. Inside the complex of invariants , the element Φ_{mn+a} differs by a coboundary δF_{mn+a} from Φ_n, a generator of $(\wedge^3\mathcal{G})^{\mathcal{G}}$. By twisting with the $\mathcal{G}$ invariant $F = 1 \otimes 1 + F_{mn+a}h^{mn+a}$ we can preserve the condition $\Delta = \Delta(h)$ and transform Φ_{mn+a} to $\Phi_{mn+a} + \delta(F_{mn+a}) = c\Phi_n$. Then applying the change of parameters $h' = h - \frac{c}{n}h^{(m-1)n+a+1}$ subtracts an appropriate multiple of Φ_n and cancels the term in question not affecting any of the preceding terms.

Now we can prove Theorem 2. More specifically

Proposition 11.5.3 *Let $\mathcal{G}$ be simple over k. Up to $k[[h]]$ isomorphism, twisting, and change of parameter there is at most one nontrivial Drinfeld algebra deformation of $(U(\mathcal{G}),\ m,\ \Delta,\ 1 \otimes 1 \otimes 1)$ and it has the form $(U_h(\mathcal{G}),\ m,\ \Delta,\ \Phi(h))$ with Φ_2 a non-zero element of $(\wedge^3\mathcal{G})^{\mathcal{G}}$, that is the multiplication and comultiplication are undeformed and Φ_2 is a multiple of the element u in Koszul's lemma.*

Proof. Assume that there exists a non-trivial deformation $U_h(\mathcal{G})$ and that it has been normalized according to Propositions 11.5.1 and 11.5.2 and such that $\Phi_2 = u$, where u is a chosen basis for $(\wedge^3\mathcal{G})^{\mathcal{G}}$. Let $U_h(\mathcal{G})'$ be a second deformation, similarly normalized. Thus both deformations have only even powers of h and agree through order 2. If $2k$ be the first index at which Φ_{2k} and Φ'_{2k} differ, then their difference defines a $\mathcal{G}$ invariant cocycle. There is a $\mathcal{G}$ invariant cochain $F_{2k} \in (U(\mathcal{G}) \otimes U(\mathcal{G}))^{\mathcal{G}}$ such that

$$\Phi_{2k} - \Phi'_{2k} = \delta_c F_{2k} + c\Phi_2.$$

Twisting $U_h(\mathcal{G})'$ by $1 \otimes 1 + F_{2k}h^{2k}$ and changing the parameter from h to $h + (c/2)h^{2k-1}$ transforms it to a deformation agreeing with $U_h(\mathcal{G})$ to order $2k + 2$. Iterating this procedure proves the theorem.

Throughout the previous discussion, if we assume that the cochains are normalized, then the twistings will be of the form $1 \otimes 1 + F_n h^n$ where $F_n \in C^{2,0}_{N,N}$. Then the total twisting will satisfy the condition $\epsilon \otimes id F = id \otimes F = 1 \otimes 1$. In general we can always assume this condition by considering only deformations involving normalized cochains.

11.6 Quasitriangular structures

In this section we investigate the role of the hexagon identities in the deformation theory of $U(\mathcal{G})$. The material is selected from Drinfeld's papers [321][322][324] and we follow his exposition fairly closely. Throughout this section $\mathcal{G}$ will be a Lie algebra, not necessarily semisimple, defined over a field k of characteristic 0, the formal power series algebra over such a field, $k_h = k[[h]]$, or some truncation thereof, $k_{(n)} = k_h/h^{n+1}k_h$. We will suppress the explicit dependence on the deformation parameter h in many formulae. All tensor products of objects defined over k_h are completed with respect to the power series topology. The symbol 1 will be used interchangeably with $1 \otimes 1 \otimes \cdots \otimes 1$ for the identity in the tensor product of algebras. No explicit reference is made to the unit or counit, which are assumed undeformed, in listing the deformation data.

Definition A *quasitriangular Drinfeld quantized universal enveloping algebra (qtDque)* $(U_h(\mathcal{G}), m(h), \Delta(h), \Phi(h), R(h))$ is a Drinfeld algebra deformation of $U(\mathcal{G})$ defined over k_h with coassociativity operator $\Phi = \Phi(h) \in U_h(\mathcal{G})^{\otimes 3}$ satisfying the pentagon identity, and cocommutativity operator $R = R(h) \in U_h(\mathcal{G})^{\otimes 2}$ satisfying the two hexagon identities. $\Phi \equiv 1 \otimes 1 \otimes 1 \mod h^2$ and $R \equiv 1 \otimes 1 \mod h$. The fact that the deformation of Φ is assumed begin at order 2 arises from the interpretation of the leading terms in Δ and Φ as defining a structure of quasi-Lie-bialgebra as noted above and explained at the end of the chapter. The full list of identities for Φ and R is

$$\Phi((\Delta \otimes id)\Delta(u)) = ((id \otimes \Delta)\Delta(u))\Phi \tag{11.6.1}$$
$$R\Delta(u) = \Delta^{\mathrm{op}}(u)R \tag{11.6.2}$$
$$(id \otimes \epsilon \otimes id)\Phi = 1 \otimes 1 \otimes 1 \tag{11.6.3}$$
$$(1 \otimes \Phi)((id \otimes \Delta \otimes id)\Phi)(\Phi \otimes 1) = ((id \otimes id \otimes \Delta)\Phi)((\Delta \otimes id \otimes id)\Phi) \tag{11.6.4}$$
$$(\Delta \otimes id)R = \Phi^{312}R^{13}(\Phi^{132})^{-1}R^{23}\Phi \tag{11.6.5}$$
$$(id \otimes \Delta)R = (\Phi^{231})^{-1}R^{13}\Phi^{213}R^{12}\Phi^{-1} \tag{11.6.6}$$

Recall that twisting by $F \in U_h \otimes U_h$ with $(\epsilon \otimes id)F = (id \otimes \epsilon)F = 1$ sends

$$\Delta \mapsto F\Delta F^{-1}$$

changes R by

$$R \mapsto F^{21}RF^{-1}. \tag{11.6.7}$$

and

$$\Phi \mapsto F^{23}(id \otimes \Delta)(F)\Phi(\Delta \otimes id)(F^{-1})(F^{12})^{-1}.$$

We saw in Chapter 8 that the pentagon identity is preserved under twisting. The formulae for the transformed elements in the first of the hexagon identities are

$$\begin{aligned}
(\tilde{\Delta} \otimes id)(\tilde{R}) &= F^{12}(\Delta \otimes id)(F^{21})(\Delta \otimes id)(R))(\Delta \otimes id)(F^{-1})(F^{12})^{-1} \\
\tilde{\Phi}^{312} &= F^{12}(\Delta \otimes id)(F^{21})\Phi^{312}(\Delta \otimes id)(F^{-1})^{312}(F^{31})^{-1} \\
\tilde{R}^{13} &= (F^{31})R^{13}(F^{13})^{-1} \\
(\tilde{\Phi}^{132})^{-1} &= F^{13}((\Delta \otimes id)(F))^{132}(\Phi^{132})^{-1}(id \otimes \Delta^{\mathrm{op}}(F))^{-1}(F^{32})^{-1} \\
\tilde{R}^{23} &= F^{32}R^{23}(F^{23})^{-1} \\
\tilde{\Phi} &= F^{23}(id \otimes \Delta)(F)\Phi(\Delta \otimes id)(F^{-1})F^{12}.
\end{aligned}$$

A sequence of cancellations and commutations of Δ and R show that the first hexagon identity is invariant. The calculation for the second hexagon identity is essentially the same.

One way of constructing deformations of $U(\mathcal{G})$ is from deformations of $\mathcal{G}$ as a Lie algebra, as we now explain.

Let $\mathcal{G}_h$ be a Lie algebra deformation of $\mathcal{G}$ defined over k_h. As usual we assume that $\mathcal{G}_h \cong \mathcal{G} \otimes k_h$ as a k_h module. If we pick a k_h basis for $\mathcal{G}_h$ in $\mathcal{G}$ and use this basis to construct the Poincaré-Birkhoff-Witt basis for the standard universal enveloping algebra, $U(\mathcal{G}_h)$, we get an isomorphism of k_h modules

$$U(\mathcal{G}_h) \cong U(\mathcal{G}) \otimes k_h.$$

Any universal enveloping algebra is cocommutative and coassociative, so if we take $R = 1$ and $\Phi = 1$ and get a qtDque $(U(\mathcal{G}_h), m_h, \Delta, 1, 1)$. We use the notation m_h in place of $m(h)$ to indicate that the dependence on h in m_h comes from the fact that the structure constants of $\mathcal{G}_h$ are in k_h. This may seem rather trivial, but Drinfeld proves the rather amazing result any quasitriangular deformation of $U(\mathcal{G})$ satisfying the additional condition

$$R^{21}(h)R(h) = 1 \tag{11.6.8}$$

is equivalent under twisting alone to a deformation obtained by deforming $\mathcal{G}$ as we have just described. A qtDque satisfying (11.6.8) is called *triangular*, (tDque). The condition is equivalent to the fact that the category of representations is a *symmetric monoidal category*, that is, the commutativity operator is an involution. Note that twisting preserves the triangular condition.

Let $U(\mathcal{G}_h)^F$ be the tDque given by twisting $(U(\mathcal{G}_h), m_h, \Delta, 1, 1)$ by F. In this section we give Drinfeld's proof [321] of

Theorem 11.6.1 Triangular Classification Theorem *Any triangular Drinfeld quantized universal enveloping algebra has the form $U(\mathcal{G}_h)^F$. Any morphism between $U(\mathcal{G}_h)^F$ and $U(\mathcal{G}'_h)^{F'}$ has the form $Ad(u) \circ \bar{\lambda}$ where $\bar{\lambda}$ is the extension to $U(\mathcal{G}_h)$ of a k_h Lie algebra homomorphism $\lambda : \mathcal{G}_h \to \mathcal{G}'_h$, $Ad(u)$ is conjugation by $u \in U(\mathcal{G}'_h)$ with $\epsilon(u) = 1$ and $u \equiv 1 \bmod h$, and $F' = (u \otimes u) \cdot (\bar{\lambda} \otimes \bar{\lambda})(F)\Delta(u)^{-1}$, where Δ is the untwisted comultiplication.*

In fact Drinfeld proves a far more general result about the reduction of qtDque $U_h(\mathcal{G})$ to the form $U(\mathcal{G}_h)$.

Theorem 11.6.2 Quasitriangular Structure Theorem. *Any quasitriangular Drinfeld quantized universal enveloping algebra can by a suitable twist be brought into the form of $U(\mathcal{G}_h)$ with $R = exp(ht)$ for $t \in Sym^2(\mathcal{G}_h)^{\mathcal{G}_h}$ and $\Phi = \varphi(ht^{12}, ht^{23})$ a universal formula valid for any $\mathcal{G}_h$.*

In contrast to the triangular case, given a Lie algebra deformation, $\mathcal{G}_h$, and a symmetric invariant $t \in \mathrm{Sym}^2(\mathcal{G}_h)^{\mathcal{G}_h}$, it is not clear that we can find a Φ which completes the data for a quasitriangular Drinfeld deformation of $U(\mathcal{G})$.

Theorem 11.6.3 Quasitriangular Existence Theorem *Such a Φ exists and is unique up to twisting by a symmetric $\mathcal{G}$ invariant element $F \in U_h \otimes U_h$.*

Drinfeld gives two proofs of the existence theorem. In the next chapter we show how the function the Φ can be constructed from the Knizhnik-Zamolodchikov equation. However, Drinfeld points out that the coefficients involve terms $\zeta(2m+1)/(2\pi i)^{2m+1}$ which are imaginary and probably transcendental. He shows how to define a modified formula valid for any field of characteristic 0. The second proof uses deformation cohomology. Both proofs are too long to present in detail here.

In the remainder of this section we will give Drinfeld's proof of the first theorem on triangular structures. In the next section we introduce a third type of structure called coboundary which allows us to use a deformation cohomology to prove the general quasitriangulalr structure theorem. In the final section we discuss the Lie algebra interpretation of the classical limit of the three types of structures and with that introduction direct the interested reader to Drinfeld's truly remarkable papers.

A result basic to the whole approach states that any formal deformation of $U(\mathcal{G})$ has the form of a universal enveloping algebra of some deformation of $\mathcal{G}$ once we know that it is cocommutative and coassociative (which is not to require that $R = 1$ and $\Phi = 1$).

Proposition 11.6.1 *Any cocommutative coassociative quantized universal enveloping algebra has the form $U(\mathcal{G}_h)$ for some Lie algebra deformation $\mathcal{G}_h$ of $\mathcal{G}$.*

Proof. Let B be such a deformation with $B/hB \cong U(\mathcal{G})$. Define the map $\rho = m \circ \Delta : B \to B$. The same map $m \circ \Delta$ on $U(\mathcal{G})$ has eigenvalues 2^m with the corresponding eigenspace, W_m, spanned by symmetric products of m basis vectors from $\mathcal{G}$. (These eigenvectors are the image of the monomials under the Weyl quantization of the polynomial algebra on $\mathcal{G}^*$.) The subspaces W_m for $m > 0$ give a direct sum decomposition of the augmentation ideal of $U(\mathcal{G})$. Let $I = \mathrm{Ker}\epsilon : B \to k[[h]]$ be the augmentation ideal of B. Then $\Delta : I \to B \otimes I + I \otimes B$ and therefore $\rho : I \to I$. For all p

$$I/h^p I = \oplus_{0<m} W_{m,p}$$

where $W_{m,p} = \mathrm{Ker}(\rho - 2^m id)^p$. The projection $B/h^{p+1}B \to B/h^p B$, maps $W_{m,p+1} \to W_{m,p}$. Taking the inverse limit defines $k[[h]]$ submodules W_m which give a topological direct sum decomposition of I. (In the topological direct sum there are infinite sums, but they are convergent in the Krull topology.) Define $W_0 = k[[h]]$. The coassociativity of Δ implies

$$\begin{aligned}(\Delta \otimes \Delta) \circ \Delta &= (id \otimes id \otimes \Delta) \circ (\Delta \otimes id) \circ \Delta \\ &= (id \otimes id \otimes \Delta) \circ (id \otimes \Delta) \circ \Delta \\ &= (id \otimes \Delta \otimes id) \circ (id \otimes \Delta) \circ \Delta.\end{aligned}$$

Then using the expression in the last line, cocommutativity implies that

$$s_{23} \circ (\Delta \otimes \Delta) \circ \Delta = (\Delta \otimes \Delta) \circ \Delta.$$

Therefore

$$\begin{aligned}\rho \otimes \rho(\Delta(u)) &= (m \otimes m) \circ (\Delta \otimes \Delta)(\Delta(u)) \\ &= (m \otimes m) \circ s_{23} \circ (\Delta \otimes \Delta) \circ \Delta(u) = \Delta \circ m \circ \Delta(u) \\ &= \Delta\rho(u).\end{aligned}$$

Since $\rho - 2^m id$ is topologically nilpotent on W_m this implies that $\rho \otimes \rho - 2^m id$ restricted to $\Delta(W_m)$ is is topologically nilpotent, but $\rho \otimes \rho - 2^m id$ is invertible on $W_i \otimes W_j$ if $i + j \neq m$. Therefore, $\Delta W_m \subset \bigoplus(W_p \otimes W_{m-p})$, and $\Delta W_1 \subset W_0 \otimes W_1 + W_1 \otimes W_0$. If $u \in W_1$ then $\Delta(u) = 1 \otimes v + w \otimes 1$ for $v, w \in W_1$. Applying $\epsilon \otimes id$ and $id \otimes \epsilon$ gives $u = v = w$, so $\Delta(u) = u \otimes 1 + 1 \otimes u$, and u is primitive. Conversely if u is primitive then $\rho(u) = 2u$ and $u \in W_1$. Thus W_1 is precisely the primitive elements of B and as such is a Lie algebra. As a direct summand W_1 is a free $k[[h]]$ module and, since $W_1/hW_1 = \mathcal{G}$, it is a Lie algebra deformation of $\mathcal{G}$. The natural homomorphism $U(W_1) \to B$ is an isomorphism since its reduction mod h is.

Before proving the Theorem 11.6.1, we describe the structure of general qtDque in a series of lemmas. In order to simplify the notation, we will denote the qtDque simply by U_h.

Lemma 11.6.1 *For any qtDque there exists an $F \in U_h \otimes U_h$ of the form $F = \prod_{\leftarrow}(1 + F_n h^n)$ which twists R into symmetric form $R^{21} = R$.*

Proof. As usual we solve the required equation recursively in h. In this case we have

$$(F^{21}RF^{-1})^{21} = F^{21}RF^{-1}. \tag{11.6.9}$$

Assume that $R^{21} = R \bmod h^n$ and try to find $F_{(n)} = 1 + fh^n$ such that 11.6.9 is true modulo h^{n+1}. The equation for f is (modulo h^{n+1})

$$\begin{aligned}(1 + f^{21}h^n)^{-1}(1 + fh^n)R^{21}(1 + f^{21}h^n)^{-1}(1 + fh^n)\\ = R^{21} + 2(f - f^{21})h^n = R.\end{aligned}$$

A solution is given by

$$f = \frac{1}{4}(R_n - R_n^{21}).$$

Lemma 11.6.2 *Let U_h be a qtDque with symmetric R, then Δ is cocommutative.*

Proof. The element R^2 commutes with $\Delta(u)$ for all u, since

$$R^2(\Delta(u))R^{-2} = R(\Delta^{\text{op}}(u))R^{-1} = R^{21}(\Delta^{\text{op}}(u))(R^{21})^{-1} = \Delta(u).$$

Then $R^2 - 1 = hS$ and S commutes with $\Delta(u)$ for all $u \in U_h$. For any element of the form $1 + hS$ we can use the binomial series as a formal power series to extract a unique square root with leading term 1. This allows us to express R as a series in S, therefore it also commutes with all $\Delta(u)$. Then $\Delta^{\text{op}} = R\Delta R^{-1} = \Delta$ and Δ is cocommutative as asserted.

In further normalizations we will use another hexagon identity.

Lemma 11.6.3 *The following identity holds in any qtDque.*

$$\Phi^{321}R^{23}(id\otimes\Delta)(R)\Phi = R^{12}(\Delta\otimes id)(R) \tag{11.6.10}$$

Proof. Multiplying $(\Delta\otimes id)R$ on the left by R^{12} and using the formula for Δ^{op}

$$R^{12}(\Delta\otimes id)(R) = (\Delta^{\mathrm{op}}\otimes id)(R)R^{12}$$

Compute this expression by applying a transposition to positions 1, 2 in the first hexagon identity and multiplying on the right by R^{12}.

$$(\Delta^{\mathrm{op}}\otimes id)(R)R^{12} = \Phi^{321}R^{23}(\Phi^{231})^{-1}R^{13}\Phi^{213}R^{12}$$

Multiply the second hexagon identity on the left by $\Phi^{321}R^{23}$ and on the right by Φ gives

$$\Phi^{321}R^{23}(id\otimes\Delta)(R)\Phi = \Phi^{321}R^{23}(\Phi^{231})^{-1}R^{13}\Phi^{213}R^{12}.$$

This sequence of equations proves the lemma.

Lemma 11.6.4 *If U_h is a qtDque with symmetric R and, as a consequence of Lemma 2, cocommutative Δ, then Φ satisfies the identity*

$$\Phi^{321} = \Phi^{-1}. \tag{11.6.11}$$

Proof. Under these assumptions, the transposition of places 1, 3 transforms $R^{12}(\Delta\otimes id)R$ into $R^{23}(id\otimes\Delta)R$. Then 11.6.10 becomes

$$\Phi(R^{12}(\Delta\otimes id)R)\Phi^{321} = R^{23}(id\otimes\Delta)R.$$

Multiplying on the right by $R^{23}(id\otimes\Delta)R$ gives

$$\Phi((R^{12}(\Delta\otimes id)R)^2)\Phi^{-1} = (R^{23}(id\otimes\Delta)R)^2.$$

Since all the elements involved in these equations have the form $1+hu$ for some some $u\in U_h^{\otimes 3}$ we can use the binomial series as a formal power series to extract a unique square root with leading term 1. This gives the equation

$$\Phi((R^{12}(\Delta\otimes id)R))\Phi^{-1} = (R^{23}(id\otimes\Delta)R).$$

Then

$$\Phi^{321}\Phi R^{12}(\Delta\otimes id)R = \Phi^{321}R^{23}(id\otimes\Delta)R\Phi = R^{12}(\Delta\otimes id)R$$

Cancelling the invertible element $R^{12}(\Delta \otimes id)R$ proves the lemma.
Proof of the theorem. If $R^{21} = R$ and $R^{21}R = 1$, then $R^2 = 1$ and extracting the unique square root with leading term 1 gives $R = 1$. We have $\Phi \equiv 1 \bmod h^2$ by assumption. Suppose that we have defined a twisting such that $\Phi = 1 + \phi h^n \bmod h^{n+1}$. The pentagon identity implies that $\delta\phi = 0$. Then there is an $f \in C^{2,0} = U(\mathcal{G})^{\otimes 2}$ and an $a \in \wedge^3\mathcal{G}$ such that

$$\phi + \delta f = a. \tag{11.6.12}$$

A simple calculation shows that for any f, $(\delta f)^{321} = -\delta(f^{21})$. The equation 11.7.1 implies that $\phi^{321} = -\phi$. Applying the transposition (13) to 11.6.12 gives

$$-\phi - \delta(f^{21}) = -a.$$

Thus we can replace f with the symmetric element $f' = 1/2(f + f^{21})$ in 11.6.12. Twisting by $1 + f'h^n$ leaves the condition $R = 1$ unchanged and transforms Φ to $1 + ah^n$. Substituting this into the hexagon equation with $R = 1$ gives

$$1 = 1 + h^n(a^{312} - a^{132} + a) = 1 + 3ah^n$$

mod h^{n+1}, so $a = 0$ and we have normalized Φ to one higher order. We can define in this way a infinite product of equivalences which converges in the formal power series topology to the desired equivalence. Thus we reduce any tDque U_h to a form with $R = 1$, $\Phi = 1$, which implies that it is cocommutative and coassociative and by Proposition 11.6.2, it is the universal algebra of a deformation of $\mathcal{G}$.

To prove the second part of the theorem suppose that $f : U(\mathcal{G}_h)^F \to U(\mathcal{G}'_h)^{F'}$ is a morphism. Then

$$\begin{aligned} f \otimes f(R) &= R', \\ f \otimes f \otimes f(\Phi) &= \Phi', \\ (f \otimes f)(F)\Delta(a)(f \otimes f)(F^{-1}) &= F'\Delta'(f(a))F'^{-1} \end{aligned} \tag{11.6.13}$$

where Δ and Δ' are the comultiplications of $U(\mathcal{G}_h)$ and $U(\mathcal{G}'_h)$ respectively. Recall

$$R = F^{21}F^{-1}, \quad R' = F'^{21}F'^{-1},$$

and

$$\begin{aligned} \Phi &= F^{23}(id \otimes \Delta)F(\Delta \otimes id)F^{-1}(F^{12})^{-1}, \\ \Phi' &= F'^{23}(id \otimes \Delta')F'(\Delta' \otimes id)F'^{-1}(F'^{12})^{-1}. \end{aligned}$$

From 11.6.13 we conclude that f is also a morphism between $U(\mathcal{G}_h)$ (untwisted) and $U(\mathcal{G}'_h)^{F''}$, where

$$F'' = (f \otimes f)(F^{-1})F' \quad R'' = 1 \quad \text{and} \quad \Phi'' = 1.$$

In order to complete the proof of the theorem we need the following lemma.

Lemma 11.6.5 *If $F \in U(\mathcal{G}_h) \otimes U(\mathcal{G}_h)$ is symmetric, $(\epsilon \otimes id)F = 1$ and*

$$F^{23}(id \otimes \Delta)F(\Delta \otimes id)F^{-1}(F^{12})^{-1} = 1 \tag{11.6.14}$$

then

$$F = (u \otimes u)\Delta(u)^{-1} \tag{11.6.15}$$

for $u \in U(\mathcal{G}_h)$ $\epsilon(u) = 1$ $u \equiv 1 \bmod h$

Proof. Using the isomorphism of $k[[h]]$ modules $U(\mathcal{G}_h) \cong U(\mathcal{G}) \otimes k[[h]]$, we can expand F in powers of h with coefficients in $\mathcal{G}$. The proof follows the standard pattern of finding the element u as an infinite product of terms $1 + u_n h^n$, where the existence of u_n follows from the fact that something is a coboundary. Suppose that we have established 11.6.15 up to order n, then we consider $\hat{F} = (u \otimes u)^{-1} F \Delta(u) = 1 + \hat{F}_n h^n \bmod h^n$. The following calculation shows that $\hat{F}$ also satisfies 11.6.14.

$$\begin{aligned}
& 1 \otimes (u \otimes u)^{-1} F^{23}(1 \otimes \Delta(u))(u \otimes \Delta(u))^{-1}(id \otimes \Delta)F(id \otimes \Delta)\Delta(u) \\
= \; & (u \otimes u \otimes u)^{-1} F^{23}(id \otimes \Delta)F(id \otimes \Delta)\Delta(u) \\
= \; & (u \otimes u \otimes u)^{-1} F^{12}(\Delta \otimes id)F(\Delta \otimes id)\Delta(u) \\
= \; & (u \otimes u)^{-1} \otimes 1 F^{12}(\Delta(u) \otimes 1)(\Delta(u) \otimes u)^{-1}(\Delta \otimes id)F(\Delta \otimes id)\Delta(u)
\end{aligned}$$

which shows that $\hat{F}_n$ is a δ cocycle. Both F and u being normalized, so is $\hat{F}$, $(\epsilon \otimes id)\hat{F}_n = 0$, and thus there exists a u_n with $\epsilon(u) = 0$ such that $\hat{F}_n - \delta u_n \in \wedge^2 \mathcal{G}$. Since $\hat{F}_n$ is symmetric we conclude that, setting $\hat{u}_n = \frac{1}{2}(u_n + u_n^{21})$, we have $\hat{F}_n = \delta \hat{u}_n$. Then $\hat{F} = ((1 + \hat{u}_n h^n) \otimes (1 + \hat{u}_n h^n))\Delta(1 + \hat{u}_n h^n) \bmod h^{n+1}$ and the inductive step is established proving the lemma.

Now we finish the proof of the theorem. The lemma shows that

$$F'' = (u \otimes u)\Delta'(u)^{-1}.$$

Then

$$\bar{\lambda} = Ad(u^{-1})f : U(\mathcal{G}_h) \to U(\mathcal{G}'_h)$$

is a morphism of bialgebras and in particular a map of coalgebras and therefore carries primitives into primitives, that is, it defines a map

$$\lambda : \mathcal{G}_h \to \mathcal{G}'_h.$$

Being induced by an (associative) algebra map of universal enveloping algebras, λ is a map of Lie algebras and $\bar{\lambda}$ is its extension to $U(\mathcal{G}_h)$.

$$f = Ad(u) \circ \bar{\lambda}$$

proving the theorem.

11.7 Coboundary deformations

In order to extend the structure theorem for triangular deformations to the quasitriangular deformations, Drinfeld introduces an sort of intermediate structure called *coboundary* or *cDque* as an acronym in the case of coboundary Drinfeld quantized universal enveloping algebra. The "universal R matrix" still satisfies the condition $R^{21}R = 1$ but the two hexagon identities are replaced by the special hexagon identity in Lemma 11.6.3. The intersection of the class of coboundary structures and the quasitriangular structures is the triangular structures. The first result for coboundary structures is

Lemma 11.7.1 *Any cDque U_h is equivalent by twisting to one with Δ cocommutative and in fact $R = 1$.*

Proof. Just as in the triangular case, we can find an F which twists R to a symmetric form, in which case Δ is cocommutative and since $R^2 = R^{21}R = 1$, we also have $R = 1$.

By lemma 11.6.4 we have

$$\Phi^{321} = \Phi^{-1}.$$

This is not strong enough to prove directly that Δ is coassociative, and then to invoke lemma 11.6.2. Nonetheless, just as in the triangular case, any coboundary Drinfeld deformation of $U(\mathcal{G})$ is a universal enveloping algebra of some Lie algebra deformation.

The significance of this result on cDque in the general theory of qtDque comes from the fact that the operation $R \to R(R^{21}R)^{-1/2}$ which is called *unitarizing* converts any quasitriangular deformation into a coboundary one. The multiplication, comultiplication and coassociativity operator are unchanged. The first and second (fundamental) hexagon identities are not preserved under unitarizing but the third one given in lemma 11.6.3 is.

Proposition 11.7.1 *Let $(U_h, m(h), \Delta(h), R(h), \Phi(h))$ be a qtDque, then unitarizing R defines a cDque.*

Proof. Let $\tilde{R} = R(R^{21}R)^{-1/2}$. Calculating the commutator

$$R^{21}R\Delta(u) = R^{21}\Delta^{\mathrm{op}}(u)R = (R\Delta(u))^{21}R = (\Delta^{\mathrm{op}}(u)R)^{21}R = \Delta(u)R^{21}R$$

one sees that $R^{21}R$ commutes with Δ, and so does $(R^{21}R)^{1/2}$ computed from the binomial series. Furthermore $R(R^{21}R)^{-1/2}R^{-1} = (RR^{21})^{-1/2}$ implies that

$$\begin{aligned}(\tilde{R})(\tilde{R})^{21} &= (R(R^{21}R)^{-1/2})(R^{21}(RR^{21})^{-1/2}) \\ &= (RR^{21})^{-1/2}RR^{21}(RR^{21})^{-1/2} = 1\end{aligned}$$

proving the symmetry condition. Finally we have to verify 11.6.10. There are three basic identities that we will need

$$\begin{aligned}(\tilde{R})^{12}(\Delta\otimes id)(\tilde{R}) &= R^{12}(\Delta\otimes id)R[R^{21}((\Delta^{\mathrm{op}}\otimes idR^{21}))R^{12}(\Delta\otimes id)R]^{-1/2} \\ (\tilde{R})^{23}(id\otimes\Delta)(\tilde{R}) &= R^{23}(id\otimes\Delta)R[R^{32}((id\otimes\Delta^{\mathrm{op}})R^{23})R^{23}(id\otimes\Delta)R]^{-1/2} \\ R^{21}(\Delta\otimes id)R &= (\Phi)^{-1}R^{23}[(id\otimes\Delta)R]^{321}(\Phi^{321})^{-1}\end{aligned}$$

The first identity is proven as follows

$$\begin{aligned}(\tilde{R})^{12}(\Delta\otimes id)\tilde{R} &= R^{12}(R^{21}R^{12})^{-1/2}(\Delta\otimes id)R[(\Delta\otimes id)R^{21}(\Delta\otimes id)R]^{-1/2} \\ &= R^{12}(\Delta\otimes id)R[R^{21}R^{12}(\Delta\otimes id)R^{21}(\Delta\otimes id)R]^{-1/2} \\ &= R^{12}(\Delta\otimes id)R[R^{21}(\Delta^{\mathrm{op}}\otimes id)R^{21})R^{12}(\Delta\otimes id)R]^{-1/2}.\end{aligned}$$

We have used repeatedly the fact that $R^{21}R$ commutes with $\Delta(u)$. The second identity is proved similarly and the final identity follows from the fact that $[R^{21}(\Delta^{\mathrm{op}}\otimes id)R^{21}]^{321} = R^{23}(id\otimes\Delta)R$. The identity 11.6.10 then follows easily.

The cohomology theory that Drinfeld uses to study these deformation problems depends only on the coalgebra structure. For any coalgebra C we can define a differential on the sum $\oplus_{p\geq 0}C^{\otimes p}$

$$\begin{aligned}\delta(a_1\otimes a_2\otimes\cdots\otimes a_p) &= 1\otimes a_1\otimes\cdots\otimes a_p \\ &+ \sum_{1\leq i\leq p}(-1)^i a_1\otimes\cdots\otimes\Delta(a_i)\otimes\cdots\otimes a_p \\ &+ (-1)^{p+1}a_1\otimes\cdots\otimes a_p\otimes 1.\end{aligned}$$

For a bialgebra B we can identify this complex with the bottom row of the Gerstenhaber-Schack complex $\mathrm{Hom}_{B-4}(\mathrm{Bar}^0(B), \mathrm{Cob}(B))$ with the horizontal differential δ. Define the subcomplex of normalized cochains $C_N(B)$ with component in degree p

$$C_N^p(B) = \mathrm{Hom}_{B-4}(\mathrm{Bar}_N^0(B), \mathrm{Cob}_N^p(B)) \subset \mathrm{Hom}(k, B^{\otimes p}) \cong B^{\otimes p}.$$

The elements in this subcomplex satisfy the condition $\epsilon_j(a) = 0$ for all j. On $C_N(B)$ there is an involution

$$\tau(a_1 \otimes \cdots \otimes a_p) = (-1)^{\frac{p(p+1)}{2}}(a_p \otimes \cdots \otimes a_1)$$

and one easily checks that τ is a cochain map. Let $\bar{C}(B)$ be the -1 eigenspace of τ acting on $C_N(B)$ and $\bar{C}'(B)$ the $+1$ eigenspace. Since τ is a cochain map these are subcomplexes.

Lemma 11.7.2 *The cohomology $\bar{H}(U(\mathcal{G}))$ of the complex $\bar{C}(U(\mathcal{G}))$ is $\wedge^p(\mathcal{G})$ if p is odd and 0 if p is even. If a $\mathcal{G}$ invariant element of $\bar{C}^n$ cobounds then it cobounds an invariant element of $\bar{C}^{n-1}$*

Proof. We from Chapter 10, and Proposition 11.1.3 that the cohomology of the full complex $C_N(U(\mathcal{G}))$ is $\wedge\mathcal{G}$. Since τ is a cochain map it determines an involution on the cohomology.

The projection $\frac{1}{2}(1-\tau)$ projects $C_N(B)$ on the subcomplex $\bar{C}(B)$ and the induced map in cohomology projects the cohomology of $C_N(B)$ on the cohomology of $\bar{C}(B)$. In the case $B = U(\mathcal{G})$ the projection applied to $\wedge\mathcal{G}$ is the identity in odd dimensions and 0 in even dimensions.

To prove the second result we use the fact that as coalgebras the universal enveloping algebra is isomorphic to the symmetric algebra on $\mathcal{G}$ and thus depends only on the vector space structure. So let us write $\mathrm{Sym}(V)$ instead of $U(\mathcal{G})$ and consider the complexes which we denote by $C_N(\mathrm{Sym}(V))$ and $\bar{C}((\mathrm{Sym}(V))$. The latter splits into a sum

$$\bar{C}((\mathrm{Sym}(V)) = (\bigoplus_p \wedge^{2p+1}(V)) \oplus \hat{C}.$$

where $\hat{C}$ is acyclic. The exact sequence

$$\hat{C}^{p-1} \overset{\delta^{p-1}}{\to} \hat{C}^p \overset{\delta^p}{\to} \hat{C}^{p+1}$$

is an exact sequence of $sl(V)$ modules. In characteristic 0 $sl(V)$ is simple and hence reductive, therefore the sequence is split exact. The splitting defines an equivariant homotopy such that

$$\delta^{p-1}h^p + h^{p+1}\delta^p = id.$$

Returning to the case $\bar{C}(U(\mathcal{G}))$ the homotopy operator which we have just constructed is equivariant with respect to $sl(\mathcal{G})$ hence, a fortiori, with respect to the adjoint action of $\mathcal{G}$. If a p-cocycle $u \in C_N(U(\mathcal{G}))$ cobounds then it is in the subcomplex $\hat{C}(U(\mathcal{G}))$ and it cobounds $h^p(u)$. If u is $\mathcal{G}$ invariant so is $h(u)$, proving the lemma.

Since the cohomology in dimension 4 vanishes, any 4 cocycle cobounds and there is equivariant homotopy defining a right inverse for δ^3:

$$id : \bar{Z}^4(U(\mathcal{G})) \xrightarrow{h^4} \bar{C}^3(U(\mathcal{G})) \xrightarrow{\delta^3} \bar{Z}^4(U(\mathcal{G})).$$

At this point it is important to note that these results remain true if we extend the coefficient ring. We can consider Lie algebras defined over a ring L extending k, of the form $\mathcal{G}_L = \mathcal{G} \otimes_k L$ as L modules. (For deformations we use the notation $\mathcal{G}_h$ for what, in the notation just introduced, would be $\mathcal{G}_{k[[h]]}$.) As a coalgebra $U(\mathcal{G}_L)$ is isomorphic to $U(\mathcal{G}) \otimes_k L$ and the complexes of $C_N(U\mathcal{G}_L)$ (defined using $\otimes_L$) and $\bar{C}((U\mathcal{G}_L))$ are isomorphic to $C_N(U\mathcal{G}) \otimes_k L$ and $\bar{C}(U(\mathcal{G})) \otimes_k L$, respectively. The coboundary operator and the equivariant homotopy extend

$$\delta_L = \delta \otimes id_L \quad h_L = h \otimes id_L.$$

The cohomology is $\wedge\mathcal{G} \otimes_k L \cong \wedge_L(\mathcal{G}_L)$.

The obstruction theory, i.e, the deformation cohomology, of $\mathcal{G}_L$ and $\mathcal{G}$ are the same up to an extension of coefficient ring from k to L.

In particular, this applies to $\mathcal{G}_n$), an nth order deformation of $\mathcal{G}$, that is, a Lie algebra defined over $k_{(n)}$, which, as a module over $k_{(n)}$

$$\mathcal{G}_{(n)} \cong \mathcal{G} \otimes_k k_{(n)}.$$

We will assume that the equivariant homotopy, which is initially defined for $\mathcal{G}$ a Lie algebra over k has been extended by tensoring with $k_{(n)}$ to $\bar{C}(U(\mathcal{G}_{(n)}))$. The axioms for a coboundary or quasitriangular structure make sense for Drinfeld algebras defined over $k_{(n)}$ and and for deformations of such algebras so we will use those terms in these contexts as well.

Preliminary remark. Now we can describe the idea behind the proof that any coboundary deformation is the universal enveloping algebra of a Lie algebra deformation. As in all the previous cases, there is an induction on powers of h. Assume that we are given a coboundary deformation B_h such that the truncation at order $n+1$, $B_{(n)} = B_h/h^{n+1}B_h$, is isomorphic to the universal enveloping algebra of an n th order deformation of a Lie algebra, $\mathcal{G}_n$.

Suppose further that for any δ cocycle $\phi_{(n)} \in U(\mathcal{G}_{(n)})^{\otimes 3}$ such that $\phi_{(n)} \equiv 0 \bmod h$ there is a universal formula, denoted $\mathcal{E}(s\phi_{(n)})$ for generating $\Phi(s)$ in a coboundary deformation depending on the (additional) deformation parameter s. $\mathcal{E}(s\phi_{(n)})$ is defined by applying the equivariant homotopy to successive obstruction classes.

Denote this deformation, with $R = 1$,

$$(U(\mathcal{G}_{(n)})[[s]], m_{(n)}, \Delta, \mathcal{E}(s\phi_{(n)}))$$

where

$$\Phi(s) = \mathcal{E}(s\phi_{(n)}) = 1 + s\phi_{(n)} + \cdots =\in U(\mathcal{G}_{(n)})[[s]]$$

satisfying $\Phi^{321} = \Phi^{-1}$ and the pentagon identity.

(Note that setting $s = 1$ in the formula $\mathcal{E}(s\phi_{(n)})$ makes sense since the assumption $\phi_{(n)} \equiv 0 \bmod h$ guarantees that the series is actually finite.) One can compare this universal construction evaluated at $s = 1$ to the existing deformation $(B_{(n)}, m_{(n)}, \Delta_{(n)}, R_{(n)}, \Phi_{(n)})$. Suppose as an induction assumption that they are isomorphic. One would like to prove the same is true to one higher order for some $\mathcal{G}_{(n+1)}$. A clever use of universal formula $\mathcal{E}$ applied to a lifting of $\phi_{(n)}$ proves the induction step.

The next proposition shows that the deformation theory of cDque is unobstructed, more precisely, there are no obstructions to extending an infinitesimal cocycle ϕ such that $\phi^{321} + \phi = 0$ to $\Phi(s)$ satisfying $\Phi^{321}\Phi = 1$ and the pentagon identity and thus to define a coboundary deformation $(U(\mathcal{G})[[s]], m, \Delta, \Phi(s), 1)$.

Proposition 11.7.2 *Given any arbitrary but fixed n. Let $\mathcal{G}_{(n)}$ be a Lie algebra over $k_{(n)}$ which is a free $k_{(n)}$ module and $B_s = U(\mathcal{G}_{(n)})[[s]]$ the trivial bialgebra deformation of the corresponding universal enveloping algebra. Let $\Phi = \Phi(s) \in B_s^{\otimes 3}$ satisfy the conditions*

1. *Φ is $\mathcal{G}_{(n)}$ invariant.*

2. *$\Phi \equiv 1 \bmod s$*

3. $\Phi^{321}\Phi \equiv 1 \bmod s^m$, *(no relation to the order, n, of truncation)*

4. $1 \otimes \Phi(id \otimes \Delta \otimes id)\Phi\Phi \otimes 1 \equiv (id \otimes id \otimes \Delta)\Phi(\Delta \otimes id \otimes id)\Phi \bmod s^m$

Then there exists a $\Phi'(s) \cong \Phi(s) \bmod s^m$ *and satisfying the same conditions modulo* s^{m+1}.

Proof. As usual we look for the extension in the form $\Phi' = \Phi + \phi s^m$, where ϕ cobounds the obstruction cocycle. The two projection operators $\frac{1}{2}(1+\tau)$ and $\frac{1}{2}(1-\tau)$ into the subcomplexes $\bar{C}'$ and $\bar{C}$ respectively, decompose ϕ into a sum $\phi_+ + \phi_-$. Since $\Phi \equiv 1 \bmod s$ condition 3 uniquely determines $\phi_+ = \frac{1}{2}(\phi^{321} + \phi)$ as a $\mathcal{G}$ invariant element of the subcomplex $\bar{C}'$. Using the vanishing of the cohomology of $\bar{C}$ in dimension 4, and the existence of a $\mathcal{G}$ equivariant inverse to δ^3 the proof is complete once we show that the obstruction cochain to the pentagon identity is actually in $\bar{Z}^4$. The cocycle condition has been verified in a more general setting in Chapter 9. The fact that it is a -1 eigenvector of τ follows from

$$\begin{aligned}&\tau(id \otimes id \otimes \Delta)(\Phi)^{-1}(1 \otimes \Phi)(id \otimes \Delta \otimes id)(\Phi)(\Phi \otimes 1)(\Delta \otimes id \otimes id)(\Phi)^{-1})\\ &= (\Delta \otimes id \otimes id)(\Phi^{321})^{-1}(\Phi^{321} \otimes 1)(id \otimes \Delta \otimes id)(\Phi^{321})\\ &\times(1 \otimes \Phi^{321})(id \otimes id \otimes \Delta)(\Phi^{321})^{-1}\end{aligned}$$

This calculation uses the cocommutativity of Δ. The first expression is $\tau(1+s^m\beta^{3,1}) = (1+s^m\tau\beta^{3,1})$ and the second is $(1+s^m\beta^{3,1})^{-1} = 1 - s^m\beta^{321}$ mod s^{m+1}, proving that $\tau\beta^{3,1} = -\beta^{3,1}$. Therefore the obstruction is in $\bar{Z}^4$, completing the proof.

For any such cocycle the element $1 + s\phi$ satisfies the conditions in the proposition mod s^2, and there are no obstructions to extending the infinitesimal Drinfeld algebra deformation to a full deformation

$$(U(\mathcal{G}_{(n)})[[s]], m_{(n)}, \Delta_{(n)}, \Phi(s))$$

where multiplication and comultiplication are independent of s. Since the construction of the extension uses a fixed homotopy operator defined for all Lie algebras over either $k, k_{(n)}$ or k_h we really do have a universal formula $\mathcal{E}(s\phi)$ generating the coboundary Drinfeld deformation beginning with an invariant cocycle ϕ such that $\phi^{321} + \phi = 0$. So we have also proved the proposition

Proposition 11.7.3 *For any* $\mathcal{G}_L$ *invariant* δ *cocycle* $\phi \in U(\mathcal{G}_L)^{\otimes 3}$ *satisfying* $\phi^{321} + \phi = 0$ *there is a universal formula* $\mathcal{E}(s\phi)$ *for generating an invariant solution to the pentagon identity* $\Phi(s)$, *satisfying* $\Phi^{321}\Phi = 1$ *and* $\Phi(s) \equiv 1 + s\phi \bmod s^2$.

This defines a coboundary deformation of $U(\mathcal{G}_L)$ with Δ cocommutative and coassociative where the conjugacy condition on Φ and Δ in the definition of Drinfeld algebra deformation is fulfilled because $\Phi(s)$ is $\mathcal{G}_L$ invariant. It is important to note that the homotopy we used to define the formula $\mathcal{E}(s\phi)$ makes sense even if ϕ is not invariant and, in the next proposition, we will apply it in such a situation.

Proposition 11.7.4 *Let $\mathcal{G}_{(n)}$ be a Lie algebra over $k_{(n)}$ which is a free $k_{(n)}$ module, and $\phi_{(n)} \in (\wedge^3 \mathcal{G}_{(n)})^{\mathcal{G}_{(n)}}$ an invariant element such that $\phi_{(n)} \equiv 0$ mod h. Let (B, m, Δ, Φ) be a coboundary Drinfeld algebra defined over $k_{(n+1)}$ normalized so that $R = 1$ and such that reducing mod h^{n+1} we have $B/h^{n+1}B \cong U(\mathcal{G}_{(n)})$ and $\Phi = \mathcal{E}(\phi_{(n)})$ mod h^{n+1}. Then after twisting by a normalized symmetric $F = 1 + fh^{n+1}$ there exist a Lie algebra, $\mathcal{G}_{(n+1)}$ over $k_{(n+1)}$ and invariant $\phi_{(n+1)} \in (\wedge^3 \mathcal{G}_{(n+1)})^{\mathcal{G}_{(n+1)}}$ with $\mathcal{G}_{(n+1)}/h^{n+1}\mathcal{G}_{(n+1)} \cong \mathcal{G}_{(n)}$ and $\phi_{(n+1)} \equiv \phi_{(n)}$ mod h^{n+1} such that*

$$(B, m, \Delta, \Phi) = (U(\mathcal{G}_{(n+1)}), m_{(n+1)}, \Delta_{(n+1)}, \mathcal{E}(\phi_{(n+1)})).$$

Proof. We first pick a not necessarily invariant lifting of $\phi_{(n)}$ as follows. Choose a $\phi' \in \wedge^3(\mathcal{G}_{(n)})$ such that $\phi_{(n)} = h\phi'$. The imbedding

$$B/h^{n+1}B \to B$$

given by multiplication by h defines an imbedding of $\mathcal{G}_{(n)}$ in B which in turn defines an imbedding of $\wedge^3\mathcal{G}_{(n)} \to \wedge^3 B$. Let ψ be the image of ϕ' under this imbedding.

The complex that we use to construct $\mathcal{E}(s\phi_{(n)})$ is $\bar{C}(U(\mathcal{G}_{(n)})) \cong \bar{C}(U(\mathcal{G})) \otimes_k k_{(n)}$, where $\mathcal{G}_{(n)} = \mathcal{G} \otimes_k k_{(n)}$ The deformation cohomology complex for B is

$$\bar{C}(B) \cong \bar{C}(U(\mathcal{G})) \otimes_k k_{(n+1)} \supset C(U(\mathcal{G}_{(n)})) \otimes_{k_{(n)}} hk_{(n+1)}.$$

If ψ and ψ' are two liftings of $\phi_{(n)}$ then $\psi - \psi' = h^{n+1}\xi$ and since both ψ and ψ' are congruent to 0 modulo h, we have $0 \equiv \psi(h^{n+1}\xi) = \psi(\psi - \psi') = \psi\psi - \psi\psi'$ in B (mod h^{n+2}). Together with the analogous equation with ψ' replacing ψ, we see that $\psi\psi = \psi'\psi'$. Since calculating the obstruction cocycle corresponding to the pentagon identity for the infinitesimal deformation $1 + s\psi$ involves only products, the result is independent of the choice of the lifting ψ. Similarly for the higher order obstructions. This fact together with the isomorphism of complexes given above shows that $\mathcal{E}(s\psi) - s\psi \in B$ is a well defined element reduces to $\mathcal{E}(s\phi_{(n)}) - s\phi_{(n)}$ modulo h^{n+1}. By assumption

$\Phi = \mathcal{E}(\phi_{(n)})$. (As we noted above, it makes sense to evaluate at $s = 1$). Define $\Psi = \mathcal{E}(\psi) \in B^{\otimes 3}$. Then $\Phi = \Psi + h^{n+1}\xi$. Since both Φ and Ψ satisfy the pentagon identity in B and they both have leading term 1, ξ is a cocycle in $\bar{C}(U(\mathcal{G}_{(0)}))$. Then $\xi = \delta f + \chi$ with $\chi \in \wedge^3\mathcal{G}_0$ and f symmetric, because it is in $\bar{C}^2$. Twisting in B by $1 + h^{n+1}f$ gives an equivalent coboundary structure (B, m, Δ', Φ') satisfying all the same conditions only now $\Phi' = \Psi + h^{n+1}\chi$. Using the inclusion described above we see that the the element $\chi \in h^{n+1} \wedge \mathcal{G}_{(0)}$ can be considered as a element of $\wedge^3(B)$.

Finally we show that in this situation, Δ' is coassociative and the element $\psi + h^{n+1}\chi \in B^{\otimes 3}$ is invariant as a consequence of the equation

$$\Phi'(\Delta \otimes id)\Delta(a) = (id \otimes \Delta)\Delta(a)\Phi'. \tag{11.7.1}$$

Write $\mathcal{E}(\psi) = 1 + \psi + \mathcal{E}'(\psi)$. All the terms in $\mathcal{E}'(\psi)$ involve product of at least two ψ and, as we saw above, these products are independent of the extension, depending only on $\phi_{(n)}$. Furthermore they sit in the subspace $h^2B^{\otimes 3}$ which is a $\mathcal{G}_{(n)}$ module. The action of $\mathcal{G}_{(n)}$ on any of these products is well defined and independent of the lifting. Therefore they are invariant, for any $a \in B$ reducing modulo h^{n+1} to $\alpha \in \mathcal{G}_{(n)}$ we have

$$\mathcal{E}'(\psi)(\Delta' \otimes id)\Delta'(a) - (id \otimes \Delta')\Delta'(a)\mathcal{E}'(\psi) = 0. \tag{11.7.2}$$

Substituting $\Phi' = 1 + \psi + \mathcal{E}'(\psi) + h^{n+1}\chi$ in 11.7.1 and using 11.7.2, we have the equation

$$(\Delta' \otimes id)\Delta'(a) - (id \otimes \Delta')\Delta'(a) = [a \otimes 1 \otimes 1 + 1 \otimes a \otimes 1 + 1 \otimes 1 \otimes a, \psi + h^{n+1}\chi].$$

Since we have assumed that Δ and hence Δ' are cocommutative, antisymmetrizing annihilates the left side. But the elements ψ and χ are in $\wedge^3(B)$ so the left side is invariant under antisymmetrization. Therefore both sides are 0. This means that Δ' is coassociative and that $\psi + h^{n+1}\chi$ is invariant. Now we can apply Proposition 6.3 to conclude that B is a universal enveloping algebra of Prim(B) relative to Δ' which form a Lie algebra $\mathcal{G}_{(n+1)}$. Since Δ' defined on B reduces modulo h^{n+1} to the standard comultiplication on the universal enveloping algebra $U(\mathcal{G}_{(n)})$ the primitive elements are liftings to B of the elements of $\mathcal{G}_{(n)} \subset B/h^{n+1}B$. If $\psi = h\sum X_i \wedge Y_i \wedge Z_i \in B$ then there exist primitives $X_i' \equiv X_i$, $Y_i' \equiv Y_i$, $Z_i' \equiv Z_i$ modulo h^{n+1} and $\psi = h\sum X_i' \wedge Y_i' \wedge Z_i' \in \wedge^3(\mathcal{G}_{(n+1)})$ and similarly for $h^{n+1}\chi$. Together with the invariance proved above

$$\psi + h^{n+1}\chi \in (\wedge^3\mathcal{G}_{(n+1)})^{\mathcal{G}_{(n+1)}}.$$

Finally $\Phi' = \mathcal{E}(\psi) + h^{n+1}\chi = \mathcal{E}(\psi + h^{n+1}\chi)$ since the term $h^{n+1}\chi$ appears only linearly in the formula.

For a Lie algebra deformation $\mathcal{G}_h$ of $\mathcal{G}$ and $\phi \in (\wedge^3(\mathcal{G}_h)^{\mathcal{G}_h}$ let $(U(\mathcal{G}_h), \phi, F)$ be the cDque given by twisting $(U(\mathcal{G}_h), m, \Delta, \mathcal{E}(\phi), 1)$ by the symmetric invariant element $F \in U(\mathcal{G}_h)^{\otimes 2}$. The formula $\mathcal{E}$ is a fixed but not necessarily unique universal formula dependent on the choice of right inverse for δ^3.

Applying Proposition 11.7.3 recursively to the truncations $B = U_h/h^{n+1}U_h$ proves the first two assertions in the following theorem.

Theorem 11.7.1 Coboundary Classification

1. *For any Lie algebra deformation $\mathcal{G}_h$ of $\mathcal{G}$ and $\phi \in (\wedge^3(\mathcal{G}_h)^{\mathcal{G}_h}$ there exists a well defined coboundary Drinfeld deformation $(U(\mathcal{G}_h), \phi, F)$.*

2. *Any coboundary Drinfeld quantized universal enveloping algebra is equivalent by twisting to some $(U(\mathcal{G}_h), \phi, F)$.*

3. *Any morphism between $(U(\mathcal{G}_h), \phi, F)$ and $(U(\mathcal{G}'_h), \phi', F')$ has the form $Ad(u) \circ \bar{\lambda}$ where $\bar{\lambda}$ is the extension to $U(\mathcal{G}_h)$ of a k_h Lie algebra homomorphism $\lambda : \mathcal{G}_h \to \mathcal{G}'_h$, $Ad(u)$ is conjugation by $u \in U(\mathcal{G}'_h)$ with $\epsilon(u) = 1$ and $u \equiv 1 \mod h$, $\phi' = Ad(u \otimes u \otimes u)(\lambda \otimes \lambda \otimes \lambda)\phi$, and $F' = (u \otimes u) \cdot (\bar{\lambda} \otimes \bar{\lambda})(F)\Delta(u)^{-1}$, where Δ is the untwisted comultiplication.*

The proof of the last claim is almost the same as in the triangular case, so we omit the details.

Turning now to quasitriangular structures proper. In this case instead of starting with an invariant alternating three tensor, the initial data is an invariant symmetric two tensor $t \in \mathrm{Sym}^2(\mathcal{G}_h)^{\mathcal{G}_h}$. Define $R = e^{ht}$ and assume that we have $\Phi(h)$ defined by some universal formula $\varphi(ht^{12}, ht^{23})$, which can be generated, for example, by applying the same proceedure as in the coboundary case to the appropriate cohomology theory, which now includes consideration of the hexagon identity as well as the pentagon identity. See [322]. As we mentioned earlier this formula may come from a **Q** modification of the Knizhnik-Zamolodchikov solution over **C** or from a cohomological procedure involving a fixed homotopy operator. We describe the K-Z solution in the next chapter, but will not enter further into the details of the construction of φ. As regards uniqueness, we have a similar proposition to the one for coboundary structures. For the purpose of the following proposition [324], we absorb the factor h into the definition of $t_{(n)}$.

Proposition 11.7.5 *Let (B, m, Δ, R, Φ) be a quasitriangular deformation of $U(\mathcal{G})$ defined over $k_{(n+1)}$ such that the reduction modulo h^{n+1} has the form $(U(\mathcal{G}_{(n)}), m_{(n)}, \Delta_{(n)}, R = e^{t_{(n)}}, \Phi = \varphi(t^{12}_{(n)}, t^{23}_{(n)})$ for $t_{(n)} \in Sym^2(\mathcal{G}_{(n)})^{\mathcal{G}}_{(n)}$, which we abreviate $(U(\mathcal{G}_{(n)})t_{(n)})$ Then there exists a symmetric $F = 1 + fh^{n+1}$ and an invariant symmetric two tensor in primitive elements, $t_{(n+1)}$ such that twisting by F transforms to the form $(U(\mathcal{G}_{(n+1)}), t_{(n+1)})$.*

Proof. With no loss of generality, we can assume that R is symmetric. Then Φ satisfies $\Phi^{321}\Phi = 1$. Consider $(B, m, \Delta, 1, \Phi)$ as a coboundary structure. By the theorem for coboundary structures, we know that there exists an $F = 1 + fh^{n+1}$ such that twisting by F transforms B to the form of a universal enveloping algebra $U(\mathcal{G}_{(n+1)})$. Let π be the projection of $U(\mathcal{G}_{(n+1)}) = \bigoplus W_m$ on the primitive elements W_1. Define $t_{(n+1)} = \pi \otimes \pi(lnR)$. By assumption lnR reduces modulo h^{n+1} to $t_{(n)}$, furthermore, π commutes with reduction modulo h^{n+1} so $t_{(n+1)}$ reduces modulo h^{n+1} to $t_{(n)}$ and differs from lnR by an element which is a multiple of h^{n+1} and in the kernel of $\pi \otimes \pi$. Thus $R = e^{t_{(n+1)}} + h^{n+1}a$ where $a \in \mathrm{Sym}^2(U(\mathcal{G}_{(0)}))$ and $(\pi \otimes \pi)(h^{n+1}a) = 0$. Similarly $\Phi = \varphi(t^{12}_{(n+1)}, t^{23}_{(n+1)}) + h^{n+1}b$. From the pentagon identities and $\Phi^{321}\Phi = 1$ we conclude

$$\delta b = 0 \quad \text{and} \quad b^{321} + b = 0$$

and from the hexagon identities

$$(\Delta \otimes id)a - a^{23} - a^{13} = b^{312} - b^{132} + b \qquad (11.7.3)$$
$$(id \otimes \Delta)a - a^{12} - a^{13} = b^{213} - b^{231} - b. \qquad (11.7.4)$$

We know that there exists a symmetric $f \in U(\mathcal{G}_{(0)})^{\otimes 2}$ such that

$$b = \delta f + b'$$

where b' is in $\wedge^3\mathcal{G}_{(0)}$. Assume we have performed the twisting by $1+fh^{n+1}$ so that b is antisymmetric. Then the right side of 11.7.3 is $3b$. Antisymmetrizing the equation leaves $3b$ invariant and annihilates the left side, so $b = 0$. The left side of 11.7.3 implies that $a \in \mathcal{G}_{(0)} \otimes U(\mathcal{G}_{(0)})$. Since a is symmetric $a \in \mathcal{G}_{(0)} \otimes \mathcal{G}_{(0)}$. But the $a = (\pi \otimes \pi)(a) = 0$, and we have proven that R and Φ have the desired form, concluding the proof.

11.8 Lie bialgebras and Poisson-Lie groups

In Proposition 11.3.2 we proved that given a bialgebra deformation of the universal enveloping algebra $U(\mathcal{G}_h)$ for $\mathcal{G}$ semisimple, it was always possible to find an equivalent deformation in which the infinitesimal comultiplication has the form

$$\Delta^{(1)}(a) = [\Delta(a), r] \quad r \in \wedge^2 \mathcal{G},$$

and therefore defines a map

$$\Delta^{(1)} : \mathcal{G} \to \wedge^2 \mathcal{G}.$$

Dualizing, we get a map

$$[\ ,\ ]^* : \wedge^2 \mathcal{G}^* \to \mathcal{G}^*.$$

The condition for this to satisfy the Jacobi identity is precisely that $[r_{12}, r_{13} + r_{23}] + [r_{13}, r_{23}]$ be invariant. Proposition 11.3.3 implies that the condition for $\Delta^{(1)}$ to be the dual of a Lie bracket on $\mathcal{G}^*$ is that the first obstruction cocycle cobound.

More generally, given any quasitriangular bialgebra deformation, with $R(h) \equiv 1 + hr \mod h^2$ and $\Delta(h) \equiv \Delta + h\Delta^{(1)} \mod h^2$, the condition

$$R(h)\Delta(h) = \Delta(h)^{\mathrm{op}} R(h)$$

implies that

$$(\Delta^{(1)} - \Delta^{(1)\mathrm{op}})(a) = [\Delta(a), r] = dr(a), \tag{11.8.1}$$

where d is the Hochschild differential. The hexagon identities at order h give

$$(\Delta \otimes id)r = r^{13} + r^{23} \quad (id \otimes \Delta)r = r^{13} + r^{12},$$

which shows that $r \in \mathcal{G} \otimes \mathcal{G}$. At order h^2 the hexagon identity gives

$$((\Delta^{(1)} - \Delta^{(1)\mathrm{op}}) \otimes id)r = [r^{13}, r^{23}].$$

Equation (11.8.1) and the identity

$$(dr \otimes id)r = -[r^{12}, r^{13} + r^{23}]$$

together imply that r satisfies the classical Yang Baxter equation

$$[r^{12}, r^{13} + r^{23}] + [r^{13}, r^{23}] = 0.$$

Since the left side of (11.8.1) is antisymmetric, the symmetric part of r is invariant and writing $r = \bar{r} + s$ where $\bar{r}$ is the antisymmetric part, we get

$$[\bar{r}^{12}, \bar{r}^{13} + \bar{r}^{23}] + [\bar{r}^{13}, \bar{r}^{23}] = -[s^{13}, s^{23}],$$

which is invariant because s is. Therefore, once again, $\Delta^{(1)} - \Delta^{(1)\mathrm{op}} = d\bar{r}$ is dual to a Lie bracket on $\mathcal{G}^*$.

Proposition 11.8.1 *Let $\Delta^{(1)}$ be the infinitesimal deformation of the comultiplication in a quasitriangular bialgebra deformation of $U(\mathcal{G})$. The antisymmetrization $\psi = \Delta^{(1)} - \Delta^{(1)\mathrm{op}}$ is a Hochschild coboundary whose restriction to $\mathcal{G}$ is dual to a Lie algebra structure on $\mathcal{G}^*$.*

In general, without the quasitriangular assumption, if we assume that the infinitesimal deformation of the multiplication vanishes, $m^{(1)} = 0$, then $\Delta^{(1)}$ is a Hochschild cocycle. The condition $\delta(\Delta^{(1)}) = 0$ implies that $\Delta^{(1)} : \mathcal{G} \longrightarrow \mathcal{G} \otimes \mathcal{G}$ and therefore $\Delta^{(1)} - \Delta^{(1)\mathrm{op}}$ is potentially dual to a Lie bracket on $\mathcal{G}$. The Jacobi identity on this bracket is satisfied if the obstruction class, which in this case has the form $(0, 0, \beta^{3,1})$ is cohomologous to zero. This motivates the following definition.

Definition. A *Lie bialgebra* is a Lie algebra $\mathcal{G}$ together with a structure of Lie algebra on $\mathcal{G}^*$, where the Lie bracket on $\mathcal{G}^*$ is dual to a $\mathcal{G}$ cocyle with values in $\wedge^2\mathcal{G}$.

We have just shown how to associate a Lie bialgebra to the infinitesimal data of either quasitriangular bialgebra deformation of $U(\mathcal{G})$ or a deformation with $m^{(1)} = 0$ which is unobstructed at second order. In this sense we can think of Lie bialgebras as the classical limit of a quantized universal enveloping algebra. The Lie algebra structure on $\mathcal{G} \oplus \mathcal{G}^*$ is a Lie algebra analog of the Drinfeld double.

The usual context for quantization is Poisson algebras as defined in section 9.8. Drinfeld proved that the cocycle appearing in the definition of a Lie bialgebra defines a group cocycle on the associated connected and simply connected Lie group, and gives rise to a very special kind of Poisson structure. We quote the definition and theorem from [317]. Let $\mathcal{O}(X)$ be one of the standard algebras of functions (smooth, analytic, etc.) on a manifold X.

Definition. A Poisson group is a group G with a Poisson bracket on $\mathcal{O}(G)$ for which the multiplication $\mu : G \times G \to G$ defines a map of Poisson algebras $\mu^* : \mathcal{O}(G) \to \mathcal{O}(G \times G)$, The definition of the bracket of $p, q \in \mathcal{O}(G \times G)$ is

$$\{p, q\}(a, b) = \{p(a, \cdot), q(a, \cdot)\}(b) + \{p(\cdot, b), q(\cdot, b)\}(a).$$

So for $f, g \in \mathcal{O}(G)$ the condition in the definition is

$$\{f, g\} \circ \mu(a, b) = \{f \circ \mu(a, \cdot), g \circ \mu(a, \cdot)\}(b) + \{f \circ \mu(\cdot, b), g \circ \mu(\cdot, b)\}(a).$$

Theorem 11.8.1 *The category of connected and simply connected Poisson-Lie groups is equivalent to the category of finite dimensional Lie bialgebras.*

We have barely scratched the surface of the subject Lie bialgebras and their associated Poisson-Lie groups. They have been studied extensively by Y. Kosmann-Schwarzbach, J. H. Lu, A. Weinstein, M. A. Semenov-Tian-Shansky, and others. The reader can browse through the bibliography for further references. Another interesting subject that we do not have room to include is the theory of deformations of Lie-bialgebras along the lines suggested by a short paragraph in [317]. This area has been developed by P. Lecomte and C. Roger [757].

The Fadeev-Reshetikhin-Takhtadjan (FRT) approach to quantum groups, the theory of quantization of the Poisson algebra of a Poisson-Lie group, is dealt with masterfully in a beautiful set of lectures by Takhtadjan, [1154] For a purely algebraic approach to the FRT theory see Manin's lecture notes [884].

There is a very interesting connection between the twisting proceedure and the quantization of Poisson-Lie groups which Takhtadjan discovered and which now sketch. Starting from the existence of the Drinfeld-Jimbo quantized universal enveloping algebra $U_h(\mathcal{G})$, [317] [576], [579], for any simple classical Lie algebra over a field of characteristic zero, we know Proposition by 11.3.1 that there is an equivalent bialgebra deformation with undeformed multiplication. We also know from the existence theorem for Drinfeld deformations that there is a deformation with both the multiplication and comultiplication undeformed but nontrivial coassociativity operator $\Phi = \Phi(h) = \varphi(ht^{12}, ht^{23})$. From the uniqueness theorem Proposition 11.5.3 it follows that the two deformations are equivalent by a twisting, that is, there exists a twisting $F = F(h) \in U(\mathcal{G})^{\otimes 2}[[h]]$ such that

$$((id \otimes \Delta)F(h))F^{23}(h) = \Phi(h)((\Delta \otimes id)F(h))F^{12}(h), \tag{11.8.2}$$

$$F(h)\Delta(h)F(h)^{-1} = \Delta(h). \tag{11.8.3}$$

Let $F^{\lambda}(h)$ ($F^{\rho}(h)$) be the formal sum of left invariant (respectively right invariant) bidifferential operators on the connected and simply connected Lie group associated to $\mathcal{G}$ defined by identifying the universal enveloping algebra with the left invariant (respectively right invariant) differential operators on the associated Lie group. If we use $F^{\lambda}(h)$ or $F^{\rho}(h)$ to define a $*$ product the condition for associativity is

$$((id \otimes \Delta)F(h))F^{23}(h) = ((\Delta \otimes id)F(h))F^{12}(h).$$

The condition (11.8.2) is weaker but if $\Phi(h)$ is $\mathcal{G}$ invariant, it is enough to prove associativity for the $*$ product defined by $F^{\rho}(h) \circ (F^{\lambda})^{-1}(h)$. In fact we have

Proposition 11.8.2 *[1153] The formal sum of bidifferential operators given by $F^\rho(h) \circ (F^\lambda)^{-1}(h)$ defines an associative $*$ product quantization of the Poisson-Lie group which is the classical limit of the the Drinfeld-Jimbo quantization $U_h(\mathcal{G})$. Together with the standard comultiplication on $\mathcal{O}(G)$ this defines a bialgebra quantization.*

To conclude this chapter we would like to indicate briefly how the Lie bialgebra picture changes when we deal with quasitriangular Drinfeld deformations. For this purpose, it is useful to have a third description of Lie bialgebras.

Proposition 11.8.3 *$\mathcal{G}$ is a Lie bialgebra if and only if there is a Lie algebra structure on the vector space sum $\mathcal{G} \oplus \mathcal{G}^*$ for which both $\mathcal{G}$ and $\mathcal{G}^*$ are imbedded as subalgebras and isotropic subspaces relative to the inner product defined by natural pairing and furthermore this pairing is ad invariant.*

Before discussing the classical limit of a qtDque, we normalize the deformation data. For a qtDque with $\Phi(h) \equiv 1 + h\phi_1 \mod h^2$ the hexagon identity implies that the full antisymmetrization of ϕ_1 equals 0. But the pentagon identity implies that ϕ_1 is a δ cocycle, so it is cohomologous to an element of $a \in \wedge^3\mathcal{G}$. Writing $\phi_1 + \delta f = a$, and antisymmetrizing, the left side is mapped to zero. We conclude that $a = 0$ and $\phi_1 = -\delta f$. Twisting by $1 - hf$ transforms $\Phi(h)$ to $\Phi'(h) \equiv 1 \mod h^2$, so we can assume from the start that $\Phi(h)$ has no term linear in h and $\Phi(h) \equiv 1 + h^2\phi \mod h^3$.

The antisymmetrized infinitesimal $\psi^* = \Delta^{(1)} - \Delta^{(1)\mathrm{op}}$ is still equal to dr and, when dualized, defines a map which we denote $\psi : \wedge^2\mathcal{G}^* \to \mathcal{G}^*$. At order h^2 the hexagon identity now gives

$$[r^{12}, r^{13} + r^{23}] + [r^{13}, r^{23}] = Alt(\phi),$$

where Alt is the full antisymmetrization. In this case ψ defines a Lie bracket, i.e. satisfies the Jacobi identity, if and only if $Alt(\phi)$ is *ad* invariant. However, even if this in not so, we can still define a Lie algebra structure on $\mathcal{G} \oplus \mathcal{G}^*$ provided we relax the requirement that $\mathcal{G}^*$ be a subalgebra. The component of the bracket $\wedge^2\mathcal{G}^* \to \mathcal{G}$ is given by interior product with ϕ. Consider ψ as an element of $\mathcal{G}^* \otimes \wedge^2\mathcal{G}$. Let $\mu \in \wedge^2\mathcal{G}^* \otimes \mathcal{G}$ be the 3 tensor representing the Lie bracket on $\mathcal{G}$. Then the full Lie algebra structure is described by

$$(\mu, \psi, \phi) \in \bigoplus_{i+j=3} \wedge^i\mathcal{G}^* \otimes \wedge^j\mathcal{G}.$$

The Jacobi identity can be expressed as the vanishing of the bracket $[[\cdot,\cdot]]$ defined in [704]. For the general theory see [952], [695], [697], and [698].

$$[[(\mu,\psi,\phi),(\mu,\psi,\phi)]] = ([[\mu,\mu]], 2[[\mu,\psi]], 2[[\mu,\phi]] + [[\psi,\psi]], 2[[\psi,\phi]]).$$

The various components do, in fact, all vanish. The first component vanishes because of the Jacobi identity in $\mathcal{G}$. The second component vanishes because ψ is a the antisymmetrization of a Hochschild coboundary and $[[\mu,\cdot]]$ is just the Chevalley Eilenberg differential which agrees with the Hochschild differential in degree zero. The vanishing of the third component follows from antisymmetrizing

$$d\phi - \delta\Delta^{(2)} = (\Delta^{(1)} \otimes id)\Delta^{(1)} - (id \otimes \Delta^{(1)})\Delta^{(1)},$$

which is just the h^2 component of $\Phi(h)(\Delta(h) \otimes id)\Delta(h) = (id \otimes \Delta(h))\Delta(h)\Phi(h)$. Finally the last component vanishes because of the vanishing of the h^3 component of the pentagon identity. This proves

Proposition 11.8.4 *The classical limit of a qtDque $U_h(\mathcal{G})$ is Lie algebra structure on $\mathcal{G} \oplus \mathcal{G}^*$ with Lie bracket given by (μ,ψ,ϕ) defined above. Such an object is called a quasi-Lie bialgebra.*

To the element $r \in \mathcal{G} \otimes \mathcal{G}$ we associate the subspace $\mathcal{W} \subset \mathcal{G}^* \oplus \mathcal{G}$ which is the graph of the mapping

$$\lambda \mapsto (id \otimes \lambda)r.$$

The subspace $\mathcal{W}$ is transverse to $\mathcal{G}$ and the canonical inner product defines a nondegenerate pairing

$$\mathcal{G} \times \mathcal{W} \longrightarrow k,$$

but $\mathcal{W}$ is no longer isotropic. The relation of the classical limit to the type of deformation is given in the following proposition. Recall that the triangular condition is the intersection of the quasitriangular and coboundary conditions.

Proposition 11.8.5 *[321]*

1. *For a qtDque $\mathcal{W}$ is an ideal.*

2. *For a cDque $\mathcal{W}$ is a $\mathcal{G}$ invariant Lagrangian subspace, that is, $\mathcal{W}^\perp = \mathcal{W}$.*

The inner product on $\mathcal{W}$ is defined by the symmetric part of r, the infinitesimal of the universal R matrix, and is invariant under twisting.

Suppose that we have two deformations B_h and B'_h with the associated subspaces $\mathcal{W}$ and $\mathcal{W}'$. If the first order truncations $B_{(1)}$ and $B'_{(1)}$ are related by twisting then the projection along $\mathcal{G}$ defines an isometry of the subspaces $\mathcal{W}$ and $\mathcal{W}'$.

Call a Lie algebra with *ad* invariant inner product a *metric Lie algebra.* Let $\mathcal{P}$ be a metric Lie algebra and $\mathcal{G}$ a Lagrangian subalgebra. Drinfeld calls the pair $(\mathcal{P}, \mathcal{G})$ a *Manin pair* . The subspace $\mathcal{W}$ in the preceding proposition is a third piece of information for the Manin pair $(\mathcal{G} \oplus \mathcal{G}^*, \mathcal{G})$.

Proposition 11.8.6 *[321] Consider triples $(\mathcal{P}, \mathcal{G}, \mathcal{W})$, where $(\mathcal{P}, \mathcal{G})$ is a Manin pair. Specifying the subspace $\mathcal{W}$ in $\mathcal{P}$ to satisfy certain conditions is equivalent to specifying corresponding data on $\mathcal{G}$.*

1. *$\mathcal{W}$ being an ideal is equivalent to specifying an $\mathcal{G}$ invariant symmetric two tensor t.*

2. *$\mathcal{W}$ being $\mathcal{G}$ invariant Lagrangian subspace is equivalent to specifying a $\mathcal{G}$ invariant antisymmetric three tensor ϕ.*

11.9 Quantum $sl(2)$

After more than 300 pages of a book on quantum groups, it's about time for an example. We pick the simplest example which demonstrates the different structures related to bialgebra deformations. To our knowledge there are no nice explicit examples for the Drinfeld algebra deformations although some explicit formulas for solutions to the Knizhnik-Zamolodchikov equation (see Chapter 12) have been given recently by Schechtman using Chen type integrals.

The example of quantum $sl(2)$ in the quantized universal enveloping algebra form is due to Kulish, Reshetikhin, and Sklyanin[317]. The presentation given here follows Majid [842].

Let X_+, X_-, and H be the standard basis for $sl(2)$ with bracket relations

$$[X_+, X_-] = H, \quad [H, X_\pm] = \pm 2X_\pm.$$

By the PBW theorem the monomials $X_+^a X_-^b H^c$ form a basis for $U(sl(2))$ hence the same is true for $U_h(sl(2))$ by the assumption that it is isomorphic as $k[[h]]$ module to $U(sl(2))$. Although, by Proposition 11.3.1 it is possible to present the deformation is a form in which the multiplication is undeformed, in the standard presentation that is not the case. The commutation relations in $U_h(sl(2))$ are

$$[X_+, X_-] = \frac{e^{\frac{1}{2}hH} - e^{-\frac{1}{2}hH}}{e^{h/2} - e^{-h/2}} \tag{11.9.1}$$

$$[H, X_\pm] = \pm 2X_\pm \tag{11.9.2}$$

It is convenient to introduce the variable

$$q = e^{\frac{1}{4}h} \in k[[h]].$$

Identifying $U_h(sl(2))$ with $U(sl(2))[[h]]$, we define

$$K_\pm = e^{\pm h/4H} = q^H \in U_h(sl(2)).$$

Then

$$K_\pm K_\mp = 1 \tag{11.9.3}$$

$$K_+ X_\pm K_- = q^{\pm 2} X_\pm \tag{11.9.4}$$

$$[X_+, X_-] = \frac{K_+^2 - K_-^2}{q^2 - q^{-2}} \tag{11.9.5}$$

Expanding (11.9.1, we see that

$$[X_+, X_-] = H \mod h^2.$$

Thus the infinitesimal deformation of the multiplication vanishes, $m^{(1)} = 0$. Even if there are higher order deformations of the multiplication and we can apply the arguments in the previous section.

The comultiplication is an algebra map so it is enough to specify it on the algebra generators, $X_\pm, H$ or $X_\pm, K_\pm$. On H it is undeformed, so the Hopf subalgebra generated by H is in fact undeformed, meeting one of the requirements for the deformation in Theorem 11.4.1.

Since H is still primitive, $K_\pm$ is grouplike.

$$\Delta(K_\pm) = K_\pm \otimes K_\pm \tag{11.9.6}$$

The remaining comultiplications are

$$\begin{aligned} \Delta(X_\pm) &= X_\pm \otimes K_+ + K_- \otimes X_\pm & (11.9.7) \\ &= X_\pm \otimes 1 + 1 \otimes X_\pm + h/4(X_\pm \otimes H - X_\pm \otimes H) \mod h^2. \end{aligned}$$

It is immediate to check that this comultiplication satisfies

$$[\Delta(Y), \Delta(Z)] = \Delta[Y, Z]$$

for generators, consistent with the defining relations.

The infinitesimal comultiplication deformation is

$$\Delta^{(1)}(X_\pm) = X_\pm \otimes H - H \otimes X_\pm, \quad \Delta^{(1)}(H) = 0.$$

A straightforward calculation shows that $\Delta^{(1)} - \Delta^{(1)\mathrm{op}}$ is the coboundary of $r = X_+ \wedge X_-$ as in Proposition 11.4.1.

$$\Delta^{(1)} - \Delta^{(1)\mathrm{op}} = d(X_+ \wedge X_-).$$

Computing the classical Yang Baxter expression

$$\langle r, r\rangle = [r^{12}, r^{13} + r^{23}] + [r^{13}, r^{23}] = 2H \wedge X_+ \wedge X_-.$$

The symmetric invariant is

$$X_+ \otimes X_- + X_- \otimes X_+ + 1/2H \otimes H.$$

Then $r' = r + 1/2t$ satisfies the classical Yang Baxter equation, $\langle r', r'\rangle = 0$.

The R matrix is a complicated expression involving "q numbers" $[n]_q = \frac{q^n - q^{-n}}{q - q^{-1}}$, see [842].

$$R = q^{1/2H\otimes H} \sum \frac{(1 - q^{-2})^n q^{1/2n(n-1)}}{[n]_q!} (K_+X_+ \otimes K_-X_-)^n, \tag{11.9.8}$$

where $q^{H\otimes H} = e^{\frac{h}{4}H\otimes H}$. The verification that this element satisfies the required equation $R\Delta = \Delta^{\mathrm{op}}R$, is a long but straightforward calculation. The expansion to first order in powers of h is

$$R = 1 + h(\frac{1}{4}H \otimes H + X_+ \otimes X_-) + O(h^2),$$

so it has the desired infinitesimal, $r + 1/2t$.

The fundamental representation of $U_h(sl(2))$ on $V = k[[h]] \oplus k[[h]]$ is given by

$$X_+ \mapsto \begin{pmatrix} 0 & 1 \\ 0 & 0 \end{pmatrix} \tag{11.9.9}$$

$$X_- \mapsto \begin{pmatrix} 0 & 0 \\ 1 & 0 \end{pmatrix} \tag{11.9.10}$$

$$K_\pm \mapsto \begin{pmatrix} q^{\pm 1} & 0 \\ 0 & q^{\mp 1} \end{pmatrix} \tag{11.9.11}$$

One checks immediately that the commutation relations are satisfied. Let x^i_j for $i,j \in \{1,2\}$ be the coordinate functions of the representation, that is, projection on the matrix entries. As elements of $U_h(sl(2))^*$ they can be multiplied by the convolution product

$$x * y(u) = x \otimes y(\Delta(u)).$$

Then

$$y * x = y \otimes x(\Delta(u)) = x \otimes y(\Delta^{\mathrm{op}}(u)) = x \otimes y(R\Delta(u)R^{-1}), \qquad (11.9.12)$$

showing that the commutation rules are governed R, or rather its image in $\mathrm{End}(V \otimes V)$ under the representation. Using the algebra homomorphism property of the representation, equation (11.9.12) gives

$$x^i_j * x^k_l = x^k_l \otimes x^i_j(R\Delta(u)R^{-1}) = R^{ki}_{pq} x^p_r * x^q_s (R^{-1})^{rs}_{lj}, \qquad (11.9.13)$$

which is equivalent to

$$R^{lj}_{rs} x^i_j * x^k_l = R^{ki}_{pq} x^p_r * x^q_s. \qquad (11.9.14)$$

The image of the R in $\mathrm{End}(V \otimes V)$ is

$$q^{-1/2}\begin{pmatrix} q & 0 & 0 & 0 \\ 0 & 1 & q-q^{-1} & 0 \\ 0 & 0 & 1 & 0 \\ 0 & 0 & 0 & q \end{pmatrix}.$$

This gives the following commutation relations for $a = x^1_1$, $b = x^1_2$, $c = x^2_1$, $d = x^2_2$

$$\begin{aligned} a * b &= qb * a & a * c &= qc * a \\ b * c &= c * b & b * d &= qd * b \\ c * d &= qd * c & a * d - d * a &= (q - q^{-1}) b * c. \end{aligned} \qquad (11.9.15)$$

These are the defining relations in the FRT quantization. To save space we express the comultiplication rule using x^i_j, the formulae with a, b, c, d appear at the beginning of section 1.2.

$$\Delta(x^i_j) = \sum x^i_k \otimes x^k_j, \qquad (11.9.16)$$

which is simply dual to the algebra homomorphism property

$$x^i_j(uv) = \sum x^i_k(u) x^k_j(v).$$

Being a $*$ product homomorphism, the comultiplication is determined on the $*$ algebra generated by the x^i_j once it defined on the x^i_j. The counit on $U_h(sl(2))$ is given by

$$\epsilon(x) = x(1),$$

so

$$\epsilon(a) = \epsilon(d) = 1, \quad \epsilon(b) = \epsilon(d) = 0. \tag{11.9.17}$$

The FRT bialgebra has algebra generators $\{a,\ b,\ c,\ d\}$ satisfying (11.9.15) with comultiplication on generators (11.9.16), and counit (11.9.17). We have just shown

Proposition 11.9.1 *There is a bialgebra map of the FRT bialgebra into $U_h(sl(2))^*$ which is injective on generators.*

We conclude with a few comments about the relation of the FRT bialgebra to the Poisson-Lie group associated to the infinitesimal $\Delta^{(1)} = d(X_+ \wedge X_-)$. The corresponding group coboundary is

$$g \mapsto Ad(g)(X_+ \wedge X_-) - X_+ \wedge X_-.$$

The corresponding Poisson-Lie structure is defined by the bitensor field

$$Ad(g)(X_+)^\lambda) \wedge Ad(g)(X_-)^\lambda - X_+^\lambda \wedge X_-^\lambda = X_+^\rho \wedge X_-^\rho - X_+^\lambda \wedge X_-^\lambda$$

where $X_\pm^\lambda$ and $X_\pm^\rho$ are the left and right invariant vector fields corresponding to $X_\pm$ and $X_\pm$ respectively.

These vectorfields, expressed in terms of the coordinates $\{a,\ b,\ c,\ d\}$ are

$$X_+^\lambda = a\frac{\partial}{\partial b} + c\frac{\partial}{\partial d} \tag{11.9.18}$$

$$X_-^\lambda = b\frac{\partial}{\partial a} + d\frac{\partial}{\partial c} \tag{11.9.19}$$

$$X_+^\rho = c\frac{\partial}{\partial a} + d\frac{\partial}{\partial b} \tag{11.9.20}$$

$$X_-^\rho = a\frac{\partial}{\partial c} + b\frac{\partial}{\partial d} \tag{11.9.21}$$

This gives the following bracket relations

$$\{a,b\} = ab, \quad \{a,c\} = ac, \quad \{b,d\} = bd, \quad \{c,d\} = cd, \quad \{a,d\} = 2bc$$

which are exactly of the form

$$1/h(x * y - y * x) \quad \mathrm{mod} \quad h^2,$$

for the FRT bialgebra.

Proposition 11.9.2 *The FRT algebra quantizes the Poisson-Lie algebra associated to the Lie bialgebra which is the classical limit of* $U_h(sl(2))$.

Chapter 12

The Knizhnik Zamolodchikov equations

12.1 Introduction

The purpose of this chapter is to describe the key geometric ideas in the remarkable papers [321][322] in which Drinfeld proves Kohno's theorem relating the representations of the braid groups given by the universal R matrices and those given by monodromy matrices.

Let $\mathcal{G}$ be a Lie algebra defined over a field of characteristic zero which contains the complex field $\mathbf{C}$. Starting with a symmetric element

$$t \in \mathcal{G} \otimes \mathcal{G} \tag{12.1.1}$$

which is $\mathcal{G}$ invariant, he constructs a deformation, $(U(\mathcal{G}), \Delta, R(h), \Phi(h))$ of the bialgebra $U(\mathcal{G})$, the universal enveloping algebra of $\mathcal{G}$, with its standard comultiplication (which is coassociative and cocommutative). The multiplication and the comultiplication are undeformed, but the $\Phi = \Phi(h)$, defining the coassociativity, and the $R = R(h)$ determining the cocommutativity are not trivial, and in particular,

$$R(h) = \exp \pi iht \tag{12.1.2}$$

so that

$$R'(0) = \pi it \tag{12.1.3}$$

gives the infinitesimal deformation of R. Drinfeld constructs this deformation by analyzing the monodromy associated to a system of differential equations which arises in conformal field theory, the Knizhnik- Zamolodchikov equations (see the next section) and which had been used to construct representations of the braid group.

Recall that for a Lie algebra, $\mathcal{G}$, the comultiplication is determined by the fact that

$$\Delta\xi = \xi \otimes 1 + 1 \otimes \xi, \quad \xi \in \mathcal{G}$$

and that $\mathcal{G}$ consists precisely of the primitive elements.

Let B be a coassociative and cocommutative bialgebra over $\mathbf{C}$. For any element

$$t = \sum t_{(1)} \otimes t_{(2)} \in B \otimes B$$

we set

$$t^{ij} = \sum 1 \otimes \cdots \otimes t_{(1)} \otimes 1 \otimes \cdots \otimes t_{(2)} \otimes 1 \otimes \cdots \in B^{\otimes n},$$

where there are 1's in all the positions other than the i–the and j–th tensor factor.

If $t \in \mathrm{Prim}(B) \otimes \mathrm{Prim}(B)$ as we are assuming, then

$$(\Delta \otimes id)t = t^{13} + t^{23}, \quad \text{and} \quad (id \otimes \Delta)t = t^{12} + t^{13}. \tag{12.1.4}$$

If t is invariant, so that

$$\Delta(a)t = t\Delta(a) \;\; \forall a \in B, \tag{12.1.5}$$

then

$$[t^{12}, t^{13} + t^{23}] = 0$$

since the left hand side is $[t^{12}, (\Delta \otimes id)t]$.

Thus in the algebra $B^{\otimes n}$ we have

$$[t^{ij}, t^{ik} + t^{jk}] = 0 \;\; \text{for } i, j, k \;\; \text{distinct} \tag{12.1.6}$$

and

$$[t^{ij}, t^{kl}] = 0, \quad if \;\; \{i, j\} \cap \{k, l\} = \emptyset. \tag{12.1.7}$$

These will turn out to be the integrabilty conditions for the KZ equations in the next section.

In what follows we will take t to be a symmetric element of $B \otimes B$, so

$$t^{12} = t^{21}.$$

Then it follows from (12.1.6) and(12.1.7) that

$$[t^{kl}, T] = 0 \tag{12.1.8}$$

where

$$T = \sum_{i \neq j} t^{ij}, \quad T \in A^{\otimes n}. \tag{12.1.9}$$

So we want to start with a (coassociative and) cocommutative bialgebra and an invariant symmetric element, $t \in \mathrm{Prim}(B) \otimes \mathrm{Prim}(B)$. We want to produce a

$$\Phi \in B^{\otimes 3}[[h]]$$

satisfying Drinfeld's pentagon identity, (11.6.4) and such that Φ together with R given by (12.1.2) satisfy the Drinfeld hexagon identities (11.6.5) and (11.6.6).

12.2 The Knizhnik Zamolodchikov equations

In this section A will be an associative algebra over $\mathbf{C}$ and $t \in A \otimes A$ is a symmetric element satisfying $[t^{12}, t^{13} + t^{23}] = 0$. Let

$$Y_n = \mathbf{C}^n \setminus multidiagonal$$

and consider the trivial vector bundle

$$Y_n \times A^{\otimes n} \to Y_n.$$

We let $A^{\otimes n}$ act on itself by right multiplication, so we can think of the t^{ij} as elements of End $A^{\otimes n}$. Thus

$$\tilde{\Omega} = -h \sum_{i \neq j} t^{ij} \frac{d(z_i - z_j)}{z_i - z_j} \tag{12.2.1}$$

is a one form on Y_n with values in End $A^{\otimes n}$, where, for the moment, h is a complex parameter entering into the definition of $\tilde{\Omega}$. Since

$$d\frac{d(z_i - z_j)}{z_i - z_j} = 0,$$

we have

$$d\tilde{\Omega} = 0,$$

and, as we are assuming (12.1.6)and (12.1.7), we have

$$\tilde{\Omega} \wedge \tilde{\Omega} = 0.$$

Thus if we think of $\tilde{\Omega}$ as defining a connection on $Y_n \times A^{\otimes n}$ relative to the global section **1** given by

$$\mathbf{1}(\mathbf{z}) = \mathbf{z} \times 1_n, \qquad 1_n = 1 \otimes \cdots \otimes 1,$$

then this connection has zero curvature.

Let W be a function on Y_n with values in $A^{\otimes n} \subset \text{End } A^{\otimes n}$. Then $\mathbf{1}W$ is a section of our vector bundle, and its covariant differential relative to our connection is given by

$$D(\mathbf{1}W) = \mathbf{1}(\tilde{\Omega}W + dW).$$

In particular, the section is covariant constant (flat) section if and only if

$$\frac{\partial W}{\partial z_i} = h\left(\sum_{i \neq j} \frac{t^{ij}}{z_i - z_j}\right) W. \tag{12.2.2}$$

These equations are known as the Knizhnik Zamolodchikov equations, or simply as the KZE. When we want to specify n, we will write KZE$_n$.

The symmetric group, S_n, acts freely on Y_n by

$$\sigma\mathbf{z} = \sigma(z_1, \ldots, z_n) = (z_{\sigma^{-1}(1)}, \ldots, z_{\sigma^{-1}(n)})$$

and on $A^{\otimes n}$ by

$$\sigma(a_1 \otimes \cdots \otimes a_n) = a_{\sigma^{-1}(1)} \otimes \cdots \otimes a_{\sigma^{-1}(n)}.$$

So we get an action of S_n on the vector bundle $Y_n \times A^{\otimes n}$. This action clearly preserves **1** and the connection determined by $\tilde{\Omega}$. So everything descends down to

$$X_n = Y_n / S_n.$$

We get a vector bundle over X_n with flat connection and preferred section such that the form, Ω, of this connection relative to the preferred section pulls back to $\tilde{\Omega}$ under the projection $Y_n \to X_n$. If we choose a base point, x, of X_n, then according to the theorem of Artin, the fundamental group $\pi_1(X_n, x)$ is isomorphic to the Artin braid group, B_n. Therefore the monodromy relative to our flat connection gives us representations

$$\rho = \rho_n : B_n \to \text{Aut } A^{\otimes n}.$$

Drinfeld shows how to use the KZ equations to give a categorical description of this representation.

For this purpose, it will be more convenient to regard the h occurring in (12.2.2) as a formal parameter rather than a complex number. So we will think of W as taking values in

$$A^{\otimes n}[[h]] = A^{\otimes n} \otimes \mathbf{C}[[h]],$$

the algebra of formal powers series with coefficients in $A^{\otimes n}$. We will look for a solution of (12.2.2) of the form

$$W(z_1, \ldots, z_n) = \sum_{m=0}^{\infty} w_m(z_1, \ldots, z_n) h^m, \tag{12.2.3}$$

where each w_n is a holomorphic function on Y_n taking values in a finite dimensional subspace of $A^{\otimes n}$.

We now use the symmetry of t and the integrability condition (12.1.6) to show that (12.2.2) can be reduced to a system of partial differential equations in $n-2$ variables. In particular, for $n = 3$ it becomes a system of ordinary differential equations. Observe that the symmetry of t^{ij} implies that

$$\sum_{i \neq j} \frac{t^{ij}}{z_i - z_j} = 0,$$

and hence that

$$\sum_{i=1}^{n} \frac{\partial W}{\partial z_i} = 0, \tag{12.2.4}$$

if W is a solution of (12.2.2). So we may change coordinates to

$$w_1 = z_1 + \cdots + z_n, \quad w_i = z_i - z_1, \quad i > 1,$$

and write

$$W(z_1, \ldots, z_n) = \tilde{W}(w_2, \ldots, w_n),$$

and W will be a solution of (12.2.2) if and only if $\tilde{W}$ is a solution of

$$\frac{\partial \tilde{W}}{\partial w_i} = h \left(\frac{t^{i1}}{w_i} + \sum_{j \neq 1,i} \frac{t^{ij}}{w_i - w_j} \right) \tilde{W} \quad i > 1. \tag{12.2.5}$$

If we apply the Euler operator in $n-1$ variables to a solution of this equation, we obtain

$$\sum_{2}^{n} w_i \frac{\partial \tilde{W}}{\partial w_i} = hT\tilde{W}, \tag{12.2.6}$$

where

$$T = \sum_{i \neq j} t^{ij}.$$

Notice that our assumptions about t imply that

$$[T, t^{ij}] = 0, \quad \forall i, j,$$

so we can think of (12.2.6) as saying that $\tilde{W}$ is a homogeneous function of $(w_2, \ldots, w_n)$ with exponent hT, i.e. that

$$\tilde{W}(uw_2, \ldots, uw_n) = u^{hT} \tilde{W}(w_2, \ldots, w_n)$$

where

$$u^{hT} \stackrel{\text{def}}{=} \exp(hT \ln u)$$

for some local choice of natural logarithm. Therefore if we make some cut in the u plane (say along the negative real axis) so that $\exp(hT \ln u)$ is defined, then, setting

$$u_i = w_i / u, \quad i = 2, \ldots, n-1$$

we have, for $w_n = u$ avoiding the cut,

$$\tilde{W}(w_2, \ldots, u) = u^{hT} \tilde{W}(u_2, \ldots, u_{n-1}, 1) \stackrel{\text{def}}{=} u^{hT} V(u_2, \ldots, u_{n-1}).$$

Then $\tilde{W}$ will be a solution of (12.2.5) if and only if V is a solution of

$$\frac{\partial V}{\partial u_i} = h \left(\frac{t^{i1}}{u_i} + \sum_{2 \leq i,j \leq n-1, j \neq i} \frac{t^{ij}}{u_i - u_j} + \frac{t^{in}}{u_i - 1} \right) V. \tag{12.2.7}$$

12.3 KZE$_3$ and the associativity

We will now assume that the algebra A of section 1 is a bialgebra with a comultiplication which is cocommutative and coassociative. For the case $n = 3$, there is only one variable, u_2, left in (12.2.7). Let us call it v. So (12.2.7) becomes an ordinary differential equation

$$\frac{dV}{dv} = h \left(\frac{t^{12}}{v} + \frac{t^{23}}{v - 1} \right) V. \tag{12.3.1}$$

This is an equation with regular singular points at 0 and at 1. Solving it will involve some logarithms, so we cut the v plane along the negative real axis, and along the positive axis with $v \geq 1$. In other words we will look for a solution in the region

$$\mathbf{C} \setminus \{(-\infty, 0] \cup [1, \infty)\}.$$

We recall from the preceding section that a solution of (12.3.1) gives a solution of KZE_3 given by

$$W(z_1, z_2, z_3) = (z_3 - z_1)^{h(t^{12}+t^{13}+t^{23})} V\left(\frac{z_2 - z_1}{z_3 - z_1}\right), \tag{12.3.2}$$

where the z's are restricted by the conditions that $z_3 - z_1$ not lie on the negative real axis, and that the ratio

$$v = \frac{z_2 - z_1}{z_3 - z_1}$$

satisfy the constraints given above.

The key idea of Drinfeld is to observe that any two solutions of the first order linear differential equation (12.3.1) should differ by a constant factor in the admissible region. Near $v = 0$ the first term in the parenthesis on the right of (12.3.1) should predominate, and hence there should be a solution, V_0, of (12.3.1) defined for $0 < v < 1$ and asymptotic to

$$v^{ht^{12}}$$

at $v = 0$, and similarly a solution, V_1 of (12.3.1) also defined on $0 < v < 1$ and asymptotic to

$$(1 - v)^{ht^{23}}$$

at $v = 1$. (We will give the proof of the existence of these asymptotic solutions in a moment.) Then the quotient

$$\Phi = V_1(v)^{-1} V_0(v) \tag{12.3.3}$$

will be a constant element of $A^{\otimes 3}[[h]]$, and this will be the coassociativity operator. Going back to the expression of v as a ratio, we see that v close to zero means that z_2 is much closer to z_1 than it is to z_3. We can express this configuration by the parenthesis insertion

$$(\bullet\ \bullet)\ \bullet$$

while the opposite configuration, where z_2 is much closer to z_3 than to z_1, corresponding to v close to 1 would be symbolized by

$$\bullet\ (\bullet\ \bullet).$$

In this way, associativity is connected to geometrical configurations in $\mathbf{CP}^3$.

Let us now turn to the existence of these asymptotic solutions. We look for V_0 of the form

$$V_0(v) = (1 + \sum_{n=1}^{\infty} a_n(h)v^n)v^{ht^{12}}.$$

Here the coefficients $a_n(h)$ are elements of $A^{\otimes 3}[[h]]$ to be determined recursively by (12.3.1). The coefficient of

$$v^{n-1}v^{ht^{12}}$$

in V_0' is

$$na_n + a_n ht^{12}$$

while the coefficient of $v^{n-1}v^{ht^{12}}$ on the right hand side of (12.3.1) is

$$ht^{12}a_n - ht^{23}\sum_{j=1}^{n-1} a_j,$$

the second term coming from the geometric series expansion of $1/(1-v)$. Thus the equations for the coefficients can be written as

$$(n - ad\ ht^{12})a_n = -ht^{23}\sum_{j=1}^{n-1} a_j.$$

Now since we are in a ring of formal power series the operator

$$n - ad\ ht^{12} = n(1 - n^{-1}ad\ ht^{12})$$

is invertible by a geometric series:

$$n^{-1}(1 - n^{-1}ad\ ht^{12})^{-1} = \frac{1}{n}\sum_{m=0}^{\infty}\left(\frac{1}{n}ad\ ht^{12}\right)^m.$$

We thus have solved for $a_n = a_n(h)$ as

$$a_n(h) = \sum_{m=0}^{\infty}\left(\frac{1}{n}ad\ ht^{12}\right)^m\left(\frac{-ht^{23}}{n}\right)\left(1 + \sum_{j=1}^{n-1} a_j\right). \tag{12.3.4}$$

If we write

$$a_n(h) = \sum_{m=1}^{\infty} a_{n,m} h^m,$$

then (12.3.4) implies that

$$a_{n,1} = -\frac{1}{n} t^{23}, \quad a_{n,2} = -\frac{1}{n^2}[t^{12}, t^{23}] + \frac{1}{n}\left(\sum_{j=1}^{n-1} \frac{1}{j}\right) t^{23}, \tag{12.3.5}$$

and by induction that $a_{n,m}$ lies in the subspace of $A^{\otimes 3}$ spanned by m-fold products of t^{12} and t^{23}. In particular, all the $a_{n,m}$ with m fixed, lie in the same finite dimensional subspace of $A^{\otimes 3}$. So if we expand V_0 out in a power series in both v and h:

$$V_0(v) = 1 + \sum_{m=1}^{\infty} V_{0,m}(v) h^m,$$

the power series $V_{0,m}$ in v lies in a finite dimensional vector space, and solves a linear differential equation derived from (12.3.1) by comparing coefficients of powers of h. Hence in fact the the $V_{0,m}$ are analytic functions of v.

The same argument, replacing v by $v - 1$ gives the solution V_1. We now define Φ by (12.3.3) so that we have

$$V_1(v)\Phi = V_0(v).$$

If we now substitute the V_0 and V_1 into (12.3.2) we get two solutions of KZE$_3$,

$$W_0(z_1, z_2, z_3) = V_0(\frac{z_2 - z_1}{z_3 - z_1}) \left(\frac{z_2 - z_1}{z_3 - z_1}\right)^{ht^{12}} (z_3 - z_1)^{h(t^{12}+t^{13}+t^{23})}$$

and

$$W_1(z_1, z_2, z_3) = V_1(\frac{z_2 - z_1}{z_3 - z_1}) \left(\frac{z_3 - z_2}{z_3 - z_1}\right)^{ht^{23}} (z_3 - z_1)^{h(t^{12}+t^{13}+t^{23})}$$

where

$$V_0(0) = 1, \quad V_1(1) = 1.$$

Since $T = t^{12}+t^{13}+t^{23}$ commutes with both t^{12} and t^{23}, $(z_3-z_1)^{hT}$ commutes with Φ and we have

$$W_1\Phi = W_0. \tag{12.3.6}$$

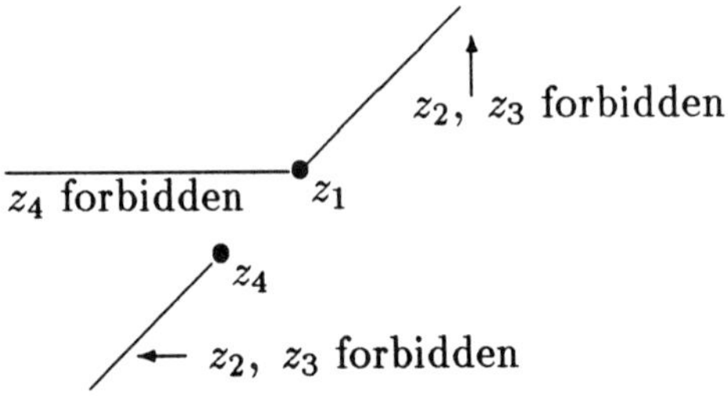

Figure 12.1:

12.4 KZE$_4$ and the pentagon constraint.

We consider the reduction (12.2.7) of KZE$_4$ to two variables:

$$\frac{\partial V}{\partial u_2} = h\left(\frac{t^{12}}{u_2} + \frac{t^{23}}{u_2 - u_3} + \frac{t^{24}}{u_2 - 1}\right)V \tag{12.4.1}$$

and

$$\frac{\partial V}{\partial u_3} = h\left(\frac{t^{13}}{u_3} + \frac{t^{23}}{u_3 - u_2} + \frac{t^{34}}{u_3 - 1}\right)V \tag{12.4.2}$$

where

$$u_2 = \frac{z_2 - z_1}{z_4 - z_1}, \quad u_3 = \frac{z_3 - z_1}{z_4 - z_1}.$$

The variable $z_4 - z_1$ was restricted by having to avoid the negative real axis, and then u_2 and u_3 are restricted by having to avoid the negative real axis and also all points > 1 on the positive real axis. We can draw the forbidden zones as in Figure 12.1.

Near the point $(u_2, u_3) = (0, 1)$ the first term predominates in (12.4.1) and the last term predominates in (12.4.2). So we would like to find a solution of the form

$$V(u_2, u_3) = U(u_2, u_3)u_2^{ht^{12}}(1 - u_3)^{ht^{34}} \tag{12.4.3}$$

with U analytic in a bidisc

$$|u_2| < \epsilon, \quad |u_3 - 1| < \epsilon$$

and with

$$U(0, 1) = 1.$$

We can do this by the methods of the preceding section, since near the point $(a, b) = (0, 1)$ the pair of equations (12.4.1) and (12.4.2) take the form

$$\frac{\partial V}{\partial u} = h\left(\frac{A}{u-a} + P\right)V, \quad \frac{\partial V}{\partial v} = h\left(\frac{B}{v-b} + Q\right)V \tag{12.4.4}$$

where

$$P = P(u, v),\ Q = Q(u, v)$$

are holomorphic in a neighborhood of (a, b) and integrability conditions are satisfied, so that in particular

$$[A, B] = 0.$$

Under such conditions, the analysis given in the preceding section establishes the existence of a solution of the form

$$V(u, v) = U(u, v)(u - a)^{hA}(v - b)^{hB}, \quad U(a, b) = 1. \tag{12.4.5}$$

A neighborhood of the point $(0, 1)$ corresponds to the configuration where z_1 and z_2 are close together but far from z_3 and z_4 which are also close together. We might picture this configuration as

$$(\bullet\bullet)(\bullet\bullet),$$

describing a vertex of the Stasheff pentagon, K_4. So let us denote the solution of KZE_4 corresponding to (12.4.3) by $W_{(\bullet\bullet)(\bullet\bullet)}$ so

$$W_{(\bullet\bullet)(\bullet\bullet)} = W_{(\bullet\bullet)(\bullet\bullet)}(z_1, z_2, z_3, z_4)$$

is the solution of the Knizhnik Zamolodchikov equations given by

$$\begin{aligned} W_{(\bullet\bullet)(\bullet\bullet)} &= V\left(\frac{z_2 - z_1}{z_4 - z_1}, \frac{z_3 - z_1}{z_4 - z_1}\right)(z_4 - z_1)^{hT} \\ &= U\left(\frac{z_2 - z_1}{z_4 - z_1}, \frac{z_3 - z_1}{z_4 - z_1}\right)\left(\frac{z_2 - z_1}{z_4 - z_1}\right)^{ht^{12}}\left(\frac{z_4 - z_3}{z_4 - z_1}\right)^{ht^{34}}(z_4 - z_1)^{hT} \\ &\sim (z_2 - z_1)^{ht^{12}}(z_4 - z_3)^{h^{34}}(z_4 - z_1)^{h(t^{13} + t^{14} + t^{23} + t^{24})} \end{aligned}$$

as we approach the forbidden region.

A neighborhood of the point $(0, 0)$ corresponds to the situation where z_1, z_2 and z_3 are close to one another but far from z_4. This would correspond to only one parenthesis insertion, so to an edge, not a vertex of K_4. Also, near $(0,0)$, the equations (12.4.1) and (12.4.2) do not have the desired form (12.4.4). Both of these troubles can be remedied by blowing up the point $(0,0)$, replacing it by a projective line describing the direction of approach: So near $(0,0)$ introduce coordinates

$$u = u_3, v = \frac{u_2}{u_3}$$

and define

$$\hat{V}(v,u) = V(uv,u).$$

Then

$$\begin{aligned}
\frac{\partial \hat{V}}{\partial u} &= \frac{\partial V}{\partial u_2} \cdot v + \frac{\partial V}{\partial u_3} \\
&= hv\left(\frac{t^{12}}{u_2} + \frac{t^{23}}{u_2 - u_3} + \frac{t^{24}}{u_2 - 1}\right) V \\
&\quad +h\left(\frac{t^{13}}{u_3} + \frac{t^{23}}{u_3 - u_2} + \frac{t^{34}}{u_3 - 1}\right) V \\
&= h\left(\frac{t^{12} + t^{13} + t^{23}}{u} + \frac{t^{24}}{uv - 1} + \frac{t^{34}}{u - 1}\right)\hat{V}
\end{aligned}$$

and

$$\begin{aligned}
\frac{\partial \hat{V}}{\partial v} &= \frac{\partial V}{\partial u_2} \cdot u \\
&= hu_3\left(\frac{t^{12}}{u_2} + \ fract^{23} u_2 - u_3 + \frac{t^{24}}{u_2 - 1}\right) V \\
&= h\left(\frac{t^{12}}{v} + \frac{t^{23}}{v - 1} + \frac{t^{24} u}{uv - 1}\right)\hat{V}
\end{aligned}$$

which have the desired normal form (12.4.4)

$$\frac{\partial \hat{V}}{\partial u} = h\left(\frac{t^{12} + t^{13} + t^{23}}{u} + \dots\right)\hat{V}, \ \frac{\partial \hat{V}}{\partial v} = h\left(\frac{t^{12}}{v} + \dots\right)\hat{V} \tag{12.4.6}$$

near $(u,v) = (0,0)$ and the desired normal form

$$\frac{\partial \hat{V}}{\partial u} = h\left(\frac{t^{12} + t^{13} + t^{23}}{u} + \dots\right)\hat{V}, \quad \frac{\partial \hat{V}}{\partial v} = h\left(\frac{t^{23}}{v - 1} + \dots\right)\hat{V} \tag{12.4.7}$$

near $(u,v)=(0,1)$.

The first of these determines a distinguished solution

$$\hat{V}_{((\bullet\bullet)\bullet)\bullet}(v,u_3)=U_{((\bullet\bullet)\bullet)\bullet}(v,u_3)v^{ht^{12}}u_3^{h(t^{12}+t^{13}+t^{23})},\quad U_{((\bullet\bullet)\bullet)\bullet}(0,0)=1$$

and we get the corresponding solution

$$W_{((\bullet\bullet)\bullet)\bullet}(z_1,z_2,z_3,z_4)=\hat{V}_{((\bullet\bullet)\bullet)\bullet}\left(\frac{z_2-z_1}{z_3-z_1},\frac{z_3-z_1}{z_4-z_1}\right)(z_4-z_1)^{hT}$$

of KZE$_4$. The second equation, (12.4.7) determines a distinguished solution

$$\hat{V}_{(\bullet(\bullet\bullet))\bullet}(v,u_3)=U_{((\bullet(\bullet\bullet)\bullet}(v,u_3)(1-v)^{ht^{23}}u_3^{h(t^{12}+t^{13}+t^{23})},\quad U_{(\bullet(\bullet\bullet))\bullet}(1,0)=1,$$

and a corresponding solution $W_{((\bullet(\bullet\bullet)\bullet}$ of KZE$_4$.

A neighborhood of $(u_2,u_3)=(1,1)$ corresponds to a region where z_2,z_3 and z_4 are close together, a whole edge of K_4. So once again we blow up, introducing the variable

$$w=\frac{1-u_3}{1-u_2}=\frac{z_4-z_3}{z_4-z_2}$$

and the function

$$\check{V}(u_2,w)=V(u_2,1+(u_2-1)w).$$

Then equations (12.4.1) and (12.4.2) become

$$\begin{aligned}\frac{\partial\check{V}}{\partial u_2} &= \frac{\partial V}{\partial u_2}+\frac{\partial V}{\partial u_3}\cdot w\\ &= h\left(\frac{t^{12}}{u_2}+\frac{t^{23}+t^{24}+t^{34}}{u_2-1}+\frac{t^{13}}{1+(u_2-1)w}\right)\check{V}\end{aligned}$$

and

$$\begin{aligned}\frac{\partial\check{V}}{\partial w} &= \frac{\partial V}{\partial u_3}\cdot(u_2-1)\\ &= h\left(\frac{t^{13}u_2-1)}{1+(u_2-1)w}+\frac{t^{23}}{w-1}+\frac{t^{34}}{w}\right)\check{V}\end{aligned}$$

which have the desired normal form near $u_2=1, w=1$ and near $u_2=1, w=0$. The point $(u_2,w)=(1,1)$ corresponds to the vertex $\bullet((\bullet\bullet)\bullet)$ of K_4 and gives rise to a solution

$$\begin{aligned}\check{V}_{\bullet((\bullet\bullet)\bullet)}(u_2,w) &= U_{\bullet((\bullet\bullet)\bullet)}(u_2,w)(1-u_2)^{h(t^{23}+t^{24}+t^{34})}(1-w)^{ht^{23}}\\ U_{\bullet((\bullet\bullet)\bullet)}(1,1) &= 1\end{aligned}$$

and a corresponding solution $W_{\bullet((\bullet\bullet)\bullet)}$ of KZE_4.

Finally, the point $(u_2, w) = (1, 0)$ corresponds to the vertex $\bullet(\bullet(\bullet\bullet))$ of K_4, a solution

$$\check{V}_{\bullet(\bullet(\bullet\bullet))}(u_2, w) = U_{\bullet(\bullet(\bullet\bullet))}(u_2, w)(1-u_2)^{h(t^{23}+t^{24}+t^{34})}w^{ht^{34}}, \quad u_{\bullet(\bullet(\bullet\bullet))}(1,0) = 1$$

and a corresponding solution $W_{\bullet(\bullet(\bullet\bullet))}$ of KZE_4.

Let us summarize what we have achieved so far, restricting the values of the z's temporarily to be real and in the order

$$z_1 < z_2 < z_3 < z_4$$

so that the u's are real and (u_2, u_3) is restricted to lie in the simplex

$$\Sigma : 0 < u_2 < u_3 < 1.$$

We have blown up the vertices $(0,0)$ and $(1,1)$ of Σ, so that Σ is replaced by the Stasheff pentagon K_4, as in Chapter 5. To each vertex of K_4 we have associated a solution $W_{\mathbf{vertex}}$ of KZE_4 defined on the region corresponding to Σ, uniquely determined by its asymptotic behaviour in the asymptotic domain determined by that vertex. We will now show how each edge of the pentagon gives a relation between the solutions associated to its vertices. Consider for example the edge $(\bullet\bullet)\bullet\bullet$ which joins the vertices $(\bullet\bullet)(\bullet\bullet)$ and $((\bullet\bullet)\bullet)\bullet$. Each of the functions

$$\tilde{W}_{(\bullet\bullet)(\bullet\bullet)}u_1^{-hT}u_2^{-ht^{12}} \quad \text{and} \quad \tilde{W}_{((\bullet\bullet)\bullet)\bullet}u_1^{hT}u_2^{-ht^{12}}$$

are solutions of (12.4.2). We may pass to the limit $u_2 \to 0$ and so obtain two solutions,

$$Z_{\circ(\circ\circ)} \stackrel{def}{=} U_{(\bullet\bullet)(\bullet\bullet)}(0, u_3)u_3^{ht^{34}} \text{ and } Z_{(\circ\circ)\circ} = U_{((\bullet\bullet)\bullet)\bullet}(u_3, 0)u_3^{h(t^{13}+t^{23})}$$

to the equation we obtain when we set $u_2 = 0$ in (12.4.2) which is

$$\frac{dZ}{du_3} = h\left(\frac{t^{13}+t^{23}}{u_3} + \frac{t^34}{u_3 - 1}\right)Z, \quad Z \in A^{\otimes 4}[[h]]. \tag{12.4.8}$$

But in view of (12.1.4) equation (12.4.8) is just $\Delta\otimes id\otimes id$ applied to (12.3.1). Hence if V_0 and V_1 are the two solutions of (12.3.1) described in the preceding section, then

$$Z_{\circ(\circ\circ)} = (\Delta\otimes id\otimes id)V_1 \quad \text{and} \quad Z_{(\circ\circ)\circ} = (\Delta\otimes id\otimes id)V_0.$$

But then the results of the preceding section imply that

$$Z_{\circ(\circ\circ)}(\Delta \otimes id \otimes id)\Phi = Z_{(\circ\circ)\circ}.$$

We claim that this is enough to prove that

$$W_{(\bullet\bullet)(\bullet\bullet)}(\Delta \otimes id \otimes id)\Phi = W_{((\bullet\bullet)\bullet)\bullet} \tag{12.4.9}$$

in the region described by Σ. Indeed, the functions $\tilde{W}_{(\bullet\bullet)(\bullet\bullet)}$ and $\tilde{W}_{((\bullet\bullet)\bullet)\bullet}$ both satisfy the system of equations

$$\begin{aligned} \frac{\partial \tilde{W}}{\partial u_2} &= h\left(\frac{\operatorname{ad} t^{12}}{u_2} + \frac{t^{23}}{u_2 - u_3} + \frac{t^{24}}{u_2 - 1}\right)\tilde{W} \\ \frac{\partial \tilde{W}}{\partial u_3} &= h\left(\frac{t^{13}}{u_3} + \frac{t^{23}}{u_3 - u_2} + \frac{t^{34}}{u_3 - 1}\right)\tilde{W}. \end{aligned}$$

This system has solutions which are analytic up to the boundary $u_2 = 0$ for $0 < u_3 < 1$. Two solutions of this system in the region $0 \le u_2 < u_3 < 1$ which agree on the segment $u_2 = 0$ agree in the whole region. Hence

$$\tilde{W}_{(\bullet\bullet)(\bullet\bullet)}(\Delta \otimes id \otimes id)\Phi = \tilde{W}_{((\bullet\bullet)\bullet)\bullet}.$$

Since $(\Delta \otimes id \otimes id)\Phi$ depends on $t^{13} + t^{23}$ and t^{34} and these both commute with T and t^{12}, the equality (12.4.9) follows from the preceding equation. We have attached an identity between functions to the edge $(\bullet\bullet)\bullet\bullet$ of K_4. A similar analysis associates an identity to every edge of K_4. We list the edges and the corresponding identities:

$(\bullet\bullet)\bullet\bullet$	$W_{(\bullet\bullet)(\bullet\bullet)}(\Delta \otimes id \otimes id)\Phi$	$=$	$W_{((\bullet\bullet)\bullet)\bullet}$
$(\bullet\bullet\bullet)\bullet$	$W_{(\bullet(\bullet\bullet))\bullet}(\Phi \otimes 1)$	$=$	$W_{((\bullet\bullet)\bullet)\bullet}$
$\bullet(\bullet\bullet)\bullet$	$W_{\bullet((\bullet\bullet)\bullet)}(id \otimes \Delta\otimes)\Phi$	$=$	$W_{(\bullet(\bullet\bullet))\bullet}$
$\bullet(\bullet\bullet\bullet)$	$W_{\bullet((\bullet\bullet\bullet))}1 \otimes \Phi$	$=$	$W_{\bullet((\bullet\bullet)\bullet)}$
$\bullet\bullet(\bullet\bullet)$	$W_{\bullet(\bullet(\bullet\bullet))}(id \otimes id \otimes \Delta)\Phi$	$=$	$W_{(\bullet\bullet)(\bullet\bullet)}$

Combining these equations we get

$$W_{\bullet(\bullet(\bullet\bullet))}(\Delta \otimes id \otimes id)\Phi(id \otimes id \otimes \Delta)\Phi = W_{\bullet(\bullet(\bullet\bullet))}1 \otimes \Phi[(id \otimes \Delta \otimes id)\Phi]\Phi \otimes 1.$$

Since $W_{\bullet(\bullet(\bullet\bullet))}$ does not vanish identically and Φ is a constant, we can cancel the $W_{\bullet(\bullet(\bullet\bullet))}$ from both sides of the above equation to obtain the pentagon identity

$$(\Delta \otimes id \otimes id)\Phi(id \otimes id \otimes i\Delta)\Phi = 1 \otimes \Phi[(id \otimes \Delta \otimes id)\Phi]\Phi \otimes 1.$$

For the convenience of the reader who wishes to verify the identities attached to the various edges of the pentagon listed above, we list the form and asymptotic behavior of the five solutions using three of the five auxiliary coordinates u_1, u_2, u_3, v, and w.

$$
\begin{array}{lcl}
W_{(\bullet\bullet)(\bullet\bullet)} & \sim & u_2^{ht^{12}}(1-u_3)^{ht^{34}}u_1^{hT} \\
W_{((\bullet\bullet)\bullet)\bullet} & \sim & u_2^{ht^{12}}u_3^{h(t^{13}+t^{23})}u_1^{hT} \\
W_{(\bullet(\bullet\bullet))\bullet} & \sim & u_3^{h(t^{12}+t^{13})}(u_3-u_2)^{ht^{23}}u_1^{hT} \\
W_{\bullet((\bullet)\bullet)\bullet)} & \sim & (1-u_2)^{h(t^{24}+t^{34})}(u_3-u_2)^{ht^{23}}u_1^{hT} \\
W_{\bullet(\bullet(\bullet\bullet))} & \sim & (1-u_2)^{h(t^{23}+t^{24})}(1-u_3)^{ht^{34}}u_1^{hT}
\end{array}
$$

12.5 KZE$_2$ and the R matrix.

Let $A = B$ be a cocommutative bialgebra. According to (12.1.2) we will set

$$R = \exp \pi i h t \in B \otimes B.$$

Then condition (12.1.5) implies that

$$\Delta(a)R = R\Delta(a) \quad \forall a \in B$$

and the symmetry of t implies that

$$sR = Rs$$

as elements of $\mathrm{End}(B \otimes B)$. So

$$R\Delta^{\mathrm{op}}R^{-1} = \Delta.$$

We would like to show that R can be interpreted in terms of a monodromy matrix.

So let us fix a base point $p \in X_2$. Recall that X_2 consists of unordered pairs of unequal complex numbers. Let us choose p to be the unordered pair

$$p = \prec 1, 2 \succ .$$

For any closed path γ starting and ending at p we must parallel translate the flat bundle along the curve γ. This will be given as multiplication by an element of $\mathrm{Aut}A^{\otimes 2}$ (let us think of h as a real number for psychological purposes) which is the monodromy. It is, of course, enough to do the computation for a generator of the fundamental group. Let us choose our generator as follows. Rotate the points 2 and 1 counterclockwise in the complex plane around their mid point through angle $0 \leq \theta \leq \pi$ so as to reach the point $\prec 2,1 \succ = \prec 1,2 \succ$. This will be our curve, γ. Let us assume that γ is parametrized as a map $\gamma : [0,1] \to X_2$, $\gamma(t) =$ rotation through angle πt so that $\gamma(0) = \gamma(1) = p$.

To compute the monodromy along γ and relate it to the Knizhnik-Zamolodchikov equation KZE_2 we must do the following: We must lift γ to a curve, Γ, in Y_2. This lifted curve need not (and in our case will not) be a closed curve, but its end point, $\Gamma(1)$ will differ from its origin, $\Gamma(0)$ by the action of an element of $g \in S_2$. We must parallel transport along Γ to map the fiber over $\Gamma(0)$ to the fiber over $\Gamma(1)$, then apply the group g and compare.

Now Y_2 consists of ordered (unequal) pairs, $\langle z_1, z_2 \rangle$. If we pick $\langle 1,2 \rangle$ as $\Gamma(0)$, then the unique lift, $\Gamma(t)$ starting at $\langle 1,2 \rangle$ will end up at $\langle 2,1 \rangle$. The permutation $g = (12) \in S_2$ maps the point $\langle 1,2 \rangle$ into $\langle 2,1 \rangle$.

Recall how a group element, g, acts on sections of a homogeneous vector bundle: If f is a section, the $g(f)$ is given by

$$(g(f))(x) = g[f(g^{-1}x)]$$

where the g^{-1} acting on x is the action on the base while the outside g maps the fiber over $g^{-1}x$ to the fiber over x. In particular, if $\mathbf{1}$ is an invariant section, and

$$f = \mathbf{1}W$$

where W is an $A^{\otimes 2}$ valued function, then

$$g(f) = \mathbf{1}gW(g^{-1}\cdot).$$

In our situation, the points $[z_1, z_2]$ near $[2,1]$ satisfy

$$Re\ z_1 > Re\ z_2$$

and hence the points $g^{-1}[z_1, z_2] = [z_2, z_1]$ avoid cut along the negative real axis. The KZ equations reduce to the Euler equation (12.2.6) with $T = t$. The application of g carries our locally constant section

$$\mathbf{1}W, \quad W(z_1, z_2) = (z_2 - z_1)^{ht}$$

into the locally constant section

$$\mathbf{1}W^{\sim}, \quad W^{\sim}(z_1, z_2) = (z_1 - z_2)^{ht}.$$

In this expression

$$(z_1 - z_2)^{ht} = \exp(ht \ln(z_1 - z_2))$$

where the principal branch of the logarithm is chosen, the one that is real on the positive real axis, and we have used the fact that $st = t$.

Starting along our path from $\Gamma(0)$, the locally constant section is given by

$$(z_2 - z_1)^{ht} = \exp(ht \ln(z_2 - z_1)).$$

The rotation rotates $z_2 - z_1$ from the point 1 on the positive real axis to the point -1 on the cut, and our locally constant solution given by parallel transport differs from the preceding one by

$$R = e^{\pi iht}.$$

To actually compute the monodromy as an element of $\mathrm{Aut} A^{\otimes 2}$ we have to fix an identification of the fiber over p with $A^{\otimes 2}$. This involves two choices: the choice of $q = \Gamma(0)$ as $\langle 1, 2\rangle$ and an isomorphism

$$a \mapsto \mathbf{1}(q)W(q)a, \quad A^{\otimes 2} \to \text{ fiber over } q.$$

So we still need to apply the element $(12) \in S_2$ to identify the fiber over $\Gamma(1)$ back to the fiber over q. Combined with the preceding equation we see that the monodromy operator is given by

$$T(\gamma)a = \rho((12))[Ra] \tag{12.5.1}$$

where, in general, the representation, ρ, of S_n on $A^{\otimes n}$ is

$$\rho(\sigma)(a_1 \otimes a_2 \otimes \cdots \otimes a_n) = a_{\sigma^{-1}(1)} \otimes a_{sigma^{-1}(2)} \otimes \cdots a_{\sigma^{-1}(n)} \tag{12.5.2}$$

as usual.

12.6 Braiding and the hexagons.

In this section we will see how the monodromy considerations of the preceding paragraph together with the associativity combine to yield the (11.6.5)

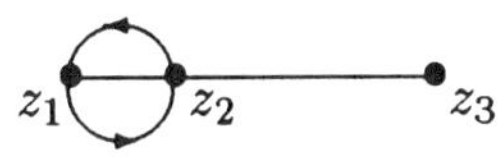

Figure 12.2:

and (11.6.6). Consider the initial configuration describing a point $p = \prec z_1, z_2, z_3 \succ \in X_3$ with the z_i real and

$$z_1 < z_2 < z_3$$

with z_2 close to z_1 and far from z_3, so as in the initial configuration in Figure 12.2.

We will consider the (closed) path γ in X_3 along which the points z_1 and z_2 are rotated counterclockwise around their midpoint until they interchange positions, and the corresponding arc Γ in Y_3 which joins the distinct points $\Gamma(0)$ and $\Gamma(1)$. Suppose that we start with the solution of KZE$_3$ corresponding to the function V_0 at the initial configuration, and so has the form

$$W(z_1, z_2, z_3) = V_0(v)v^{ht^{12}}(z_3 - z_1)^{hT}, \quad v = \frac{z_2 - z_1}{z_3 - z_1}$$

near the initial configuration, $\Gamma(0)$. Let us continue this solution along the path Γ to the point $\Gamma(1)$. As we have seen in the preceding section, the asymptotic behaviour of this solution as the two close points coalesce will be

$$\left(\frac{(z_1 - z_2)}{z_3 - z_1}\right)^{ht^{12}} R^{12}(z_3 - z_1)^{h(t^{12}+t^{13}+t^{23})} = R^{12} \tag{12.6.1}$$

where

$$R = \exp \pi i h t, \text{ and } R^{12} = R \otimes 1$$

as usual.

On the other hand, elements of the group S_3 act to carry solutions of KZE$_3$ into solutions. Applying the element $g = (12)$ (which acts as s_{12} on $A^{\otimes 3}$) we see that

$$W_{\sim}(z_1, z_2, z_3) = s_{12}W(z_2, z_1, z_3)$$

is a solution of KZE_3 with asymptotics

$$(z_1 - z_2)^{ht^{12}}(z_3 - z_1)^{h(t^{13}+t^{23})}. \tag{12.6.2}$$

By the uniqueness of solutions we can then say the following: Let $\mathbf{s}_{12}$ denote the operation consisting of analytic continuation along Γ from $\Gamma(0)$ to $\Gamma(1)$. Then

$$\mathbf{s}_{12}W_0 = W_{\sim}R^{12}. \tag{12.6.3}$$

It is important to emphasize that the operator $\mathbf{s}_{12}$ depends on the homotopy class of the path γ. For example, if we were to traverse γ in the opposite direction the R occuring on the right of (12.6.3) would have to be replaced by R^{-1}.

In order to keep track of the various local sections, let us use (12)3 to denote the configuration where the z_1 are real and $z_1 < z_2 < z_3$ with z_2 close to z_1 and far from z_3, and let us use $W_{(12)3}$ to denote the function we have called W_0 above. Similarly we would let (21)3 denote the configuration with $z_2 < z_1 << z_3$ and $W_{(21)3}$ the function we called $W_{\sim}$ above, so we will write (12.6.3) as

$$\mathbf{s}_{12}W_{(12)3} = W_{(21)3}R^{12}.$$

Similarly, the configuration with $z_1 << z_2 < z_3$ will be denoted by 1(23) and the function that we denoted by W_1 in section 12.2 we will now denote by $W_{1(23)}$ so we will write the associativity relation (12.3.6) as

$$W_{1(23)}\Phi = W_{(12)3}.$$

So we will associate a local solution of KZE_3 to each of the twelve possible associations and permutations of the letter $1, 2, 3$. In particular, consider the functions tabulated below with the configurations where they are defined and their asymptotic behavior.

(12)3	$z_1\ z_2$	z_3	$W_{(12)3}$	$\sim \quad (z_2 - z_1)^{ht^{12}}(z_3 - z_1)^{h(t^{13}+t^{23})}$
1(23)	z_1	$z_2\ z_3$	$W_{1(23)}$	$\sim \quad (z_3 - z_2)^{ht^{23}}(z_3 - z_1)^{h(t^{12}+t^{13})}$
1(32)	z_1	$z_3\ z_2$	$W_{1(32)}$	$\sim \quad (z_2 - z_3)^{ht^{23}}(z_2 - z_1)^{h(t^{12}+t^{13})}$
(13)2	$z_1\ z_3$	z_2	$W_{(13)2}$	$\sim \quad (z_3 - z_1)^{ht^{13}}(z_2 - z_1)^{h(t^{12}+T^{23})}$
(31)2	$z_3\ z_1$	z_2	$W_{(31)2}$	$\sim \quad (z - 1 - z_3)^{ht^{13}}(z_2 - z_3)^{h(t^{13}+t^{23})}$
3(12)	z_3	$z_1\ z_2$	$W_{3(12)}$	$\sim \quad (z_2 - z_1)^{ht^{12}}(z_2 - z_3)^{h(t^{13}+t^{23})}$

Of course we want to think of these six configurations arranged as vertices of Mac Lane's hexagon. As such, the functions attached to vertices related by an associativity are related to one another by action of a Φ:

$$W_{(12)3} = W_{1(23)}\Phi, \quad W_{(13)2} = W_{1(32)}\Phi^{132}, \text{and} \quad W_{(31)2} = W_{3(12)}\Phi^{312}.$$

On the other hand, we also have the relations

$$W_{1(32)}R^{23} = \mathbf{s}_{23}W_{1(23)}$$

and

$$W_{(31)2}R^{13} = \mathbf{s}_{13}W_{(13)2}.$$

Thus

$$\begin{aligned}
\mathbf{s}_{23}W_{(12)3}\Phi^{-1} &= W_{1(32)}R^{23} \\
&= W_{(13)2}(\Phi^{132})^{-1}R^{23} \\
\mathbf{s}_{12}\mathbf{s}_{23}W_{(12)3}\Phi^{-1} &= W_{(31)2}R^{13}(\Phi^{132})^{-1}R^{23} \\
&= W_{3(12)}\Phi^{312}R^{13}(\Phi^{132})^{-1}R^{23}
\end{aligned}$$

so

$$\mathbf{s}_{12}\mathbf{s}_{23}W_{(12)3} = W_{3(12)}\Phi^{312}R^{13}(\Phi^{132})^{-1}R^{23}\Phi. \tag{12.6.4}$$

But the path involved in $\mathbf{s}_2$ is homotopic to rotating z_3 around z_2 counterclockwise through 180 degrees (followed by a simultaneous translation of z_2 and z_3), and similarly, the path involved in $\mathbf{s}_{12}$ is rotating the second point around the first. So the composite path is homotopic to the path γ consisting of rotating the point z_3 through 180 degrees counterclockwise around the pair z_1, z_2. So we may replace $\mathbf{s}_{12} \circ \mathbf{s}_{23}$ by analytic continuation along γ. Along γ $\arg(z_3 - z_1)$ increases by πi while $\arg(z_2 - z_1)$ is constant. We conclude that

$$\mathbf{s}_{12}\mathbf{s}_{23}W_{(12)3} = W_{3(12)}e^{\pi i h(t^{13}+t^{23})} = W_{3(12)}(\Delta \otimes id)R. \tag{12.6.5}$$

Comparing (12.6.4) and (12.6.5) and using the fact that $W_{3(12)}$ is invertible, we get the first hexagon identity (11.6.5)

$$(\Delta \otimes id)R = \Phi^{312}R^{13}(\Phi^{132})^{-1}R^{23}\Phi. \tag{12.6.6}$$

The second hexagon (11.6.6) identity

$$(id \otimes \Delta)R = (\Phi^{231})^{-1}R^{13}\Phi^{213}R^{12}\Phi^{-1} \tag{12.6.7}$$

follows by following the clockwise paths.

12.7 The Drinfeld-Yang-Baxter equations.

Let $p = \prec z_1, z_2, z_3 \succ$ be a point of X_3 that we choose as our base point. Let us consider the paths γ_1 rotating z_1 and z_2 as in the preceding section, so $\gamma_1(0) = \gamma_1(1) = p$. Similarly, let γ_2 be the path beginning and ending at p which rotates z_2 and z_3. Then we have the Artin braid relation expressing the homotopy

$$\gamma_1\gamma_2\gamma_1 \sim \gamma_2\gamma_1\gamma_2.$$

A solution of KZE_3 gives a representation of this relation by assigning a reresentation

$$\gamma \mapsto T(\gamma) \in \mathrm{Aut}(A^{\otimes 3})$$

given by monodromy, once we have chosen an identification of the fiber over p with $A^{\otimes 3}$. The expression for T will involve Φ and R (as we shall see), and hence the equation

$$T(\gamma_1)T(\gamma_2)T(\gamma_1) = T(\gamma_2)T(\gamma_1)T(\gamma_2) \tag{12.7.1}$$

will give an equation involving Φ and R which will be Drinfeld's generalized Yang-Baxter equation. For the case $\Phi = 1 \otimes 1 \otimes 1$ it will reduce to the old Yang Baxter equation.

To get the form of Drinfeld's generalized Yang Baxter equation we need an explicit form for the monodromy operators $T(\gamma_1)$ and $T(\gamma_2)$.

So let us choose our base point $p \in X_3$ to be the image of a point $q \in Y_3$ with z_i real and $z_1 < z_2 < z_3$. To fix the ideas we might pick $p = \prec 1, 2, 3 \succ$ and $q = \langle 1, 2, 3 \rangle$. Identify $A^{\otimes 3}$ with the fiber over q via

$$a \mapsto \mathbf{1}(q)W_0(q)a \quad a \in A^{\otimes 3}. \tag{12.7.2}$$

Recall that there are six points in Y_3 which lie over p and the group S_3 permutes them, and acts as automorphisms of the fiber bundle over them. The fiber over p is obtained by identifying the fibers over these six points by the action of S_3. In particular, having fixed q as one of these six points, (12.7.2) gives an identification of $A^{\otimes 3}$ with the fiber of our bundle over $p \in X_3$.

Let

$$\sigma_1 = (12) \text{ and } \sigma_2 = (23) \quad \in S_3.$$

Letting Γ_1 be the lift of γ_1 starting at q we see that parallel transport along Γ_1 will take the point $\bullet 1(q)W_0(q)a$ to the point

$$\mathbf{1}(\Gamma(t))W_0(\Gamma(t))a \to \sigma_1[\mathbf{1}(q)W_0(q)]R^{12}a$$

as $t \to 1$. We can write the expression on the right as

$$\sigma_1[\mathbf{1}(q)W_0(q)\rho(\sigma)(R^{12}a)]$$

where ρ is the representation of S_3 on $A^{\otimes 3}$, cf. (12.5.2). So applying $\sigma_1^{-1} = \sigma_1$ to bring the fiber over $\Gamma_1(1)$ shows that

$$T(\gamma_1)a = \rho(\sigma_1)[R^{12}a] \tag{12.7.3}$$

in accordance with (12.5.1).To compute the monodromy $T(\gamma_2)$ we write

$$W_1(q)\Phi a = W_0(q)a$$

so parallel transport along Γ_2 carries this point to

$$\begin{aligned} \mathbf{1}(\Gamma_2(t)W_1(\Gamma_2(t)\Phi a &\longrightarrow \sigma_2[\mathbf{1}(q)W_1(q)]R^{23}\Phi a \\ &= \sigma_2[\mathbf{1}(q)W_0(q)\Phi^{-1}]R^{23}\Phi a \\ &= \sigma_2[\mathbf{1}(q)W_1(q)\Phi^{-1}\rho(\sigma_2)(R^{23}\Phi a)]. \end{aligned}$$

Applying $\sigma_2^{-1} = \sigma_2$ to bring the fiber back to q shows that

$$T(\gamma_2)(a) = \Phi^{-1}\rho(\sigma_2)(R^{23}\Phi a) = \rho(\sigma_2)[(\Phi^{132})^{-1}R^{23}\Phi a]. \tag{12.7.4}$$

Substituting (12.7.3) and (12.7.4) into (12.7.1), moving all the $\rho(\sigma_i)$ to the left (say) and using the braid relation

$$\sigma_1\sigma_2\sigma_1 = \sigma_2\sigma_1\sigma_2$$

in S_3 gives Drinfeld's non-associative Yang Baxter equation

$$\phi^{321}R^{23}(\Phi^{-1})^{231}R^{13}\Phi^{213}R^{12} = R^{12}\Phi^{312}R^{13}(\Phi^{-1})^{132}R^{23}\Phi. \tag{12.7.5}$$

12.8 Solution by Chen integrals

In section 3.9 we saw that Chen's iterated integrals can be used to solve the initial value problem

$$G'(s) = (f_1(s)X_1 + \cdots + f_n(s)X_n)G(s), \quad G(a) = 1. \tag{12.8.1}$$

The solution can be expressed as

$$G(s) = \sum_J \int_a^s J(f)X^J. \tag{12.8.2}$$

We can't use this method for equation (12.3.1) on the interval $[0,1]$ since we are given asymptotics and not initial values. If we consider a system of the form (12.8.1) on the interval $[\epsilon, 1-\epsilon]$ with

$$f_1(s) = \frac{1}{s}, \quad f_2(s) = \frac{1}{1-s}, \quad X_1 = ht^{12}, \quad X_2 = -ht^{23},$$

and $a = \epsilon$ or $a = 1-\epsilon$, that is,

$$G(\epsilon) = (\epsilon)^{ht^{12}} \quad \text{or} \quad G(1-\epsilon) = (1-\epsilon)^{ht^{23}},$$

then we have the problem of integrals which diverge as $\epsilon \to 0$.

However most of the Chen integrals $\int_\epsilon^s J(f)$ and $\int_s^\eta J(f)$, where $\eta = 1-\epsilon$ are not divergent.

Lemma 12.8.1 *Let* $f_1(s) = \frac{1}{s}, \quad f_2(s) = \frac{1}{1-s}$ *and* $0 < s < 1$.

1. *If J is not of the form $(K,1)$ then the limit $\lim_{\epsilon\to 0}\int_\epsilon^s J(f)$ converges uniformly in s on closed subintervals of $(0,1)$ to a differentiable function, which we denote $g_J(s)$.*

2. *If $J=(i,J')$ then $g_J(s)'=f_i(s)g_{J'}(s)$ for $0<s<1$. Setting $g_J(0)=0$ defines a continous extension to the interval $[0,1)$.*

3. *If J is not of the form $(K,2)$ then the limit $\lim_{\eta\to 1}\int_s^\eta J(f)$ converges uniformly in s on closed subintervals of $(0,1)$ to a differentiable function, which we denote $h_J(s)$.*

4. *If $J=(i,J')$ then $h_J(s)'=f_i(s)h_{J'}(s)$ for $0<s<1$. Setting $h_J(1)=0$ defines a continous extension to the interval $(0,1]$.*

Clearly it is enough to prove items 1 and 2 of the lemma. Items 3 and 4 are the corresponding statements near the other endpoint of the interval. Let $g_{J,\epsilon}(s)=\int_\epsilon^s J(f)$. The proof is based on a bound for $g_{J,\epsilon}(s)$ and an estimate for $g_{J\epsilon}(s)-g_{J\epsilon'}(s)$ for $\epsilon<\epsilon'$. Note that $g_{J,\epsilon}(s)>g_{J,\epsilon'}(s)$ for $\epsilon<\epsilon'$.
Claim For any finite sequence $j\neq(K,1)$ and any $0<a<1$ there exist constants $M_{J,a}$ and $m_{J,a}$ such that

$$g_{J,\epsilon}(s)<M_{J,a}s \quad \text{for} \quad \epsilon\leq s\leq a,$$

and

$$g_{J,\epsilon}(s)-g_{J,\epsilon'}(s)<m_{J,a}(\epsilon'-\epsilon)^{1/2^j} \quad \text{for} \quad \epsilon<\epsilon' \quad \text{and} \quad s\leq a,$$

where j is the length of J.

The second estimate proves uniform convergence on any closed subinterval of $(0,1)$. The expression for the derivatives relative to s is given by a continous function times an expression of the same form; therefore, if the sequence converges uniformly the the sequence of derivatives does also. We can then conclude that the limit is differentiable since the following well know theorem on sequences applies equally well to limits depending on a continuous parameter.

Given a sequence of continuously differentiable functions which converges uniformly together with the sequence of derivatives, the limit is differentiable and the derivative of the limit is the limit of the derivatives.

The first estimate proves that the limit has a continous extension to $[0,a]$.

We now prove the first estimate. If J is not of the form $(K,1)$ then either $J=(2)$ or $J=(i,J')$ with J' also not of the form $(K,1)$. The proof is by induction on the length of J. If $J=(2)$ then the Chen integral is the ordinary integral

$$\begin{aligned}\int_\epsilon^s f_2(s)ds &= \int_\epsilon^s \frac{1}{1-s}ds \\ &\leq \int_\epsilon^s \frac{1}{1-a}dt = \frac{s-\epsilon}{1-a} \\ &< (\frac{1}{1-a})s.\end{aligned}$$

Assume that $J=(i,J')$ and we have proven the estimate for J'. Then

$$g_{J,\epsilon}(s) = \int_\epsilon^s f_i(t) g_{J,,\epsilon}(t)dt.$$

If $i=1$ the integrand is less that $M_{J'a}$ so the integral is less that $M_{Ja}(s-\epsilon) < M_{Ja}s$. If $i=2$, the integrand is less than $M_{J'a}\frac{a}{1-a}$ and the integral is less than $M_{J'a}\frac{a}{1-a}s$.

Now to the second estimate.

$$\begin{aligned}\int_\epsilon^s f_i(t) g_{J'\epsilon}(t)dt &- \int_{\epsilon\prime}^s f_i(t) g_{J'\epsilon\prime}(t)dt \\ = \int_\epsilon^{\epsilon\prime} f_i(t) g_{J'\epsilon}(t)dt &+ \int_{\epsilon\prime}^s f_i(t)(g_{J\prime\epsilon}(t) - g_{J\prime\epsilon\prime}(t))dt \\ &\leq M_{J'a}(\epsilon'-\epsilon) + \int_{\epsilon'}^s f_i(t)(\epsilon'-\epsilon)^{1/2^{j-1}}dt.\end{aligned}$$

If $i=2$ the estimate follows because $(\epsilon'-\epsilon) < (\epsilon'-\epsilon)^{1/2^{j-1}} < (\epsilon'-\epsilon)^{1/2^j}$ and the integral of $\int_{\epsilon\prime}^s f_2(t)dt$ is bounded by $\frac{1}{1-a}$. If $i=1$ the estimate follows from the fact that $-(\epsilon'-\epsilon)^{1/2^{j-1}} ln(\epsilon')$ is bounded for $0<\epsilon<\epsilon'<1$. This proves the estimate and the lemma follows.

Substituting $G_0(s) = H_0(s)s^{hX_1}$ in

$$G'(s) = h(\frac{X_1}{s} + \frac{X_2}{1-s})G(s)$$

we find that $H_0(s)$ solves

$$H'(s) = h(\frac{adX_1}{s} + \frac{X_2}{1-s})H(s), \tag{12.8.3}$$

where $ad(X)Y = XY - YX$. Let F be the noncommutative formal power series in X_1, X_2 as defined in section 3.8, only now with coefficients in $\mathbf{C}[[h]]$. Let ρ_0 be the representation of F into $\mathrm{End}_{\mathbf{C}[[h]]}(F)$ given by

$$\rho_0(X_1)(Y) = ad(hX_1)(Y), \qquad \rho_0(X_2)(Y) = -hX_2\cdot Y.$$

Define

$$H_{u\epsilon}(s) = \sum_J \int_\epsilon^s J(f)\rho_0(X^J)(u). \tag{12.8.4}$$

$H_{u\epsilon}(s)$ solves the equation (12.8.3) with initial conditions

$$H_{u\epsilon}(\epsilon) = u.$$

For all $n > 0$ we can interpret (12.8.3) as an equation on $F_{(n)} = F/h^{n+1}F$ with initial condition $H(\epsilon) = u_{(n)}$, where $u_{(n)}$ is the image in $F_{(n)}$ of $u \in F$ under the canonical quotient map. Since the term $\rho(X^J)$ has a factor of h^j, (12.8.4) when reduced modulo h^{n+1} is a finite sum which gives a solution to (12.8.3). Now note that if we pick initial conditions u such that $ad(X_1)u = 0$ then the only integrals appearing in the sum have $J \neq (K, 1)$, so the limit as $\epsilon \to 0$ exists uniformly in s and gives the solution to (12.8.3) for $0 \leq s < 1$. Letting $u = 1$ this proves the first half of the following proposition, the proof of the second half is completely parallel.

Proposition 12.8.1

1. *The solution $G_0(s)$ to (12.8.3) with asymptotics $G_0(s) \sim s^{hX_1}$ at $s = 0$ can be expressed*

$$G_0(s) = (\sum_{J\neq(K,1)} \int_0^s J(f)(\rho_0(X))^J)(1) \cdot s^{hX_1},$$

where

$$f_1(s) = \frac{1}{s}, \quad f_2(s) = \frac{1}{1-s}$$

and

$$\rho_1(X_1)(Y) = ad(hX_1), \quad \rho_0(X_2)(Y) = -hX_2 \cdot Y.$$

2. *The solution $G_1(s)$ to (12.8.3) with asymptotics $G_1(s) \sim (1-s)^{hX_2}$ at $s = 1$ can be expressed*

$$G_1(s) = -(\sum_{J\neq(K,2)} \int_s^1 J(f)(\rho_0 X)^J)(1) \cdot (1-s)^{hX_2},$$

where $\rho_1(X_1)(Y) = X_1 \cdot Y, \qquad \rho_0(X_2)(Y) = ad(-hX_2)(Y).$

If we define $\Phi = (G_1(s))^{-1}G_0(s)$ the considerations in section 3.8 and 3.9 together with the fact that s^{hX_1} and $(1-s)^{hX_2}$ are themselves exponentials and the standard Baker-Campbell-Hausdorff formula shows that Φ has the form $expZ$ for some Z in $\mathcal{L}(X_1, X_2)[[h]]$, the free Lie algebra generated over $\mathbf{C}[[h]]$ by X_1, X_2. Substituting $X_1 = t^{12}$ and $X_2 = t^{23}$ shows that

Proposition 12.8.2 *The solution Φ to the pentagon equation given by solving KZE_3 has the form $exp(\phi(ht^{12}, ht^{23}))$ for some canonical function $\varphi(X_1, X_2$ taking values in $\mathcal{L}(X_1, X_2)[[h]]$.*

Chapter 13

Deformation of Drinfeld algebras

We saw in Chapter 9 that the Gerstenhaber-Schack complex defined using the bar and cobar complex gave a good cohomology theory for the study of those deformations of Drinfeld algebras in which the initial structure has a trivial coassociativity operator, $\Phi_0 = 1$. When $\Phi_0 \neq 1$ a number of complications arise. First of all the definition of the cobar complex must be modified to include the operator Φ_0. From one point of view this can be understood as a way of compensating for the nonassociativity of Δ_0. Another way of understanding the modification is that, as long as $\Phi_0 \neq 1$, it must appear explicitly in the differential so that the cocycle and coboundary conditions will correspond to the structure equations. In [1095], we gave a rather ad hoc definition of the modified differential based on the idea that the tensor product of modules (left, right, or bi-) over a Drinfeld algebra is a nonassociative operation and one must keep track of the placement of parentheses. Markl and Stasheff, [896], independently, gave an equivalent definition in which they replaced the tensor algebra with a non-associative algebra. Once the principal of a nonassociative tensor structure is recognized, the remaining details follow very naturally. The resulting modified cobar complex defines a cohomology theory suited to the study of deformations of the pentagon identity in the same way as the complex $\hat{C}$ or $\tilde{C}$ which we used repeatedly in the last chapter.

The next complication comes when we want to build the double complex using the $B-4$ homomorphisms from the bar complex to the (modified) cobar complex. In order to define a $B-4$ structure on the bar complex we need to use the tensor product of comodules, but the coassociativity operator enters into the definition in a way that creates problems with the definition of the tensor product of comodules. In fact the naive definition, mimicking the definition for comodules over bialgebras, does not work. (There is a definition of tensor product over the bialgebra, but this does not give us what we want in the case of the bar complex.) Markl and Shnider [895] found a way around this problem by generalizing the definition of a comodule over a Drinfeld algebra to include terms of higher order. The new operations are similar (dually) to the ternary and higher "multiplications" in the definition of Stasheff's homotopy associative algebras; therefore, the structure which these higher order operations define is called a homotopy associative comodule. They show that the usual bar complex of a Drinfeld algebra is a homotopy associative bicomodule, with a compatible B bimodule structure. The complex which replaces $\mathrm{Hom}(\mathrm{Bar}(B), \mathrm{Cob}(B))$ is still bigraded but no longer a bicomplex. The differential has components $D_n : C^{p,q} \to C^{p+n,q-n+1}$ for $n > 1$. These higher order operators are exactly the ones needed to adapt the cohomology theory to the study of deformations of Drinfeld algebras.

In section 1 we give our original definition of the modified cobar complex relative to a preferred placement of parentheses in the tensor product, i.e., a preferred vertex in the Stasheff polyhedra. In section 2 we study the deformations of the pentagon identity following closely the paper [1095]. If the modified cobar complex alone was sufficient for the full deformation theory then the choice between the construction in section 1 and that of Markl and Stasheff would be a matter of taste. The formulae are no more complicated and have the advantage that they describe all the operators explicitly on a fixed vector space. However, in the constructions that are necessary for the full deformation theory the explicit appearance of all the associativity operators becomes very awkward, and an alternative approach is clearly preferable. Therefore, in section 3 we describe the nonassociative tensor algebra of a Drinfeld algebra bimodule, following very closely, in some places verbatim, the paper by Markl and Stasheff. This is the approach we adopt in the rest of the chapter. In section 4 we discuss the problems that arise in defining the tensor product of comodules and define homotopy comodules. In section 5 we define operators $M_n[\sigma]$ for simplices σ in K_n which are the building blocks of the higher D_{n-1} in the homotopy bicomodule structure on the bar complex. In section 6 we give the explicit definitions of the operators D_{n-1} and prove that using them we can define a cohomology theory generalizing the Gerstenhaber-Schack theory. In section 7 we show that this theory reduces to the standard one when $\Phi_0 = 1$ and show how this cohomology theory can be used to study Drinfeld algebra deformations.

13.1 The modified cobar complex

Let (B, m, Δ, u, Φ) be a Drinfeld algebra. By ${}_B\mathcal{M}$ we mean the category of left B modules, $\mathcal{M}_B$ the category of right modules, and ${}_B\mathcal{M}_B$ the category of B bimodules. Each of these categories has a monoidal structure. As a vector space, the tensor product is the usual tensor product over k. In order to fix notation we will remind the reader of the definitions of the module structure and the associativity operators.

For $M, N \in {}_B\mathcal{M}$ with $m \in M$, $n \in N$ the action of $a \in A$ is given by

$$a(m \otimes n) = \sum a_{(1)}m \otimes a_{(2)}n.$$

When Δ is not coassociative in order to have a notation suited to tensor products with more than two factors we modify the Sweedler notation and include in the subscript an indication of which particular iteration of Δ is being applied. We give a preferred status to the composition with Δ always in the first position,

$$(\Delta \otimes id)\Delta(a) = \sum a_{(1)} \otimes a_{(2)} \otimes a_{(3)}.$$

For the other composition

$$(id \otimes \Delta)\Delta(a) = \sum a_{(1)'} \otimes a_{(2)'} \otimes a_{(3)'}.$$

Denoting the coassociativity operator for B (which determines the associativity operator for the category of B modules) by

$$\Phi = \sum \Phi^{(1)} \otimes \Phi^{(2)} \otimes \Phi^{(3)} \in B \otimes B \otimes B,$$

we have the identity

$$\sum \Phi^{(1)}a_{(1)} \otimes \Phi^{(2)}a_{(2)} \otimes \Phi^{(3)}a_{(3)} = \sum a_{(1)'}\Phi^{(1)} \otimes a_{(2)'}\Phi^{(2)} \otimes a_{(3)'}\Phi^{(3)}.$$

The left (right)B module action of an element $a \in B$ on $(M \otimes N) \otimes P$ is given by the left (right) multiplication by $(\Delta \otimes id)\Delta(a)$. The left (right) module action on $M \otimes (N \otimes P)$ is given by left (right) multiplication by $(id \otimes \Delta)\Delta(a)$.

The associativity operator between the two distinct left modules

$$\alpha^l_{M,N,P} : (M \otimes N) \otimes P \to M \otimes (N \otimes P)$$

is given by

$$\alpha^l_{M,N,P}((m\otimes n)\otimes p)=\sum \Phi^{(1)}m\otimes(\Phi^{(2)}n\otimes\Phi^{(3)}m)$$

and the associativity operator between the two distinct right modules is

$$\alpha^r_{M,N,P}((m\otimes n)\otimes p)=\sum m(\Phi^{-1})^{(1)}\otimes(n(\Phi^{-1})^{(2)}\otimes p(\Phi^{-1})^{(3)}).$$

For ${}_B\mathcal{M}_B$ the associativity operator is the composition

$$\alpha_{M,N,P}=\alpha^l_{M,N,P}\circ\alpha^r_{M,N,P}=\alpha^r_{M,N,P}\circ\alpha^l_{M,N,P}.$$

Define the maps

$$\Delta_{i,p}:B^{\otimes p}\to B^{\otimes(p+1)},$$

by

$$\Delta_{i,p}(a_1\otimes\cdots\otimes a_p)=a_1\otimes\cdots\otimes\Delta(a_i)\otimes\cdots\otimes a_p. \tag{13.1.1}$$

Let $\mathcal{V}_p$ denote the vertices of the Stasheff polyhedron K_p, $p\geq 2$, that is, the set of all bracketings of p nonassociative indeterminates. Let $\mathcal{V}_1=\{(\bullet)\}$ and $\mathcal{V}_2=\{(\bullet\bullet)\}$ each consist of one element. For $v\in\mathcal{V}_p$ and $1\leq i\leq p$, let $v_{[i]}\in\mathcal{V}_{p+1}$ denote the bracketing obtained from v by the replacement $\bullet\mapsto(\bullet\bullet)$ at the i-th place. Notice that any bracketing can be obtained from the "trivial bracketing" $(\bullet)\in\mathcal{V}_1$ by a successive application of this operation (for example, $(\bullet\bullet)\bullet=((\bullet)_{[1]})_{[1]}$, $\bullet(\bullet\bullet)=((\bullet)_{[1]})_{[2]}$ and so on). For $v\in\mathcal{V}_p$ let $M_v^{\otimes p}$ be the p-fold tensor product of M with the parentheses placed in conformity with the bracketing v. The following two conditions clearly define, for any $v\in\mathcal{V}_p$, a map $\Delta^v:B\to B^{\otimes p}$ by means of which the B module structure on the tensor product $M_v^{\otimes p}$ is defined.

1. $\Delta^\bullet=id$ and
2. $\Delta^{v_{[i]}}=\Delta_{i,p}\Delta^v$, for $v\in\mathcal{V}_p$ and $1\leq i\leq p$.

We reiterate the general result for any tensor product of modules (not necessarily the same).

The left (right) B module structure on a tensor product of p modules with parentheses placed according to the formula determined by the vertex v is left (right) multiplication by $\Delta^v(a)$.

As mentioned above, we give a preferred status to the tensor product in which the parentheses expand out from the left. For example, when all the modules are equal

$$(((((M\otimes M)\otimes M)\otimes M)\otimes M)\otimes M)\otimes M.$$

We will denote the vertex in K_p corresponding to this bracketing by $[p]$ and let $M^{[p]}$ denote $M_{[p]}^{\otimes p}$ with the bimodule structure given above.

We will adopt the convention

$$\Delta^{[0]} = \epsilon. \tag{13.1.2}$$

Also we note the additional identity

$$\Delta^{[p]} = (\Delta^{[p-1]} \otimes id)\Delta$$

which could have been used to define $\Delta^{[p]}$ inductively beginning with $\Delta^{[0]}$.

In particular all the $B^{[p]}$ are B bimodules. We would like to build the cobar resolution out of the operators $\Delta_{i,p}$ but unfortunately,

$$\Delta_{i,p} : B^{[p]} \to B_{[p]_i}^{\otimes(p+1)}$$

instead of $B^{[p]} \to B^{[p+1]}$. The issue is the placement of the parentheses in ${}_B\mathcal{M}_B$. For example, the image of $\Delta_{5,6}$ will lie in

$$((((B \otimes B) \otimes B) \otimes B) \otimes (B \otimes B)) \otimes B$$

while we would like all the images of the $\Delta_{i,6}$ to lie in

$$(((((B \otimes B) \otimes B) \otimes B) \otimes B) \otimes B) \otimes B.$$

To remedy this we must apply the appropriate associativity operator. For instance, in the example above, we would want to apply the associativity operator

$$a^{-1}_{B^{[4]},B,B} \otimes id$$

which is given by conjugation by the element

$$(\Delta^{[4]} \otimes id \otimes id)(\Phi^{-1}) \otimes 1.$$

More generally, let $1_j = 1 \otimes \cdots \otimes 1$, $j-$times. Then conjugation by

$$\Psi_{i,p+1} = (\Delta^{[i-1]} \otimes id \otimes id)(\Phi^{-1}) \otimes 1_{p-i} \tag{13.1.3}$$

gives the associativity in ${}_B\mathcal{M}_B$ from $M_i^{[p]}$ to $M^{[p+1]} = M_1^{[p]}$ where $M_i^{[p]} = M_{[p]_i}^{\otimes(p+1)}$ in our previous notation.

$$\Psi_{1,p+1} = 1_{p+1}$$

and

$$\Psi_{2,p+1} = \Phi^{-1} \otimes 1_{p-2}.$$

Define

$$\delta_p : B^{[p]} \to B^{[p+1]}$$

by

$$\delta_p = \sum_{i=1}^{p} (-1)^{i-1} Ad\ (\Psi_{i,p+1})\Delta_{i,p}. \tag{13.1.4}$$

The we have the following propositions

Proposition 13.1.1 $\delta_{p+1} \circ \delta_p = 0$.

Proposition 13.1.2 *δ_p is an B bimodule morphism for all p.*

Proposition 13.1.3 *For each p the module $B^{[p]}$ has the structure of an B bicomodule, and the maps δ_p are B bicomodule morphisms.*

Note. We are **not** dealing here with a tensor product of comodules. As in the standard cobar complex the left comodule structure involves only the extreme left factor in the tensor product.

Proposition 13.1.4 *The B bimodule maps $\epsilon_{1,p} : B^{[p]} \to B^{[p-1]}$ given by $\epsilon_{1,1} = 0$ and*

$$\epsilon_{1.p} a_1 \otimes \cdots \otimes a_p = \epsilon(a_1) a_2 \otimes \cdots \otimes a_p, \quad p \geq 2 \tag{13.1.5}$$

define a contracting homotopy, that is, they satisfy the equations

$$\delta_{p-1}\epsilon_{1,p} + \epsilon_{1,p+1}\delta_p = id. \tag{13.1.6}$$

For $p = 1$ this is understood to mean

$$\epsilon_{1,2}\delta_1 = id.$$

The content of these four propositions is thus the assertion that

$$C(B) = \bigoplus_{p\geq 1} B^{[p]}, \quad \delta = \bigoplus \delta_p$$

is an acyclic complex of B bimodules and bicomodules.

We now turn to the proofs of these propositions. For Prop.1, write the double sum occurring in $\delta_{p+1} \circ \delta_p$ as the sum of three terms

$$\delta_{p+1} \circ \delta_p = \sum{}_1 + \sum{}_2 + \sum{}_3$$

where

$$\begin{aligned}
\sum{}_1 &= \sum_{i<j} (-1)^{i+j} Ad\Psi_{i,p+2}\Delta_{i,p+1} \circ Ad\Psi_{j,p+1}\Delta_{j,p} \\
\sum{}_2 &= \sum_{i>j+1} (-1)^{i+j} Ad\Psi_{i,p+2}\Delta_{i,p+1} \circ Ad\Psi_{j,p+1}\Delta_{j,p} \\
\sum{}_3 &= \sum_i [Ad(\Psi_{i,p+2}\Delta_{i,p+1}(\Psi_{i,p+1}))\Delta_{i,p+1}\Delta_{i,p} - \\
&\qquad Ad(\Psi_{i+1,p+2}\Delta_{i+1,p+1}(\Psi_{i,p+1}))\Delta_{i+1,p+1}\Delta_{i,p}].
\end{aligned}$$

We claim that we get term by term cancellation between $\sum_1$ and $\sum_2$ and that the summands in $\sum_3$ vanish. To see this, observe that although these expressions look formidable, this is because of the presence of the associativity operators.

For example, without the presence of the associativity operators, then for $i < j$, we would have a simple term $\Delta_{i,p+1} \circ \Delta_{j,p}$ occurring with sign $(-1)^{i+j}$ in $\sum_1$ and we could consider this as a map, from $B^{[p]}$ to $B^{[p]}_{i,j}$ where the subscript i,j means that we have replaced both the i–th and the j–th factor B by $(B \otimes B)$. Let $\Delta^{ij} : B^{[p]} \to B^{[p]}_{ij}$ denote this operator. The actual operator occurring in $\sum_1$ has a couple of associativities intervening. But the fact that the associativity is a natural transformation, and the coherence property of the associativity, implies that operator occurring in $\sum_1$ is $a \circ \Delta^{ij}$ where $a : V^{[p]}_{ij} \to V^{[p+2]}$ is, in fact, the appropriate (unique) associativity. For the same reason the corresponding term occurring in $\sum_2$ is also $a \circ \Delta^{ij}$, but now with sign $(-1)^{i+j+1}$. Thus $\sum_1$ and $\sum_2$ cancel. (Alternatively, one can argue directly from the definitions that these sums cancel, hence verifying the required coherence in this case without appeal to the MacLane theorem. The same goes for the vanishing of $\sum_3$ to which we now turn.)

We claim that the summands in $\sum_3$ vanish because of the naturality of the associativity and MacLane's pentagon constraint. It is easiest to explain this argument diagrammatically. For the i–th summand, we can clearly ignore all the tensor factors of B occurring in $B^{[p]}$ which are to the right of the i–th position, as they just get carried along. Similarly, the $B^{[i-1]}$ which occurs to the left of the i–th position we can treat as a single factor as far as the associativities are concerned. So we can represent the starting position as $\bullet\ \bullet$ where the left hand $\bullet$ stands for $B^{[i-1]}$ and the right $\bullet$ stands for the i–the factor, B. Then one of the operators occurring in the i–th summand (the one where the second Δ occurs at the right of the two newly created $B \otimes B$ factors, (the term containing $\Delta_{i+1}\Delta_i$) can be represented as

$$\bullet\ \bullet \rightarrow \bullet\ (\bullet\ \bullet) \rightarrow (\bullet\ \bullet)\ \bullet \rightarrow (\bullet\ \bullet)(\bullet\ \bullet) \rightarrow ((\bullet\ \bullet)\ \bullet)\ \bullet\ ,$$

where we have included arrows for the associativities. By the naturality of the associativity, we can replace this by

$$\bullet\ \bullet \rightarrow \bullet\ (\bullet\ \bullet) \rightarrow \bullet\ (\bullet\ (\bullet\ \bullet)) \rightarrow (\bullet\ \bullet)(\bullet\ \bullet) \rightarrow ((\bullet\ \bullet)\ \bullet)\ \ \bullet\ .$$

Notice that the last two arrows are two sides to Mac Lane's pentagon. Similarly, the other operator (occuring with the opposite sign) in which the second Δ is applied to the left hand B in the newly created $(B \otimes B)$ (the term containing $\Delta_i\Delta_i$) is equivalent by naturality to

$$\bullet\ \bullet \rightarrow \bullet\ (\bullet\ \bullet) \rightarrow \bullet\ ((\bullet\ \bullet)\ \bullet) \rightarrow (\bullet\ (\bullet\ \bullet))\ \bullet \rightarrow ((\bullet\ \bullet)\ \bullet)\ \ \bullet\ .$$

The two right arrows are again two edges in Mac Lane's pentagon.The remaining edge of Mac Lane's pentagon, corresponding to the associativity

$$\bullet\ (\bullet\ (\bullet\ \bullet)) \rightarrow \bullet\ ((\bullet\ \bullet)\ \bullet)$$

together with the commutativity of the diagram

$$\begin{array}{ccccc} \bullet\ \bullet \rightarrow \bullet\ (\bullet\ \bullet) & & \longrightarrow & & \bullet\ ((\bullet\ \bullet)\ \bullet) \\ & \searrow & & \nearrow & \\ & & \bullet\ (\bullet\ (\bullet\ \bullet)) & & \end{array}$$

is provided exactly by our axiom asserting the equality up to associativity of $(\Delta \otimes id) \circ \Delta$ and $(id \otimes \Delta) \circ \Delta$. This completes the proof of Prop. 1.

Prop. 2 follows from the identities, for all p and $i \leq p$

$$\begin{aligned} Ad\Psi_{i,p+1}\Delta_{i,p}(\Delta^{[p]}(a)\sigma) &= Ad\Psi_{i,p+1}(\Delta_{i,p}(\Delta^{[p]}(a))Ad\Psi_{i,p+1}\Delta_{i,p}\sigma \\ &= \Delta^{[p+1]}(a)Ad\Psi_{i,p+1}\Delta_{i,p}\sigma \end{aligned}$$

For the proof of Prop. 3, let $\alpha_{[p]}$ denote the associativity operator from $B^{[p+1]}$ to $B \otimes B^{[p]}$. Then

$$\lambda_p = \alpha_{[p]} \circ \Delta_{1,p} : B^{[p]} \rightarrow B \otimes B^{[p]}$$

defines a left comodule structure on $B^{[p]}$. Similarly, $\mathrm{Ad}\Psi_{p,p+1}$ is the associativity operator from $B^{[p-1]} \otimes (B \otimes B)$ to $B^{[p+1]}$ so

$$\rho_p = Ad\Psi_{p,p+1} \circ \Delta_{p,p} : B^{[p]} \to B^{[p+1]} = B^{[p]} \otimes B$$

defines a right comodule structure. The compatibility conditions for these structures as well as the identities

$$\lambda_{p+1}\delta_p = (id \otimes \delta_p)\lambda, \tag{13.1.7}$$

and

$$\rho_{p+1}\delta_p = (\delta_p \otimes id)\rho_p \tag{13.1.8}$$

follow from Mac Lane coherence, or directly from the definitions.

For the proof of Prop. 4, we use the following identities which follow immediately from the definitions:

$$\begin{aligned} \epsilon_{1,p}\Psi_{j,p} &= \Psi_{j-1,p-1}, \quad j \geq 2 && (13.1.9)\\ \epsilon_{1,p+1}\Delta_{j,p} &= \Delta_{j-1,p-1}\epsilon_{1,p}, \quad j \geq 2 && (13.1.10)\\ \epsilon_{1,p+1}\Delta_{1,p} &= id. && (13.1.11) \end{aligned}$$

This last equation, for $p = 1$ reduces to the case $p = 1$ of (13.1.6) For $p \geq 2$ we have

$$\begin{aligned} \epsilon_{1,p+1}\delta_p &= \epsilon_{1,p+1}(\Delta_{1,p} + \sum_{1=2}^{p}(-1)^{i-1} Ad\ \Psi_{i,p+1}\Delta_{i,p}) \\ &= id + \sum_{i=2}^{p}(-1)^{i-1} Ad\ (\epsilon_{1,p+1}(\Psi_{i,p+1}))\epsilon_{1,p+1}\Delta_{i,p} \\ &= id - \sum(-1)^{i-2} Ad\ \Psi_{i-1,p}\Delta_{i-1,p-1}\epsilon_{1,p}, \end{aligned}$$

proving Prop.4.

At this stage in the usual theory, one would normally consider the functor $\mathrm{Hom}_{B-\mathrm{bicomod}}(k, \bullet)$ applied to the cobar complex that we have just constructed and use an adjoint relation to pass to something like the "reduced cobar complex" which we used in the last section of Chapter 11. However,

$$\mathrm{Hom}_{B-\mathrm{bicomod}}(M, B \otimes (N \otimes B) \neq \mathrm{Hom}_k(M, N).$$

The adjoint functor relation fails because the various compatability conditions for the left comodule structure λ or right comodule structure or the compatibility conditions between the left comodule structure λ and right comodule structure ρ on M involve the associativity operators defined by Φ. For example

$$\Phi[(\lambda \otimes id) \circ \rho] = [(id \otimes \rho) \circ \lambda]\Phi.$$

The formula for the operator γ which generates a bicomodule map from a k linear map must now be consistent with the placement of parentheses. If we define

$$\gamma(\psi) = (id \otimes \psi \otimes id) \circ (id \otimes \rho) \circ (\lambda).$$

then the conditions for $\gamma(\psi)$ to be a bicomodule map involve commuting $id \otimes \psi \otimes id$ past an associativity operator, and this obviously places restrictions on the linear map ψ. For example, it would be sufficient for ψ to be a B bimodule map, but the B bimodule structure on k is given by the augmentation map. Bimodule maps with domain k have range in a submodule annihilated by the augmentation ideal so this approach clearly will not work. If we try to use Bar_0 instead of k as the domain of our homomorphisms, and consider an adjoint relation like

$$\mathrm{Hom}_{B-4}(M, B \otimes (N \otimes B)) = \mathrm{Hom}_{B-\mathrm{bimod}}(M, N)$$

then in order to set $M = \mathrm{Bar}_0(B)$ we have the problem of defining a B bicomodule structure on the bar complex.

At the initial stage of the deformation theory, when we are dealing only with the pentagon identity, we can avoid all these complications and simply consider a subcomplex of $(C(B), \delta)$. In the full theory these considerations will be essential.

Define the subspaces

$$B^{[p]\flat} = \{\sigma \in B^{[p]} | a\sigma = \sigma a, \quad \forall a \in B\}.$$

Let

$$\eta = \alpha^{-1}_{[p+2]} : B \otimes B^{[p+1]} \to B^{[p+2]}$$

and define

$$j_p : B^{[p]\flat} \to B^{[p+2]}$$

by

$$j_p(\sigma) = \eta(1 \otimes (\sigma \otimes 1)). \tag{13.1.12}$$

Define

$$\delta^\flat : B^{[p]\flat} \to B^{[p+1]\flat}$$

by

$$\delta_p^\flat(\sigma) = \eta(1\otimes\sigma) - \delta_p\sigma + (-1)^{p+1}\sigma\otimes 1. \qquad (13.1.13)$$

Then we have

Proposition 13.1.5 *(i)* $\delta_{p+2}\jmath_p(\sigma) = \jmath_{p+1}(\delta_p^\flat\sigma)$, *(ii)* $\delta^\flat$ *preserves* $B^\flat$.

Proof. By definition

$$\delta_{p+2}\jmath_p(\sigma) = \sum_{i=1}^{p+2}(-1)^{i-1} Ad\ \Psi_{i,p+3}\Delta_{i,p+2}\eta(1\otimes(\sigma\otimes 1)).$$

The first term is

$$\Delta_1\circ\eta(1\otimes(\sigma\otimes 1)) = \xi((1\otimes 1)\otimes(\sigma\otimes 1))$$

where ξ is the appropriate associativity operator, and

$$\xi((1\otimes 1)\otimes(\sigma\otimes 1)) = \eta(1\otimes((1\otimes\sigma)\otimes 1)$$

since in this equation the two sides differ by the action of $Ad((1\otimes\Phi^{-1})\Delta_{3,3}\Phi)$ on $1\otimes 1\otimes\sigma\otimes 1$, which is trivial by the invariance hypothesis. So this yields the first term in $\jmath_{p+1}(\delta_p^\flat\sigma)$. By a similar argument,

$$\begin{aligned} Ad\ \Psi_{p+2,p+3}\Delta_{p+2,p+2}\circ\eta(1\otimes(\sigma\otimes\Delta(1))) &= \xi(1\otimes(\sigma\otimes(1\otimes 1))) \\ &= \eta(1\otimes((\sigma\otimes 1)\otimes 1))) \end{aligned}$$

giving the last term in $\jmath_{p+1}(\delta_p^\flat\sigma)$. The remaining terms in the sum add up to $-\jmath_{p+1}(\delta_p\sigma)$ by the naturality of associativity.

To prove the second assertion, note that for any $\sigma\in B^{[p]}$:

$$\begin{aligned} a\eta(1\otimes\sigma) &= \Delta^{[p+1]}(a)\times\eta(1\otimes\sigma) \\ &= \sum\eta((a_{(1)}\otimes\Delta^{[p]}a_{(2)})\times(1\otimes\sigma)) \\ &= \sum\eta(a_{(1)}\otimes a_{(2)}\sigma) \end{aligned}$$

where the second equality follows from the naturality of associativity. Similarly for the right action, therefore, $\sigma\in B^{[p]\flat}$ implies $\eta(1\otimes\sigma)\in B^{[p+1]\flat}$. The same holds for $\sigma\otimes 1$. Finally, since δ_p is an B bimodule map, $\delta_p\sigma\in B^{[p+1]\flat}$, completing the proof.

13.2 Deformation theory.

Let (B, m, Δ, Φ) be a Drinfeld algebra defined over k . A *restricted deformation* $(B_h, m, \Delta, \Phi(h))$ is one in which the deformed multiplication and comultiplication which we normally write, $m(h)$ and $\Delta(h)$ respectively, are the (undeformed) $k[[h]]$ linear extensions of m and Δ, so that all that is being deformed is Φ. For restricted deformations the equivalence relation is restricted to twisting in which Δ remains undeformed

$$AdF(h) \circ \Delta = \tilde{\Delta} \tag{13.2.1}$$
$$F^{23}(id \otimes \Delta(F))\Phi(h)(\Delta \otimes id)(F^{-1})(F^{12})^{-1} = \tilde{\Phi}(h). \tag{13.2.2}$$

The first equation says that F is invariant. We will show that the complex $(C^\flat(B), \delta^\flat)$ where

$$C^\flat(B) = \bigoplus_{p \geq 2} B^{[p]\flat}, \quad \delta^\flat = \bigoplus_p \delta_p^\flat$$

describes the restricted deformation theory with strong equivalence in the usual sense. Any infinitesimal restricted deformation is a cocycle, two infinitesimals are strongly equivalent if and only if they are cohomologous, we can define the notion of restricted deformations up to "order n" and the obstruction to going one step further is a cocycle, and finally, two such extensions differ by a cocycle which is a coboundary if and only if they are strongly equivalent up to "order $n+1$".

To start the theory off, let us write the deformation of Φ multiplicatively:

$$\Phi(h) = \Phi(1 + h\phi_1 + h^2\phi_2 + \ldots)$$

and look at the coefficient of h in the conjugacy relation which we write as

$$\Phi(h)(\Delta \otimes id)\Delta(a) = (id \otimes \Delta)\Delta(a)\Phi(h).$$

We get

$$\phi_1(\Delta \otimes id)\Delta(a) = \Phi^{-1}(id \otimes \Delta)\Delta(a)\Phi\ \phi_1 = (\Delta \otimes id)\Delta(a)\phi_1.$$

In other words, considering ϕ_1 as an element of $B^{[3]}$

$$a\phi_1 = \phi_1 a \quad \forall a \in B.$$

So calling ϕ_1 the infinitesimal of the deformation, we have proved

Proposition 13.2.1 *Infinitesimals belong to* $B^{[3]\flat}$.

Let us write the pentagon identity in the form

$$\Delta_{1,3}\Phi(h)^{-1}\Delta_{3,3}\Phi(h)^{-1}\ (1\otimes\Phi(h))\Delta_{2,3}\Phi(h)(\Phi(h)\otimes 1)=1.$$

Using $(1+h\Phi_1+\ldots)^{-1}=(1-h\phi_1+\ldots)$ the coefficient of h on the left becomes (after pulling all the zero-th order terms to the right and using the pentagon identity for Φ)

$$\begin{aligned}\Delta_{1,3}\phi_1-Ad(\Delta_{1,3}\Phi^{-1})\Delta_{3,3}\phi_1+Ad(\Delta_{2.3}\Phi(\Phi\otimes 1))^{-1}\ 1\otimes\phi_1+&\\ +Ad(\Phi\otimes 1)^{-1}\Delta_{2,3}\phi_1+\phi_1\otimes 1 &=\\ \eta(1\otimes\phi_1)-\delta_3\phi_1+\phi_1\otimes 1 &=\\ &=\delta_3^\flat\phi_1.\end{aligned}$$

Thus

Proposition 13.2.2 *The pentagon identity is satisfied to order h^2 if and only if the infinitesimal deformation , $\phi_1\in B^{[3]\flat}$ is a $\delta_3^\flat$ cocycle.*

Let $(B_{(n-1)},m,\Delta,\Phi_{(n-1)})$ be an $(n-1)$-truncated restricted deformation. We shall now define the obstruction to extending a $(n-1)$–truncated restricted deformation to a n–truncated one. Write the $(n-1)$–truncated deformation as

$$\Phi_{(n-1)}=\Phi(1+h\phi_1+\ldots+h^{n-1}\phi_{n-1}).$$

Define the *obstruction cochain*, u_n by the congruence mod h^{n+1}

$$1+h^n u_n\equiv(\Delta_{1,3}\Phi_{(n-1)}^{-1})(\Delta_{3,3}\Phi_{(n-1)}^{-1})(1\otimes\Phi_{(n-1)})(\Delta_{2,3}\Phi_{(n-1)})(\Phi_{(n-1)}\otimes 1) \tag{13.2.3}$$

We will consider u_n as an element of $B^{[4]}$ and prove

Proposition 13.2.3 *The obstruction cochain, u_n, is a $\delta^\flat$ cocycle in $A^{[p]\flat}$, i.e.*

$$\delta_4^\flat u_n=0.$$

Proof. The proof of invariance follows from the following five identities, for the first of which we give two intermediate equations,

$$\begin{aligned}\Delta^{[4]}(a)\Delta_1\Phi_{(n-1)}^{-1}=\Delta_1(\Delta^{[3]}(a)\Phi_{(n-1)}^{-1}) &=\Delta_1(\Phi_{(n-1)}^{-1}\Delta_2\Delta(a))\\ &=\Delta_1\Phi_{(n-1)}^{-1}(\Delta_3\Delta^{[3]}(a))\\ (\Delta_3\Delta^{[3]}(a))\Delta_3\Phi_{(n-1)}^{-1} &=\Delta_3\Phi_{(n-1)}^{-1}(\Delta_3\Delta_2\Delta(a))\\ (\Delta_3\Delta_2\Delta(a))1\otimes\Phi_{(n-1)} &=1\otimes\Phi_{(n-1)}(\Delta_2^2\Delta(a))\\ (\Delta_2^2\Delta(a))\Delta_2\Phi_{(n-1)} &=\Delta_2\Phi_{(n-1)}(\Delta_2\Delta_1\Delta(a))\\ (\Delta_2\Delta_1\Delta(a))(\Phi_{(n-1)}\otimes 1) &=(\Phi_{(n-1)}\otimes 1)\Delta^{[4]}(a).\end{aligned}$$

Then the proof of the cocycle condition consists of looking at the coefficient of h^n in the product of six terms listed below. This coefficient is exactly $\delta_4^\flat u_n$. So if we can show that the product equals 1 mod h^{n+1}, this will complete the proof of the proposition. Here are the six terms:

$$\begin{array}{rcl} A & = & \eta(1_5 + h^n(1 \otimes u_n)) = \eta(1 \otimes (1_4 + h^n u_n)) \\ B & \equiv & (1_5 - h^n \Delta_{1,4} u_n) \equiv \Delta_{1,4}(1_4 + h^n u_n)^{-1} \qquad mod\ h^{n+1} \\ C & = & Ad(\Phi^{-1} \otimes 1_2)(1_5 + h^n \Delta_{2,4} u_n) = Ad(\Phi^{-1} \otimes 1_2)\Delta_{2,4}(1_4 + h^n u_n) \\ D & \equiv & Ad\ \Psi_{3,5}(1_5 - h^n \Delta_{3,4} u_n) \equiv Ad\ \Psi_{3,5}[\Delta_{3,4}(1_4 + h^n u_n)^{-1}] \quad mod\ h^{n+1} \\ E & = & Ad\ \Psi_{4,5}(1_5 + h^n \Delta_{4,5} u_n) = Ad\ \Psi_{4,5}[\Delta_{4,5}(1_4 + h^n u_n)] \\ F & \equiv & (1_5 - h^n u_n \otimes 1) \equiv (1_4 + h^n u_4)^{-1} \otimes 1\ mod\ h^{n+1}. \end{array}$$

The multiplication is to take place in $B_h^{\otimes 5}$ and since we are computing mod h^{n+1}, we may take the product in any order and we may replace Φ by $\Phi_{(n-1)}$.

We need to be explicit about the associativity operators occurring in the definitions of these terms. The η occurring in the definition of A can be written as the sequence of associativities

$$\bullet\,(((\bullet\,\bullet)\,\bullet)\,\bullet) \to (\bullet\,((\bullet\,\bullet)\,\bullet))\,\bullet \to ((\bullet\,(\bullet\,\bullet))\bullet)\bullet \to (((\bullet\,\bullet)\,\bullet)\,\bullet)\,\bullet$$

and so

$$\eta = Ad(\Phi^{-1} \otimes 1_2) \circ Ad(\Delta_2 \Phi^{-1} \otimes 1) \circ Ad\ \Delta_2^2 \Phi^{-1}$$

where we write Δ_i instead of $\Delta_{i,p}$ so that the formulas can fit on the line. We also recall that

$$\Psi_{3,5} = \Delta_1 \Phi^{-1} \otimes 1, \quad \Psi_{4,5} = \Delta_1 \Delta_1 \Phi^{-1}.$$

We will replace Φ by $\Phi_{(n-1)}$ throughout, and use $\equiv$ to denote equality or equivalence mod h^{n+1}. So it follows from the definition (13.2.3) that

$$F \equiv (\Phi_{(n-1)}^{-1} \otimes 1_2)\,(\Delta_2 \Phi_{(n-1)}^{-1} \otimes 1)\,(1 \otimes \Phi_{(n-1)}^{-1} \otimes 1)\,(\Delta_3 \Phi_{(n-1)} \otimes 1)\,(\Delta_1 \Phi_{(n-1)} \otimes 1).$$

If we compare this expression for F with the one for η given above (with Φ replaced by $\Phi_{(n-1)}$) we see that the last two terms in the expression for A will cancel the first two terms in the expression for F if we form the product AF. The last five factors in the product AF will then be

$$(1 \otimes \Phi_{(n-1)} \otimes 1)\,(\Delta_2^2 \Phi_{(n-1)})\,(1 \otimes \Phi_{(n-1)}^{-1} \otimes 1)\,(\Delta_3 \Phi_{(n-1)} \otimes 1)(\Delta_1 \Phi_{(n-1)} \otimes 1), \tag{13.2.4}$$

where the term $(1\otimes\Phi_{(n-1)}\otimes 1)$ comes from the last factor in $1\otimes(1_4+h^n u_n)$, the next term is what is left over from applying η and the last three factors are what are left over from F. Now

$$\Delta_2^2 = (id\otimes\Delta\otimes id\otimes id)(id\otimes\Delta\otimes id) = id\otimes(\Delta\otimes id)\Delta\otimes id.$$

But conjugation by $\Phi_{(n-1)}$ carries $(\Delta\otimes id)\Delta$ into $(id\otimes\Delta)\Delta$ by our assumption that we have a truncated restricted deformation up to order $n-1$. (If we are dealing with a situation in which, in spite of the fact that $\Phi\neq 1$, nonetheless $(\Delta\otimes id)\Delta = (id\otimes\Delta)\Delta$ this is an invariance condition.) Thus (13.2.4) simplifies to

$$\begin{aligned}(\Delta_3\Delta_2\Phi_{(n-1)})\,(\Delta_3\Phi_{(n-1)}\otimes 1)\,(\Delta_1\Phi_{(n-1)}\otimes 1) &= \\ \Delta_3[(\Delta_2\Phi_{(n-1)})\,(\Phi_{(n-1)}\otimes 1)]\,(\Delta_1\Phi_{(n-1)}\otimes 1).\end{aligned}$$

Now from (13.2.3) we see that the next to the last term in $1\otimes(1_4+t^n u_n)$ is

$$1\otimes\Delta_2\Phi_{(n-1)} = \Delta_3(1\otimes\Phi_{(n-1)}).$$

So AF is $\equiv$ to

$$\begin{aligned}&(\Phi^{-1}_{(n-1)}\otimes 1_2)\;\Delta_2(\Phi^{-1}_{(n-1}\otimes 1)\;\Delta_2^2\Phi^{-1}_{(n-1)}\;\Delta_2(1\otimes\Phi^{-1}_{(n-1)})\;\Delta_4(1\otimes\Phi^{-1}_{(n-1)})\\ &\quad\times 1_2\otimes\Phi_{(n-1)}\;\Delta_3[1\otimes\Phi_{(n-1)}\;\Delta_2\Phi_{(n-1)}\;\Phi_{(n-1)}\otimes 1]\;\Delta_1\Phi_{(n-1)}\otimes 1.\end{aligned}$$

Now let us write out the full expression for C which is:

$$\begin{aligned}(\Phi^{-1}_{(n-1)}\otimes 1_2)\Delta_2\Delta_1\Phi^{-1}_{(n-1)} \quad &\times\quad \Delta_2\Delta_3\Phi^{-1}_{(n-1)}\Delta_2(1\otimes\Phi_{(n-1)})\Delta_2^2\Phi_{(n-1)}\\ &\times\quad \Delta_2(\Phi_{(n-1)}\otimes 1)(\Phi_{(n-1)}\otimes 1_2).\end{aligned}$$

We see that the last four terms in this expression for C cancel the first four terms we obtained above for AF so we obtain, for CAF, the expression

$$\begin{aligned}\Phi^{-1}_{(n-1)}\otimes 1_2\;\Delta_2\Delta_1\Phi^{-1}_{(n-1)}\Delta_2\Delta_3\Phi^{-1}_{(n-1)}\;\Delta_4(1\otimes\Phi^{-1}_{(n-1)})\;1_2\otimes\Phi_{(n-1)}\quad&\times\\ \times\Delta_3[1\otimes\Phi_{(n-1)}\Delta_2\Phi_{(n-1)}\;\Phi_{(n-1)\otimes 1}]\;\Delta_1\Phi_{(n-1)}\otimes 1\quad&=\\ \Delta_1^2\Phi^{-1}_{(n-1)}\;\Phi^{-1}_{(n-1)}\otimes 1_2\;\Delta_2\Delta_3\Phi^{-1}_{(n-1)}\;\Delta_4(1\otimes\Phi^{-1}_{(n-1)})\;1_2\otimes\Phi_{(n-1)}\quad&\times\\ \times\Delta_3[1\otimes\Phi_{(n-1)}\;\Delta_2\Phi_{(n-1)}\;\Phi_{(n-1)}\otimes 1]\;\Delta_1\Phi_{(n-1)}\otimes 1\quad&=\\ \Delta_1^2\Phi^{-1}_{(n-1)}\;\Delta_4[\Phi^{-1}_{(n-1)}\otimes 1\;\Delta_2\Phi^{-1}_{(n-1)}\;1\otimes\Phi^{-1}_{(n-1)}]1_2\otimes\Phi_{n-1}\quad&\times\\ \times\Delta_3[1\otimes\Phi_{(n-1)}\;\times\Delta_2\Phi_{(n-1)}\;\Phi_{(n-1)}]\;\Delta_1\Phi_{(n-1)}\otimes 1,&\end{aligned}$$

the last equality following from

$$\Delta_2\Delta_3 = \Delta_4\Delta_2$$

and the first, from

$$\Phi^{-1}_{(n-1)}\Delta_2\Delta_1 = \Delta_1^2\Phi^{-1}_{(n-1)}$$

We now use E which when written out becomes

$$\Delta_1^2\Phi^{-1}_{(n-1)}\ \Delta_4[\Delta_1\Phi^{-1}_{(n-1)}\Delta_3\Phi^{-1}_{(n-1)}\ 1\otimes\Phi_{(n-1)}\ \Delta_2\Phi_{(n-1)}\Phi_{(n-1)}\otimes 1]\Delta_1^2\Phi_{(n-1)}.$$

The last four terms in this expression cancel the first four terms in CAF so $ECAF$ becomes

$$\Delta_1^2\Phi^{-1}_{(n-1)}\Delta_4[\Delta_1\Phi^{-1}_{(n-1)}\ \Delta_3\Phi^{-1}_{(n-1)}] \ \times$$
$$\times\ 1_2\otimes\Phi_{(n-1)}\Delta_3[1\otimes\Phi_{(n-1)}\ \Delta_2\Phi_{(n-1)}\Phi_{(n-1)}\otimes 1]\Delta_1\Phi_{(n-1)}\otimes 1.$$

We use $\Delta_4\Delta_1 = \Delta_1\Delta_3$ to rewrite the first expression inside the $\Delta_4[\,]$ and multiply on the left by B which is given by

$$\Delta_1(\Phi^{-1}_{(n-1)}\otimes 1)\Delta_1\Delta_2\Phi^{-1}_{(n-1)}\ 1_2\otimes\Phi^{-1}_{(n-1)}\Delta_1\Delta_3\Phi_{(n-1)}\Delta_1^2\Phi_{(n-1)}$$

and cancel the first two terms and giving $BECAF$ as

$$\Delta_1[\Phi^{-1}_{(n-1)}\otimes 1\ \Delta_2\Phi^{-1}_{(n-1)}]\ \{1_2\otimes\Phi^{-1}_{(n-1)}\Delta_4\Delta_3\Phi^{-1}_{(n-1)}\ 1_2\otimes\Phi_{(n-1)}\} \ \times$$
$$\times\Delta_3[1\otimes\Phi_{(n-1)}\ \Delta_2\Phi_{(n-1)}\Phi_{(n-1)}\otimes 1]\Delta_1\Phi_{(n-1)}\otimes 1.$$

The middle expression in braces simplifies to $\Delta_3^2\Phi_{(n-1)}$, and if we use $\Delta_1\Delta_2 = \Delta_3\Delta_1$ we find that

$$BECAF \equiv D^{-1}.$$

This completes the proof of Prop. 7.

Proposition 13.2.4 *The obstruction cocycle is a coboundary if and only if the (n-1)- truncated restricted deformation $\Phi_{(n-1)}$ extends to one of order n, $\Phi_{(n)}$.*

Proof. The same calculation as in the proof of Proposition 6 shows that if we calculate the coefficient of h^n in the pentagon identity for $\Phi_{(n)}$ the contribution of the term ϕ_n is $\delta_3^\flat\phi_n$. By definition, the remaining terms combine to give $u_{(n)}$. Thus the complete expression is $\delta_3^\flat\phi_n + u_{(n)}$. This shows that to satisfy the pentagon identity to the next order it is necessary and sufficient that $u_{(n)}$ be a coboundary.

Suppose now that $u_{(n)}$ is a coboundary and that ϕ_n and $\tilde{\phi_n}$ are two distinct choices of cochains cobounding $u_{(n)}$, with $\Phi_{(n)}$ and $\tilde{\Phi}_{(n)}$ the associated deformations, which, by construction, agree to order $n-1$. The difference $\phi_n - \tilde{\phi_n}$ is a cocycle which we call the difference cocyle. The next proposition shows that two extensions are equivalent under twisting if and only if the difference cocycle is cohomologous to zero.

Proposition 13.2.5 *Two n-truncated deformations extending a given $(n-1)$-truncated deformation are equivalent under twisting by an element $F_{(n)} = 1 \otimes 1 + f_n h^n$ if and only if the difference cocycle is the coboundary of f.*

Proof. For $F_{(n)}$ as above we calculate the coefficient of h^n in (25).

$$(f^{23} + (id \otimes \Delta)f)\Phi - \Phi(f^{12} + (\Delta \otimes id)f) = \Phi\tilde{\phi_n}$$

Collecting terms and multiplying on the left by Φ^{-1}

$$\phi_n - \tilde{\phi_n} = f \otimes 1 + \Delta_1 f - Ad\Phi^{-1}(1 \otimes f + \Delta_2 f) = \delta_2^{\flat} f$$

proving the proposition.

13.3 Nonassociative tensor algebra of a bimodule

Except for a few minor changes, the following section is taken directly from the paper [896] of Markl and Stasheff. Let $F(M) = \sum F^k(M)$ be the free unitary nonassociative algebra on M which for the time being is simply a vector space and let $\star$ denote the product on $F(M)$.

$$\begin{aligned}
F^0(M) &= \mathbf{k}, \\
F^1(M) &= M, \\
F^2(M) &= M \otimes M = M^{\otimes 2}_{(\bullet,\bullet)}, \\
F^3 = M^{\otimes 3}_{\bullet(\bullet\bullet)} &\oplus M^{\otimes 3}_{(\bullet\bullet)\bullet}, \\
F^4 = M^{\otimes 4}_{\bullet(\bullet(\bullet\bullet))} \oplus M^{\otimes 4}_{((\bullet\bullet)\bullet)\bullet} &\oplus M^{\otimes 4}_{(\bullet\bullet)(\bullet\bullet)} \oplus M^{\otimes 4}_{\bullet((\bullet\bullet)\bullet)} \oplus M^{\otimes 4}_{(\bullet(\bullet\bullet))\bullet}, \cdots
\end{aligned}$$

.

Notice first that when B is a Drinfeld algebra and M is a left B module $F(M)$ admits a natural left action of B by the following two rules:

1. for $f \in F^0 = \mathbf{k}$, $a \bullet f = \epsilon(a)f$,

2. for $f \in F^1(M) = M$, $a \bullet f = m(a, f)$,

3. $a \bullet (f \star g) = \sum (a_{(1)} \bullet f) \star (a_{(2)} \bullet g)$.

A right action $(f, b) \mapsto f \bullet b$ is defined by similar rules when M is a right B module.

When M is a B bimodule these operations define on $F(M)$ the structure of a B-bimodule, i.e. that $a \bullet (b \bullet f) = (a \cdot b) \bullet f$, $a \bullet (f \bullet b) = (a \bullet f) \bullet b$ and $(f \bullet a) \bullet b = f \bullet (a \cdot b)$ for $a, b \in B$, $a \cdot b$ the product in B and $f \in F(M)$. From now on we will assume that all our modules are B bimodules.

There are a number of operations which we will be introducing, and to avoid bizarre symbols we have chosen to follow Markl and Stasheff and use $\bullet$ for the module structure as well as a place holder in the various bracketings.

Now for a pair of B bimodules M, N we can do the same construction on the B bimodule $M \oplus N$, and isolate the terms with one factor N. For example we define the subspace

$$F^3(M, N) \subset F^3(M \oplus N) = (M \oplus N)^{\oplus 3}_{\bullet(\bullet\bullet)} \oplus (M \oplus N)^{\oplus 3}_{(\bullet\bullet)\bullet}$$

as the sum

$$\begin{gathered}(N \otimes M^{\otimes 2})_{\bullet(\bullet\bullet)} \oplus (M \otimes N \otimes M)_{\bullet(\bullet\bullet)} \oplus (M^{\otimes 2} \otimes N)_{\bullet(\bullet\bullet)} \\ \oplus (N \otimes M^{\otimes 2})_{(\bullet\bullet)\bullet} \oplus (M \otimes N \otimes M)_{(\bullet\bullet)\bullet} \oplus (M^{\otimes 2} \otimes N)_{(\bullet\bullet)\bullet}\end{gathered}$$

The vector space constructed by carrying out the same procedure on F^k will be denoted $F^k(M, N)$. The same considerations as before show that $F(M, N)$ has the structure of a B bimodule. In addition, there is are two bilinear mappings :

$$F(M) \times F(M, N) \longrightarrow F(M, N), \quad (f, u) \mapsto f \star u,$$

$$F(M, N) \times F(M) \longrightarrow F(M, N), \quad (u, f) \mapsto u \star f$$

which make $F(M, N)$ into a $F(M)$ bimodule, where once again we simplify notation by using the same symbol $\star$ for the $F(M)$ algebra structure and for the $F(M)$ module structure.

The associativity operators

$$\alpha_{v', v} : M_v^{\otimes p} \to M_{v'}^{\otimes p}.$$

are given by conjugation with $\Psi_{v',v} \in B^{\otimes p}$. We define an equivalence relation on $F(M)$ as follows

$$M_v^{\otimes p} \ni x \sim y \in M_{v'}^{\otimes p}$$

if and only if, with respect to the $B^{\otimes p}$ bimodule structure

$$y = \Psi_{v',v} x \Psi_{v,v'} = \Psi_{v',v} x \Psi_{v',v}^{-1}.$$

Markl and Stasheff have a particularly nice way of describing this equivalence relation. Equivalence to 0 in $F(M)$ means belonging to the $\star$-multiplicative ideal, I_M generated by expressions of the form below; and equivalence to 0 in $F(M,N)$ means belonging to the $F(M)$ $\star$-module generated by expressions of the form

$$\Phi \bullet (x \star (y \star z)) - ((x \star y) \star z) \bullet \Phi, \tag{13.3.1}$$

where $x, y, z \in F(M)$ or two of the three are in $F(M)$ and one of the three in $F(M,N)$. Expanding Φ as $\sum \Phi^{(1)} \otimes \Phi^{(2)} \otimes \Phi^{(3)}$, $\Phi \bullet (x \star (y \star z))$ means, simply $\sum (\Phi^{(1)} \bullet x) \star ((\Phi^{(2)} \bullet y) \star (\Phi^{(3)} \bullet z))$, the meaning of the right-hand expression being similar. Finally, let $\mathcal{C}(M)$ be the graded $\mathbf{k}$-module $F^*(M)/\sim$ and $\mathcal{C}(M,N)$ the graded $\mathbf{k}$-module $F^*(M,N)/\sim$. We omit the proof of the following straightforward but useful result from [896]

Proposition 13.3.1

1. *The left and right actions $\bullet$ of B on $F(M)$ induce on $\mathcal{C}(M)$ the structure (denoted again by $\bullet$) of a graded B-bimodule.*

2. *The left and right actions $\bullet$ of B on $F(M,N)$ induce on $\mathcal{C}(M,N)$ the structure (denoted again by $\bullet$) of a graded B-bimodule.*

3. *The (nonassociative) multiplication $\star$ on $F(M)$ induces on $\mathcal{C}(M)$ a (nonassociative) multiplication $\odot$ satisfying*

$$\Phi \bullet [p \odot (q \odot r)] = [(p \odot q) \odot r] \bullet \Phi, \tag{13.3.2}$$

where again $\Phi \bullet [p \odot (q \odot r)]$ abbreviates $\sum (\Phi^{(1)} \bullet p) \odot ((\Phi^{(2)} \bullet q) \odot (\Phi^{(3)} \bullet r))$ and similarly for the second expression.

4. *The bilinear mappings $F(M) \times F(M,N) \mapsto F(M)$ and $F(M,N) \times F(M) \mapsto F(M,N)$ induce on $\mathcal{C}(M,N)$ right and left $\mathcal{C}(M)$ module structures which we also denote by $\odot$ satisfying*

$$\Phi \bullet [p \odot (q \odot r)] = [(p \odot q) \odot r] \bullet \Phi, \tag{13.3.3}$$

where one of the three elements p, q, r is in $\mathcal{C}(M,N)$.

We take the conditions in item 4 of the preceding proposition as the defining property of a $\mathcal{C}(M)$ bimodule replacing the usual "associativity" conditions $a(bm) = (ab)m$, $a(mb) = (am)b$, $(ma)b = m(ab)$. Since the definition requires the use of the B bimodule structure, we must assume condition 2 as well, therefore *when we refer to a $\mathcal{C}(M)$ bimodule we assume the existence of a compatible B bimodule structure.*

For each vertex $v \in K_p$ there is a section of the equivalence relation

$$\eta[v] : \mathcal{C}^p(M) \to M_v^{\otimes p}$$

and an inverse projection, satisfying the identities

$$\xi[v] : M_v^{\otimes p} \to \mathcal{C}^p(M)$$

$$\xi[v]\eta[v] = id_{\mathcal{C}^p(M)} \tag{13.3.4}$$

$$\eta[v]\xi[v] = id_{M_v^{\otimes p}} \tag{13.3.5}$$

$$\eta[v']\xi[v](\hat{b}) = \Psi_{v',v} \cdot \hat{b} \cdot \Psi_{v',v}^{-1} \tag{13.3.6}$$

$$\xi[v'](\Psi_{v',v} \cdot \hat{b}) = \xi[v](\hat{b} \cdot \Psi_{v,v'}) \tag{13.3.7}$$

$$\eta[v'](\tilde{b}) \cdot \Psi_{v,v'} = \Psi_{v',v} \cdot \eta[v](\tilde{b}) \tag{13.3.8}$$

$$\xi[v](\Delta^{[v]}(a) \cdot \hat{b}) = a \bullet (\xi[v](\hat{b})) \tag{13.3.9}$$

$$\Delta^{[v]}(a) \cdot (\eta[v](\tilde{b})) = \eta[v](a \bullet \tilde{b}) \tag{13.3.10}$$

$$\xi[v](\hat{b} \cdot \Delta^{[v]}(a)) = (\xi[v](\hat{b})) \bullet a \tag{13.3.11}$$

$$(\eta[v](\tilde{b})) \cdot \Delta^{[v]}(a) = \eta[v](\tilde{b} \bullet a) \tag{13.3.12}$$

where $\hat{b} \in M_v^{\otimes p}$, $\tilde{b} \in \mathcal{C}^p(M)$. This list of equations is redundant, (13.3.7) and (13.3.8) follow from (13.3.6), given the inverse relations of η and ξ. In the same way (13.3.12) and (13.3.10) follow from (13.3.11)and (13.3.9) respectively.

The maps η and ξ for the preferred vertex $[p] \in K_p$ will be denoted simply $\eta[p]$ and $\xi[p]$. They establish an isomorphism between the B bimodules $M^{[p]}$ defined in section 1 and $\mathcal{C}^p(M)$. In the case $M = B$ the direct sum maps $\xi = \oplus\xi[p]$ and $\eta = \oplus\eta[p]$ define a B bimodule isomorphism between the modified cobar complex and $\mathcal{C}(B)$.

The operators $\Delta_{i,p} : B_v^{\otimes p} \to B_{v[i]}^{\otimes(p+1)}$ induce operators $\tilde{\Delta}_{i,p}$ on $\mathcal{C}^p(B)$. Define a differential

$$\delta_{\mathcal{C}} = \sum(-1)^i\tilde{\Delta}_{i,p}.$$

With respect to this differential on $\mathcal{C}(B)$ the maps η and ξ are isomorphisms of between the chain complex defined in section 1 and $(\mathcal{C}(B), \delta_{\mathcal{C}})$.

$$\xi[p+1]\delta\eta[p] = \delta_{\mathcal{C}}.$$

Henceforth we will use the symbol $\mathcal{C}(B)$ for the modified cobar complex of a Drinfeld algebra B.

Proposition 13.3.2 *There is an operator $\delta_{\mathcal{C}}$ on $\mathcal{C}(B)$ which satisfies the following conditions*

1. *$\delta_{\mathcal{C}}$ is a B bimodule map relative to $\bullet$: $\delta_{\mathcal{C}}(a\bullet x) = a\bullet(\delta_{\mathcal{C}}x)$ and $\delta_{\mathcal{C}}(x\bullet a) = (\delta_{\mathcal{C}}x)\bullet a$, where $a \in B$ and $x \in \mathcal{C}(B)$,*
2. *$\delta_{\mathcal{C}}$ has degree 1: $\delta_{\mathcal{C}} : \mathcal{C}^i(B) \to \mathcal{C}^{i+1}(B)$,*
3. *$\delta_{\mathcal{C}}$ is a derivation relative to $\odot$: $\delta_{\mathcal{C}}(x \odot y) = \delta_{\mathcal{C}}x \odot y + (-1)^{|x|}x \odot \delta_{\mathcal{C}}y$*
4. *$\delta_{\mathcal{C}}^2 = 0$.*

13.4 Comodules over a Drinfeld algebra

The standard definition of a (left) comodule over a coalgebra, which by now is very familiar, has an alternative formulation using the cobar complex. This will be the key to our definition of homotopy associative comodules over a Drinfeld algebra. For the purposes of the current discussion we need to extend the cobar differential to the full tensor algebra. This is the same as including the base field as a component of degree -2, $\mathrm{Cob}^{-2}(C) = k$, and extending the differential to be 0 in degree -2. In this way we give the tensor algebra the structure of an acyclic complex. We will denote by $T(C)$ the tensor algebra with the usual grading and the extension of the cobar differential just defined. Relative to the multiplication induced by the tensor product it is a differential graded algebra with unit.

Let M be a left B comodule with structure map λ. The tensor product $T(C) \otimes M$ is a free left module over $T(C)$. Extend λ to $T(C) \otimes M$ as a δ derivation, which we will denote D. That is $D(m) = \lambda(m)$ for $m \in M$ and

$$D(x \otimes y) = \delta(x) \otimes y + (-1)^p x \otimes \lambda(y) \tag{13.4.1}$$

for $x \in \mathcal{C}^p(B)$ and $y \in M$.

The operator D has three important properties which allow us to give an alternative definition of left comodules

Proposition 13.4.1 *M is a left comodule over a coalgebra C if and only if there exists an operator D on $T(C) \otimes M$ satisfying the conditions*

1. *D is of graded degree 1.*

2. *D is a derivation relative to the left $T(C)$ module structure.*

$$D(x \otimes y) = \delta(x) \otimes y + (-1)^p x \otimes D(y) \tag{13.4.2}$$

   where $x \in T^p(C)$, $\quad y \in T^q(C) \otimes M$.

3. *$D^2 = 0$*

Proof. Assume that M is a left C comodule. The first two conditions follow immediately from the definitions. The last one is, in fact, equivalent to the compatibility condition. Since $T(C) \otimes M$ is a free $T(C)$ module, and $\delta^2 = 0$, it is enough to check condition 3 on M. There we have

$$D^2(m) = D(\lambda(m)) = (\Delta \otimes id)\lambda(m) - (id \otimes \lambda)\lambda(m) = 0,$$

since $\delta = \Delta$ on $T^0(C) = C$.

Conversely given a D satisfying the three conditions we can define a left C comodule structure with the structure map coming from restricting D to M. Condition 1 guarantees that this takes values in $C \otimes M$. Conditions 2 and 3 imply the structure equation.

This makes $T(C) \otimes M$ a differential graded $T(C)$ module and all properties of comodules and comodule maps can be reformulated in terms of this structure. For example there is a bijective correspondence between comodule maps $\phi : M \to N$ and maps of differential graded modules (commuting with D) $\tilde{\phi} : T(C) \otimes M \to T(C) \otimes N$ of graded degree 0.

In the case of a B bicomodule, M. The structure maps

$$\lambda : M \to B \otimes M \qquad \rho : M \to M \otimes B$$

together define a mapping

$$\lambda - \rho : N \to (B \otimes M) \oplus (M \otimes B) \subset T(B) \otimes M \otimes T(B).$$

In this case we have the same proposition only the operator D is a $T(B)$ bimodule derivation extending $\lambda - \rho$.

Proposition 13.4.2 *M is a bimodule over a coalgebra C if and only if there exists an operator D on $T(C) \otimes M \otimes T(C)$ satisfying the conditions*

1. *D is of graded degree 1.*

2. *D is a derivation relative to the $T(C)$ bimodule structure.*

$$D(x\otimes y\otimes z)=\delta(x)\otimes y\otimes z+(-1)^{p}x\otimes D(y)\otimes z+(-1)^{p+q+r}x\otimes y\otimes\delta(z) \tag{13.4.3}$$

where $x\in T^{p}(C),\quad y\in T^{q}(C)\otimes M\otimes T^{r}(C)$.

3. $D^2=0$

For a Drinfeld algebra, we want to consider comodules with values in the category ${}_B\mathcal{M}_B$. Generalizing the standard definition of a comodule to the case of a comodule over a Drinfeld algebra leads to the compatibility diagram

$$\begin{array}{rlcll}
 & M & \xrightarrow{\lambda} & B\otimes M & \\
\lambda & \downarrow & & | & \\
 & B\otimes M & & | & id\otimes\lambda \\
\Delta\otimes id & \downarrow & & \downarrow & \\
 & (B\otimes B)\otimes M & \xrightarrow{\alpha_{B,B,M}} & B\otimes(B\otimes M) &
\end{array}$$

The diagram above becomes the equation

$$\Phi(\Delta\otimes id)\lambda\Phi^{-1}=(id\otimes\lambda)\lambda.$$

What we are really defining is a comodule in a monoidal category. Shahn Majid was one of the first people to formulate various aspects of quantum group theory in these terms and he has developed this and related ideas quite extensively in his papers on braided groups, braided Lie algebras etc.

In fact we will want to consider bicomodules, so there will be two more compatibility diagrams and the corresponding equations.

If we start with this definition and try to define the tensor product of bicomodules using the formula

$$\lambda_{M\otimes N}:M\otimes N\xrightarrow{\lambda_M\otimes\lambda_N}B\otimes M\otimes B\otimes N\xrightarrow{(m\otimes id\otimes id)\circ s_{23}}B\otimes M\otimes N,$$

we find that the diagram is not commutative, that is, the compatibility conditions are not satisfied for either the standard bimodule structure on the tensor product, nor for the one in which the left module action is on M and the right module action is on N. (We need to use the second action when $M=N=B$ in $\mathrm{Bar}_0(B)$ and then inductively for $M=B$ and $N=\mathrm{Bar}_p(B)$.)

This failure is basically because we have product which is a mixture of algebra multiplications and tensor product multiplications and conjugation using either bimodule structure does not act as an automorphism for this mixed product.

Thus we are lead to consider the alternative definition of a bicomodule given in the last proposition and consider possible modifications in the case of Drinfeld algebras. One might think that the only change necessary should be the use of $\mathcal{C}(B, M)$ in place of $T(B) \otimes M \otimes T(B)$, keeping the equation (13.4.3), but taking into consideration the nonassociativity. However, this is just the incorporation of the associativity operators of the category ${}_B\mathcal{M}_B$, and a simple calculation shows that this does not work.

For the case $M = \mathrm{Bar}(B)$, which is the main example in which we are interested, we will define operators

$$D_k : \mathrm{Bar}_i(B) \to \mathcal{C}^{k+1}(B, \mathrm{Bar}(B)_{i+k}) \tag{13.4.4}$$

to compensate for the failure of $\mathrm{Bar}(B)$ to be a "standard" bicomodule. These operators give $\mathcal{C}(B, \mathrm{Bar}(B))$ what we will call a homotopy associative bicomodule structure The operator D will now have the form $D = D_0 + D_1 + \sum D_k$. The first term D_0 is the bar complex differential. The second term $D_1 = \sum D_{1,i}$ where

$$D_{1,i} : \mathrm{Bar}_i(B) \to \mathcal{C}(B \otimes \mathrm{Bar}_i(B))$$

is the extension of

$$D_{1,i} = \lambda - (-1)^i \rho \tag{13.4.5}$$

as a $\delta_{\mathcal{C}}$ derivation defined using the formula for the "standard" bicomodule structures λ and ρ on a tensor product $M^{\otimes i}$. The definition of the higher D_k will be given in the next two sections.

In section 6 we will prove

Theorem 13.4.1 *There exists an operator D on the bigraded $\mathcal{C}(B)$- bimodule and B-bimodule $\mathcal{C}(B, \mathrm{Bar}(B))$ satisfying the following conditions*

1. *D is a B bimodule map: $D(a \bullet x) = a \bullet Dx$, $D(x \bullet a) = Dx \bullet a$*

2. *D is a $\delta_{\mathcal{C}}$ derivative relative to the right and left $\mathcal{C}(B)$ module structures:*

$$D((x \odot w) \odot y) = (\delta_{\mathcal{C}} x \odot w) \odot y + (-1)^{|x|}(x \odot D(w)) \odot y + (-1)^{|x|+|w|}(x \odot w) \odot \delta_{\mathcal{C}} y.$$

The corresponding equation with the alternative placement of parentheses follows from this one and the fact that D is a B bimodule map.

3. D has square 0: $D^2 = 0$.

Define $\mathcal{C}^i_q(B) = \mathcal{C}^i(B, \mathrm{Bar}_q(B)$. Given $\alpha \in Hom_{B-\mathrm{bimod}}(\mathrm{Bar}_q(B), \mathcal{C}^p(B))$ we extend it to $\alpha \in \mathrm{Hom}_{\mathcal{C}(B)-\mathrm{bimod}}(\mathcal{C}(B, \mathrm{Bar}(B)), \mathcal{C}(B))$ requiring $\alpha(\mathcal{C}^i_q(B) \subset \mathcal{C}^{p+i}$ and

$$\alpha(x \odot y) = (-1)^{|\alpha|i} x \odot \alpha(y)$$
$$\alpha(y \odot x) = \alpha(y) \odot x$$

for $x \in \mathcal{C}^i(B)$ and $y \in \mathcal{C}(B, \mathrm{Bar}_q(B))$. We say that α has bidegree (p, q) and total degree $|\alpha = p + q$ and denote the vector space of such homomorphisms by $\mathcal{E}^{p,q}$. The parity is equal to the total degree reduced mod 2. We are giving the extension α the parity of its total degree.

As we stated earlier, the definition of $\mathcal{C}(B)$ bimodule assumes the existence of a compatible B bimodule structure so we do not explicitly include it in the subscript of the Hom functor. Once again we have an adjoint functor relation.

Proposition 13.4.3 *For a $\mathcal{C}(B)$ bimodule N and a B bimodule M*

$$Hom_{\mathcal{C}(B)-bimod}(\mathcal{C}(B, M), N) \;\cong\; \mathrm{Hom}_{B-bimod}(M, N).$$

Note that because of the question of parity, there are distinct extensions of B bimodule homomorphisms to $\mathcal{C}(B)$ bimodule homomorphisms depending on parity for the left and right $\mathcal{C}(B)$ module structures and we have chosen one. Each choice of parity for the extension defines a distinct adjoint relation.

In the last section we discuss the proof of the following theorem.

Theorem 13.4.2 *Deformations of Drinfeld algebras are classified in the sense of Gerstenhaber by the complex*

$$\mathrm{Hom}_{\mathcal{C}(B)-bimod}(\mathcal{C}(B, \mathrm{Bar}(B)), \mathcal{C}(B))$$

with differential $\tilde{D}(\alpha) = \delta_{\mathcal{C}} \circ \alpha - (-1)^{|\alpha|} \alpha \circ D$.

13.5 Interpolation operators M_k

The definition of the operators D_k involves several stages and is rather complicated. In this section we describe operators M_k which contain all the fundamental structure of the D_k. These operators give a way of translating the combinatorial information contained in the Stasheff polyhedra into operators on the bar complex interpolating between the different maps that arise from the compatibility diagrams for comodules. The basic insight is straightforward. We want to use the boundary operator in the bar complex to "compensate for the deviation" (failure to commute) in the compatibility diagram for the standard definition of the product of comodules. The failure of this diagram to commute says that if we define the derivation, which we will call D_1 of $\mathcal{C}(B, \mathrm{Bar}_q(B))$ using the standard definition of the comodule structure then $D_1^2 \neq 0$. Let D_0 be the $\mathcal{C}(B)$ module extension of the bar complex differential. By "compensating for the deviation using the cobar differential" we mean considering $D = D_0 \pm D_1$ and trying to add higher order terms so that the sum has square 0. In fact, one could formulate the proceedure as a deformation problem for the Gerstenhaber-Schack complex, because in the bialgebra case, $D^2 = (D_0 \pm D_1)^2 = 0$ and the formula in Theorem 13.4.2 defines the standard Gerstenhaber-Schack differential.

In the case of $\mathrm{Bar}_2(B)$ the deviation for the left comodule structure $\lambda_{B\otimes B}$ is

$$((\Delta \otimes id)\lambda_{B\otimes B}) - ((id \otimes \lambda_{B\otimes B})\lambda_{B\otimes B})\Phi =$$

$$= \sum \Phi^{(1)}a_{(1)}b_{(1)} \otimes \Phi^{(2)}a_{(2)}b_{(2)} \otimes \Phi^{(3)}a_{(3)} \otimes b_{(3)}$$
$$-a_{(1)'}b_{(1)'}\Phi^{(1)} \otimes a_{(2)'}b_{(2)'}\Phi^{(2)} \otimes a_{(3)'} \otimes b_{(3)'}\Phi^{(3)}$$

$$= \sum a_{(1)'}\Phi^{(1)}b_{(1)} \otimes a_{(2)'}\Phi^{(2)}b_{(2)} \otimes a_{(3)'}\Phi^{(3)} \otimes b_{(3)}$$
$$-a_{(1)'}\Phi^{(1)}b_{(1)} \otimes a_{(2)'}\Phi^{(2)}b_{(2)} \otimes a_{(3)'} \otimes \Phi^{(3)}b_{(3)}$$

$$= \sum [a_{(1)'}\Phi^{(1)}b_{(1)} \otimes a_{(2)'}\Phi^{(2)}b_{(2)}] \otimes [a_{(3)'}\Phi^{(3)} \otimes b_{(3)} - a_{(3)'} \otimes \Phi^{(3)}b_{(3)}].$$

The expression $a_{(3)'}\Phi^{(3)} \otimes b_{(3)} - a_{(3)'} \otimes \Phi^{(3)}b_{(3)}$ inside the right brackets in last line is exactly $D_0(a_{(3)'} \otimes \Phi^{(3)} \otimes b_{(3)})$ therefore the deviation is $id \otimes id \otimes D_0$ acting on

$$\sum [a_{(1)'}\Phi^{(1)}b_{(1)} \otimes a_{(2)'}\Phi^{(2)}b_{(2)}] \otimes [a_{(3)'} \otimes \Phi^{(3)} \otimes b_{(3)}]. \tag{13.5.1}$$

This suggests the form for the higher operators D_2. Consider the diagram for K_3

$$(\bullet\bullet)\bullet \xrightarrow{\Phi} \bullet(\bullet\bullet) \tag{13.5.2}$$

and the comultiplications associated to the two vertices

$$\Delta^{(\bullet\bullet)\bullet}(a) = \sum a_{(1)} \otimes a_{(2)} \otimes a_{(3)} \qquad \Delta^{\bullet(\bullet\bullet)}(a) = \sum a_{(1)'} \otimes a_{(2)'} \otimes a_{(3)'}.$$

Then 13.5.1 is a kind of product of three terms

$$\Delta^{\bullet(\bullet\bullet)}(a) \quad \Phi \quad \Delta^{(\bullet\bullet)\bullet}(b)$$

corresponding to the diagram 13.5.2 but reversing the order because in the correspondence between functional notation and arrow notation,

$$\begin{aligned} a &\xrightarrow{T} b \quad \text{becomes} \\ b &= T(a). \end{aligned}$$

All the operators D_{n-1} will correspond to the geometry of K_n in precisely this way. For any k simplex $\sigma = [v_0, v_1, \ldots, v_k]$ spanned by $k+1$ of the vertices of K_n we will define an operator $M_n[\sigma]$ which is a product of terms $\Delta^{v_i}(a)$ and $\Psi_{v_i, v_{i+1}}$, the associativity operator corresponding to the edges with reversed orientation from v_{i+1} to v_i.

The map $M[\ ]$ from simplices to operators extended linearly to the simplicial chain complex will eventually be applied to the fundamental class of a simplicial decomposition of the Stasheff polyhedra. Simplices appear here because the differential for the bar complex has exactly the form of an alternating sum of face operators for a simplicial complex. It is interesting to consider the possibility of developing a new combinatorial (co)homology theory in which the K_n replace simplices σ_{n-2} and which would have the same relation to the (co)homology of non(co)associative (co)algebras that simplicial theory has to (co)associative (co)algebra (co)homology. For the present we do not have such a theory and must make do with simplices. With this introduction we proceed to the actual definition.

Let $[v_0, \cdots, v_k]$ be an oriented k-simplex spanned by vertices of K_n, with the orientation of the edges going from v_i to v_{i-1}. We chose a preferred base vertex $[n] \in K_n$ for each n and we will use it in the following definition, but the resulting definition will, in fact, be independent of the choice of vertex.

Let $P(q,k)$ be the partitions of the segment of $q+2$ elements into $k+1$ subsegments such that the first and last are nonempty. For $\pi \in P(q,k)$ the segments are

$$(0, \ldots, \pi(1)-1)(\pi(1), \ldots, \pi(2)-1) \ldots (\pi(k), \ldots, q+1).$$

For convenience in some formula, we define $\pi(0) = 0$, $\pi(k+1) = q+1$. Define $\ell(\pi) = \sum_{0 \le j \le k+1} \pi(j)$ and $o(\pi) = \sum_{0 \le j \le [k/2]} \pi(2j+1) - \pi(2j)$, that is the sum of the lengths of the odd segments. Then for $\pi \in P(q,k)$

$$\begin{aligned}(-1)^{o(\pi)} &= (-1)^{\ell(\pi)} \quad k \quad \text{even} \\ &= (-1)^{q+1}(-1)^{\ell(\pi)} \quad k \quad \text{odd.}\end{aligned} \tag{13.5.3}$$

For $\hat{a} = a_0 \otimes a_1 \otimes \cdots \otimes a_{q+1} \in \mathrm{Bar}_q(B)$ and $\pi \in P(q,k)$ let

$$\pi_i(\hat{a}) = a_{\pi(i)} \otimes \cdots \otimes a_{\pi(i+1)-1}.$$

Finally let $\overline{\otimes}$ be the product on the external tensor product of tensor algebras with identity.

$$\begin{aligned}(a_1 \otimes \cdots \otimes a_r)\overline{\otimes}(b_1 \otimes \cdots \otimes b_s) &= (a_1 \otimes b_1) \otimes \cdots (a_r \otimes b_r) \otimes b_{r+1} \otimes \cdots b_s \\ &\quad \text{when} \quad s > r \\ &= (a_1 \otimes b_1) \otimes \cdots (a_s \otimes b_s) \otimes a_{s+1} \otimes \cdots a_r \\ &\quad \text{when} \quad r \geq s\end{aligned}$$

By analogy to (13.5.1), but before multiplying any terms, define

$$\Psi[v_0, \cdots, v_k] =$$

$$\sum_{\pi \in P(q,k)} (-1)^{o(\pi)} \Delta^{v_0}(\pi_0(\hat{a}))\overline{\otimes}\Psi_{v_0,v_1}\overline{\otimes}\Delta^{v_1}(\pi_1(\hat{a}))\overline{\otimes} \cdots \Psi_{v_{k-1},v_k}\overline{\otimes}\Delta^{v_k}(\pi_k(\hat{a})).$$

The expression above takes values in $\otimes^n(B^{\otimes(q+k+2)}) = \otimes^n \mathrm{Bar}_{q+k}(B)$ not in $\mathcal{C}(B, \mathrm{Bar}(B))$ as we would like. The last stage of the definition is basically to apply the multiplication in B to $n-1$ factors and to project into $\mathcal{C}(B, \mathrm{Bar}(B))$, however we also need to use the $\otimes^n B$ module structure on $\otimes^n \mathrm{Bar}_{q+k}(B)$ (which we denote $\cdot$) premultiplying by $\Psi_{[n],v_0}$ and postmultiplying by $\Psi_{v_k,[n]}$. To simplify notation we introduce $A = B^{\otimes(q+k+2)}$ and for $1 \leq r \leq n$ define the operation

$$\Box^r = \tilde{m}^{\otimes r} \otimes id \otimes \tilde{m}^{\otimes n-r} : A^{\otimes n} \longrightarrow \otimes^{r-1} B \otimes A \otimes^{n-r} B$$

where

$$\tilde{m} : (B^{\otimes(q+k+2)}) \rightarrow B$$

is multiplication in B. To get a map with values in $\mathcal{C}(B, \mathrm{Bar}(B))$, we define

$$\bar{M}_n[v_0, \cdots, v_k](\hat{a}) = \sum_{1 \leq r \leq n} \eta[n]\Box^r[\Psi_{[n],v_0} \cdot \Psi[v_0, \ldots, v_k](\hat{a}) \cdot \Psi_{v_k,[n]}]. \tag{13.5.4}$$

For a k simplex $\sigma \in K_n$

$$\bar{M}_n[\sigma] : \mathrm{Bar}_q(B) \to \mathcal{C}^{n-1}(B, \mathrm{Bar}_{k+q}(B))$$

and we extend the definition to $\mathcal{C}(B, \mathrm{Bar}_q(B))$ by

$$M_n[\sigma]((x \odot y) \odot z) = (-1)^{(n-1-k)(|x|+|y|}((x \odot \bar{M}_n[\sigma](y)) \odot z)$$

for $y \in \mathrm{Bar}_q(B)$ and $x, z \in \mathcal{C}(B)$. Then $M_n[\sigma]$ has parity $n-1-k$.

Proposition 13.5.1 Boundary Formula

Let D_0 be the $\mathcal{C}(B)$ bimodule extension of the bar differential. For $\hat{a} \in \mathrm{Bar}_q(B)$

$$\begin{aligned} D_0 M_n[v_0, \cdots, v_k](\hat{a}) \quad &- \quad (-1)^{n+k-1} M_n[v_0, \cdots, v_k] D_0(\hat{a}) \\ = \sum (-1)^i \quad &M_n \quad [v_0, \cdots, v_{i-1}, v_{i+1}, \cdots, v_k](\hat{a}). \end{aligned}$$

In more compact notation, as operators on $\mathrm{Bar}_q(B)$

$$D_0 M_n[\sigma] + (-1)^{n+k} M_n[\sigma] D_0 = M_n[d_S \sigma]$$

where d_S is the simplicial differential.

Proof. The operator $\bar{M}[\sigma]$ is a sum of n terms corresponding to the $\square^r$ for distinct r. We will prove that the boundary formula is true for each summand separately. Using a modified Sweedler notation,

$$\Delta^v(a) = \sum a_{(v,1)} \otimes a_{(v,2)} \otimes \cdots \otimes a_{(v,n)}$$

for any $v \in K_n$. In the term corresponding to M^r we need to consider the bar differential ∂ acting on the expression

$$\begin{aligned} a_{0(v_0,r)} \otimes \cdots \otimes a_{\pi(1)-1(v_0,r)} \otimes \Psi^r_{v_0,v_1} \otimes a_{\pi(1)(v_1,r)} \otimes \quad \cdots \\ \otimes a_{\pi(k)-1(v_k,r)} \otimes \Psi^r_{v_k,v_{k+1}} \otimes a_{\pi(k)(v_{k+1},r)} \otimes \quad \cdots \quad \otimes a_{q+1(v_{k+1},r)} \end{aligned}$$

Since ∂ acts by multiplying adjacent terms, there are four kinds of products $aa, a\Psi, \Psi a, \Psi\Psi$ that can arise in $D_0 \bar{M}_n$ but only one type, aa, can arise in $M_n D_0$. We claim that all but the $a\Psi$ on the extreme left, the Ψa on the extreme right and the$\Psi\Psi$ terms cancel, and that these give the right hand side of the equation in the statement of the proposition.

Every term in $\bar{M}_n D_0$ appears as an aa term in $D_0 \bar{M}_n$. Suppose it is a product of the type $a_{j(v_c,r)} a_{j+1(v_c,r)}$ of terms in the $c+1$st segment, then the factor in $D_0\bar{M}_n$ is $(-1)^{o(\pi)+j+c}$ because there are c additional factors Ψ to the left of the product. The coefficient of the same term in $\bar{M}_n D_0$ is $(-1)^{j+o(\pi')}$ where π' is a partition with one less element in the $c+1$st segment. If the parity of $o(\pi)+c$ and $o(\pi')$ is opposite these terms cancel. But if c is even $o(\pi) = o(\pi')+1$ and if c is odd $o(\pi') = o(\pi)$ so the parity is indeed opposite and the terms do cancel.

All but the extreme right Ψa terms and the extreme left $a\Psi$ terms cancel in pairs. When the algebra product $\Psi^r_{v_i,v_{i+1}} a_{j(v_{i+1},r)}$ occurs in the factor of the n fold tensor product of algebras in which the bar complex appears we get terms of the type $u \otimes \Psi_{v_i,v_{i+1}} \Delta^{v_{i+1}}(a_j) \otimes v$ and a matching term $u \otimes \Delta^{v_i}(a_j)\Psi_{v_i,v_{i+1}} \otimes v$ where $u \otimes v \in B^{\otimes(n+q+k-3)}$ represent the remaining factors coming from the operation of Δ^{v_i} on the other components a_i and the other associativity operators. We can apply the formula

$$\Psi_{v_i,v_{i+1}} \Delta^{v_{i+1}}(a) = \Delta^{v_i}(a)\Psi_{v_i,v_{i+1}}$$

to cancel terms provided the coefficients have opposite sign. The only difference in the coefficients of the term of type $a\Psi$ and the term of type Ψa is due to the change in the partition. The transformation of one partition into the other involves two adjacent segments which change in length by one. Since one of the segments is odd, and this causes a change of sign, so these terms cancel. This is true for every product except those involving $a\Psi$ on the extreme left and Ψa on the extreme right, in which case the exclusion of partitions with the first or last segment empty eliminates the matching term. The terms of type $a\Psi$ on the left can still be rewritten as Ψa terms as before. Doing so allows us to rewrite then as the left $B^{\otimes n}$ action of Ψ_{v_0,v_1}. Since we already have the term $\Psi_{[n],v_0}$ acting on the left, we can use the MacLane coherence theorem and write

$$\Psi_{[n],v_0}\Psi_{v_0,v_1} = \Psi_{[n],v_1}.$$

These terms are precisely the terms in $\bar{M}_n[v_1,\ldots,v_k]$, corresponding to the face opposite v_0. Once again we only need to check the sign. The partition has changed by combining the first two segments to give a partition involving k segments instead of $k+1$ and changing the parity of the position of a given segment, the odd segments become even and conversely. According to 13.5.3 the difference in sign is $(-1)^q$ because $\ell(\pi)$ has also changed by 1.

We deal with the terms Ψa on the right by a similar argument. Here the signs differ by $(-1)^{q+k}$ because of the sign $(-1)^{q+k-1}$ in the bar differential and another factor of -1 coming from the change in the term $(-1)^{o(\pi)}$. After substituting the expression on the right we get precisely the terms appearing in the expression for $(-1)^{q+k}\bar{M}_n[v_0, \cdots, v_{k-1}]$.

Finally we have the $\Psi\Psi$ terms. By the MacLane coherence theorem

$$\Psi_{v_{i-1},v_i}\Psi_{v_i,v_{i+1}} = \Psi_{v_{i-1},v_{i+1}}.$$

Substituting the term on the right in the appropriate $\Psi\Psi$ products gives the terms in $\bar{M}_n[v_0, \cdots v_{i-1}, v_{i+1}, \cdots v_k]$. Consider the two partitions involved here, the original partition π before applying ∂ has the $i+1$st segment empty. The other π' after applying ∂ has one less segment. In $o(\pi)$ the term $\pi(i)+\pi(i+1) = 2\pi(i)$ has even parity and drops out of the expression $(-1)^{o(\pi)}$. The sign coming from ∂ is $(-1)^{\pi(i)+i-1}$ because there are $\pi(i)+i-1$ elements to the left of Ψ_{v_{i-1},v_i}. Then there is the additional fact of the change of the parity of the total number of segments which introduces a factor of $(-1)^{q+1}$. Putting this all together we get

$$\sum(-1)^{q+i}M[v_0, \cdots, v_{i-1}, v_{i+1}, \cdots v_k].$$

This proves the formula

$$\begin{aligned} &D_0\bar{M}_n[v_0, \cdots, v_k](\hat{a}) \quad + \quad \bar{M}_n[v_0, \cdots, v_k]D_0(\hat{a}) \\ &= (-1)^q\sum(-1)^i \quad \bar{M}_n \quad [v_0, \cdots, v_{i-1}, v_{i+1}, \cdots, v_k](\hat{a}). \end{aligned}$$

Multiplying both sides by $(-1)^{(n-1-k)q}$ and taking into account that $D_0(\hat{a})$ has degree $q-1$ and $[v_0, \cdots, v_{i-1}, v_{i+1}, \cdots, v_k]$ is a $k-1$ simplex proves the theorem.

13.6 Homotopy operators D_k

The operators D_n are defined as a sum $\sum M_{n+1}[\sigma]$ where $\sum\sigma$ is the fundamental class of K_n.

Let K_n be, for $n \geq 2$, the Stasheff polyedron defined in Chapter 5. The vertices $\mathcal{V}_n$ of K_n are indexed by the set Br_n of full bracketings of n indeterminates. For $a, b \geq 2$ and $1 \leq t \leq b$, there is a map $(-,-)_t$: $\mathrm{Br}_a \times \mathrm{Br}_b \to \mathrm{Br}_{a+b-1}$, given by inserting the first argument into the t-th position in the second argument. This map defines the inclusions ι_t : $K_a \times K_b \hookrightarrow \partial K_{a+b-1}$ and it is known that

$$\partial K_n = \bigcup\nolimits_J \iota_t(K_a \times K_b), \tag{13.6.1}$$

where $J = \{(a, b, t);\ a+b = n+1, a, b \geq 2 \text{ and } 1 \leq t \leq b\}$.

The following lemma is an immediate consequence of the definitions.

Lemma 13.6.1 *For $u, u' \in K_a$ and $v, v' \in K_b$*

1.
$$(id^{\otimes(t-1)} \otimes \Delta^u \otimes id^{\otimes(b-t)})\Delta^v = \Delta^{\iota_t(u,v)} \tag{13.6.2}$$

2.
$$\Psi_{\iota_t(u,v),\iota_t(u,v')} = (id^{\otimes(t-1)} \otimes \Delta^u \otimes id^{\otimes(b-t)})\Psi_{v,v'} \tag{13.6.3}$$

3.
$$\Psi_{\iota(u,v),\iota(u',v)} = id^{\otimes(t-1)} \otimes \Psi_{u,u'} \otimes id^{\otimes(b-t)} \tag{13.6.4}$$

Let $\prec$ be the partial order on Br_n defined by saying that $u \prec v$ if and only if v is obtained from u by the substitution $(w_1, w_2)w_3 \mapsto w_1(w_2, w_3)$ with some $w_i \in \mathrm{Br}_{n_i}$, $i = 1, 2, 3$, $n_1 + n_2 + n_3 \leq n$. The preferred vertex $[n]$ is the (unique) minimal element of Br_n.

Proposition 13.6.1

$$K_n = \mathrm{Cone}(\textstyle\bigcup_{J'} \iota_t(K_a \times K_b); [n]),$$

where $J' = \{(a, b, t) \in J;\ t \geq 2\}$.

The cone structure above induces on K_n inductively the triangulation $\mathcal{T}_n$ as follows:

1. $\mathcal{T}_2$ is the only possible triangulation of $K_2 = \{\bullet\bullet\}$ and
2. for $n \geq 2$, $\mathcal{T}_n$ is induced by the cone structure of Proposition 13.6.1 from the triangulations $\mathcal{T}_a$ and $\mathcal{T}_b$ on K_a and K_b which are already defined, by the induction assumption.

It is possible to show that $\mathcal{T}_n$ has the property that $\prec$ induces a total order on the vertices of any simplex of $\mathcal{T}_n$.

We draw two triangulations of the pentagon K_4. There are only two triangulations compatible with this order, but only the first has the correct cone structure.

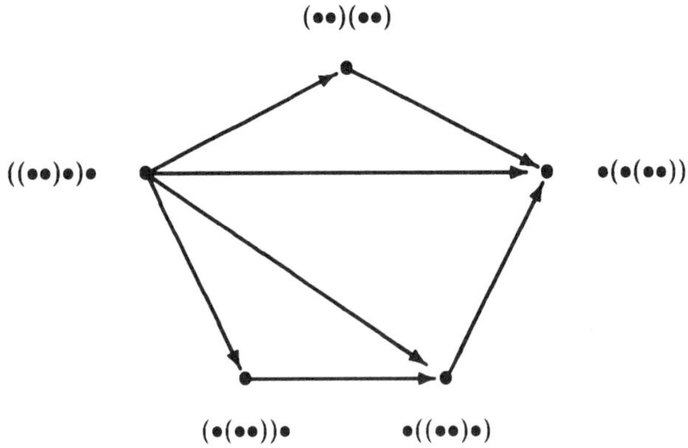

and

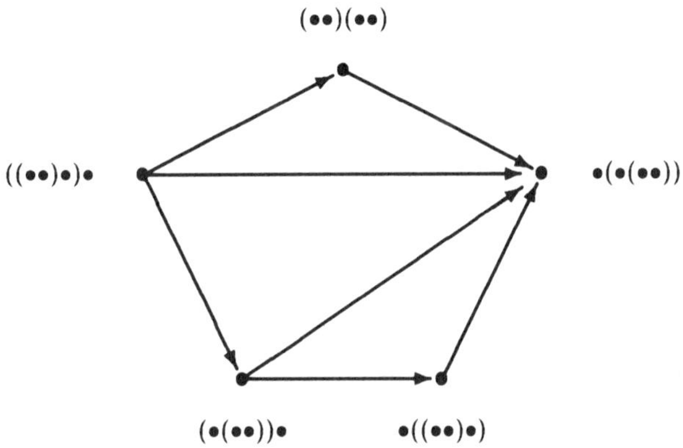

Let $C_i(K_n)$ denote the set of i-dimensional oriented simplicial chains of $\mathcal{T}_n$ with coefficients in $\mathbf{k}$ and let $d_S : C_i(K_n) \to C_{i-1}(K_n)$ be the simplicial differential. For $\sigma \in C_r(K_a)$ and $\tau \in C_s(K_b)$, $r, s \leq 0$, $a, b \leq 2$, set

$$\sigma \diamond \tau := (-1)^{rs} \sum_{1 \leq t \leq b} (-1)^{(a+1)(s+t+1)} (\iota_t)_*(\sigma \times \tau) \in C_{r+s}(K_{a+b-1}).$$

Let $C(K)^n := \bigotimes_{n=i+r-1} C_r(K_i)$ and extend $\diamond$ to a bilinear map (denoted by the same symbol) $\diamond : C(K)^* \otimes C(K)^* \to C(K)^*$.

The product $\times$ is the natural chain transformation given by the Eilenberg-Zilber Theorem. It is determined by the triangulating the cartesian product of simplices using (r, s) shuffles. We quote briefly from the discussion in MacLane's book [826].

The simplices in the cartesian product are indexed by (r, s) shuffles. An (r, s) shuffle is a partition of the set $0, r + s - 1$ of integers into 2 disjoint sets $i_1 < i_2 < \cdots i_r$ and $j_1 < j_2 < \cdots < j_s$. A shuffle may be pictured as a sequence of moves in the lattice of points (m,n) in the plane with integral coordinates: Start at $(0,0)$ at time 0 ; at time k move to the right if k is one of $i_1,\ldots,i_r$ and up if k is one of $j_1,\ldots,j_s$; the result is a staircase from $(0,0)$ to (r,s). Consider the product of simplices $\sigma = [u_0,u_1,\ldots,u_r]$ and $\tau = [v_0,v_1,\ldots,v_s]$. The $r+s$ simplices in the triangulation of the product have vertices $[(u_0,v_0),\ldots(u_{m(i)},v_{n(i)}),\ldots(u_r,v_s)]$, where $(m(i),n(i))$ $0 \leq i \leq r+s$ is any one of the paths corresponding to an (r,s) shuffle.

Proposition 13.6.2 *[895]* $(C(K)^*,\diamond,d_S)$ *is an associative differential graded algebra.*

Observe that the bimodule endomorphisms $\mathrm{End}_{B-\mathrm{bimod}}(\mathcal{C}(B,\mathrm{Bar}(B)))$ of the B bimodule $\mathcal{C}(B,\mathrm{Bar}(B))$ form a bigraded vector space where θ is of bidegree r,s if

$$\theta(\mathrm{Bar}_q(B)) \subset \mathcal{C}^r(B,\mathrm{Bar}_{q+s}(B)).$$

Once again $|theta| = r+s$ is the total degree.

Define a differential by

$$\nabla(\theta) = (D_0 \circ \theta - (-1)^{|\theta|}\theta \circ D_0)$$

The **Boundary formula** says

$$\nabla M[\sigma] = M[d_S\sigma].$$

Let $\mathbf{e}_n \in C_n(K_{n+2})$, $n \geq 0$, be the fundamental classes in a suitable triangulation of K_n chosen in such a way that $\mathbf{e}_0 = 1\cdot(\bullet,\bullet)$ and that

$$d_S\mathbf{e}_n + \sum_{i+j=n-1} (\mathbf{e}_i \diamond \mathbf{e}_j) = 0. \tag{13.6.5}$$

It follows from (13.6.1) and the fact which we proved in Chapter 5 that such a sequence exists.

We would like to know the way that the operators $\bar{M}_n$ behave in relation to the product structure defined above. From the definition 13.5.4, it follows that given an r simplex $\sigma \in K_a$ and an s simplex $\tau \in K_b$ the rule for the composition $M_r[\sigma]\circ M_s[\tau](\hat{a})$ on an element $\hat{a} \in \mathrm{Bar}_q(B)$ follows immediately once we know the rule for the composition

$$(id^{\otimes(t-1)(q+s+2)} \otimes \Psi[\sigma] \otimes id^{\otimes(b-t)(q+s+2)}) \circ \Psi[\tau](\hat{a}).$$

The tensor products of id reflect the fact that M_b takes values in $\mathcal{C}^{b-1}(B,\mathrm{Bar}_{q+s+2})$ and M_a is a $\mathcal{C}(B)$ linear extension of the definition on

Bar_{q+s+2}. This composition of Ψ's involves a double sum over partitions of $q+2$ elements into $s+1$ sets and of $q+s+2$ elements into $r+1$ sets. Let $\sigma = [u_0, \cdots, u_r]$ and $\tau = [v_0, \cdots, v_s]$. A typical composite operation corresponding to one pair of partitions, π a partition of $q+2$ elements into $s+1$ sets and ρ a partition of $q+s+2$ elements into $r+1$ sets can be represented

$$\overbrace{\underbrace{1, \cdots, \pi(1)-1}_{v_0} \Psi_{v_0,v_1} \cdots \Psi_{v_{j-1},v_j} \underbrace{\pi(j) \cdots \rho(1)}_{v_j}}^{u_0} \Psi_{u_0,u_1} \cdot$$

$$\cdot \overbrace{\underbrace{\rho(1) \cdots \pi(j+1)-1}_{v_j} \Psi_{v_j,v_{j+1}} \cdots}^{u_1}$$

This corresponds to the $r+s$ simplex with vertices

$$\iota_t(u_0, v_0), \cdots \iota_t(u_0, v_j), \iota_t(u_1, v_j) \cdots \iota_t(u_1, v_{j+1}) \iota_t(u_1, v_{j+2}) \cdots$$

which is one of the simplices in the triangulation of the cartesian product as described above.

The comultiplications are

$$(id^{\otimes(t-1)} \otimes \Delta^{u_i} \otimes id^{\otimes(b-t)}) \circ \Delta^{v_j} = \Delta^{\iota_t(u_i,v_j)}$$

The transition operators are

$$(id^{\otimes(t-1)} \otimes \Delta^{u_i} \otimes id^{\otimes(b-t)}) \circ \Psi_{v_j,v_{j+1}} = \Psi_{\iota_t(u_i,v_j)\iota_t(u_i,v_{j+1})}$$

and

$$id^{\otimes(t-1)} \otimes \Psi_{u_i,u_{i+1}} \otimes id^{\otimes(b-t)} = \Psi_{\iota_t(u_i,v_j),\iota_t(u_{i+1},v_j)}.$$

This leads us to the following theorem:

Theorem 13.6.1 *The k linear map*

$$\tilde{M} : (C(K)^*, \diamond, d_S) \to (End_{B-bimod}(C(B, \mathrm{Bar}(B))), \circ, \nabla)$$

defined on $C_i(K_n)$ by $\sigma \mapsto M_n[\sigma]$ for $n > 2$ and on $C_0(K_2)$ by $M(\mathbf{e}_0) = D_1$. is a map of differential graded algebras.

$$\begin{aligned} \tilde{M}[\sigma \diamond \tau] &= \tilde{M}[\sigma] \circ \tilde{M}[\tau], \\ \tilde{M}[d_S \sigma] &= \nabla \tilde{M}[\sigma]. \end{aligned}$$

Furthermore,

$$\tilde{M}_n(C_p(K_n))(\mathcal{C}^i(B, \mathrm{Bar}_q(B)) \subset \mathcal{C}^{i+n-1}(B, \mathrm{Bar}_{p+q}(B)),$$

$i, p \geq 0,\ n \geq 2,\ q \geq 1.$

Together with the *Boundary Formula* we get:

Theorem 13.6.2 *Let $D_k := M(\mathbf{e}_{k-1})$, $k \geq 1$. Then $D := D_0 + \sum_{k\geq 1} D_k$ is a homotopy comodule structure on $\mathcal{C}(B, \mathrm{Bar}\, B)$*

Proof. Looking at the homogeneous components of D^2 we see that $D^2 = 0$ is equivalent to $[D_0, D_k] + \sum_{i+j=k} D_i \circ D_j = 0$.

$$\begin{aligned} \nabla(D_k) + \sum_{i+j=k} D_i \circ D_j &= \nabla\tilde{M}[e_{k-1}] + \sum_{(i-1)+(j-1)=k-2} \tilde{M}[\mathbf{e}_{i-1}] \circ M[\mathbf{e}_{j-1}] \\ &= \tilde{M}[d_S(\mathbf{e}_{k-1}) + \sum_{i'+j'=k-2} \mathbf{e}_{i'} \diamond \mathbf{e}_{j'}] \\ &= 0. \end{aligned}$$

13.7 Relation of the D_k to deformation theory

In this final section we want to discuss the relation of the complex

$$\mathrm{Hom}_{\mathcal{C}(B)-\mathrm{bimod}}(\mathcal{C}(B, \mathrm{Bar}(B)), \mathcal{C}(B))$$

with differential

$$\tilde{D}(\alpha) = \delta_{\mathcal{C}} \circ \alpha - (-1)^{|\alpha|}\alpha \circ D$$

to the deformation theory of Drinfeld algebras. First of all we show that the cohomology theory reduces to the Gerstenhaber-Schack theory when $\Phi_0 = 1$. Then we give some examples to show that the coboundary formula agree with the terms in the various structure equations. The discussion here is just a sketch. The full details appear in the paper [895].

Recall the definition $\mathcal{C}^i_q = \mathcal{C}^i(B, \mathrm{Bar}_q)$ and $\mathcal{C}^j = \mathcal{C}^j(B)$. The operators D_n defined in the previous section satisfy

$$D_n = M_{n+1}[\mathbf{e}_{n-1}] : \mathcal{C}^i_q \longrightarrow \mathcal{C}^{i+n}_{q+n-1}.$$

If $\alpha \in \mathcal{E}^{p,q}$ the homomorphisms of bidegree (p, q) then $\alpha(\mathcal{C}^i_q) \subset \mathcal{C}^{i+p}$.

Thus $\alpha \in \mathcal{E}^{p,q}$ implies $\alpha \circ D_n \in \mathcal{E}^{p+n,q-n+1}$, since

$$C^i_{q-n+1} \xrightarrow{D_n} C^{i+n}_q \xrightarrow{\alpha} C^{p+i+n}.$$

Define operators

$$\tilde{D}_i(\alpha) = -(-1)^{|\alpha|}\alpha \circ D_i$$

for $i \neq 1$ and

$$\tilde{D}_1(\alpha) = \delta_{\mathcal{C}} \circ \alpha - (-1)^{|\alpha|}\alpha \circ D_1.$$

Our deformation cohomology complex is $\mathcal{E} = \oplus \mathcal{E}^{p,q}$ with differential $\tilde{D} = \sum_{0 \leq n} \tilde{D}_n$. Note that that this agrees with the definition given in the statement of Theorem 13.4.2.

The complex $\mathcal{E}$ has components $\{\mathcal{E}^{p,q}; p, q \geq 0\}$, and maps $\{\tilde{D}_1^{p,q} : \mathcal{E}^{p,q} \to \mathcal{E}^{p+1,q}; p, q \geq 0\}$, $\{\tilde{D}_0^{p,q} : \mathcal{E}^{p,q} \to \mathcal{E}^{p,q+1}; p, q \geq 0\}$ and $\{\tilde{D}_l^{p,q} : \mathcal{E}^{p,q} \to \mathcal{E}^{p+l,q-l}; p, q \geq 0, l \geq 2\}$. So the maps $\tilde{D}_0$ and $\tilde{D}_1$ form the bicomplex-like pattern just like in the Gerstenhaber- Schack complex. Arrows corresponding to the operators in the full complex which are relevant to the deformation cohomology are shown in the following diagrams, which should be compared to the diagrams in the preface.

We prove below that when $\Phi = 1$ the complex with differential $\tilde{D}_0 + \tilde{D}_1$ is isomorphic to the Gerstenhaber-Schack complex and its cohomology is isomorphic to the full cohomology of the Drinfeld deformation cohomology that we have just defined.

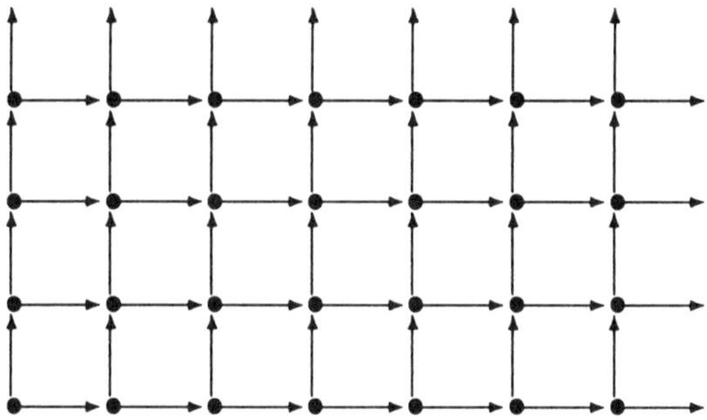

the maps D_2 form the following pattern:

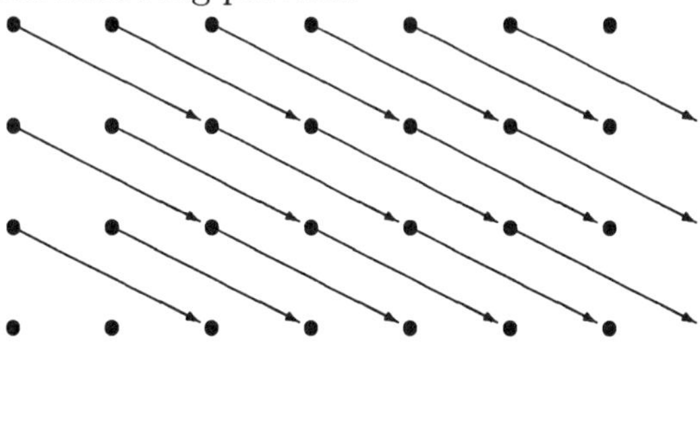

the maps D_3 form the following pattern:

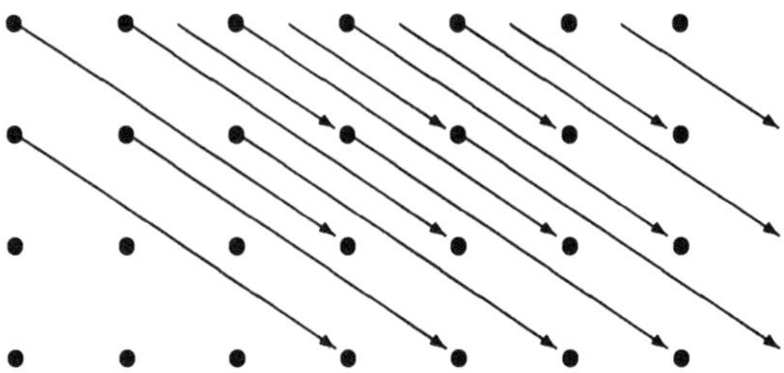

the maps D_4 form the following pattern:

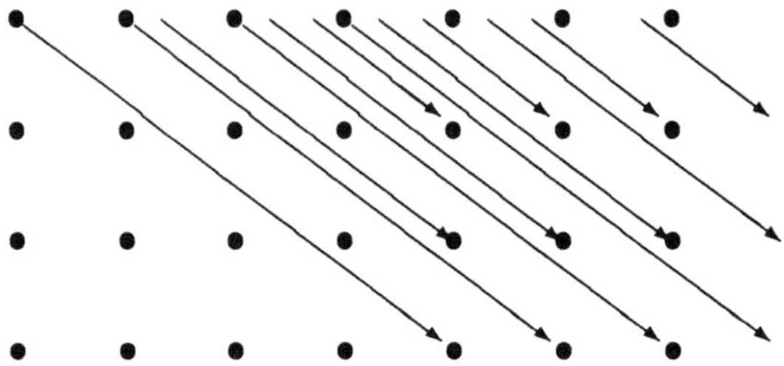

and so on.

The component $\mathcal{E}^{p,q}$ is represented by the lattice point in the p st column and the qst row, where the labelling of rows and columns begins at 0. The triple (m_n, Δ_n, Φ_n) representing the coefficients of h^n in a deformation belong to the components in positions $(1,2),(2,1)$ and $(3,0)$. The obstruction cocycle representing (associativity of m, compatibility of m and Δ, coassociativity of Δ, pentagon identity) is in positions $(1,3),(2,2),(3,1)$, and $(4,0)$. Looking at the diagrams we see that the operators D_i for $0 \leq i \leq 4$ all come into the general deformation cohomology.

Proposition 13.7.1 *The differential complex for deformations of the Drinfeld algebra (B, m, Δ, Φ)*

$$\mathcal{E} = Hom_{\mathcal{C}(B)-bimod}(\mathcal{C}(B, \mathrm{Bar}(B)), \mathcal{C}(B))$$

with differential

$$\tilde{D}(\alpha) = \delta_{\mathcal{C}} \circ \alpha - (-1)^{|\alpha|} \alpha \circ D = \sum_{0 \leq n} \tilde{D}_n$$

has the same cohomology as the Gerstenhaber-Schack complex when $\Phi = 1$.

Proof.

Note first that in this formulation we do not consider the Hom functor relative to any explicit comodule structures. The bicomodule homomorphism condition on the Hom functor has been replaced by the $\mathcal{C}(B)$ bimodule condition. As before, we have an adjoint functor relation relative to the module structures.

$$\begin{aligned}
&\mathrm{Hom}_{\mathcal{C}(B)-\mathrm{bimod},B-\mathrm{bimod}}(\mathcal{C}(B,\mathrm{Bar}(B)),\mathcal{C}(B)) \\
&\cong \ \mathrm{Hom}_{B-\mathrm{bimod}}(\mathrm{Bar}(B),\mathcal{C}(B)) \\
&\cong \ \mathrm{Hom}_k(\otimes B,\mathcal{C}(B))
\end{aligned}$$

As a graded vector space $\mathcal{C}(B)$ is isomorphic to the tensor algebra. (There is no shift in degree by 2.) Since $\Phi = 1$, the two are in fact isomorphic as algebras. Thus

$$\mathcal{E}^{p,q} \cong \mathrm{Hom}_k(B^{\otimes q}, B^{\otimes p}).$$

The boundary operator $\tilde{D}_0$ is the standard differential on the bar complex and reduces to the Hochschild coboundary when we apply the adjoint functor relation for B bimodules just as before. The formula for the $\tilde{D}_1(\alpha)$ is

$$\delta_{\mathcal{C}} \circ \alpha - (-1)^{|\alpha|}\alpha \circ D_1.$$

If we conjugate the standard Gerstenhaber-Schack differential by the k isomorphism which equals $(-1)^{\frac{1}{2}q(q+1)}$ on $C^{p,q}_{GS}$ then we get an equivalent differential with

$$\sum D^{p,q} \qquad D^{p,q} = (-1)^{p+q} d_H + \delta$$

for $\alpha \in C^{p,q}_{GS}$. We identify both $C^{p,q}_{GS}$ and $\mathcal{E}^{p,q}$ as vector spaces with $\mathrm{Hom}_k(B^{\otimes q}, B^{\otimes p})$. Then

$$\begin{aligned}
((-1)^{p+q}d_H + \delta)(\alpha) &= (-1)^{p+q}d_H(\alpha) + (id \otimes \alpha) \circ \lambda \\
&\quad -\delta_{\mathrm{cobar}} \circ \alpha + (-1)^{p+1}(\alpha \otimes id) \circ \rho \\
\text{for} \alpha \in \mathcal{E}^{p,q} &= (-1)^{|\alpha|}\alpha \circ D_0 - \delta_{\mathcal{C}} \circ \alpha + (-1)^{|\alpha|}\alpha(\lambda(u) - (-1)^q \rho(u)) \\
&= (-1)^{|\alpha|}\alpha(D_0 + D_1) - \delta_{\mathcal{C}} \circ \alpha \\
&= -(\tilde{D}_0 + \tilde{D}_1)\alpha
\end{aligned}$$

In passing to the second equation we used equation (13.4.5) and the fact that since $\alpha \in \mathcal{E}^{p,q}$ has parity $p+q$,

$$\alpha \circ \lambda = (-1)^{p+q} id \otimes \alpha \circ \lambda.$$

The higher operators $\tilde{D}_j$ for $j \geq 2$ are, up to a sign, composition with D_j, and because all the associativity operators are tensor products of 1's they map $\mathrm{Bar}_q(B)$ into $C^i(B, \mathrm{Bar}_S(B)) \subset C^i(B, \mathrm{Bar}_{q+i})$ where Bar_S is the complex we used to define the normalized bar resolution, which is spanned by terms with 1 in some slot of the tensor product other than the first or last. That means that the for any normalized cochain $\tilde{D}_j(\alpha) = 0$ for $j \geq 2$.

At this point we have to invoke a spectral sequence argument to prove that the cohomology of the complex with differential $\tilde{D}' = \tilde{D}_0 + \tilde{D}_1$ is isomorphic to the cohomology with the full differential $\sum \tilde{D}_i = D' + \sum_{i \geq 2} \tilde{D}_i$. This is similar to the double complex arguments that we have seen earlier but with one big exception. Although the differential is a sum of two operators only one, D' has square 0, therefore we cannot consider the cohomology of the second separately. We can only consider it as a differential when it is applied to the cohomology $H^*(\mathcal{E}, D')$. The theory of spectral sequences says that using this proceedure, and assuming certain finiteness conditions which exist here, the cohomology of the total complex can be computed from an iterated sequence of cohomologies. Without going into the details, we need only note that each $\tilde{D}_j$ is the zero operator on the cohomology $H(\mathcal{E}, D')$ because any cohomology class is represented by a normalized cochain and the $\tilde{D}_j$ annihilate normalized cochains. Therefore the cohomology in the limit, which is the cohomology of $\tilde{D}$, is the cohomology of the second term in the sequence

$$H^*(\mathcal{E}, \tilde{D}_0 + \tilde{D}_1) = H^*_{G.S.}(\mathrm{Hom}(\otimes B, \otimes B)).$$

Now we describe those of $D_*^{*,*}$'s which are relevant for the infinitesimal cocycle condition and for determining when a deformation is equivalent to the trivial one. A deformation B_h of a Drinfeld algebra $(B, m_0, \Delta_0, \Phi_0)$ with

$$m(h) = m_0 + mh + o(h^2), \Delta(h) = \Delta_0 + \Delta h + o(h^2), \Phi(h) = \Phi_0 + \Phi h + o(h^2))$$

as above determines an infinitesimal cocycle $(m, \Delta, \Phi) \in \mathcal{E}^2$. The $(0,3)$ and $(1,2)$ components are

$$D_0^{1,2}(m) = 0, \; D_1^{1,2}(m) + D_0^{2,1}(\Delta) = 0,$$
$$\text{the}(2,1)\text{component is} \quad D_2^{1,2}(m) + D_1^{2,1}(\Delta) + D_0^{3,0}(\Phi) = 0$$
$$\text{and the}(3,1)\text{component is} \quad D_3^{1,2}(m) + D_2^{1,1}(\Delta) + D_1^{3,0}(\Phi) = 0.$$

Let us look at the $(3,1)$ component of the cocycle condition, which arises from the h coefficient of the the conjugate coassociativity equation. For $a, b \in B^{\otimes n}$, $a \circ b$ denotes the product on $B^{\otimes n}$ induced by $m_0 + tm$. Let then

$a \bullet_m b$ denotes the coefficient at t in $a \circ b$. So, for example, for $n = 1$ we have $a \bullet_m b = m(a, b)$, for $n = 2$ we have $a \bullet_m b = \{(m_0 \odot m) + (m \odot m_0)\} T_{1324}(a \odot b)$, etc. Then

$$\begin{aligned}
D_2^{1,2}(m) &= (id \odot \Delta_0)\Delta_0 \bullet_m \Phi_0 - \Phi_0 \bullet_m (\Delta_0 \odot id)\Delta_0 \text{ and} \\
D_1^{2,1}(\Delta) &= (id \odot \Delta)(\Delta_0) \cdot \Phi_0 - \Phi_0 \cdot (\Delta_0 \odot id)\Delta + (id \odot \Delta_0)\Delta \cdot \Phi_0 \\
&\quad - \Phi_0 \cdot (\Delta \odot id)\Delta_0,
\end{aligned}$$

$$D_0^{3,0}(\Phi) = (id \odot \Delta_0)\Delta_0 \cdot \Phi - \Phi \cdot (\Delta_0 \odot id)\Delta_0,$$

The $(4, 0)$ component of the cocycle condition comes from the pentagon equation.

$$\begin{aligned}
D_3^{1,2}(m) &= (1 \odot \Phi_0) \bullet_m \{(id \odot \Delta_0 \odot id)(\Phi_0) \cdot (\Phi_0 \odot 1)\} \\
&+ (1 \odot \Phi_0) \cdot \{id \odot \Delta_0 \odot id)(\Phi_0) \bullet_m (\Phi_0 \odot 1)\} \\
&- (id \odot id \odot \Delta_0)(\Phi_0) \bullet_m (\Delta_0 \odot id \odot id)(\Phi_0)
\end{aligned}$$

$$\begin{aligned}
D_2^{2,1}(\Delta) &= -(id \odot id \odot \Delta_0)(\Phi_0)(\Delta \odot id \odot id)(\Phi_0) \\
&\quad +(1 \odot \Phi_0)(id \odot \Delta \odot id)(\Phi_0)(\Phi_0 \odot 1) \\
&\quad -(id \odot id \odot \Delta)(\Phi_0)(\Delta_0 \odot id \odot id)(\Phi_0)
\end{aligned}$$

$$\begin{aligned}
D_1^{3,0}(\Phi) &= (1 \odot \Phi)(id \odot \Delta_0 \odot id)(\Phi_0)(\Phi_0 \odot 1) \\
&\quad -(id \odot id \odot \Delta_0)(\Phi_0)(\Delta_0 \odot id \odot id)(\Phi) \\
&\quad +(1 \odot \Phi_0)(id \odot \Delta_0 \odot id)(\Phi)(\Phi_0 \odot 1) \\
&\quad -(id \odot id \odot \Delta_0)(\Phi)(\Delta_0 \odot id \odot id)(\Phi_0) \\
&\quad +(1 \odot \Phi_0)(id \odot \Delta_0 \odot id)(\Phi_0)(\Phi \odot 1),
\end{aligned}$$

Such an infinitesimal deformation is equivalent to the trivial one if and only if there are some $F \in B^{\otimes 2}$ and $\phi \in \mathrm{Hom}_k(B, B)$ such that

$$m = D_0^{1,1}(\phi),\ \Delta = D_1^{1,1}(\phi) + D_0^{2,0}(F) \text{ and } \Phi = D_2^{1,1}(\phi) + D_1^{2,0}(F).$$

The new terms here are

$$\begin{aligned}
D_1^{1,1}(\phi) &= (id \odot \phi)\Delta_0 - \Delta_0\phi + (\phi \odot id)\Delta_0, \\
D_1^{2,0}(F) &= (1 \odot F)\Phi_0 - \Phi_0(\Delta_0 \odot id)(F) + (id \odot \Delta_0)(F)\Phi_0 - \Phi_0(F \odot 1), \\
D_2^{1,1}(\phi) &= (\phi \odot 1 \odot 1)(\Phi_0) + (1 \odot \phi \odot 1)(\Phi_0) + (1 \odot 1 \odot \phi)(\Phi_0)
\end{aligned}$$

All these formulae, which come from isolating the h term in an expansion of the structure equations for the deformation agree with the operators $\tilde{D}_k$. The operators $D_1^{i,j}$ come directly from the modified cobar complex. The operators $D_2 i, j$ come from K_3 using formula 13.5.1.

Finally we describe how the expression $D_3^{1,2}(m)$ comes from the first triangulation of K_4. The vertex on the extreme left is the preferred vertex [4], label the other vertices, going counter clockwise from [4], v_3, v_2, v_0, v_1. The triangles in the triangulation are $[v_2, v_3, [4]], [v_0, v_2, [4]]$ and $[v_0, v_1, [4]]$. The corresponding associativity operators are

$$\begin{aligned}
\Psi_{v_0,v_2} &= 1 \otimes \Phi, \\
\Psi_{v_2,v_{[4]}} &= ((id \otimes \Delta \otimes id)\Phi)\Phi \otimes 1, \\
\Psi_{v_2,v_3} &= (id \otimes \Delta id)\Phi \\
\Psi_{v_3,[4]} &= \Phi \otimes 1 \\
\Psi_{v_0,v_1} &= (id \otimes \Delta \otimes id)\Phi \\
\Psi_{v_1,[4]} &= (\Delta \otimes id \otimes id)\Phi
\end{aligned}$$

Let $m^3 = (m \otimes m_0 \otimes m_0 + m_0 \otimes m \otimes m_0 + m_0 \otimes m_0 \otimes m)$. The three summands of the formula in order of appearance correspond to the triangles in order listed above. Before projecting into $C(B)$, the first summand is $m^3(\Psi_{v_0,v_2} \overline{\otimes} \Psi_{v_2,[4]})$, the second summand is $\Psi_{v_0,v_2} \cdot m^3(\Psi_{v_2,v_3} \overline{\otimes} \Psi_{v_3,[4]})$, the last summand is $m^3(\Psi_{v_0,v_1} \overline{\otimes} \Psi_{v_1,[4]})$. In the definition of $M_3[v_0, v_1, v_4]$ there is a left multiplication by $\Psi_{v_i[4]}$ which does not appear in the formula but up to this common factor the two expressions agree.

To do the full obstruction theory we need to know one more operator, D_4 which comes from a triangulation of the polyhedron, K_5. This is obviously much more complicated. The details appear in [895].

Bibliography

[1] E.Abe, *Hopf algebras*, Cambridge University Press, Cambridge-New York, 1980. [MR83a:16010]

[2] M.L.Glasser, D.B.Abraham and E.H.Lieb, *Analytic properties of the free energy for the "ice" models*, J. Math. Phys. **13** (1972), 887–900. [MR55#7227]

[3] G.S.Agrawal and E.Wolf, *Calculus for functions of noncommuting operators and general phase-space methods in quantum mechanics. I. Mapping theorems and ordering of functions of noncommuting operators*, Phys. Rev. D **2** (1970), 2161–2170.

[4] M.Wadati, T.Deguchi and Y.Akutsu, *Yang-Baxter relation, exactly solvable models and link polynomials*, Quantum groups (Leningrad 1990), Lect. Notes Math. **1510** (1992), 373–388. [MR93e:57016]

[5] G.Albertini, B.M.McCoy and J.H.H.Perk, *Eigenvalue spectrum of the superintegrable chiral Potts model*, Integrable Systems in Quantum Field Theory and Statistical Mechanics, Adv. Stud. Pure Math. **19** (1989), 1–55. [MR91e:82022]

[6] G.Albertini,
B.M.McCoy and J.H.H.Perk, *Commensurate-incommensurate transition in the ground state of the superintegrable chiral Potts model*, Phys. Lett. **135A** (1989), 159–166.

[7] F.C.Alcaraz, *Parafermions and surface exponents of self-dual $Z(N)$ spin models from conformal invariance*, J. Phys. A **20** (1987), L623–L628.

[8] F.C.Alcaraz, *The critical behaviour of self-dual $Z(N)$ spin-systems: finite-size scaling and conformal invariance*, J. Phys. A **20** (1987), 2511–2526.

[9] F.C.Alcaraz and A.Lima-Santos, *Conservation laws for $Z(N)$ symmetric quantum spin models and their exact ground state energies*, Nucl. Phys. **B275** (1986), 436–458.

[10] A.Yu.Alekseev, *From geometric quantization to conformal field theory*, Problems of modern quantum field theory (Alushta 1989), 22–42, Springer, Berlin, 1989. [MR92e:81216]

[11] A.Yu.Alekseev and L.D.Faddeev, *$(T^*G)_t$: A toy model for conformal field theory*, Comm. Math. Phys. **141** (1991), 413–422.

[12] A.Yu.Alekseev and S.Shatashvili, *Quantum groups and WZNW models*, Comm. Math. Phys. **133** (1990), 353–368. [MR92b:81070]

[13] T.V.Alekseevskaya, I.M.Gelfand and A.V.Zelevinsky, *Distribution of real hyperplanes and the partition function connected with it*, Soviet Math. Dokl. **36** (1988), 589–593. [MR89m:05014]

[14] L.Alvarez-Gaumé, C.Gomez and G.Sierra, *Duality and quantum groups*, Nucl. Phys. **B330** (1990), 347–398. [MR91c:81122]

[15] J.Alev and M.Chamarie, *Derivations et automorphismes de quelques algèbres quantiques*, Comm. Algebra **20** (1992), 1787–1802. [MR93f:16036]

[16] D.Altschüler and A.Coste, *Quasi-quantum groups, knots, three-manifolds, and topological field theory*, Comm. Math. Phys. **150** (1992), 83–107.

[17] R.Aminou and Y.Kosmann-Schwarzbach, *Bigèbres de Lie, doubles et carrés*, Ann. Inst. H. Poincaré **49** (1988), 461–478. [MR91a:17019]

[18] H.H.Andersen, P.Polo and W.Kexin, *Injective modules for quantum algebras*, Amer. J. Math. **114** (1992), 571–604.

[19] H.H.Andersen, P.Polo and K.Wen, *Representations of quantum algebras*, Invent. Math. **104** (1991), 1–59. [MR92e:17011]

[20] G.E.Andrews and R.J.Baxter, *Lattice gas generalization of the hard hexagon model. II. The local densities as elliptic functions*, J. Statist. Phys. **44** (1986), 713–728. [MR88i:82063]

[21] G.E.Andrews and R.J.Baxter, *Lattice gas generalization of the hard hexagon model. III. q-trinomial coefficients*, J. Statist. Phys. **47** (1987), 297–330. [MR88h:82069]

[22] G.E.Andrews and R.J.Baxter, *A motivated proof of the Rogers-Ramanujan identities*, Amer. Math. Monthly **96** (1989), 401–409. [MR90e:11147]

[23] G.E.Andrews and R.J.Baxter, *SCRATCHPAD explorations for elliptic theta functions*, Computers in mathematics (Stanford, CA, 1986), 17–33, Dekker, New York, 1990. [MR91i:11188]

[24] G.E.Andrews, R.J.Baxter, D.M.Bressoud, W.H.Burge, P.J.Forrester and G.Viennot, *Partitions with prescribed hook differences*, European J. Combin. **8** (1987), 341–350. [MR89g:05014]

[25] N.Andruskiewitsch, *Some exceptional compact matrix pseudogroups*, Bull. Soc. Math. France. **120** (1992), 297–325.

[26] B.Aneva, *A deformed quantum* SU(2) *superalgebra*, J. Phys. A **24** (1991), L455–L458. [MR92e:17012]

[27] J.C.Anglès-d'Auriac, D.Hansel and J.M.Maillard, *On the phase diagram of the chiral Potts model*, Preprint PAR LPTHE 88-17.

[28] K.Aomoto, *On the acyclicity of free cobar constructions. I*, Proc. Japan Acad. Ser. A **53** (1977), 35–36. [MR58#13101a]

[29] K.Aomoto, *A construction of integrable differential system associated with braid groups*, Braids (Santa Cruz, CA, 1986), Contemp. Math. **78** (1988), 1–11.

[30] H.Aratyn, *A geometric approach to the string BRS cohomology*, Differential geometric methods in theoretical physics (Davis, CA, 1988), 431–443, Plenum, New York, 1990.

[31] H.Aratyn, E.Nissimov and S.Pacheva, *Symplectic geometry and infinite-dimensional symmetry groups*, Proc. XXth Intern. Conf. Diff. Geom. Methods Theor. Phys., World Sci. Publ., 1991.

[32] H.Aratyn, E.Nissimov and S.Pacheva, *Classical r-matrices and Poisson bracket structures on infinite-dimensional groups*, Phys. Lett. B **284** (1992), 273–282.

[33] H.Aratyn, E.Nissimov, S.Pacheva and I.Vaysburd, *R-matrix formulation of KP hierarchies and their gauge equivalence*, Phys. Lett. B **294** (1992), 167–176.

[34] H.Aratyn, E.Nissimov, S.Pacheva and A.Zimerman, *Symplectic actions on coadjoint orbits*, Phys. Lett. B **240** (1990), 127–132. [MR91d:58066]

[35] D.Arnal, *Simultaneous deformations of a Lie algebra and its modules*, Differential geomety and mathematical physics, 3–15, Reidel, Dordrecht-Boston, 1983. [MR85b:17011]

[36] D.Arnal, M.Cahen and S.Gutt, *Deformations on coadjoint orbits*, J. Geom. Phys. **3** (1986), 327–351. [MR88k:58047]

[37] D.Arnal, M.Cahen and S.Gutt, *Representation of compact Lie groups and quantization by deformation*, Acad. Roy. Belg. Bull. Cl. Sci. **74** (1988), no. 4-5, 123–141. [MR91a:58068]

[38] D.Arnal and J.C.Cortet, *Geometrical theory of contractions of groups and representations*, J. Math. Phys. **20** (1979), 556–563. [MR80j:58034]

[39] D.Arnal and J.C.Cortet, *Déformations algébriques sur le dual g^* d'une algèbre de Lie*, Actés du colloque Jean Braconnier (Lyon 1986), Publ. Dep. Math. Univ. Claude Bernard, Lyon, 1987, no. 18, 147–162. [MR89a:17019]

[40] D.Arnal and J.C.Cortet, *Représentations $*$ des groupes exponentiels*, J. Funct. Anal. **92** (1990), 103–135. [MR91h:22014]

[41] D.Arnal, J.C.Cortet, M.Flato and D.Sternheimer, *Star-product: quantization and representations without operators*, Field theory, quantization and statistical physics, 85–111, Reidel, Dordrecht-Boston, 1981. [MR82e:81019]

[42] D.Arnal, J.C.Cortet, P.Molin and G.Pinczon, *Covariance and geometrical invariance in $*$-quantization*, J. Math. Phys. **24** (1983), 276–283. [MR84h:58061]

[43] D.Arnaudon and A.Chakrabarti, *Flat periodic representations of $U_q(g)$*, Comm. Math. Phys. **139** (1991), 605–617. [MR92k:17018]

[44] P.Roche and D.Arnaudon, *Irreducible representations of the quantum analogue of* SU(2), Lett. Math. Phys. **17** (1989), 295–300. [MR90f:17013]

[45] V.I.Arnold, *Remarks on Poisson structures on a plane and on other powers of volume elements*, Trudy Sem. Petrovsk. (1987), no. 12, 37–46. [MR89e:58040]

[46] V.I.Arnold and A.B.Givental, *Symplectic geometry*, Current problems in mathematics. Fundamental directions **4**, 5–139, Akad. Nauk SSSR, Vsesoyuz. Inst. Nauchn. i Tekhn. Inform. Moscow, 1985. [MR88b:58044]

[47] V.A.Artamonov, *The structure of Hopf algebras*, Algebra. Topology. Geometry **29**, 3–63, Akad. Nauk SSSR, Vsesoyuz. Inst. Nauchn. i Tekhn. Inform., Moscow, 1991. [MR93b:16064]

[48] E.Artin, *Theorie der Zöpfe*, Hamb. Abh. **4** (1925), 47–72; Collected Papers, 416–441, Addison-Wesley, 1965.

[49] E.Artin, *Theory of braids*, Ann. Math. **48** (1947), 101–126; Collected Papers, 446–471, Addison-Wesley, 1965.

[50] E.Artin, *Braids and permutations*, Ann. Math. **48** (1947), 643–649; Collected Papers, 472–478, Addison-Wesley, 1965.

[51] M.Artin and C.Rotthaus, *A structure theorem for power series rings*, Algebraic geometry and commutative algebra, Vol. I, 35–44, Kinokuniya, Tokyo, 1988. [MR90b:14006]

[52] M.Artin, W.Schelter and J.Tate, *Quantum deformations of* GL_n, Comm. Pure Appl. Math. **44** (1991), 879–895. [MR92i:17014]

[53] M.Artin, J.Tate and M.Van den Bergh, *Some algebras associated to automorphisms of elliptic curves*, The Grothendieck Festschrift, Vol. I, 33–85, Birkhäuser, Boston, MA, 1990. [MR92e:14002]

[54] F.J.Bloore, M.Assimakopoulos and I.R.Ghobrial, *Quantization as deformation theory*, J. Math. Phys. **17** (1976), 1034–1038. [MR54#12006]

[55] M.F.Atiyah, *The impact of physics on geometry*, Differential geometrical methods in theoretical physics (Como 1987), 1–9, Kluwer Acad. Publ., Dordrecht, 1988. [MR90d:00055]

[56] M.F.Atiyah, *Topological quantum field theories*, Publ. Math. I.H.E.S. **68** (1988), 175–186. [MR90e:57059]

[57] M.F.Atiyah, *New invariants of* 3- *and* 4-*dimensional manifolds*, Proc. Symposia Pure Math. **48** (1988), 285–299. [MR91i:57020]

[58] M.F.Atiyah, *The geometry and physics of knots*, Miniconference on Geometry and Physics (Canberra 1989), 1–17, Austral. Nat. Univ., Canberra, 1989. [MR91b:57003]

[59] M.F.Atiyah, *The frontier between geometry and physics*, Jahresber. Deutsch. Math. Verein. **91** (1989), 149–158. [MR91d:57001]

[60] M.F.Atiyah, *The Jones-Witten invariants of knots*, Astérisque **189-190** (1990), 7–16. [MR92k:57005]

[61] M.F.Atiyah, *The geometry and physics of knots*, Cambridge University Press, Cambridge, 1990. [MR92b:57008]

[62] M.F.Atiyah, *On the work of Edward Witten*, Proceedings of the International Congress of Mathematicians (Kyoto 1990), 31–35, Math. Soc. Japan, Tokyo, 1991. [MR93e:81001b]

[63] M.F.Atiyah, *Magnetic monopoles and the Yang-Baxter equations*, Internat. J. Modern Phys. A **6** (1991), 2761–2774. [MR92g:82029]

[64] R.J.Baxter, J.H.H.Perk and H.Au-Yang, *New solutions of the star-triangle relations for the chiral Potts model*, Phys. Lett. A **128** (1988), 138–142. [MR89g:82062]

[65] H.Au-Yang, B.M.McCoy, J.H.H.Perk and S.Tang, *Solvable models in statistical mechanics and Riemann surfaces of genus greater than one*, Algebraic Analysis, Vol. 1, 29–39, Academic Press, Boston, 1988. [MR90b:82059]

[66] H.Au-Yang and J.H.H.Perk, *Critical correlations in a Z-invariant inhomogeneous Ising model*, Physica **144A** (1987), 44–104. [MR88k:821119]

[67] H.Au-Yang and J.H.H.Perk, *Onsager's star-triangle equation: master key to integrability*, Integrable Systems in Quantum Field Theory and Statistical Mechanics, Adv. Stud. Pure Math. **19** (1989), 57–94. [MR91e:82023]

[68] H.Au-Yang, J.H.H.Perk, B.M.McCoy, S.Tang and M.L.Yan, *Commuting transfer matrices in the chiral Potts models: solutions of star-triangle equations with genus* > 1, Phys. Lett. A **123** (1987), 219–223. [MR89g:82061]

[69] A.Avez, *Représentations de Dirac de l'algèbre de Lie des polynômes en les coordonnées canoniques munie du crochet de Poisson*, C.R.A.S. Paris **290** (1980), A301–A303. [MR81i:17008]

[70] S.Ouvry, R.Stora and P. van Baal, *On the algebraic characterization of Witten's topological Yang-Mills theory*, Phys. Lett. B **220** (1989), 159–163.

[71] O.Babelon, *The role of the Yang-Baxter equation in conformal field theory*, Knots, topology and quantum field theories (Florence 1989), 447–468, World Sci. Publishing, River Edge, NJ, 1989. [MR92m:81227]

[72] O.Babelon and D.Bernard, *Dressing transformations and the origin of the quantum group symmetries*, Phys. Lett. B **260** (1991), 81–86. [MR92k:81067]

[73] J.Balog, L.Dabrowski and L.Feher, *A new quantum deformation of* SL(3), Phys. Lett. B **257** (1991), 74–78. [MR92d:17013]

[74] P.Bamberg and S.Sternberg, *A course in mathematics for students of physics. 2*, Cambridge University Press, 1991. [MR92j:00003b]

[75] M.Bangoura and Y.Kosmann-Schwarzbach, *Double of a Jacobian quasi-bialgebra*, Lett. Math. Phys. B **28** (1993), 13–29.

[76] D.Bar-Natan, *On the Vassiliev knot invariants*, Harvard Preprint, July 1992.

[77] D.Bar-Natan and E.Witten, *Perturbative expansion of Chern-Simons theory with noncompact gauge group*, Comm. Math. Phys. **141** (1991), 423–440. [MR93c:81244]

[78] M.N.Barber, *Perturbation expansions of the axial next-nearest neighbour Ising and of asymmetric four-state clock models*, J. Phys. A **15** (1982), 915–938.

[79] H.Basart, M.Flato, A.Lichnerowicz and D.Sternheimer, *Deformation theory applied to quantization and statistical mechanics*, Lett. Math. Phys. **8** (1984), 483–494. [MR86j:58038]

[80] H.Basart and A.Lichnerowicz, *Déformations d'un star-produit sur une variété symplectique*, C.R.A.S. Paris **293** (1981), 347–350. [MR82j:58053]

[81] H.Basart and A.Lichnerowicz, *Conformal symplectic geometry, deformations, rigidity and geometrical (KMS) conditions*, Lett. Math. Phys. **10** (1985), 167–177. [MR87h:58058]

[82] Yu.A.Bashilov and S.V.Pokrovsky, *Conservation laws in the quantum version of N-dimensional Potts model*, Comm. Math. Phys. **76** (1980), 129–141.

[83] S.V.Pokrovsky and Yu.A.Bashilov, *Star-triangle relations in the exactly solvable statistical models*, Comm. Math. Phys. **84** (1982), 103–132.

[84] I.A.Batalin and E.S.Fradkin, *Operatorial quantization of dynamical systems subject to constraints. A further study of the construction*, Ann. Inst. H. Poincaré Phys. Theor. **49** (1988), 145–214. [MR90m:81037]

[85] H.J.Baues, *Geometry of loop spaces and the cobar construction*, Mem. Amer. Math. Soc. **25** (1980), no. 230. [MR81m:55010]

[86] G.Baxter, *Weight factors for the two-dimensional Ising model*, J. Math. Phys. **6** (1965), 1015–1021. [MR34#1054]

[87] R.J.Baxter, *Colorings of a hexagonal lattice*, J. Math. Phys. **11** (1970) 784–789. [MR42#1457]

[88] R.J.Baxter, *Eight-vertex model in lattice statistics*, Phys. Rev. Lett. **26** (1971), 832–833.

[89] R.J.Baxter, *Generalized ferroelectric model on a square lattice*, Studies Appl. Math. **50** (1971), 51–69. [MR46#6766]

[90] R.J.Baxter, *Partition function of the eight-vertex lattice model*, Ann. Physics **70** (1972), 193–228. [MR44#7912]

[91] R.J.Baxter, *One-dimensional anisotropic Heisenberg chain*, Ann. Phys. **70** (1972), 323–337. [MR45#8147]

[92] R.J.Baxter, *Ising model on a triangular lattice with three-spin interactions. II. Free energy and correlation length*, Austral. J. Phys. **27** (1974), 369–381. [MR53#10050]

[93] R.J.Baxter, *Corner transfer matrices of the eight-vertex model. II. The Ising model case*, J. Statist. Phys. **17** (1977), 1–14. [MR58#32674]

[94] R.J.Baxter, *Solvable eight-vertex model on an arbitrary planar lattice*, Philos. Trans. Roy. Soc. London Ser. A **289** (1978), 315–346. [MR57#18660]

[95] R.J.Baxter, *Exactly solved models*, Fundamental problems in statistical mechanics, Vol. 5, 109–141, North-Holland, Amsterdam, 1980.

[96] R.J.Baxter, *Free-fermion, checkerboard and* **Z***-invariant lattice models in statistical mechanics*, Proc. Roy. Soc. London Ser. A **404** (1986), 1–33. [MR87h:82078]

[97] R.J.Baxter, *The Yang-Baxter equations and the Zamolodchikov model*, Physica D **18** (1986), 321–347. [MR87i:82099]

[98] R.J.Baxter, *Chromatic polynomials of large triangular lattices*, J. Phys. A **20** (1987), 5241–5261. [MR89a:82033]

[99] R.J.Baxter, *Ramanujan's identities in statistical mechanics*, Ramanujan revisited (Urbana-Champaign 1987), 69–84, Academic Press, Boston, MA, 1988. [MR89h:82043]

[100] R.J.Baxter, *Free energy of the solvable chiral Potts model*, J. Statist. Phys. **52** (1988), 639–667. [MR89j:82081]

[101] R.J.Baxter, *The superintegrable chiral Potts model*, Phys. Lett. A **133** (1988), 185–189. [MR90e:82086]

[102] R.J.Baxter, *Solving models in statistical mechanics*, Integrable systems in quantum field theory and statistical mechanics, 95–116, Academic Press, Boston, MA, 1989. [MR91j:82020]

[103] R.J.Baxter, *A simple solvable Z_N Hamiltonian*, Phys. Lett. A **140** (1989), 155–157. [MR90k:82085]

[104] R.J.Baxter, *Superintegrable chiral Potts model: thermodynamic properties, an "inverse" model, and a simple associated Hamiltonian*, J. Statist. Phys. **57** (1989), 1–39. [MR90m:82050]

[105] R.J.Baxter, *Exactly solved models in statistical mechanics*, Academic Press, London-New York, 1989. [MR90b:82001]

[106] R.J.Baxter, *Chiral Potts model: eigenvalues of the transfer matrix*, Phys. Lett. A **146** (1990), 110–114. [MR91h:82007]

[107] R.J.Baxter, *On solving the Ising model functional relation without elliptic functions*, Phys. A **177** (1991), 101–108. [MR92m:82036]

[108] R.J.Baxter, *Corner transfer matrices of the chiral Potts model. II. The triangular lattice*, J. Statist. Phys. **70** (1993), 535–582.

[109] V.V.Bazhanov and R.J.Baxter, *New solvable lattice models in three dimensions*, J. Statist. Phys. **69** (1992), 453–485.

[110] R.J.Baxter, V.V.Bazhanov and J.H.H.Perk, *Functional relations for transfer matrices of the chiral Potts model*, Internat. J. Modern Phys. B **4** (1990), 803–870. [MR91j:82009]

[111] R.J.Baxter and T.C.Choy, *Spontaneous magnetisations of the Ising model on the bathroom tile lattice*, J. Phys. A **21** (1988), 2143–2150. [MR89i:82082]

[112] R.J.Baxter and T.C.Choy, *Local three-spin correlations in the free-fermion and planar Ising models*, Proc. Roy. Soc. London Ser. A **423** (1989), 279–300. [MR90k:82086]

[113] R.J.Baxter and G.R.W.Quispel, *Hamiltonian limit of the 3D Zamolodchikov model*, J. Statist. Phys. **58** (1990), 411–430. [MR91e:82024]

[114] R.J.Baxter and F.Y.Wu, *Ising model on a triangular lattice with three-spin interactions. I. The eigenvalue equation*, Austral. J. Phys. **27** (1974), 357–367. [MR53#10049]

[115] F.Bayen, M.Flato, C.Fronsdal, A.Lichnerowicz and D.Sternheimer, *Quantum mechanics as a deformation of classical mechanics*, Lett. Math. Phys. **1** (1975), 521–530. [MR58#32617]

[116] F.Bayen, M.Flato, C.Fronsdal, A.Lichnerowicz and D.Sternheimer, *Deformation theory and quantization. I. Deformations of symplectic structures*, Ann. Physics **111** (1978), 61–110. [MR58#14737a]

[117] F.Bayen, M.Flato, C.Fronsdal, A.Lichnerowicz and D.Sternheimer, *Deformation theory and quantization. II. Physical applications*, Ann. Physics **111** (1978), 111–151. [MR58#14737b]

[118] F.Bayen and C.Fronsdal, *Quantization on the sphere*, J. Math. Phys. **22** (1981), 1345–1349. [MR83c:58034]

[119] F.Bayen and J.M.Maillard, *Star exponentials of the elements of the inhomogeneous symplectic Lie algebra*, Lett. Math. Phys. **6** (1982), 491–497. [MR84c:58028]

[120] D.Bayer and P.Diaconis, *Trailing the dovetail shuffle to its lair*, Ann. Appl. Probab. **2** (1992), 294–313. [MR93d:60014]

[121] V.V.Bazhanov, *Trigonometric solutions of triangle equations and classical Lie algebras*, Phys. Lett. B **159** (1985), 321–324. [MR87g:82009]

[122] E.Beggs and S.Majid, *Matched pairs of topological Lie algebras corresponding to Lie bialgebra structures on* $\mathrm{diff}(\mathbf{S}^1)$ *and* $\mathrm{diff}(\mathbf{R})$, Ann. Inst. H. Poincaré Phys. Theor. **53** (1990), 15–34. [MR91k:17026]

[123] A.A.Beilinson, G.Lusztig and R.MacPherson, *A geometric setting for the quantum deformation of* GL_n, Duke Math. J. **61** (1990), 655–677. [MR91m:17012]

[124] A.A.Beilinson and V.V.Schechtman, *Determinant bundles and Virasoro algebras*, Comm. Math. Phys. **118** (1988), 651–701.

[125] A.A.Belavin and V.G.Drinfeld, *Solutions of the classical Yang-Baxter equation for simple Lie algebras*, Funct. Anal. Appl. **16** (1982), 159–180. [MR84e:81034]

[126] A.A.Belavin and V.G.Drinfeld, *The classical Yang-Baxter equation for simple Lie algebras*, Funktsional. Anal. Prilozhen. **17** (1983), no. 3, 69–70. [MR85e:58059]

[127] A.A.Belavin and V.G.Drinfeld, *Triangle equations and simple Lie algebras*, Mathematical physics reviews, Vol. 4, 93–165, Harwood, Academic Publ., 1984. [MR87h:58078]

[128] A.A.Belavin, A.M.Polyakov and A.B.Zamolodchikov, *Infinite conformal symmetry in two-dimensional quantum field theory*, Nuclear Phys. B **241** (1984), 333–380. [MR86m:81097]

[129] H.Benamor and G.Pinczon, *The graded Lie algebra structure of Lie superalgebra deformation theory*, Lett. Math. Phys. **18** (1989), 307–313. [MR91g:17021]

[130] D.B.Benson, *The shuffle bialgebra*, Mathematical foundations of programming language semantics (New Orleans, LA, 1987), Lect. Notes Comp. Sci. **298** (1988), 616–637. [MR89h:68037]

[131] D.B.Benson, *Bialgebras: some foundations for distributed and concurrent computation*, Fund. Inform. **12** (1989), 427–486. [MR91e:68044]

[132] J.Bergen and M.Cohen, *Actions of commutative Hopf algebras*, Bull. London Math. Soc. **18** (1986), 159–164. [MR87e:16052]

[133] J.Bergen and S.Montgomery, *Ideals and quotients in crossed products of Hopf algebras*, J. Algebra **152** (1992), 374–396.

[134] R.Berger, *Géométrie algébrique de Poisson*, C.R.A.S. Paris **289** (1979), 583–585.

[135] R.Berger, *Déformations et lois de groupes formels*, Actés du colloque Jean Braconnier (Lyon 1986), Publ. Dep. Math. Univ. Claude Bernard, Lyon, 1987, no. 87-1, 163–173. [MR88h:17017]

[136] R.Berger, *Algebraic quantization*, Lett. Math. Phys. **17** (1989), 275–283. [MR90e:81065]

[137] R.Berger, *The quantum Poincaré-Birkhoff-Witt theorem*, Comm. Math. Phys. **143** (1992), 215–234. [MR93c:17020]

[138] F.Bergeron and N.Bergeron, *Orthogonal idempotents in the descent algebra of B_n and applications*, J. Pure Appl. Algebra **79** (1992), 109–129. [MR93f:20054]

[139] G.M.Bergman, *Everybody knows what a Hopf algebra is*, Group actions on rings (Brunswick, Maine, 1984), Contemp. Math. **43** (1985), 25–48. [MR87e:16024]

[140] M.Kardar and A.N.Berker, *Commensurate-incommensurate phase diagrams for overlayers from a helical Potts model*, Phys. Rev. Lett. **48** (1982), 1552–1555.

[141] D.Bernard, *A remark on quasi-triangular quantum Lie algebras*, Phys. Lett. B **260** (1991), 389–393. [MR92e:17013]

[142] J.Bernstein, *Trace in categories*, Operator algebras, unitary representations, enveloping algebras, and invariant theory (Paris 1989), 417–423, Birkhäuser, Boston, MA, 1990. [MR92d:17010]

[143] J.Bernstein, Sackler Lectures, Tel-Aviv Univ., 1992, to appear.

[144] J.Bernstein and S.Gelfand, *Tensor products of finite- and infinite-dimensional representations of semisimple Lie algebras*, Compositio Math. **41** (1980), 245–285. [MR82c:17003]

[145] J.Berstel and C.Reutenauer, *Rational series and their languages*, Springer, 1988.

[146] H.Bethe, *Zur Theorie der Metalle. I. Eigenwerte und Eigenfunktionen der linearen Atomkette*, Z. Phys. **71** (1931), 205–226.

[147] K.H.Bhaskara and K.Viswanath, *Poisson algebras and Poisson manifolds*, Wiley, New York, 1988. [MR89j:58026]

[148] L.C.Biedenharn, *The quantum group* $\mathrm{SU}_q(2)$ *and a q-analogue of the boson operators*, J. Phys. A **22** (1989), L873–L878. [MR90k:17027]

[149] L.C.Biedenharn, *Tensor operator structures in quantum unitary groups*, Differential geometric methods in theoretical physics (Rapallo 1990), Lect. Notes Phys. **375** (1991), 61–69. [MR92h:81069]

[150] L.C.Biedenharn, *An overview of quantum groups*, Group theoretical methods in physics (Moscow 1990), Lect. Notes Phys. **382** (1991), 147–163. [MR92i:17015]

[151] D.E.Flath and L.C.Biedenharn, *Beyond the enveloping algebra of* sl_3, Canad. J. Math. **37** (1985), 710–729. [MR87a:17014]

[152] M.Gould and L.C.Biedenharn, *The pattern calculus for tensor operators in quantum groups*, J. Math. Phys. **33** (1992), 3613–3635.

[153] L.P.Horwitz, D.Sepunaru and L.C.Biedenharn, *Some quantum aspects of theories with hypercomplex and nonassociative structure*, Proceedings of the Third Johns Hopkins Workshop on Current Problems in High Energy Particle Theory (Florence 1979), 121–153, Johns Hopkins Univ., Baltimore, Md., 1979. [MR83c:81090]

[154] L.C.Biedenharn and M.A.Lohe, *Some remarks on quantum groups—a new variation on the theme of symmetry*, From symmetries to strings: forty years of Rochester Conferences (Rochester, NY, 1990), 189–206, World Sci. Publishing, River Edge, NJ, 1990.

[155] L.C.Biedenharn and M.A.Lohe, *Quantum groups and basic hypergeometric functions*, Quantum groups (Argonne, IL, 1990), 123–132, World Sci. Publishing, Teaneck, NJ, 1991. [MR92d:33047]

[156] L.C.Biedenharn and M.A.Lohe, *Induced representations and tensor operators for quantum groups*, Quantum groups (Leningrad 1990), Lect. Notes Math. **1510** (1992), 197–209.

[157] L.C.Biedenharn and M.A.Lohe, *An extension of the Borel-Weil construction to the quantum group* $\mathrm{U}_q(n)$, Comm. Math. Phys. **146** (1992), 483–504.

[158] L.C.Biedenharn and M.A.Lohe, *Symmetries of quantum group coupling coefficients*, Differential geometry, group representations, and quantization, Lect. Notes Phys. **379** (1991), 193–206.

[159] L.C.Biedenharn and J.D.Louck, *Angular momentum in quantum physics: Theory and applications*, Encyclopedia of Math. and its Appl. **8**, Addison-Wesley, 1981.

[160] L.C.Biedenharn and J.D.Louck, *The Racah-Wigner algebra in quantum theory*, Encyclopedia of Math. and its Appl. **9**, Addison-Wesley, 1981. [MR83d:81002]

[161] M.Nomura and L.C.Biedenharn, *On the q-symplecton realization of the quantum group* $\mathrm{SU}_q(2)$, J. Math. Phys. **33** (1992), 3636–3648.

[162] L.C.Biedenharn and M.Tarlini, *On q-tensor operators for quantum groups*, Lett. Math. Phys. **20** (1990), 271–278. [MR91k:17009]

[163] B.Binegar, *Cohomology and deformations of Lie superalgebras*, Lett. Math. Phys. **12** (1986), 301–308. [MR88a:17037]

[164] J.S.Birman, *On the Jones polynomial of closed 3-braids*, Invent. Math. **81** (1985), 287–294.

[165] J.S.Birman, *A progress report on the study of links via closed braids*, Mitt. Math. Ges. Hamburg **12** (1991), 869–895. [MR93f:57008]

[166] J.S.Birman, *Recent developments in braid and link theory*, Math. Intelligencer **13** (1991), no. 1, 52–60. [MR91k:57006]

[167] J.S.Birman, *New points of view in knot theory*, Bull. Amer. Math. Soc. **28** (1993), 253–287.

[168] J.S.Birman and H.Wenzl, *Braids, link polynomials and a new algebra*, Trans. Amer. Math. Soc. **313** (1989), 249–273. [MR90g:57004]

[169] R.J.Blattner, M.Cohen and S.Montgomery, *Crossed products and inner actions of Hopf algebras*, Trans. Amer. Math. Soc. **298** (1986), 671–711. [MR87k:16012]

[170] R.J.Blattner and S.Montgomery, *A duality theorem for Hopf module algebras*, J. Algebra **95** (1985), 153–172. [MR87h:16016]

[171] R.J.Blattner and S.Montgomery, *Crossed products and Galois extensions of Hopf algebras*, Pacific J. Math. **137** (1989), 37–54. [MR90a:16007]

[172] R.E.Block, *Commutative Hopf algebras, Lie coalgebras, and divided powers*, J. Algebra **96** (1985), 275–306. [MR87c:16012a]

[173] N.M.Bogolyubov, A.G.Izergin and N.Yu.Reshetikhin, *Finite-size effects and infrared asymptotics of the correlation functions in two dimensions*, J. Phys. A **20** (1987), 5361–5369. [MR89b:82020]

[174] N.N.Bogolyubov, I.V.Mikityuk and A.K.Prikarpatskii, *Verma modules over the quantum Lie algebra of currents on the circle*, Soviet Math. Dokl. **42** (1991), 424–428. [MR92a:17042]

[175] F.Bohnenblust, *The algebraical braid group*, Ann. Math. **48** (1947), 127–136.

[176] A.I.Bondal and M.M.Kapranov, *Representable functors, Serre functors, and reconstructions*, Izv. Akad. Nauk SSSR Ser. Mat. **53** (1989), 1183–1205. [MR91b:14013]

[177] P.Bonneau, *Cohomology and associated deformations for not necessarily co-associative bialgebras*, Lett. Math. Phys. **26** (1992), 277–283.

[178] P.Bonneau, M.Flato and G.Pinczon, *A natural and rigid model of quantum groups*, Lett. Math. Phys. **25** (1992), 75–84.

[179] M.Bordemann, *Generalized Lax pairs, the modified classical Yang-Baxter equation, and affine geometry of Lie groups*, Comm. Math. Phys. **135** (1990), 201–216. [MR91k:58049]

[180] H.Boseck, *Hopf algebras—an algebraic link between geometry and physics*, Proceedings of the 3rd Congress of Geometry (Thessaloniki 1991), 115–121, Aristotle Univ. Thessaloniki, 1992. [MR93d:16047]

[181] R.Bott and L.W.Tu, *Differential forms in algebraic topology*, Springer, New York-Berlin, 1982. [MR83i:57016]

[182] N.Bourbaki, *Groupes et algèbres de Lie*, Chap. 1-3, Hermann, Paris, 1971, 1972.

[183] M.D.Gould, R.B.Zhang and A.J.Bracken, *Lie bi-superalgebras and the graded classical Yang-Baxter equation*, Rev. Math. Phys. **3** (1991), 223–240. [MR92h:17014]

[184] R.B.Zhang, M.D.Gould and A.J.Bracken, *From representations of the braid group to solutions of the Yang-Baxter equation*, Nuclear Phys. B **354** (1991), 625–652. [MR92m:17031]

[185] W.Browder, *On differential Hopf algebras*, Trans. Amer. Math. Soc. **107** (1963), 153–176. [MR26#3061]

[186] R.Brustein, Doctoral thesis, Tel-Aviv Univ., 1988.

[187] R.Brustein, Y.Ne'eman and S.Sternberg, *Two-dimensional conformal field theory, duality and MacLane's coherence*, Fields, strings and quantum gravity (Beijing 1989), 89– , Gordon and Breach, N.Y., 1990.

[188] R.Brustein, Y.Ne'eman and S.Sternberg, *Conformal field theory and category theory*, Physics and geometry (Lake Tahoe 1989), to appear.

[189] R.Brustein, Y.Ne'eman and S.Sternberg, *Duality, crossing and MacLane's coherence*, Israel J. Math. **72** (1990), 19–37. [MR92c:81153]

[190] R.Brustein, S.Yankielowicz and J.-B.Zuber, *Factorization and selection rules of operator product algebras in conformal field theories*, Nucl. Phys. **B313** (1989), 321–347.

[191] T.Brzeziński, *Exterior bialgebras*, Preprint DAMTP/91-53, 1991.

[192] T.Brzeziński and S.Majid, *A class of bicovariant differential calculi on Hopf algebras*, Lett. Math. Phys. **26** (1992), 67–78.

[193] T.Brzeziński and S.Majid, *Quantum group gauge theory on classical spaces*, Phys. Lett. B **298** (1993), 339–343.

[194] D.Buchsbaum and D.S.Rim, *A generalized Koszul complex*, Bull. Amer. Math. Soc. **69** (1963), 382–385. [MR26#6226]

[195] D.A.Buchsbaum and D.S.Rim, *A generalized Koszul complex. II. Depth and multiplicity*, Trans. Amer. Math. Soc. **111** (1964), 197–224. [MR28#3076]

[196] D.A.Buchsbaum and D.S.Rim, *A generalized Koszul complex. III. A remark on generic acyclicity*, Proc. Amer. Math. Soc. **16** (1965), 555–558. [MR31#1285]

[197] N.Burroughs, *The universal R-matrix for* $U_q\mathrm{sl}(3)$ *and beyond!* Comm. Math. Phys. **127** (1990), 109–128. [MR91a:17017]

[198] N.Burroughs, *Relating the approaches to quantised algebras and quantum groups*, Comm. Math. Phys. **133** (1990), 91–117. [MR92c:17018]

[199] M.Cahen, *Déformations et quantification*, Physique quantique et géométrie (Paris 1986), 43–62, Hermann, Paris, 1988. [MR89k:58107]

[200] M.Cahen, M.Flato, S.Gutt and D.Sternheimer, *Do different deformations lead to the same spectrum?* J. Geom. Phys. **2** (1985), 35–48. [MR87h:58075]

[201] M.Cahen and S.Gutt, *Regular ∗-representations of Lie algebras*, Lett. Math. Phys. **6** (1982), 395–404. [MR84d:58025]

[202] M.Cahen and S.Gutt, *Theory of deformations and group representations*, Bifurcation theory, mechanics and physics, 77–98, Reidel, Dordrecht-Boston, 1983. [MR85k:58037]

[203] M.Cahen and S.Gutt, *Discrete spectrum of the hydrogen atom: an illustration of deformation theory methods and problems*, J. Geom. Phys. **1** (1984), 65–83. [MR87a:58073]

[204] M.Cahen and S.Gutt, *An algebraic construction of ∗-product on the regular orbits of semisimple Lie groups*, Gravitation and geometry, 71–82, Bibliopolis, Naples, 1987. [MR89e:22023]

[205] M.Cahen, S.Gutt, C.Ohn and M.Parker, *Lie-Poisson groups: remarks and examples*, Lett. Math. Phys. **19** (1990), 343–353. [MR91d:22012]

[206] M.Cahen, S.Gutt and J.Rawnsley, *Nonlinearizability of the Iwasawa-Poisson-Lie structure*, Lett. Math. Phys. **24** (1992), 79–83.

[207] M.Cahen and C.Ohn, *Bialgebra structures on the Heisenberg algebra*, Acad. Roy. Belg. Bull. Cl. Sci. **75** (1989), no. 10, 315–321. [MR93d:16048]

[208] A.Chodos and D.G.Caldi, *Quantum deformations of the Heisenberg equations of motion*, J. Phys. A **24** (1991), 5505–5515. [MR92g:81082]

[209] R.Cappuccio and E.Guadagnini, *On braid statistics*, Phys. Lett. B **252** (1990), 420–424. [MR92g:20058]

[210] A.L.Carey and K.C.Hannabuss, *Simple examples of conformal field theories*, Internat. J. Modern Phys. B **4** (1990), 1059–1068. [MR91k:81153]

[211] H.Cartan and S.Eilenberg, *Homological algebra*, Princeton University Press, 1956. [MR17,1040e]

[212] P.Cartier, *Modules sur une coalgèbre*, Sém. Sophus Lie 1955-1956, exp. 4.

[213] P.Cartier, *Cohomologie des coalgèbres*, Sém. Sophus Lie 1955-1956, exp. 5.

[214] P.Cartier, *Les arrangements d'hyperplans: un chapitre de géométrie combinatoire*, Sém. Bourbaki 1980, Lect. Notes Math. **901** (1982), 1–22. [MR84d:32017]

[215] P.Cartier, *Jacobiennes généralisées, monodromie unipotente et intégrales itérées*, Astérisque **161-162** (1988), 31–52. [MR90d:32024]

[216] P.Cartier, *Développements récents sur les groupes de tresses. Applications à la topologie et à l'algèbre*, Astérisque **189-190** (1990), 17–67. [MR92d:20053]

[217] E.Celeghini, *Coherence and quantum groups*, Nonlinear coherent structures in physics and biology (Dijon 1991), Lect. Notes Phys. **393** (1991), 326–331.

[218] P.Centen, V.Rittenberg and M.Marcu, *Non-universality in Z_3 symmetric spin systems*, Nucl. Phys. **B205** (1982), 585–600.

[219] A.Pressley and V.Chari, *Notes on quantum groups*, Nuclear Phys. B Proc. Suppl. **18A** (1990), 207–228. [MR92j:17014]

[220] V.Chari and A.Pressley, *Yangians and R-matrices*, Enseign. Math. **36** (1990), 267–302. [MR92h:17009]

[221] V.Chari and A.Pressley, *Introduction to quantum groups*, Proceedings of the Hyderabad Conference on Algebraic Groups (Hyderabad 1989), 81–122, Manoj Prakashan, Madras, 1991. [MR93f:17019]

[222] V.Chari and A.Pressley, *Fundamental representations of Yangians and singularities of R-matrices*, J. Reine Angew. Math. **417** (1991), 87–128. [MR92h:17010]

[223] V.Chari and A.Pressley, *Quantum affine algebras*, Comm. Math. Phys. **142** (1991), 261–283. [MR93d:17017]

[224] V.Chari and A.Pressley, *Fundamental representations of quantum groups at roots of* 1, Lett. Math. Phys. **26** (1992), 133–146.

[225] J.Q.Chen, *Group representation theory for physicists*, World Scientific Publishing, Teaneck, NJ, 1989. [MR91h:81047]

[226] K.T.Chen, *Reduced bar constructions on de Rham complexes*, Algebra, topology, and category theory (a collection of papers in honor of Samuel Eilenberg), 19–32, Academic Press, New York, 1976. [MR54#1272]

[227] K.T.Chen, *Circular bar construction*, J. Algebra **57** (1979), 446–483. [MR80j:57045]

[228] K.T.Chen, *Pullback path fibrations, homotopies and iterated integrals*, Bull. Inst. Math. Acad. Sinica **8** (1980), no. 2-3, part 1, 263–275. [MR82e:58004]

[229] Y.Cheng and K.Xue, *New solutions of Yang-Baxter equation and new link polynomials*, J. Knot Theory Ramifications **1** (1992), 31–46. [MR93d:57012]

[230] I.V.Cherednik, *On a method of constructing factorized S-matrices in terms of elementary functions*, Teor. Mat. Fiz. **43** (1980), 117–119. [MR81g:81084]

[231] I.V.Cherednik, *"Quantum" deformations of irreducible finite-dimensional representations of gl_N*, Dokl. Akad. Nauk SSSR **287** (1986), 1076–1079. [MR87j:17007]

[232] I.V.Cherednik, *A new interpretation of Gelfand-Tzetlin bases*, Duke Math. J. **54** (1987), 563–577. [MR88k:17005]

[233] I.V.Cherednik, *Quantum groups as hidden symmetries of classic representation theory*, Differential geometric methods in theoretical physics (Chester, 1988), 47–54, World Sci. Publishing, Teaneck, NJ, 1989. [MR92h:17004]

[234] I.Cherednik, *A unification of Knizhnik-Zamolodchikov and Dunkl operators via affine Hecke algebras*, Invent. Math. **106** (1991), 411–431. [MR93b:17040]

[235] I.M.Gelfand and I.V.Cherednik, *Abstract Hamiltonian formalism for classical Yang-Baxter sheaves*, Uspekhi Mat. Nauk **38** (1983), no. 3, 3–21. [MR85f:58052]

[236] C.Chevalley and S.Eilenberg, *Cohomology theory of Lie groups and Lie algebras*, Trans. Amer. Math. Soc. **63** (1948), 85–124. [MR9,567a]

[237] F.R.K. Chung and S. Sternberg, *A note on the Clebsch Gordan problem*, to appear

[238] C.Cibils, *A quiver quantum group*, Preprint, May 1992.

[239] B.A.Cipra, *An introduction to the Ising model*, Amer. Math. Monthly **94** (1987), 937–959. [MR89g:82001]

[240] M.Cohen and D.Fishman, *Hopf algebra actions*, J. Algebra **100** (1986), 363–379. [MR87i:16012]

[241] M.Cohen and S.Westreich, *From Supersymmetry to Quantum Commutativity*, J. Algebra (1992).

[242] V.Coll, M.Gerstenhaber and A.Giaquinto, *An explicit deformation formula with noncommuting derivations*, Ring theory (Ramat Gan and Jerusalem, 1988), Israel Math. Soc. Conf. Proc. **1** (1989), 396–403. [MR91e:16035]

[243] J.F.Conn, *Normal forms for analytic Poisson structures*, Ann. Math. **119** (1984), 577–601. [MR85f:58037]; Correction: **125** (1987), 433–436. [MR88b:58048]

[244] J.F.Conn, *Normal forms for smooth Poisson structures*, Ann. Math. **121** (1985), 565–593. [MR86m:58050]

[245] A.Connes, *C^*-algèbres et géométrie différentielle*, C.R.A.S. Paris **290** (1980), 599–604.

[246] A.Connes, M.Flato and D.Sternheimer, *Closed star products and cyclic cohomology*, Lett. Math. Phys. **24** (1992), 1–12. [MR93d:19003]

[247] A.Connes and M.A.Rieffel, *Yang-Mills for non-commutative two-tori*, Contemp. Math. **62** (1987), 237–266.

[248] E.Corrigan, D.B.Fairlie and P.Fletcher, *Some aspects of quantum groups and supergroups*, J. Math. Phys. **31** (1990), 776–780. [MR91b:16045]

[249] L.Corwin, Y.Ne'eman and S.Sternberg, *Graded Lie algebras in mathematics and physics (Bose-Fermi symmetry)*, Rev. Modern Phys. **47** (1975), 573–603. [MR55#11828]

[250] P.Cotta-Ramusino and M.Rinaldi, *Link-diagrams, Yang-Baxter equations and quantum holonomy*, Deformation theory and quantum groups with applications to mathematical physics (Amherst, MA, 1990), Contemp. Math. **134** (1992), 19–44.

[251] M.Couture, *New braid group representations of the B_2, B_3 and B_4 types, their associated link polynomials and quantum R-matrices*, J. Phys. A **23** (1990), 4765–4778. [MR92e:20026]

[252] M.Couture and N.C.Schmeing, *New braid group representations of the D_2 and D_3 types and their Baxterization*, J. Phys. A **23** (1990), 4751–4764. [MR92e:20025]

[253] L.Crane, *2-d physics and 3-d topology*, Comm. Math. Phys. **135** (1991), 615–640. [MR92h:57026]

[254] L.Crane, *Categorical physics*, Preprint; available at `hep-th@xxx.lanl.gov` as `hep-th/9301061`.

[255] L.Crane and D.Yetter, *A categorical construction of 4d topological quantum field theories*, Preprint; available at `hep-th@xxx.lanl.gov` as `hep-th/9301062`.

[256] A.Crumeyrolle, *Déformations d'algèbres associées a une variété symplectique, une construction effective*, Ann. Inst. H. Poincaré Sect. A **35** (1981), 175–194. [MR83h:58047]

[257] C.W.Curtis, N.Iwahori and R.Kilmoyer, *Hecke algebras and characters of parabolic type of finite groups with (B,N) pairs*, Publ. Math. I.H.E.S. **40** (1972), 81–116.

[258] T.L.Curtright and C.K.Zachos, *Deforming maps for quantum algebras*, Phys. Lett. B **243** (1990), 237–244. [MR91i:17020]

[259] Y.L.Daletskii and B.L.Tsygan, *Hamiltonian operators and Hochschild homology*, Funktsional. Anal. Prilozhen. **19** (1985), no. 4, 82–83. [MR87h:17034]

[260] E.Date and M.Jimbo, *Exactly solvable SOS models. II. Proof of the star-triangle relation and combinatorial identities*, Conformal field theory and solvable lattice models (Kyoto 1986), 17–122, Academic Press, Boston, MA, 1988.

[261] E.Date and M.Jimbo, *Path space realization of the basic representation of* $A_n^{(1)}$, Infinite-dimensional Lie algebras and groups (Luminy Marseille, 1988), 108–123, World Sci. Publishing, Teaneck, NJ, 1989. [MR91a:82028]

[262] E.Date and M.Jimbo, *One-dimensional configuration sums in vertex models and affine Lie algebra characters*, Lett. Math. Phys. **17** (1989), 69–77. [MR90c:82052]

[263] E.Date, M.Jimbo, A.Kuniba, T.Miwa and M.Okado, *Paths, Maya diagrams and representations of* $\widehat{sl}(r, \mathbf{C})$, Integrable systems in quantum field theory and statistical mechanics, 149–191, Academic Press, Boston, MA, 1989. [MR91g:17028]

[264] E.Date, M.Jimbo, K.Miki and T.Miwa, *Cyclic representations of* $U_q(sl(n+1, \mathbf{C}))$ *at* $q^N = 1$, Publ. R.I.M.S. **27** (1991), 347–366. [MR92i:17018]

[265] E.Date, M.Jimbo, K.Miki and T.Miwa, *Generalized chiral Potts models and minimal cyclic representations of* $U_q(\widehat{gl}(n, \mathbf{C}))$, Comm. Math. Phys. **137** (1991), 133–147. [MR92b:82043]

[266] E.Date, M.Jimbo and T.Miwa, *Representations of* $U_q(gl(n, \mathbf{C}))$ *at* $q = 0$ *and the Robinson-Shensted correspondence*, Physics and mathematics of strings, 185–211, World Sci. Publishing, Teaneck, NJ, 1990. [MR92h:17012]

[267] E.Date, M.Jimbo, T.Miwa and M.Okado, *Fusion of the eight vertex SOS model*, Lett. Math. Phys. **12** (1986), 209–215. [MR88a:82042]

[268] E.Date, M.Jimbo, T.Miwa and M.Okado, *Solvable lattice models*, Theta functions - Bowdoin, Part 1 (Brunswick, ME, 1987), Proc. Symposia Pure Math. **49** (1989), 295–331. [MR90i:82062]

[269] D.David and D.D.Holm, *Multiple Lie-Poisson structures, reductions, and geometric phases for the Maxwell-Bloch travelling wave equations*, J. Nonlinear Sci. **2** (1992), 241–262. [MR93e:58060]

[270] B.Davies, O.Foda, M.Jimbo, T.Miwa and A.Nakayashiki, *Diagonalization of the XXZ Hamiltonian by vertex operators*, Comm. Math. Phys. **151** (1993), 89–153.

[271] B.J.Day, *Note on monoidal monads*, J. Austral. Math. Soc. Ser. A **23** (1977), 292–311. [MR58#22237]

[272] C.De Concini and V.G.Kac, *Representations of quantum groups at roots of* 1, Operator algebras, unitary representations, enveloping algebras, and invariant theory (Paris 1989), 471–506, Birkhäuser, Boston, MA, 1990. [MR92g:17012]

[273] C.De Concini and V.G.Kac, *Representations of quantum groups at roots of* 1*: reduction to the exceptional case*, Infinite analysis (Kyoto 1991), 141–149, World Sci. Publishing, River Edge, NJ, 1992.

[274] C.De Concini, V.G.Kac and C.Procesi, *Quantum coadjoint action*, J. Amer. Math. Soc. **5** (1992), 151–189. [MR93f:17020]

[275] M.De Wilde, *Star-products on symplectic manifolds*, Rend. Circ. Mat. Palermo Suppl. no. 6 (1984), 307–312. [MR86h:58052]

[276] M.De Wilde, *Existence de déformations formelles de l'algèbre de Poisson et de star-produits sur une variété symplectique*, Physique quantique et géométrie (Paris 1986), 195–203, Hermann, Paris, 1988. [MR89m:17030]

[277] M.De Wilde and P.B.A.Lecomte, *Cohomologie 3-différentiable de l'algèbre de Poisson d'une variété symplectique*, Ann. Inst. Fourier **33** (1983), 83–94. [MR85b:58042]

[278] M.De Wilde and P.B.A.Lecomte, *Existence of star-products on exact symplectic manifolds*, Ann. Inst. Fourier **35** (1985), 117–143. [MR87c:58037]

[279] M.De Wilde and P.B.A.Lecomte, *Formal deformations of the Poisson-Lie algebra of a symplectic manifold and star-products. Existence, equivalence, derivations*, Deformation theory of algebras and structures and applications (Il Ciocco 1986), 897–960, Kluwer Acad. Publ., Dordrecht, 1988. [MR90c:58052]

[280] M.De Wilde, P.B.A.Lecomte and S.Gutt, *À propos des deuxième et troisième espaces de cohomologie de l'algèbre de Lie de Poisson d'une variété symplectique*, Ann. Inst. H. Poincaré Sect. A **40** (1984), 77–83. [MR86a:58036]

[281] M.De Wilde, P.B.A.Lecomte and D.Mélotte, *Invariant star-products on symplectic manifolds*, J. Geom. Phys. **2** (1985), 121–129. [MR88a:17027]

[282] P.A.Deift and L.C.Li, *Generalized affine Lie algebras and the solution of a class of flows associated with the QR eigenvalue algorithm*, Comm. Pure Appl. Math. **42** (1989), 963–991. [MR92a:58063]

[283] P.Deligne, *Le groupe fondamentale de la droite projective moins trois points*, Galois groups over **Q** (Berkeley, CA, 1987), M.S.R.I. Publ. **16**, Springer, 1989, 79–297. [MR90m:14016]

[284] P.Deligne, *Catégories tannakiennes*, The Grothendieck Festschrift, Vol. II, 111–195, Birkhäuser, Boston, MA, 1990. [MR92d:14002]

[285] P.Deligne, *Resumé des premièrs exposés de A. Grothendieck*, SGA 7, Lecture Notes Math. **288** (1972), 1–24.

[286] P.Deligne and J.S.Milne, *Tannakian categories*, Lect. Notes Math. **900** (1982), 101–228.

[287] J.Yeomans and B.Derrida, *Bulk and interface scaling properties of the chiral clock model*, J. Phys. A **18** (1985), 2343–2355.

[288] R.Dijkgraaf, V.Pasquier and P.Roche, *Quasi-Hopf algebras, group cohomology and orbifold models*, Nuclear Phys. B Proc. Suppl. **18B** (1990), 60–72. [MR92m:81238]

[289] R.Dijkgraaf and E.Verlinde, *Modular invariance and the fusion algebra*, Nucl. Phys. B Proc. Suppl. **5B** (1988), 87–97.

[290] R.Dijkgraaf and E.Witten, *Topological gauge theories and group cohomology*, Comm. Math. Phys. **129** (1990), 393–429. [MR91g:81133]

[291] R.Dipper and G.James, *Representations of Hecke algebras of general linear groups*, Proc. London Math. Soc. **52** (1986), 20–52.

[292] J.Distler and Z.Qiu, *BRS cohomology and a Feigin-Fuchs representation of Kac-Moody and parafermionic theories*, Nuclear Phys. B **336** (1990), 533–546. [MR91f:17022]

[293] V.K.Dobrev, *Representations of quantum groups*, Symmetries in science, V (Lochau 1990), 93–135, Plenum, New York, 1991.

[294] R.L.Dobrushin, *Gibbsian random fields for lattice systems with pairwise interactions*, Funktsional. Anal. Prilozhen. **2** (1968), no. 4, 31–43. [MR40#3862]

[295] H.D.Doebner, J.D.Hennig and W.Lucke, *Mathematical guide to quantum groups*, Quantum groups (Clausthal 1989), Lect. Notes Phys. **370** (1990), 29–63.

[296] Y.Doi, *Homological coalgebra*, J. Math. Soc. Japan **33** (1981), 31–50. [MR82g:16014]

[297] R.Dolen, D.Horn and C.Schmid, *Prediction of Regge parameters of ρ poles from low-energy πN data*, Phys. Rev. Lett. **19** (1967), 402–407.

[298] J.Donin and D.Gurevich, *Quasi-Hopf algebras and R-matrix structures in line bundles over flag manifolds*, Preprint MPI/92-69, 1992; to appear in Selecta Math. Sovietica.

[299] J.Donin and D.Gurevich, *Poisson brackets associated to R-matrices on symmetric spaces and their quantization*, to appear in Israel J. Math.

[300] J.Donin and D.Gurevich, *Some Poisson structures associated to Drinfeld-Jimbo R-matrices and their quantization*, Preprint.

[301] S.Donkin, *On the Hopf algebra dual of an enveloping algebra*, Math. Proc. Cambridge Philos. Soc. **91** (1982), 215–224. [MR83h:16014]

[302] I.Ya.Dorfman, *Deformations of Hamiltonian structures and integrable systems*, Nonlinear and turbulent processes in physics, Vol. 3 (Kiev 1983), 1313–1318, Harwood Academic Publ., 1984. [MR87e:58092]

[303] I.Ya.Dorfman, *On symplectic and Hamiltonian differential operators*, Nonlinear evolution equations and dynamical systems (Kolymbari 1989), 148–151, Springer, Berlin, 1990. [MR91m:58058]

[304] I.M.Gelfand and I.Ya.Dorfman, *Hamiltonian operators and algebraic structures associated with them*, Funktsional. Anal. Prilozhen. **13** (1979), no. 4, 13–30. [MR81c:58035]

[305] I.M.Gelfand and I.Ya.Dorfman, *Schouten bracket and Hamiltonian operators*, Funktsional. Anal. Prilozhen. **14** (1980), no. 3, 71–74. [MR82e:58039]

[306] I.M.Gelfand and I.Ya.Dorfman, *Hamiltonian operators and infinite-dimensional Lie algebras*, Funktsional. Anal. Prilozhen. **15** (1981), no. 3, 23–40. [MR82j:58045]

[307] I.M.Gelfand and I.Ya.Dorfman, *Integrable equations of KdV-H. Dym type*, Current problems in mathematical physics and numerical mathematics, 102–112, Nauka, Moscow, 1982. [MR84i:58055]

[308] I.M.Gelfand and I.Ya.Dorfman, *Hamiltonian operators and the classical Yang-Baxter equation*, Funktsional. Anal. Prilozhen. **16** (1982), no. 4, 1–9. [MR84e:58030]

[309] V.S.Dotsenko and V.A.Fateev, *Conformal algebra and multipoint correlation functions in 2D statistical modles*, Nucl. Phys. B **240** (1984), 312–348. [MR85i:82061]

[310] P.Doubilet, *A Hopf algebra arising from the lattice of partitions of a set*, J. Algebra **28** (1974), 127–132. [MR49#349]

[311] V.G.Drinfeld, *Hamiltonian structures on Lie groups, Lie bialgebras and the geometric meaning of classical Yang-Baxter equations*, Dokl. Akad. Nauk SSSR **268** (1983), 285–287. [MR84i:58044]

[312] V.G.Drinfeld, *Constant quasiclassical solutions of the Yang-Baxter quantum equation*, Soviet Math. Dokl. **28** (1983), 667–671. [MR85d:58040]

[313] V.G.Drinfeld, *Hopf algebras and the quantum Yang-Baxter equation*, Soviet Math. Dokl. **32** (1985), 254–258. [MR87h:58080]

[314] V.G.Drinfeld, *On quadratic commutation relations in the quasiclassical case*, Mathematical physics, functional analysis, 25–34, Naukova Dumka, Kiev, 1986. [MR89c:58048]

[315] V.G.Drinfeld, *Degenerate affine Hecke algebras and Yangians*, Funct. Anal. Appl. **20** (1986), 58–60. [MR87m:22044]

[316] V.G.Drinfeld, *Quantum groups*, Zap. Nauchn. Sem. LOMI **155** (1986), 18–49. [MR88f:17017]

[317] V.G.Drinfeld, *Quantum groups*, Proceedings of the International Congress of Mathematicians (Berkeley 1986), 798–820, Amer. Math. Soc., Providence, RI, 1987. [MR89f:17017]

[318] V.G.Drinfeld, *A new realization of Yangians and of quantum affine algebras*, Soviet Math. Dokl. **36** (1988), 212–216. [MR88j:17020]

[319] V.G.Drinfeld, *Quasi-Hopf algebras and Knizhnik-Zamolodchikov equations*, Problems of modern quantum field theory (Alushta 1989), 1–13, Springer, Berlin, 1989. [MR92j:17011]

[320] V.G.Drinfeld, *Almost cocommutative Hopf algebras*, Leningrad Math. J. **1** (1990), 321–342. [MR91b:16046]

[321] V.G.Drinfeld, *Quasi-Hopf algebras*, Leningrad Math. J. **1** (1990), 1419–1457. [MR91b:17016]

[322] V.G.Drinfeld, *On quasitriangular quasi-Hopf algebras and on a group that is closely connected with* Gal($\overline{\mathbf{Q}}/\mathbf{Q}$), Leningrad Math. J. **2** (1991), 829–860. [MR92f:16047]

[323] V.G.Drinfeld, *On some unsolved problems in quantum group theory*, Quantum groups (Leningrad 1990), Lect. Notes Math. **1510** (1992), 1–8.

[324] V.G.Drinfeld, *On the structure of quasitriangular quasi-Hopf algebras*, Functional Anal. Appl. **26** (1992), 63–65.

[325] V.G.Drinfeld and V.V.Sokolov, *Lie algebras and equations of Korteweg-de Vries type*, Current Problems in Math. **24**, 81–100, Akad. Nauk SSSR, Vsesoyuz. Inst. Nauchn. i Tekhn. Inform., Moscow, 1984. [MR86h:58071]

[326] M.Dubois-Violette, *On the theory of quantum groups*, Lett. Math. Phys. **19** (1990), 121–126. [MR91a:17014]

[327] B.A.Dubrovin, *Differential-geometric Poisson brackets on a lattice*, Funkts. Anal. Prilozh. **23** (1989), no. 2, 131–133. [MR90k:58063]

[328] G.Duchamp and D.Krob, *Computing with P.B.W. in enveloping algebras*, Algebraic computing in control (Paris 1991), Lect. Notes Control Inform. Sci. **165** (1991), 223–240. [MR93b:17038]

[329] J.P.Dufour, *Linearisation de certaines structures de Poisson*, J. Differential Geom. **32** (1990), 415–428. [MR91m:58139]

[330] B.Durhuus, H.P.Jakobsen and R.Nest, *A construction of topological quantum field theories from 6j-symbols*, Nuclear Phys. B Proc. Suppl. **25A** (1992), 109–121.

[331] C.Duval, J.Elhadad, M.J.Gotay, J.Sniatycki and G.M.Tuynman, *Quantization and bosonic BRST theory*, Ann. Physics **206** (1991), 1–26. [MR92h:58076]

[332] C.Duval, J.Elhadad and G.M.Tuynman, *The BRS method and geometric quantization: some examples*, Comm. Math. Phys. **126** (1990), 535–557. [MR91b:81054]

[333] I.L.Egusquiza, *Quantum mechanics on the quantum sphere*, Preprint DAMTP/92-18, 1992.

[334] A.Ehresmann and C.Ehresmann, *Multiple functors. IV. Monoidal closed structures on* Cat_n, Cahiers Topologie Géom. Différentielle **20** (1979), 59–104. [MR81g:18010]

[335] S.Eilenberg, *Automata, languages, and machines*, Vol. A, Academic Press, 1974.

[336] S.Eilenberg and S.MacLane, *General theory of natural equivalences*, Trans. Amer. Math. Soc. **58** (1945), 231–294. [MR7,109d]

[337] S.Eilenberg and J.C.Moore, *Foundations of relative homological algebra*, Mem. Amer. Math. Soc. **55** (1965). [MR31#2294]

[338] Y.Eisenberg, *Infinite dimensional algebra for all the (super-)field theories in four dimensions*, Phys. Lett. B. **233** (1989), 349–354.

[339] G.G.Emch, *Non-equilibrium quantum statistical mechanics*, Springer, Vienna, 1976. [MR58#25679]

[340] B.Enriquez, *Rational forms for twistings of enveloping algebras of simple Lie algebras*, Lett. Math. Phys. **25** (1992), 111–120.

[341] B.Enriquez, *Integrity, integral closedness and finiteness over their centers of the coordinate algebras of quantum groups at p^{κ}-th roots of unity*, Preprint no. 1020, Centre Math. École Polytechnique, December 1991.

[342] B.Enriquez, *Le centre des algèbres de cordonnées sur les groupes qauntiques aux racines p^{α}-ièmes de l'unité*, Preprint no. 1039, Centre Math. École Polytechnique, April 1992.

[343] M.Siegert and H.U.Everts, *Mean-field theory of the three-state chiral clock model*, Z. Phys. B **60** (1985), 265–281.

[344] M.Siegert and H.U.Everts, *The Bethe approximation to the three-state chiral clock model*, Z. Phys. B **66** (1987), 227–235.

[345] L.D.Faddeev, *Integrable models in* $(1+1)$ *dimensional quantum field theory*, Recent advances in field theory and statistical mechanics (Les Houches 1982), 561–608, North-Holland, Amsterdam, 1984.

[346] L.D.Faddeev, *Quantum groups*, Bol. Soc. Brasil. Mat. **20** (1989), 47–54. [MR92f:17016]

[347] L.Faddeev, *Lectures on quantum inverse scattering method*, New problems, methods and techniques in quantum field theory and statistical mechanics, 7–54, World Sci. Publishing, River Edge, NJ, 1990.

[348] L.D.Faddeev, *A mathematician's view of the development of physics*, Miscellanea mathematica, 119–127, Springer, Berlin, 1991.

[349] N.Yu.Reshetikhin and L.D.Faddeev, *Hamiltonian structures for integrable field theory models*, Teoret. Mat. Fiz. **56** (1983), 323–343. [MR85j:58079]

[350] L.D.Faddeev and N.Yu.Reshetikhin, *Integrability of a quantum model of a principal chiral field*, Proceedings of the VII international conference on the problems of quantum field theory (Alushta 1984), 37–55, Ob'ed. Inst. Yadernykh Issled., Dubna, 1984. [MR88b:81104]

[351] L.Faddeev, N.Reshetikhin and L.Takhtajan, *Quantum groups*, Braid group, knot theory and statistical mechanics, 97–110, World Sci. Publishing, Teaneck, NJ, 1989.

[352] N.Yu.Reshetikhin, L.A.Takhtajan and L.D.Faddeev, *Quantization of Lie groups and Lie algebras*, Leningrad Math. J. **1** (1990), 193–225. [MR90j:17039]

[353] L.D.Faddeev, E.K.Sklyanin and L.A.Takhatjan, *The quantum inverse problem. I*, Teor. Mat. Fiz. **40** (1979), 194–220.

[354] A.A.Slavnov and L.D.Faddeev, *Introduction to the quantum theory of gauge fields*, Frontiers in Phys. **50**, Benjamin, Reading, MA, 1980.

[355] L.A.Takhtajan and L.D.Faddeev, *The quantum method for the inverse problem and the XYZ Heisenberg model*, Uspekhi Mat. Nauk **34** (1979), no. 5, 13–63. [MR81d:82066]

[356] L.D.Faddeev and L.A.Takhtajan, *Poisson structure for the KdV equation*, Lett. Math. Phys. **10** (1985), 183–188. [MR87f:58065]

[357] V.O.Tarasov, L.A.Takhtajan and L.D.Faddeev, *Local Hamiltonians for integrable quantum models on a lattice*, Teoret. Mat. Fiz. **57** (1983), 163–181. [MR85d:81154]

[358] D.B.Fairlie, *Quantum groups or the uses of algebraic computational programs in the suggestion and verification of algebraic conjectures*, Strings '89 (College Station, TX, 1989), 334–340, World Sci. Publishing, River Edge, NJ, 1990.

[359] D.B.Fairlie, *Quantum deformations of* SU(2), J. Phys. A **23** (1990), L183–L187. [MR91f:17011]

[360] D.B.Fairlie, *q-analysis and quantum groups*, Symmetries in science, V (Lochau 1990), 147–157, Plenum, New York, 1991. [MR92k:17020]

[361] D.B.Fairlie, *Polynomial algebras with q-Heisenberg operators*, Quantum groups (Argonne, IL, 1990), 133–142, World Sci. Publishing, Teaneck, NJ, 1991. [MR92b:81076]

[362] D.B.Fairlie, P.Fletcher and C.K.Zachos, *Infinite-dimensional algebras and a trigonometric basis for the classical Lie algebras*, J. Math. Phys. **31** (1990), 1088–1094. [MR91k:17020]

[363] I.M.Gelfand and D.B.Fairlie, *The algebra of Weyl symmetrised polynomials and its quantum extension*, Comm. Math. Phys. **136** (1991), 487–499. [MR92d:17015]

[364] D.B.Fairlie and C.K.Zachos, *Multiparameter associative generalizations of canonical commutation relations and quantized planes*, Phys. Lett. B **256** (1991), 43–49. [MR92e:81092]

[365] V.A.Fateev and A.B.Zamolodchikov, *Self-dual solutions of the star-triangle relations in Z_N-models*, Phys. Lett. **92A** (1982), 37–39. [MR85j:82054b]

[366] Y.Yu and S.M.Fei, *Symplectic structures, BKS kernels and path integrals of BRST bosonic strings*, Z. Phys. C **43** (1989), 633–640. [MR91b:81164]

[367] B.Feigin and E.Frenkel, *Quantization of the Drinfeld-Sokolov reduction*, Phys. Lett. B **246** (1990), 75–81. [MR92g:17029]

[368] B.L.Feigin, V.V.Schechtman and A.N.Varchenko, *On algebraic equations satisfied by correlators in Wess-Zumino-Witten models*, Lett. Math. Phys. **20** (1990), 291–297. [MR92d:81121]

[369] G.Felder, J.Fröhlich and G.Keller, *Braid matrices and structure constants for minimal conformal models*, Comm. Math. Phys. **124** (1989), 647–664. [MR91f:81178]

[370] G.Felder, J.Fröhlich and G.Keller, *On the structure of unitary conformal field theory. II. Representation theoretic approach*, Comm. Math. Phys. **130** (1990), 1–49. [MR91f:81177]

[371] P.Feng and B.Tsygan, *Hochschild and cyclic homology of quantum groups*, Comm. Math. Phys. **140** (1991), 481–521. [MR93e:19006]

[372] A.Fialowski, *An example of formal deformations of Lie algebras*, Deformation theory of algebras and structures and applications (Il Ciocco 1986), 375–401, Kluwer Acad. Publ., Dordrecht, 1988. [MR90c:17027]

[373] D.Fischman, *Schur's double centralizer theorem: A Hopf algebra approach*, Preprint 1991; to appear in J. Algebra.

[374] D.Fischman and S.Montgomery, *A Schur double centralizer theorem for cotriangular Hopf algebras and generalized Lie algebras*, Preprint 1993; to appear in J. Algebra.

[375] M.E.Fisher, *Transformations of Ising models*, Phys. Rev. **113** (1959), 969–981.

[376] M.E.Fisher, *Walks, walls, wetting, and melting*, J. Stat. Phys. **34** (1984), 667–729.

[377] D.A.Huse and M.E.Fisher, *Commensurate melting, domain walls, and dislocations*, Phys. Rev. **B29** (1984), 239–270.

[378] D.A.Huse, A.M.Szpilka and M.E.Fisher, *Melting and wetting transitions in the three-state chiral clock model*, Physica **121A** (1983), 363–398.

[379] J.M.Yeomans and M.E.Fisher, *Many commensurate phases in the chiral Potts or asymmetric clock models*, J. Phys. C **14** (1981), L835–L839.

[380] J.M.Yeomans and M.E.Fisher, *Analysis of multiphase region in the three-state chiral clock model*, Physica **127A** (1984), 1–37.

[381] M.Flato, A.Lichnerowicz and D.Sternheimer, *Deformations of Poisson brackets, Dirac brackets and applications*, J. Math. Phys. **17** (1976), 1794–1762. [MR54#8735]

[382] M.Flato and Z.Lu, *Remarks on quantum groups*, Lett. Math. Phys. **21** (1991), 85–88. [MR92g:81084]

[383] M.Flato and D.Sternheimer, *On a possible origin of quantum groups*, Lett. Math. Phys. **22** (1991), 155–160. [MR92i:17019]

[384] P.Fletcher, *The Moyal bracket*, Quantum groups (Argonne, IL, 1990), 143–157, World Sci. Publishing, Teaneck, NJ, 1991. [MR92k:17036]

[385] M.Fliess, *Du produit de Hurwitz de deux séries formelles*, C.R.A.S. Paris **268** (1969), 535–537.

[386] M.Fliess, *Application des variables non commutatives a divers produits de séries formelles*, Séminaire Delange-Pisot-Poitou: Théorie des nombres (1969/1970), Fasc. 2, Exp. 21, Secretariat Mathématique, Paris, 1970. [MR43#6201]

[387] M.Fliess, *Séries reconnaissables, rationelles et algébriques*, Bull. Sci. Math. **94** (1970), 231–239.

[388] M.Fliess, *Sur divers produits de séries formelles*, Bull. Soc. Math. France **102** (1974), 181–191.

[389] M.Fliess and D.Normand-Cyrot, *Algèbres de Lie nilpotentes, formule de Baker-Campbell-Hausdorff et intégrales itérées de K.T.Chen*, Séminaire de Probabilités XVI, Lect. Notes Math. **920** (1982), 257–267. [MR84f:93052]

[390] A.T.Fomenko, *Algebraic structure of certain classes of completely integrable Hamiltonian systems on Lie algebras*, Geometric theory of functions and topology, 85–126, Inst. Mat. Akad. Nauk Ukrain. SSR, Kiev, 1981. [MR84c:58034]

[391] A.T.Fomenko, *Complete integrability of some classical Hamiltonian systems*, Monogenic functions and mappings, 3–19, Inst. Mat. Akad. Nauk Ukrain. SSR, Kiev, 1982. [MR84k:58115]

[392] A.T.Fomenko, *Algebraic structure of certain integrable Hamiltonian systems*, Topological and geometric methods in mathematical physics, 84–110, Voronezh. Gos. Univ., 1983. [MR84j:58061]

[393] V.V.Trofimov and A.T.Fomenko, *Integrability in the sense of Liouville of Hamiltonian systems on Lie algebras*, Uspekhi Mat. Nauk **39** (1984), no. 2, 3–56. [MR85d:58044]

[394] V.V.Trofimov and A.T.Fomenko, *Geometry of Poisson brackets and methods for integration, in the sense of Liouville, of systems in symmetric spaces*, Encyclopedia of Math. Sci. **29** (Geometry II), Springer, Berlin-New York.

[395] G.E.Uhlenbeck and G.W.Ford, *Lectures in statistical mechanics*, American Mathematical Society, Providence, RI, 1963. [MR27#1241]

[396] A.P.Fordy and A.G.Reyman, *Classical r-matrices and compatible Poisson brackets for coupled KdV systems*, Lett. Math. Phys. **17** (1989), 25–29. [MR90e:58061]

[397] M.Forger and J.Kellendonk, *Classical BRST cohomology and invariant functions on constraint manifolds. I*, Comm. Math. Phys. **143** (1992), 235–251. [MR92k:17035]

[398] M.Forger, J.Laartz and U.Schaper, *Current algebra of classical nonlinear sigma models*, Comm. Math. Phys. **146** (1992), 397–402.

[399] T.F.Fox, *Coalgebras and Cartesian categories*, Comm. Algebra **4** (1976), 665–667. [MR53#8189]

[400] T.F.Fox, *Algebraic deformations and triple cohomology*, Proc. Amer. Math. Soc. **78** (1980), 467–472. [MR81c:16040]

[401] T.F.Fox, *Algebraic deformations and bicohomology*, Canad. Math. Bull. **32** (1989), 182–189. [MR90h:18007]

[402] T.F.Fox, *An introduction to algebraic deformation theory*, J. Pure Appl. Algebra **84** (1993), 17–41.

[403] T.F.Fox, *The construction of cofree coalgebras*, J. Pure Appl. Algebra **84** (1993), 191–198.

[404] E.S.Fradkin and V.Ya.Linetsky, *Infinite-dimensional generalizations of simple Lie algebras*, Modern Phys. Lett. A **5** (1990), 1967–1977. [MR91h:17014]

[405] D.Freed, *Higher algebraic structures and quantization*, Preprint 1992.

[406] E.Frenkel, V.Kac and M.Wakimoto, *Characters and fusion rules for W-algebras via quantized Drinfeld-Sokolov reduction*, Comm. Math. Phys. **147** (1992), 295–328.

[407] I.Frenkel, H.Garland and G.J.Zuckerman, *Semi-infinite cohomology and string theory*, Proc. Nat. Acad. Sci. U.S.A. **83** (1986), 8442–8446. [MR88d:17016]

[408] I.Frenkel, J.Lepowsky and A.Meurman, *Vertex operator algebras and the Monster*, Academic Press, Boston, MA, 1988. [MR90h:17026]

[409] I.Frenkel and N.Reshetikhin, *Quantum affine algebras and holonomic difference equations*, Comm. Math. Phys. **146** (1992), 1–60.

[410] P.J.Freyd and D.N.Yetter, *Braided compact closed categories with applications to low-dimensional topology*, Adv. Math. **77** (1989), 156–182. [MR91c:57019]

[411] P.J.Freyd and D.N.Yetter, *Coherence theorems via knot theory*, J. Pure Appl. Algebra **78** (1992), 49–76. [MR93d:18013]

[412] A.Fröhlich and C.T.C.Wall, *Graded monoidal categories*, Compositio Math. **28** (1974), 229–285. [MR50#2297]

[413] J.Fröhlich, *Statistics of fields, the Yang-Baxter equation, and the theory of knots and links*, New problems, methods and techniques in quantum field theory and statistical mechanics, 93–122, World Sci. Publishing, River Edge, NJ, 1990.

[414] J.Fröhlich and T.Kevler, *Quantum groups, quantum categories and quantum field theory*, Lect. Notes Math. **1542** (1993).

[415] J.Fröhlich and C.King, *The Chern-Simons theory and knot polynomials*, Comm. Math. Phys. **126** (1989), 167–199. [MR91e:57012]

[416] A.Frölicher and A.Nijenhuis, *A theorem on stability of complex structures*, Proc. Nat. Acad. Sci. U.S.A. **43** (1957), 239–241.

[417] C.Fronsdal, *Invariant $*$-product quantization of the one-dimensional Kepler problem*, J. Math. Phys. **20** (1979), 2226–2232. [MR80m:53054]

[418] C.Fronsdal, *Some ideas about quantization*, Rep. Math. Phys. **15** (1979), 111–145. [MR81a:81014]

[419] C.Fronsdal, *Normal ordering and quantum groups*, Lett. Math. Phys. **22** (1991), 225–228. [MR92i:81128]

[420] C.Fronsdal, $\mathrm{gl}_q(n)$ *and quantum monodromy*, Lett. Math. Phys. **24** (1992), 73–78. [MR93f:58098]

[421] G.Gallavotti and H.J.F.Knops, *Block-spins interactions in the Ising model*, Comm. Math. Phys. **36** (1974), 171–184. [MR49#4505]

[422] G.Gallavotti and S.Miracle-Sole, *Statistical mechanics of lattice systems*, Comm. Math. Phys. **5** (1967), 317–323. [MR36#1173]

[423] G.Gallavotti and S.Miracle-Sole, *Correlation functions of a lattice system*, Comm. Math. Phys. **7** (1968), 274–288. [MR37#1144]

[424] F.de Liberto, G.Gallavotti and L.Russo, *Markov processes, Bernoulli schemes, and Ising model*, Comm. Math. Phys. **33** (1973), 259–282. [MR49#11633]

[425] M.Gaudin, *Un système à une dimension de fermions en interaction*, Phys. Lett. **24A** (1967), 55–56.

[426] M.Gaudin, *Algebraic aspects of exact models*, Structural elements in particle physics and statistical mechanics (Freiburg 1981), 43–60, Plenum, New York, 1983.

[427] M.Gaudin, *La fonction d'onde de Bethe*, Masson, Paris, 1983.

[428] M.Gaudin, *Matrices R de dimension infinie*, J. Physique **49** (1988), 1857–1865. [MR90a:82066]

[429] K.Gawedzki, *Conformal field theory*, Astérisque **177-178** (1989), 95–126. [MR91g:81137]

[430] K.Gawedzki, *Geometry of Wess-Zumino-Witten models of conformal field theory*, Nuclear Phys. B Proc. Suppl. **18B** (1990), 78–91. [MR92g:81170]

[431] K.Gawedzki, *Classical origin of quantum group symmetries in Wess-Zumino-Witten conformal field theory*, Comm. Math. Phys. **139** (1991), 201–213. [MR92h:81153]

[432] M.L.Ge and X.F.Liu, *Construction of a new quantum double and new solutions of the Yang-Baxter equation*, Lett. Math. Phys. **24** (1992), 197–203.

[433] M.L.Ge and A.C.T.Wu, *Quantum groups constructed from the nonstandard braid group representations in the Faddeev-Reshetikhin-Takhtajan approach*, J. Phys. A **24** (1991), L725–L732. [MR92i:17020]

[434] G.von Gehlen and V.Rittenberg, *Finite-size effects in the three-state quantum asymmetric clock model*, Nucl. Phys. **B230** (1984), 455–472.

[435] G.von Gehlen and V.Rittenberg, Z_n*-symmetric quantum chains with an infinite set of conserved charges and* Z_n *zero modes*, Nucl. Phys. **B257** (1985), 351–370. [MR87g:82010]

[436] I.M.Gelfand, *Webs, Veronese curves, and bi-Hamiltonian systems*, J. Funct. Anal. **99** (1991), 150–178. [MR93d:58070]

[437] I.M.Gelfand, R.M.Goresky, R.D.MacPherson and V.V.Serganova, *Combinatorial geometries, convex polyhedra, and Schubert cells*, Adv. Math. **63** (1987), 301–316. [MR88f:14045]

[438] I.M.Gelfand, M.M.Kapranov and A.V.Zelevinsky, *Discriminants of polynomials in several variables and triangulations of Newton polytopes*, Leningrad Math. J. **2** (1991), 449–505.

[439] I.M.Gelfand and V.V.Serganova, *Combinatorial geometries and torus strata on homogeneous spaces*, Russian Math. Surveys **42** (1987), 107–134. [MR89g:32049]

[440] I.M.Gelfand and A.V.Zelevinsky, *Models of representations of classical groups and their hidden symmetries*, Funktsional. Anal. Prilozhen. **18** (1984), no. 3, 14–31. [MR86i:22024]

[441] I.M.Gelfand and A.V.Zelevinsky, *Polyhedra in a space of diagrams and the canonical basis in irreducible representations of* gl_3, Funktsional. Anal. Prilozhen. **19** (1985), no. 2, 72–75. [MR87c:17014]

[442] S.Gelfand and D.Kazhdan, *Examples of tensor categories*, Invent. Math. **109** (1992), 595–617.

[443] S.I.Gelfand and Y.I.Manin, *Homological algebra*, Current problems in mathematics. Fundamental directions **38**, 5–240, Akad. Nauk SSSR, Vsesoyuz. Inst. Nauchn. i Tekhn. Inform. Moscow, 1989. [MR92a:18003b]

[444] M.Gerstenhaber, *The cohomology structure of an associative ring*, Ann. Math. **78** (1963), 267–288. [MR28#5102]

[445] M.Gerstenhaber, *A uniform cohomology theory for algebras*, Proc. Nat. Acad. Sci. U.S.A. **51** (1964), 626–629. [MR28#4017]

[446] M.Gerstenhaber, *On the deformation of rings and algebras*, Ann. Math. **79** (1964), 59–103. [MR30#2034]

[447] M.Gerstenhaber, *On the deformation of rings and algebras. II*, Ann. Math. **84** (1966), 1–19. [MR34#7608]

[448] M.Gerstenhaber, *The third cohomology group of a ring and the commutative cohomology theory*, Bull. Amer. Math. Soc. **73** (1967), 950–954. [MR36#240]

[449] M.Gerstenhaber, *On the deformation of rings and algebras. III*, Ann. Math. **88** (1968), 1–34. [MR39#1521]

[450] M.Gerstenhaber, *On the deformation of sheaves of rings*, Global Analysis (Papers in Honor of K. Kodaira), 149–157, Univ. Tokyo Press, Tokyo, 1969. [MR41#1833]

[451] M.Gerstenhaber, *On the deformation of rings and algebras. IV*, Ann. Math. **99** (1974), 257–276. [MR52#10807]

[452] M.Gerstenhaber, A.Giaquinto and S.D.Schack, *Quantum symmetry*, Quantum groups (Leningrad 1990), Lect. Notes Math. **1510** (1992), 9–46.

[453] M.Gerstenhaber, A.Giaquinto and S.D.Schack, *Quantum groups, cohomology and preferred deformations*, Proc. XX Intern. Conf. Diff. Geom. Methods Theor. Phys., 529–538, World Sci. Publ., Singapore, 1992.

[454] M.Gerstenhaber, A.Giaquinto and S.D.Schack, *Construction of quantum groups from Belavin-Drinfeld infinitesimals*, Israel Math. Conf. Proc. **7** (1993), editors Anthony Joseph and Steve Shnider (distributed by AMS).

[455] M.Gerstenhaber and S.D.Schack, *On the deformation of algebra morphisms and diagrams*, Trans. Amer. Math. Soc. **279** (1983), 1–50. [MR85d:16021]

[456] M.Gerstenhaber and S.D.Schack, *Relative Hochschild cohomology, rigid algebras, and the Bockstein*, J. Pure Appl. Algebra **43** (1986), 53–74. [MR88a:16045]

[457] M.Gerstenhaber and S.D.Schack, *A Hodge-type decomposition for commutative algebra cohomology*, J. Pure Appl. Algebra **48** (1987), 229–247. [MR88k:13011]

[458] M.Gerstenhaber and S.D.Schack, *Sometimes H^1 is H^2 and discrete groups deform*, Geometry of group representations (Boulder, CO, 1987), Contemp. Math. **74** (1988), 149–168. [MR89g:16034]

[459] M.Gerstenhaber and S.D.Schack, *Algebraic cohomology and deformation theory*, Deformation theory of algebras and structures and applications (Il Ciocco 1986), 11–264, Kluwer Acad. Publ., Dordrecht, 1988. [MR90c:16016]

[460] M.Gerstenhaber and S.D.Schack, *Triangular algebras*, Deformation theory of algebras and structures and applications (Il Ciocco 1986), 447–498, Kluwer Acad. Publ., Dordrecht, 1988. [MR90c:16017]

[461] M.Gerstenhaber and S.D.Schack, *The cohomology of presheaves of algebras. I. Presheaves over a partially ordered set*, Trans. Amer. Math. Soc. **310** (1988), 135–165. [MR89k:16052]

[462] M.Gerstenhaber and S.D.Schack, *Bialgebra cohomology, deformations, and quantum groups*, Proc. Nat. Acad. Sci. U.S.A. **87** (1990), 478–481. [MR90j:16062]

[463] M.Gerstenhaber and S.D.Schack, *The shuffle bialgebra and the cohomology of commutative algebras*, J. Pure Appl. Algebra **70** (1991), 263–272. [MR92e:13008]

[464] M.Gerstenhaber and S.D.Schack, *Algebras, bialgebras, quantum groups, and algebraic deformations*, Deformation theory and quantum groups with applications to mathematical physics (Amherst, MA, 1990), Contemp. Math. **134** (1992), 51–92.

[465] M.Gerstenhaber and S.D.Schack, *Quantum groups as deformations of Hopf algebras*, Preprint.

[466] A.Giaquinto, *Quantization of tensor representations and deformation of matrix bialgebras*, J. Pure Appl. Algebra **79** (1992), 169–190. [MR93b:16066]

[467] V.A.Ginzburg, *Enveloping algebras and deformations*, Uspekhi Mat. Nauk **34** (1979), no. 2, 191–192. [MR80i:17012]

[468] V.A.Ginzburg, *Lagrangian construction of the enveloping algebra* $U(\mathrm{sl}_n)$, C.R.A.S. Paris **312** (1991), 907–912. [MR92c:17017]

[469] V.Ginzburg and S.Kumar, *Cohomology of quantum groups at roots of unity*, Duke Math. J. **69** (1993), 179–198.

[470] V.Ginzburg and A.Weinstein, *Lie-Poisson structure on some Poisson-Lie groups*, J. Amer. Math. Soc. **5** (1992), 445–453. [MR92h:17022]

[471] W.M.Goldman and J.J.Millson, *Differential graded Lie algebras and singularities of level sets of momentum mappings*, Comm. Math. Phys. **131** (1990), 495–515. [MR91i:58044]

[472] H.Goldschmidt, *Duality theorems in deformation theory*, Trans. Amer. Math. Soc. **292** (1985), 1–49. [MR87d:58162]

[473] F.M.Goodman, P.de la Harpe and V.F.R.Jones, *Coxeter graphs and towers of algebras*, M.S.R.I. Publ. **14**, Springer, 1989.

[474] J.Grabowski, *The Lie structure of C^* and Poisson algebras*, Studia Math. **81** (1985), 259–270. [MR87c:17021]

[475] J.Grabowski, *Abstract Jacobi and Poisson structures. Quantization and star-products*, J. Geom. Phys. **9** (1992), 45–73. [MR93e:58073]

[476] J.Grabowski, *Isomorphisms of the Jacobi and Poisson brackets*, Preprint ESI 5, The Erwin Schrödinger Intern. Inst. Math. Phys., Vienna, March 1993.

[477] J.Grabowski, *A Poisson-Lie group structure on the diffeomorphism group of a circle*, Preprint ESI 26, The Erwin Schrödinger Intern. Inst. Math. Phys., Vienna, June 1993.

[478] J.Grabowski, G.Landi, G.Marmo and G.Vilasi, *Generalized reduction procedure: symplectic and Poisson formalism*, Preprint ESI 28, The Erwin Schrödinger Intern. Inst. Math. Phys., Vienna, June 1993.

[479] J.Grabowski, G.Marmo and A.M.Perelomov, *Poisson structures: towards a classification*, Preprint ESI 17, The Erwin Schrödinger Intern. Inst. Math. Phys., Vienna, April 1993.

[480] G.Griffing, *The dual Friedrichs' theorem*, J. Algebra **132** (1990), 446–460. [MR92h:17002]

[481] N.A.Gromov, *Contractions and analytic continuations of the irreducible representations of the quantum algebra* $\mathrm{su}_q(2)$, Group theoretical methods in physics (Moscow 1990), Lect. Notes Phys. **382** (1991), 225–228.

[482] N.A.Gromov and V.I.Manko, *Contractions of the irreducible representations of the quantum algebras* $\mathrm{su}_q(2)$ *and* $\mathrm{so}_q(3)$, J. Math. Phys. **33** (1992), 1374–1378. [MR93b:17043]

[483] R.Grossman and D.E.Radford, *A simple construction of bialgebra deformations*, Deformation theory and quantum groups with applications to mathematical physics (Amherst, MA, 1990), Contemp. Math. **134** (1992), 115–117.

[484] R.Grossman and D.E.Radford, *Bialgebra deformations of certain universal enveloping algebras*, to appear in J. Pure Appl. Algebra.

[485] A.Grothendieck, *Esquisse d'un programme*, Mimeographed notes, Univ. Montpellier, 1982.

[486] C.Gruber, A.Hintermann and D.Merlini, *Group analysis of classical lattice systems*, Lect. Notes Phys. **60** (1977). [MR58#25661]

[487] C.H.Gu and C.N.Yang, *A one-dimensional N fermion problem with factorized S-matrix*, Comm. Math. Phys. **122** (1989), 105–116. [MR91a:82018]

[488] F.Guedira and A.Lichnerowicz, *Géométrie des algèbres de Lie locales de Kirillov*, J. Math. Pures Appl. **63** (1984), 407–484.

[489] V.K.A.M.Gugenheim, *Cohomology theory in the category of Hopf algebras*, Colloque de Toplogie (Brussels 1964), 137–148, Librairie Universitaire, Louvain, 1966. [MR37#2836]

[490] V.K.A.M.Gugenheim, *On a perturbation theory for the homology of the loop-space*, J. Pure Appl. Algebra **25** (1982), 197–205. [MR83j:55021]

[491] V.K.A.M.Gugenheim, L.A.Lambe and J.D.Stasheff, *Perturbation theory in differential homological algebra. II*, Illinois J. Math. **35** (1991), 357–373. [MR93e:55018]

[492] V.Guillemin and S.Sternberg, *An algebraic model of transitive differential geometry*, Bull. Amer. Math. Soc. **70** (1964), 16–47. [MR30#533]

[493] V.Guillemin and S.Sternberg, *Deformation theory of pseudogroup structures*, Mem. Amer. Math. Soc. **64** (1966). [MR35#2302]

[494] V.Guillemin and S.Sternberg, *Geometric asymptotics*, American Mathematical Society, Providence, RI, 1977. [MR58#24404]

[495] V.Guillemin and S.Sternberg, *Some problems in integral geometry and some related problems in microlocal analysis*, Amer. J. Math. **101** (1979), 915–955.

[496] V.Guillemin and S.Sternberg, *The moment map and collective motion*, Ann. Physics **127** (1980), 220–253. [MR81g:58011]

[497] V.Guillemin and S.Sternberg, *On the method of Symes for integrating systems of the Toda type*, Lett. Math. Phys. **7** (1983), 113–115. [MR85k:58038]

[498] V.Guillemin and S.Sternberg, *A normal form for the moment map*, Differential geometric methods in mathematical physics (Jerusalem 1982), 161–175, Reidel, Dordrecht-Boston, 1984. [MR86d:58033]

[499] V.Guillemin and S.Sternberg, *Birational equivalence in the symplectic category*, Invent. Math. **97** (1989), 485–522. [MR90f:58060]

[500] V.Guillemin and S.Sternberg, *Symplectic techniques in physics*, Cambridge University Press, 1990. [MR91d:58073]

[501] V.Guillemin and S.Sternberg, *Variations on a theme by Kepler*, American Mathematical Society, Providence, RI, 1990.

[502] D.I.Gurevich, *The Yang-Baxter equation and the generalization of formal Lie theory*, Sov. Math. Dokl. **33** (1986), 758–762. [MR88b:58066]

[503] D.I.Gurevich, *Trace and determinant in algebras associated with the Yang-Baxter equation*, Funktsional. Anal. Prilozhen. **21** (1987), no. 3, 79–80. [MR89c:58050]

[504] D.I.Gurevich, *Hecke symmetries and quantum determinants*, Soviet Math. Dokl. **38** (1989), 555–559. [MR90c:17018]

[505] D.I.Gurevich, *Poisson brackets associated with the classical Yang-Baxter equation*, Functional Anal. Appl. **23** (1989), 57–59. [MR90f:58061]

[506] D.Gurevich, *Equation de Yang-Baxter et quantification des cocycles*, C.R.A.S. Paris **310** (1990), 845–848. [MR92g:17028]

[507] D.Gurevich, *Equation de Yang-Baxter et S-traces*, C.R.A.S. Paris **313** (1991), 809–812.

[508] D.Gurevich, *Cohomologie de "grand crochet" et filtration "non-holonome"*, C.R.A.S. Paris **313** (1991), 855–858. [MR93b:58171]

[509] D.I.Gurevich, *Algebraic aspects of the quantum Yang-Baxter equation*, Leningrad Math. J. **2** (1991), 801–828. [MR93e:17018]

[510] D.Gurevich, *Quantization of Poisson pairs: the R-matrix approach*, J. Geom. Phys. **9** (1992), 25–44. [MR93f:58088]

[511] D.Gurevich and S.Majid, *Braided groups of Hopf algebras obtained by twisting*, Preprint DAMTP/91-49, 1991; to appear in Pacific J. Math.

[512] D.Gurevich and V.Rubtsov, *Yang-Baxter equation and deformation of associative and Lie algebras*, Quantum groups (Leningrad 1990), Lect. Notes Math. **1510** (1992), 47–55.

[513] E.Gutkin, *Bethe ansatz and the generalized Yang-Baxter equations*, Ann. Physics **176** (1987), 22–48. [MR88g:58081]

[514] E.Gutkin, *Equivalence principle in statistical mechanics*, Physica **144A** (1987), 430–444.

[515] S.Gutt, *Second et troisième éspaces de cohomologie différentiable de l'algèbre de Lie de Poisson d'une variété symplectique*, Ann. Inst. H. Poincaré Sect. A **33** (1980), 1–31. [MR82e:58044]

[516] S.Gutt, *Deformation theory and its applications to mechanics and to group representations*, Atti Accad. Sci. Torino Cl. Sci. Fis. Mat. Natur. **117** (1983), suppl. no. 1, 149–162. [MR86d:58039]

[517] S.Gutt, *Some aspects of deformation theory and quantization*, Quantum theories and geometry (Les Treilles 1987), 77–102, Kluwer Acad. Publ., Dordrecht, 1988. [MR90b:58096]

[518] R.M.Hain, *Iterated integrals and homotopy periods*, Mem. Amer. Math. Soc. **47** (1984), no. 291. [MR85d:55015]

[519] E.Halpern, *A note on divided powers in a Hopf algebra*, Proc. Amer. Math. Soc. **11** (1960), 547–556. [MR23#A2448]

[520] J.M.Hammersley, *Self-avoiding walks*, Physica A **177** (1991), 51–57. [MR92i:82054]

[521] D.Hansel and J.M.Maillard, *Symmetries of models with genus* > 1, Phys. Lett. **133A** (1988), 11–14.

[522] H.Harari, *Duality diagrams*, Phys. Rev. Lett. **22** (1969), 562–565.

[523] T.Hayashi, *q-analogues of Clifford and Weyl algebras—spinor and oscillator representations of quantum enveloping algebras*, Comm. Math. Phys. **127** (1990), 129–144. [MR91a:17015]

[524] T.Hayashi, *Quantum deformation of classical groups*, Publ. R.I.M.S. **28** (1992), 57–81. [MR92m:17025]

[525] T.Hayashi, *Quantum groups and quantum determinants*, J. Algebra **152** (1992), 146–165.

[526] M.Hazewinkel, *Formal groups and applications*, Academic Press, New York-London, 1978. [MR82a:14020]

[527] M.Hazewinkel, *The philosophy of deformations: introductory remarks and a guide to this volume*, Deformation theory of algebras and structures and applications (Il Ciocco 1986), 1–7, Kluwer Acad. Publ., Dordrecht, 1988. [MR90c:58198]

[528] M.Hazewinkel, *Introductory recommendations for the study of Hopf algebras in mathematics and physics*, CWI Quarterly **4** (1991), 3–26. [MR92h:16035]

[529] O.J.Heilmann, D.J.Kleitman, E.H.Lieb and S.Sherman, *Some positive definite functions on sets and their application to the Ising model*, Discrete Math. **1** (1971), 19–27. [MR46#10342]

[530] A.Heller, *Adjoint functors and bar constructions*, Adv. Math. **12** (1974), 8–31. [MR48#12585]

[531] J.Helmstetter, *Algèbres de Weyl et $*$-produit*, Université des Sciences et Techniques du Languedoc, U.E.R. de Mathématiques, Montpellier, 1985. [MR87m:58057]

[532] M.Henneaux, *Duality theorems in BRST cohomology*, Ann. Physics **194** (1989), 281–302. [MR90k:17042]

[533] M.Henneaux and C.Teitelboim, *BRST cohomology in classical mechanics*, Comm. Math. Phys. **115** (1988), 213–230. [MR89h:17022]

[534] R.Hermann, *Infinite dimensional Lie algebras and current algebra*, Lect. Notes Phys. **6** (1970), 312–337. [MR42#6878]

[535] R.Hermann, *Inverse scattering technique of soliton theory, Lie algebras, the quantum mechanical Poisson-Moyal bracket, and the rotating rigid body*, Phys. Rev. Lett. **37** (1976), 1591–1593. [MR54#9451]

[536] R.Hermann, *Geometric and Lie-theoretic principles in pure and applied deformation theory*, Deformation theory of algebras and structures and applications (Il Ciocco 1986), 701–796, Kluwer Acad. Publ., Dordrecht, 1988. [MR90a:58005]

[537] R.Hermann, *Topics in physical geometry*, Math Sci Press, Brookline, MA, 1988. [MR90h:58001]

[538] P.Hilton and J.Pedersen, *Catalan numbers, their generalization, and their uses*, Math. Intelligencer **13** (1991), no. 2, 64–75. [MR93d:05006]

[539] N.J.Hitchin, *Flat connections and geometric quantization*, Comm. Math. Phys. **131** (1990), 347–380. [MR91g:32022]

[540] G.Hochschild, *On the cohomology groups of an associative algebra*, Ann. Math. **46** (1945), 58–67.

[541] G.Hochschild, *On the cohomology theory for associative algebras*, Ann. Math. **47** (1946), 568–579. [MR8,64c]

[542] G.Hochschild, *Cohomology and representations of associative algebras*, Duke Math. J. **14** (1947), 921–948. [MR9,267b]

[543] G.Hochschild, *Relative homological algebra*, Trans. Amer. Math. Soc. **82** (1956), 246–269. [MR18,278a]

[544] G.Hochschild, *Algebraic groups and Hopf algebras*, Illinois J. Math. **14** (1970), 52–65. [MR41#1742]

[545] G.Hochschild, B.Kostant and A.Rosenberg, *Differential forms on regular affine algebras*, Trans. Amer. Math. Soc. **102** (1962) 383–408. [MR26#167]

[546] G.Hochschild and J.-P.Serre, *Cohomology of group extensions*, Trans. Amer. Math. Soc. **74** (1953), 110–134. [MR14,619b]

[547] G.Hochschild and J.-P.Serre, *Cohomology of Lie algebras*, Ann. Math. **57** (1953), 591–603. [MR14,943c]

[548] P.N.Hoffman, *Hopf algebras from branching rules*, Canad. J. Math. **35** (1983), 177–192. [MR84b:16011]

[549] P.N.Hoffman and J.F.Humphreys, *Hopf algebras and projective representations of $G \wr S_n$ and $G \wr A_n$*, Canad. J. Math. **38** (1986), 1380–1458. [MR88h:20014]

[550] P.N.Hoffman and J.F.Humphreys, *Primitives in the Hopf algebra of projective S_n-representations*, J. Pure Appl. Algebra **47** (1987), 155–164. [MR88h:20015]

[551] D.D.Holm and K.B.Wolf, *Lie-Poisson description of Hamiltonian ray optics*, Phys. D **51** (1991), 189–199. [MR92j:58035]

[552] H.Hopf, *Ein topologischer Beitrag zur reellen Algebra*, Comment. Math. Helv. **13** (1941), 219–239. [MR3,61c]

[553] H.Hopf, *Einige Anwendungen der Topologie auf die Algebra*, Univ. Politecnico Torino. Rend. Sem. Mat. **11** (1952), 75–91. [MR14,720g]

[554] J.M.Hourlik and S.J.Knak Jensen, *Phase diagram of the three-state chiral clock model studied by Monte-Carlo renormalization-group calculations*, Phys. Rev. **B34** (1986), 325–329.

[555] S.F.Howes, *Commensurate-incommensurate transitions and the Lifshitz point in the quantum asymmetric clock model*, Phys. Rev. **B27** (1983), 1762–1768.

[556] S.Howes, L.P.Kadanoff and M.den Nijs, *Quantum model for commensurate-incommensurate transitions*, Nucl. Phys. **B215** (1983), 169–208.

[557] J.R.Hubbuck, *On homotopy commutative H-spaces*, Topology **8** (1969), 119–126. [MR38#6592]

[558] J.Huebschmann, *Poisson cohomology and quantization*, J. Reine Angew. Math. **408** (1990), 57–113.

[559] K.Huitu and D.Nemeschansky, *Supersymmetric Gelfand-Dickey algebra*, Modern Phys. Lett. A **6** (1991), 3179–3190. [MR92i:81098]

[560] A.Hurwitz, *Über die Composition der quadratischen Formen von beliebig vielen Variablen*, Götting. Nachrichten (1898), 309–316; Mathematische Werke, Band II, 565–571, Birkhäuser, Basel, 1963.

[561] A.Hurwitz, *Sur un théorème de M. Hadamard*, C.R.A.S. Paris **128** (1899), 350–353; Mathematische Werke, Band I, 481–484, Birkhäuser, Basel, 1962.

[562] D.A.Huse, *Simple three-state model with infinitely many phases*, Phys. Rev. **B24** (1981), 5180–5194.

[563] D.Husemoller, *Elliptic curves*, Springer, New York-Berlin, 1987. [MR88h:11039]

[564] G.B.Im and G.M.Kelly, *A universal property of the convolution monoidal structure*, J. Pure Appl. Algebra **43** (1986), 75–88. [MR87m:18011]

[565] M.S.Iotov and I.T.Todorov, *On the universal R-matrix for $U_q(sl_{r+1})$*, Lett. Math. Phys. **23** (1991), 241–250. [MR93a:17014]

[566] E.Ising, Zeitschrift für Physik **31** (1925), 253–258.

[567] H.Itoyama and A.Sevrin, *Braiding matrices of conformal blocks and coset models*, Internat. J. Modern Phys. A **5** (1990), 211–222. [MR90j:81208]

[568] A.R.Its, A.G.Izergin, V.E.Korepin and N.A.Slavnov, *Differential equations for quantum correlation functions*, Internat. J. Modern Phys. B **4** (1990), 1003–1037. [MR91k:82009]

[569] C.Itzykson, *From the harmonic oscillator to the A-D-E classification of conformal models*, Integrable systems in quantum field theory and statistical mechanics, 287–346, Academic Press, Boston, MA, 1989. [MR92g:17046]

[570] H.Saleur and C.Itzykson, *Two-dimensional field theories close to criticality*, J. Statist. Phys. **48** (1987), 449–475. [MR88m:82016]

[571] A.G.Izergin and V.E.Korepin, *The inverse scattering approach to the quantum Shabat-Mikhailov model*, Comm. Math. Phys. **79** (1981), 303–316. [MR83c:81092]

[572] A.G.Izergin, V.E.Korepin and N.Yu.Reshetikhin, *Correlation functions in a one-dimensional Bose gas*, J. Phys. A **20** (1987), 4799–4822. [MR89b:82021]

[573] A.G.Izergin, V.E.Korepin and N.Yu.Reshetikhin, *Conformal dimensions in Bethe ansatz solvable models*, J. Phys. A **22** (1989), 2615–2620. [MR90d:81113]

[574] M.Jambu, *Algèbre d'holonomie de Lie et certaines fibrations topologiques*, C.R.A.S. Paris **306** (1988), 479–482. [MR89j:32018]

[575] G.James and A.Kerber, *The representation theory of the symmetric group*, Addison-Wesley, Reading, Mass., 1981. [MR83k:20003]

[576] M.Jimbo, *A q-difference analogue of $U(g)$ and the Yang-Baxter equation*, Lett. Math. Phys. **10** (1985), 63–69. [MR86k:17008]

[577] M.Jimbo, *Quantum R-matrix related to the generalized Toda system: an algebraic approach*, Field theory, quantum gravity and strings (Meudon/Paris, 1985), Lect. Notes Phys. **246** (1986), 335–361. [MR87j:17013]

[578] M.Jimbo, *A q-analogue of $U(gl(N+1))$ Hecke algebra, and the Yang-Baxter equation*, Lett. Math. Phys. **11** (1986), 247–252. [MR87k:17011]

[579] M.Jimbo, *Quantum R-matrix for the generalized Toda system*, Comm. Math. Phys. **102** (1986), 537–547. [MR87h:58086]

[580] M.Jimbo, *The $A_n^{(1)}$ face models*, Comm. Math. Phys. **119** (1988), 543–565. [MR90h:17044]

[581] M.Jimbo, *Introduction to the Yang-Baxter equation*, Internat. J. Modern Phys. A **4** (1989), 3759–3777. [MR90j:82001]

[582] M.Jimbo, K.Miki, T.Miwa and A.Nakayashiki, *Correlation functions of the XXZ model for $\Delta < -1$*, Phys. Lett. A **168** (1992), 256–263.

[583] M.Jimbo, K.C.Misra, T.Miwa and M.Okado, *Combinatorics of representations of $U_q(\widehat{sl}(n))$ at $q = 0$*, Comm. Math. Phys. **136** (1991), 543–566. [MR93a:17015]

[584] M.Jimbo and T.Miwa, *Classification of solutions to the star-triangle relation for a class of 3- and 4-state IRF models*, Nuclear Phys. B **257** (1985), 1–18. [MR86j:82059]

[585] M.Jimbo, T.Miwa and M.Okado, *Solvable lattice models related to the vector representation of classical simple Lie algebras*, Comm. Math. Phys. **116** (1988), 507–525. [MR89j:17018]

[586] M.Johnson, *The combinatorics of n-categorical pasting*, J. Pure Appl. Alg. **62** (1989), 211–225.

[587] D.W.Jonah, *Cohomology of coalgebras*, Mem. Amer. Math. Soc. **82** (1968).

[588] V.F.R.Jones, *A polynomial invariant for knots via von Neumann algebras*, Bull. Amer. Math. Soc. **12** (1985), 103–111. [MR86e:57006]

[589] V.F.R.Jones, *On knot invariants related to some statistical mechanical models*, Pacific J. Math. **137** (1989), 311–334. [MR89m:57005]

[590] V.F.R.Jones, *Knot theory and statistical mechanics*, Scientific American **263** (1990), no. 5, 98–103. [MR91k:57007]

[591] V.F.R.Jones, *Baxterization*, Internat. J. Modern Phys. B **4** (1990), 701–713. [MR91m:82038]

[592] V.F.R.Jones, *Knots, braids, statistical mechanics and von Neumann algebras*, New Zealand J. Math. **21** (1992), 1–16. [MR93e:57009]

[593] S.A.Joni and G.C.Rota, *Coalgebras and bialgebras in combinatorics*, Umbral calculus and Hopf algebras (Norman, Okla. 1978), Contemp. Math. **6** (1982), 1–47. [MR83b:05003]

[594] A.Joseph, *Some ring theoretic techniques and open problems in enveloping algebras*, Preprint.

[595] A.Joseph and G.Letzter, *Local finiteness of the adjoint action for quantized enveloping algebras*, J. Algebra **153** (1992), 289–318.

[596] A.Joyal and R.Street, *Braided monoidal categories*, Macquaire Math. Reports #850067 and #860081, December 1985 and November 1986.

[597] A.Joyal and R.Street, *An introduction to Tannaka duality and quantum groups*, Category theory (Como 1990), Lect. Notes Math. **1488** (1991), 413–492. [MR93f:18015]

[598] A.Joyal and R.Street, *The geometry of tensor calculus. I*, Adv. Math. (1991), 55–112.

[599] A.Joyal and R.Street, *Tortile Yang-Baxter operators in tensor categories*, J. Pure Appl. Algebra **71** (1991), 43–51. [MR92e:18006]

[600] A.Joyal and R.Street, *Braided tensor categories*, to appear in Adv. Math.

[601] B.Jurco, *On coherent states for the simplest quantum groups*, Lett. Math. Phys. **21** (1991), 51–58. [MR92b:81079]

[602] I.I.Kachurik and A.U.Klimyk, *On Racah coefficients of the quantum algebra* $U_q(\mathrm{su}_2)$, J. Phys. A **23** (1990), 2717–2728. [MR92a:33011]

[603] L.P.Kadanoff and F.J.Wegner, *Some critical properties of the eight-vertex model*, Phys. Rev. **B4** (1971), 3989–3993.

[604] J.B.Kammerer, *Analysis of the Moyal product in a flat space*, J. Math. Phys. **27** (1986), 529–535. [MR87d:58065]

[605] R.M.Kane, *The homology of Hopf spaces*, North-Holland, Amsterdam-New York, 1988. [MR90f:55018]

[606] A.Tsuchiya and Y.Kanie, *Vertex operators in conformal field theory on* $\mathbf{P}^1$ *and monodromy representations of braid group*, Adv. Stud. Pure Math. **12** (1988), 297–372.

[607] I.Kaplansky, *Bialgebras*, Department of Mathematics, University of Chicago, 1975. [MR55#8087]

[608] M.M.Kapranov, *The permutoassociahedron, MacLane's coherence theorem and asymptotic zones for the KZ equation*, J. Pure Appl. Alg. **85** (1993), 119–142.

[609] M.M.Kapranov and V.A.Voevodsky, *Braided monomial 2-categories and Zamolodchikov tetrahedra equations*, Mathematics Report 1991, Harvard and Northwestern Univ., to appear in Proc. Symp. Pure Math.

[610] M.V.Karasev, *Flat Poisson manifolds and finite-dimensional pseudogroups*, Mat. Zametki **45** (1989), no. 3, 53–65.

[611] M.V.Karasev and V.P.Maslov, *Nonlinear Poisson brackets. Geometry and quantization*, AMS, Providence, RI, 1993. [MR93a:58062]

[612] Y.M.Vorobev and M.V.Karasev, *Corrections to the classical dynamics and to the quantization conditions arising in the deformation of Poisson brackets*, Soviet Math. Dokl. **36** (1988), 594–598. [MR89c:58046]

[613] Y.M.Vorobev and M.V.Karasev, *Poisson manifolds and the Schouten bracket*, Functional Anal. Appl. **22** (1988), 1–9. [MR89k:58011]

[614] Y.M.Vorobev and M.V.Karasev, *Deformation and cohomologies of Poisson brackets*, Global analysis – studies and applications IV, Lect. Notes Math. **1453** (1990), 271–289. [MR91g:58098]

[615] S.Kasangian and F.Rossi, *Some remarks on symmetry for a monoidal category*, Bull. Austral. Math. Soc. **23** (1981), 209–214. [MR82g:18007]

[616] M.Kashiwara, *On crystal bases of the* Q*-analogue of universal enveloping algebras*, Duke Math. J. **63** (1991), 465–516. [MR93b:17045]

[617] M.Kashiwara, *Global crystal bases of quantum groups*, Duke Math. J. **69** (1993), 455–485.

[618] M.Kashiwara and T.Miwa, *A class of elliptic solutions to the star-triangle relation*, Nuclear Phys. B **275** (1986), 121–134. [MR87j:82093]

[619] P.W.Kasteleyn, *Exactly solvable lattice models*, Fundamental problems in statistical mechanics, Vol. 3, 103–155, North-Holland, Amsterdam, 1975.

[620] D.Kastler and R.Stora, *Lie-Cartan pairs*, J. Geom. Phys. **2** (1985), 1–31. [MR87k:58005]

[621] L.H.Kauffman, *On knots*, Princeton University Press, 1987. [MR89c:57005]

[622] L.H.Kauffman, *State models and the Jones polynomial*, Topology **26** (1987), 395–407. [MR88f:57006]

[623] L.H.Kauffman, *Statistical mechanics and the Jones polynomial*, Braids (Santa Cruz, CA, 1986), Contemp. Math. **78** (1988), 263–297. [MR89j:57002]

[624] L.H.Kauffman, *New invariants in the theory of knots*, Amer. Math. Monthly **95** (1988), 195–242. [MR89d:57005]

[625] L.H.Kauffman, *New invariants in the theory of knots*, Astérisque **163-164** (1988), 137–219. [MR90e:57009]

[626] L.H.Kauffman, *Statistical mechanics and the Alexander polynomial*, Algebraic topology (Evanston, IL, 1988), Contemp. Math. **96** (1989), 221–231. [MR90i:57003]

[627] L.H.Kauffman, *Knots, abstract tensors and the Yang-Baxter equation*, Knots, topology and quantum field theories (Florence 1989), 179–334, World Sci. Publishing, River Edge, NJ, 1989.

[628] L.H.Kauffman, *Polynomial invariants in knot theory*, Braid group, knot theory and statistical mechanics, 27–58, World Sci. Publishing, Teaneck, NJ, 1989.

[629] L.H.Kauffman, *Statistical mechanics and the Jones polynomial*, New problems, methods and techniques in quantum field theory and statistical mechanics, 175–222, World Sci. Publishing, River Edge, NJ, 1990.

[630] L.H.Kauffman, *Spin networks and knot polynomials*, Internat. J. Modern Phys. A **5** (1990), 93–115. [MR90j:57005]

[631] L.H.Kauffman, *An integral heuristic*, Internat. J. Modern Phys. A **5** (1990), 1363–1367. [MR92a:57006]

[632] L.H.Kauffman, *State models for link polynomials*, Enseign. Math. **36** (1990), 1–37. [MR91k:57008]

[633] L.H.Kauffman, *From knots to quantum groups and back*, Quantum groups (Argonne, IL, 1990), 1–32, World Sci. Publishing, Teaneck, NJ, 1991. [MR92b:57014]

[634] L.H.Kauffman, *Knots and physics*, World Scientific Publishing, River Edge, NJ, 1991. [MR93b:57010]

[635] L.H.Kauffman and S.Lins, *Computing Turaev-Viro invariants for 3-manifolds*, Manuscripta Math. **72** (1991), 81–94.

[636] L.H.Kauffman and D.E.Radford, *A necessary and sufficient condition for a finite-dimensional Drinfeld double to be a ribbon Hopf algebra*, J. Algebra **159** (1993), 98–114.

[637] L.H.Kauffman and P.Vogel, *Link polynomials and a graphical calculus*, J. Knot Theory Ramifications **1** (1992), 59–104.

[638] D.Kazhdan and G.Lusztig, *Representations of Coxeter groups and Hecke algebras*, Inv. Math. **53** (1979), 165–184.

[639] D.Kazhdan and G.Lusztig, *Affine Lie algebras and quantum groups*, Internat. Math. Res. Notices (inside Duke Math. J.) 1991, no. 2, 21–29. [MR92g:17015]

[640] D.Kazhdan and G.Lusztig, *Tensor structures arising from affine Lie algebras. I, II*, to appear in J. Amer. Math. Soc.

[641] D.Kazhdan and H.Wenzl, *Reconstructing monoidal categories*, Preprint.

[642] G.Keller, *Fusion rules of* $U_q(\mathrm{sl}(2, \mathbf{C}))$, $q^m = 1$, Lett. Math. Phys. **21** (1991), 273–286. [MR93a:17016]

[643] G.M.Kelly, *On MacLane's conditions for coherence of natural associativities, commutativities, etc* J. Algebra **1** (1964), 397–402. [MR32#132]

[644] G.M.Kelly, *Tensor products in categories*, J. Algebra **2** (1965), 15–37. [MR31#1288]

[645] G.M.Kelly, *Many-variable functorial calculus. I*, Coherence in categories, Lect. Notes Math. **281** (1972), 66–105. [MR49#5126]

[646] G.M.Kelly, *An abstract approach to coherence*, Coherence in categories, Lect. Notes Math. **281** (1972), 106–147. [MR49#5127]

[647] G.M.Kelly and M.L.Laplaza, *Coherence for compact closed categories*, J. Pure Appl. Alg. **19** (1980), 193–213. [MR81m:18008]

[648] G.M.Kelly and S.MacLane, *Closed coherence for a natural transformation*, Coherence in categories, Lect. Notes Math. **281** (1972), 1–28. [MR51#10437]

[649] G.M.Kelly and R.Street, *Review of the elements of* 2*-categories*, Category Seminar (Sydney 1972), Lect. Notes Math. **420** (1974), 75–103. [MR50#10010]

[650] A.Kempf, *Quantum group-symmetric Fock spaces with Bargmann-Fock representation*, Lett. Math. Phys. **26** (1992), 1–12.

[651] A.E.Kennelly, *The equivalence of triangles and three-pointed stars in conducting networks*, Electrical World and Engineer **34** (1989), 413–414

[652] S.V.Kerov, *Realization of $*$-representations of Hecke algebras, and the orthogonal Young form*, Zap. Nauchn. Sem. LOMI **161** (1987), 155–172. [MR88m:20085]

[653] S.V.Kerov, A.N.Kirillov and N.Yu.Reshetikhin, *Combinatorics, the Bethe ansatz and representations of the symmetric group*, Zap. Nauchn. Sem. LOMI **155** (1986), 50–64. [MR88i:82021]

[654] A.M.Vershik and S.V.Kerov, *Characters and realizations of representations of the infinite-dimensional Hecke algebra, and knot invariants*, Soviet Math. Dokl. **38** (1989), 134–137. [MR89k:17011]

[655] A.M.Vershik and S.V.Kerov, *The Grothendieck group of infinite symmetric group and symmetric functions (with the elements of the theory of K_0-functor of AF-algebras)*, Representation of Lie groups and related topics, 39–117, Gordon and Breach, New York, 1990.

[656] S.M.Khoroshkin and V.N.Tolstoy, *The uniqueness theorem for the universal R-matrix*, Lett. Math. Phys. **24** (1992), 231–244. [MR93e:17020]

[657] R.Kielpinski, *On homological dimensions of cocommutative Hopf algebras*, Colloq. Math. **39** (1978), 229–231. [MR80f:16015]

[658] R.Kielpinski, *Polynomial type Hopf algebras*, Tsukuba J. Math. **7** (1983), 191–202. [MR84k:16009]

[659] K.Kikkawa, B.Sakita and M.A.Virasoro, *Feynman-like diagrams compatible with duality. I. Planar diagrams*, Phys. Rev. **184** (1969), 1701–1713.

[660] T.Kimura, *Prequantum BRST cohomology*, Mathematical aspects of classical field theory (Seattle, WA, 1991), Contemp. Math. **132** (1992), 439–457.

[661] T.Kimura, *Generalized classical BRST cohomology and reduction of Poisson manifolds*, Comm. Math. Phys. **151** (1993), 155–182.

[662] A.A.Kirillov, *Geometric quantization*, Encyclopedia of Math. Sci. **4** (Dynamical Systems IV), Springer, Berlin-New York, 1990.

[663] A.A.Kirillov Jr. *Semi-infinite cohomology of Lie superalgebras*, Funktsional. Anal. Prilozhen. **23** (1989), no. 3, 61–62. [MR90k:17043]

[664] A.N.Kirillov, *Clebsch-Gordan quantum coefficients*, Zap. Nauchn. Sem. LOMI **168** (1988), 67–84. [MR90c:22039]

[665] N.A.Liskova and A.N.Kirillov, *Clebsch Gordan and Racah Wigner coefficients for $U_q(\mathrm{SU}(1,1))$*, Infinite analysis (Kyoto 1991), 611–621, World Sci. Publishing, River Edge, NJ, 1992.

[666] A.N.Kirillov and N.Yu.Reshetikhin, *The Bethe ansatz and the combinatorics of Young tableaux*, Zap. Nauchn. Sem. LOMI **155** (1986), 65–115. [MR88i:82020]

[667] A.N.Kirillov and N.Yu.Reshetikhin, *Exact solution of the integrable XXZ Heisenberg model with arbitrary spin. I. The ground state and the excitation spectrum*, J. Phys. A **20** (1987), 1565–1585. [MR88i:82071a]

[668] A.N.Kirillov and N.Yu.Reshetikhin, *Exact solution of the integrable XXZ Heisenberg model with arbitrary spin. II. Thermodynamics of the system*, J. Phys. A **20** (1987), 1587–1597. [MR88i:82071b]

[669] A.N.Kirillov and N.Yu.Reshetikhin, *Representations of Yangians and multiplicities of the inclusion of the irreducible components of the tensor product of representations of simple Lie algebras*, Zap. Nauchn. Sem. LOMI **160** (1987), 211–221. [MR89b:17012]

[670] A.N.Kirillov and N.Yu.Reshetikhin, *Multiplicity of weights in irreducible tensor representations of a complete linear superalgebra*, Functional. Anal. Appl. **22** (1988), 328–330. [MR90c:17013]

[671] A.N.Kirillov and N.Yu.Reshetikhin, *Representations of the algebra $U_q(\mathrm{sl}(2))$, q-orthogonal polynomials and invariants of links*, Infinite-dimensional Lie algebras and groups (Luminy Marseille, 1988), 285–339, World Sci. Publishing, Teaneck, NJ, 1989. [MR90m:17022]

[672] A.N.Kirillov and N.Reshetikhin, *q-Weyl group and a multiplicative formula for universal R-matrices*, Comm. Math. Phys. **134** (1990), 421–431. [MR92c:17023]

[673] E.Kirkman, C.Procesi and L.Small, *A q-analog for the Virasoro algebra*, Preprint.

[674] E.Kirkman and L.Small, *q-analogs of harmonic oscillators and related rings*, Preprint.

[675] V.G.Knizhnik and A.B.Zamolodchikov, *Current algebras and Wess-Zumino model in two dimensions*, Nucl. Phys. **B247** (1984), 83–103. [MR87h:81129]

[676] Y.Kobayashi and S.Nagamachi, *The chain rule of differentiation in superspace*, Lett. Math. Phys. **11** (1986), 293–297. [MR87k:58043]

[677] K.Kodaira and D.C.Spencer, *On deformations of complex analytic structures. I, II*, Ann. Math. **67** (1958), 328–466.

[678] K.Kodaira and D.C.Spencer, *On deformations of complex analytic structures. III*, Ann. Math. **71** (1960), 43–76.

[679] T.Kohno, *One-parameter family of linear representations of Artin's braid groups*, Galois representations and arithmetic algebraic geometry (Kyoto 1985 & Tokyo 1986), 189–200, North-Holland, Amsterdam-New York, 1987. [MR90d:32025]

[680] T.Kohno, *Monodromy representations of braid groups and Yang-Baxter equations*, Ann. Inst. Fourier **37** (1987), 139–160. [MR89h:17030]

[681] T.Kohno, *Linear representations of braid groups and classical Yang-Baxter equations*, Braids (Santa Cruz, CA, 1986), Contemp. Math. **78** (1988), 339-363. [MR90h:20056]

[682] T.Kohno, *Hecke algebra representations of braid groups and classical Yang-Baxter equations*, Conformal field theory and solvable lattice models (Kyoto 1986), 255–269, Academic Press, Boston, MA, 1988. [MR90a:17016]

[683] T.Kohno, *Monodromy of braid groups in conformal field theory and positive Markov traces*, Differential geometric methods in theoretical physics (Chester 1988), 64–73, World Sci. Publishing, Teaneck, NJ, 1989. [MR92e:20027]

[684] T.Kohno, *Integrable systems related to braid groups and Yang-Baxter equation*, Braid group, knot theory and statistical mechanics, 135–149, World Sci. Publishing, Teaneck, NJ, 1989.

[685] T.Kohno, *Integrable connections related to Manin and Schechtman's higher braid groups*, Illinois J. Math. **34** (1990), 476–484. [MR91d:32051]

[686] M.Koppinen, *A duality theorem for crossed products of Hopf algebras*, J. Algebra **146** (1992), 153–174. [MR93b:16068]

[687] L.L.Vaksman and L.I.Korogodskii, *The algebra of bounded functions on the quantum group of motions of the plane and q-analogues of Bessel functions*, Soviet Math. Dokl. **39** (1989), 173–177. [MR90j:17027]

[688] L.L.Vaksman and L.I.Korogodskii, *Spherical functions on the quantum group* $\mathrm{SU}(1,1)$ *and a q-analogue of the Mehler-Fock formula*, Functional. Anal. Appl. **25** (1991), 48–49.

[689] Y.Kosmann-Schwarzbach, *Poisson-Drinfeld groups*, Topics in soliton theory and exactly solvable nonlinear equations (Oberwolfach 1986), 191–215, World Sci. Publishing, Singapore, 1987. [MR88k:58049]

[690] Y.Kosmann-Schwarzbach, *The modified Yang-Baxter equation and bi-Hamiltonian structures*, Differential geometric methods in theoretical physics (Chester 1988), 12–25, World Sci. Publishing, Teaneck, NJ, 1989. [MR92e:58093]

[691] Y.Kosmann-Schwarzbach, *Generalized symmetries, recursion operators and bi-Hamiltonian systems*, Partially integrable evolution equations in physics (Les Houches 1989), 479–489, Kluwer Acad. Publ., Dordrecht, 1990. [MR92d:58078]

[692] Y.Kosmann-Schwarzbach, *Groupes de Lie-Poisson quasitriangulaires*, Géométrie symplectique et mécanique (La Grande Motte 1988), Lect. Notes Math. **1416** (1990), 161–177. [MR91g:58116]

[693] Y.Kosmann-Schwarzbach, *Quantum and classical Yang-Baxter equations*, Modern Phys. Lett. A **5** (1990), 981–990. [MR91g:58117]; Errata: **6** (1991), 3373.

[694] Y.Kosmann-Schwarzbach, *From "quantum groups" to "quasi-quantum groups"*, Symmetries in science, V (Lochau 1990), 369–393, Plenum, New York, 1991. [MR93b:16069]

[695] Y.Kosmann-Schwarzbach, *Grand crochet, crochets de Schouten et cohomologies d'algèbres de Lie*, C.R.A.S. Paris **312** (1991), 123–126. [MR92b:17028]

[696] Y.Kosmann-Schwarzbach, *Champs affines de multivecteurs sur les groupes de Lie*, C.R.A.S. Paris **312** (1991), 233–236. [MR92d:17023]

[697] Y.Kosmann-Schwarzbach, *Quasi-bigèbres de Lie et groupes de Lie quasi-Poisson* C.R.A.S. Paris **312** (1991), 391–394.

[698] Y.Kosmann-Schwarzbach, *Jacobian quasi-bialgebras and quasi-Poisson-Lie groups*, Mathematical aspects of classical field theory (Seattle, WA, 1991), Contemp. Math. **132** (1992), 459–489.

[699] Y.Kosmann-Schwarzbach and F.Magri, *Poisson-Lie groups and complete integrability. I. Drinfeld bialgebras, dual extensions and their canonical representations*, Ann. Inst. H. Poincaré Phys. Theor. **49** (1988), 433–460. [MR91a:17018]

[700] Y.Kosmann-Schwarzbach and F.Magri, *Dualization and deformation of Lie brackets on Poisson manifolds*, Differential geometry and its applications (Brno 1989), 79–84, World Sci. Publishing, Teaneck, NJ, 1990.

[701] Y.Kosmann-Schwarzbach and F.Magri, *Poisson-Nijenhuis structures*, Ann. Inst. H. Poincaré Phys. Theor. **53** (1990), 35–81. [MR92b:17026]

[702] B.Kostant, *Quantization and unitary representations. I*, Lect. Notes Math. **170** (1970), 87–208.

[703] B.Kostant, *Quantization and representation theory*, Representation theory of Lie groups (Oxford 1977), 287–316, London Math. Soc. Lect. Notes Ser. **34**, Cambridge Univ. Press, 1979.

[704] B.Kostant and S.Sternberg, *Symplectic reduction, BRS cohomology, and infinite-dimensional Clifford algebras*, Ann. Physics **176** (1987), 49–113. [MR88m:58057]

[705] B.Kostant and S.Sternberg, *The Schwartzian derivative and the conformal geometry of the Lorentz hyperboloid*, Quantum theories and geometry (Les Treilles 1987), 113–125, Kluwer Acad. Publ., Dordrecht, 1988. [MR90j:53015]

[706] J.-L.Koszul, *Sur le troisième nombre de Betti des espaces de groupes de Lie compacts*, C.R.A.S. Paris **224** (1947), 251–253. [MR8,368a]

[707] J.-L.Koszul, *Sur les opérateurs de dérivations dans un anneau*, C.R.A.S. Paris **225** (1947), 217–219. [MR9,196h]

[708] J.-L.Koszul, *Sur l'homologie des espaces homogènes*, C.R.A.S. Paris **225** (1947), 477–479. [MR9,196i]

[709] J.-L.Koszul, *Homologie et cohomologie des algèbres de Lie*, Bull. Soc. Math. France **78** (1950), 65–127. [MR12,120g]

[710] J.-L.Koszul, *Sur un type d'algèbres différentielles en rapport avec la transgression*, Colloque de topologie (espaces fibrés) (Bruxelles 1950), 73–81, Georges Thone, Liege & Masson, Paris, 1951. [MR13,109a]

[711] J.-L.Koszul, *Homologie des complexes de formes différentielles d'ordre supérieur*, Ann. Sci. École Norm. Sup. **7** (1974), 139–153. [MR50#8583]

[712] J.-L.Koszul, *Crochet de Schouten-Nijenhuis et cohomologie*, Élie Cartan et les mathématiques d'adjourd'hui, Astérisque, Numéro Hors Série (1985), 257–271. [MR88m:17013]

[713] Yu.Yu.Kotchetkoff, *Déformations des superalgèbres de Buttin et quantification*, C.R.A.S. Paris **299** (1984), 643–645. [MR85k:17021]

[714] H.A.Kramers and G.H.Wannier, *Statistics of the two-dimensional ferromagnet. I, II*, Phys. Rev. **60** (1941), 252–276. [MR3,63i & MR3,64a]

[715] I.S.Krasilshchik, *Hamiltonian cohomology of canonical algebras*, Dokl. Akad. Nauk SSSR **251** (1980), 1306–1309. [MR81j:58043]

[716] A.M.Vinogradov and I.S.Krasilshchik, *What is Hamiltonian formalism?* Uspekhi Mat. Nauk **30** (1975), no. 1, 173–198. [MR58#31235]

[717] M.G.Krein, *A principle of duality for a bicompact group and square block algebra*, Dokl. Akad. Nauk SSSR **69** (1949), 725–728.

[718] I.M.Krichever, *The Baxter equations and algebraic geometry*, Funkts. Anal. Prilozh. **15** (1981), no. 2, 22–35. [MR82j:82024]

[719] I.M.Krichever, *Algebraic geometry methods in the theory of the Baxter-Yang equations*, Mathematical physics reviews, Vol. **3**, 53–81, Harwood Academic Publ., 1982. [MR85g:58045]

[720] C.N.Ktorides and L.Ch.Papaloucas, *Derivation properties of a deformed Poisson algebra and the quantisation problem*, J. Phys. A **15** (1982), L451–L455. [MR84a:58041]

[721] C.N.Ktorides and L.Ch.Papaloucas, *Nonassociative properties of a deformed Poisson product and the quantisation problem*, J. Phys. A **16** (1983), 3485–3492. [MR85b:58053]

[722] P.P.Kulish and N.Yu.Reshetikhin, *The quantum linear problem for the sine-Gordon equation and higher representations*, Zap. Nauchn. Sem. LOMI **101** (1981), 101–110.

[723] P.P.Kulish and N.Yu.Reshetikhin, *On* GL_3*-invariant solutions of the Yang-Baxter equation and associated quantum systems*, Zap. Nauchn. Sem. LOMI **120** (1982), 92–121. [MR84i:58058]

[724] P.P.Kulish and N.Yu.Reshetikhin, *Universal R-matrix of the quantum superalgebra* osp(2|1), Lett. Math. Phys. **18** (1989), 143–149. [MR90j:17024]

[725] P.P.Kulish, N.Yu.Reshetikhin and E.K.Sklyanin, *Yang-Baxter equations and representation theory. I*, Lett. Math. Phys. **5** (1981), 393–403. [MR83g:81099]

[726] P.P.Kulish and E.K.Sklyanin, *Solutions of the Yang-Baxter equation*, J. Soviet Math. **19** (1982), 1596–1620.

[727] P.P.Kulish and E.K.Sklyanin, *Quantum spectral transform method. Recent developments*, Integrable quantum field theories, Lect. Notes Phys. **151** (1981), 61–119.

[728] K.Kumar, *Solution of eight-vertex lattice model without elliptic functions*, Austral. J. Phys. **27** (1974), 433–456. [MR53#10051]

[729] I.A.Kunin, *Group deformations and quantizations*, Trends in applications of pure mathematics to mechanics, Vol. II (Kozubnik 1977), 171–178, Pitman, Boston-London, 1979. [MR81f:22038]

[730] R.N.Onody and V.Kurak, *Exact solution of $Z(N)$ invariant spin models on a triangle lattice*, J. Phys. A **17** (1984), L615–L619.

[731] G.Kuster, *On the Hurwitz product of formal power series and automata*, Theoret. Comput. Sci. **83** (1991), 261–273. [MR93b:68056]

[732] V.Lakshmibai and N.Reshetikhin, *Quantum flag and Schubert schemes*, Deformation theory and quantum groups with applications to mathematical physics (Amherst, MA, 1990), Contemp. Math. **134** (1992), 145–181.

[733] L.A.Lambe, *Homological perturbation theory, Hochschild homology, and formal groups*, Deformation theory and quantum groups with applications to mathematical physics (Amherst, MA, 1990), Contemp. Math. **134** (1992), 183–218.

[734] L.A.Lambe and D.E.Radford, *Algebraic aspects of the quantum Yang-Baxter equation*, J. Algebra **154** (1993), 228–288.

[735] K.H.Lameche, *Propriétés algébriques des séries formelles liées au shuffle*, Groupe de Travail en Théorie Analytique et Élémentaire des Nombres (1989/1990), 65–83, Univ. Paris XI, Orsay, 1992.

[736] M.L.Laplaza, *Coherence in nonmonoidal closed categories*, Trans. Amer. Math. Soc. **230** (1977), 293–311. [MR56#3089]

[737] R.G.Larson and E.J.Taft, *The algebraic structure of linearly recursive sequences under Hadamard product*, Israel J. Math. **72** (1990), 118–132. [MR92g:16056]

[738] R.G.Larson and J.Towber, *Two dual classes of bialgebras related to the concepts of "quantum group" and "quantum Lie algebra"*, Comm. Algebra **19** (1991), 3295–3345. [MR93b:16070]

[739] G.J.Lasinio, *Stochastic dynamics and the semiclassical limit of quantum mechanics*, Springer, Vienna, 1980. [MR82k:81019]

[740] R.J.Lawrence, *A universal link invariant using quantum groups*, Differential geometric methods in theoretical physics (Chester 1988), 55–63, World Sci. Publishing, Teaneck, NJ, 1989. [MR92e:57012]

[741] R.J.Lawrence, *A universal link invariant*, The interface of mathematics and particle physics (Oxford 1988), 151–156, Oxford Univ. Press, New York, 1990. [MR92e:57013]

[742] R.J.Lawrence, *A topological approach to representations of the Iwahori Hecke algebra*, Internat. J. Modern Phys. A **5** (1990), 3213–3219. [MR91j:32043]

[743] R.J.Lawrence, *Homological representations of the Hecke algebra*, Comm. Math. Phys. **135** (1990), 141–191. [MR92d:16020]

[744] R.J.Lawrence, *Connections between CFT and topology via knot theory*, Differential geometric methods in theoretical physics (Rapallo 1990), Lect. Notes Phys. **375** (1991), 245–254. [MR93f:20051]

[745] R.J.Lawrence, *A functional approach to the one-variable Jones polynomial*, J. Diff. Geom. **37** (1993), 689–710.

[746] R.J.Lawrence, *On algebras and triangle relations*, Harvard Preprint, 1991.

[747] P.D.Lax, *Integrals of nonlinear equations of evolution and solitary waves*, Comm. Pure Appl. Math. **21** (1968), 467–490.

[748] A.Yu.Lazarev and M.V.Movshev, *Deformations of Hopf algebras*, Uspekhi Mat. Nauk **46** (1991), no. 1, 253–254.

[749] A.Yu.Lazarev and M.V.Movshev, *Quantization of some Lie groups and algebras*, Uspekhi Mat. Nauk **46** (1991), no. 6, 215–216. [MR93e:17021]

[750] L.Le Bruyn, *Universal bialgebras associated with orders*, Comm. Algebra **10** (1982), 457–478. [MR84i:16006]

[751] J.V.Leahy, *Deformation theory of graded and filtered objects in an abelian category*, J. Fac. Sci. Univ. Tokyo Sect. I **14** (1967) 59–80. [MR36#3849]

[752] P.B.A.Lecomte, *Sur l'algèbre de Lie des sections d'un fibré en algèbres de Lie*, Ann. Inst. Fourier **30** (1980), 35–50. [MR82h:57021]

[753] P.B.A.Lecomte, *Some remarks on the cohomology of the Poisson algebra*, Rend. Circ. Mat. Palermo Suppl. no. 3 (1984), 169–175. [MR86g:17012]

[754] P.B.A.Lecomte, *Application of the cohomology of graded Lie algebras to formal deformations of Lie algebras*, Lett. Math. Phys. **13** (1987), 157–166. [MR88g:17019]

[755] P.B.A.Lecomte, D.Mélotte and C.Roger, *Explicit form and convergence of 1-differential formal deformations of the Poisson Lie algebra*, Lett. Math. Phys. **18** (1989), 275–285. [MR90i:58057]

[756] P.B.A.Lecomte and C.Roger, *Formal deformations of the associative algebra of smooth matrices*, Lett. Math. Phys. **15** (1988), 55–63. [MR89f:58013]

[757] P.B.A.Lecomte and C.Roger, *Modules et cohomologies des bigèbres de Lie*, C.R.A.S. Paris **310** (1990), 405–410. [MR91c:17013]; Erratum: **311** (1990), 893–894. [MR92e:17028]

[758] C.W.Lee, *The associahedron and triangulations of the n-gon*, European J. Combin. **10** (1989), 551–560. [MR90i:52010]

[759] A.M.Legendre, *Calculus integralis. II.*

[760] D.A.Leites, *Clifford algebras as superalgebras, and quantization*, Teoret. Mat. Fiz. **58** (1984), 229–232. [MR85k:81083]

[761] D.A.Leites and V.V.Serganova, *Solutions of the classical Yang-Baxter equation for simple superalgebras*, Teoret. Mat. Fiz. **58** (1984), 26–37. [MR85m:82020]

[762] B.Lesche, *From classical mechanics to Feynman graphs with $*$-products*, Phys. Rev. D **29** (1984), 2270–2274. [MR86b:58053]

[763] D.Levy, *The algebra of the Gauss law in d=4 massless axial QED with a WZ term*, Phys. Lett. B **201** (1988), 143–146. [MR89h:81205]

[764] D.Levy, *Temperley-Lieb algebras*, Recent developments in conformal field theories (Trieste 1989), 165–178, World Sci. Publishing, River Edge, NJ, 1990.

[765] D.Levy, *Structure of Temperley-Lieb algebras and its application to 2D statistical models*, Phys. Rev. Lett. **64** (1990), 499–502. [MR91f:82006]

[766] D.Levy, *Algebraic structure of translation-invariant spin-$\frac{1}{2}$ XXZ and q-Potts quantum chains*, Phys. Rev. Lett. **67** (1991), 1971–1974. [MR92g:82033]

[767] D.Levy, *Lattice algebras and the hidden symmetry of the 2D Ising model in a magnetic field*, Intern. J. Modern Phys. A **6** (1991), 5127–5153.

[768] D.Levy, *The structure of the affine Hecke quotient underlying the translation invariant XXZ chain*, Preprint TAUP 1986-92, Tel-Aviv Univ., June 1992.

[769] A.Lichnerowicz, *New geometrical dynamics*, Lect. Notes Math. **570** (1977), 377–395.

[770] A.Lichnerowicz, *Les variétés de Poisson et leurs algèbres de Lie associées*, J. Differential Geometry **12** (1977), 253–300. [MR58#18565]

[771] A.Lichnerowicz, *Déformations d'algèbres associées à une variété symplectique*, Rend. Sem. Mat. Univ. Politec. Torino **36** (1977), 43–58. [MR80h:58026]

[772] A.Lichnerowicz, *Les variétés de Jacobi et leurs algèbres de Lie associées*, J. Math. Pures Appl. **57** (1978), 453–488. [MR80m:58016]

[773] A.Lichnerowicz, *Feuilletages et déformations infinitesimales des algèbres de Lie associées*, Proc. IV Intern. Colloq. Diff. Geom (Santiago de Compostela 1978), 192–204, Univ. Santiago de Compostela, 1979. [MR81g:57016]

[774] A.Lichnerowicz, *Existence and equivalence of twisted products on a symplectic manifold*, Lett. Math. Phys. **3** (1979), 495–502. [MR80m:58043]

[775] A.Lichnerowicz, *Sur les algèbres formelles associees par déformation a une variété symplectique*, Ann. Mat. Pura Appl. **123** (1980), 287–330. [MR82m:58030]

[776] A.Lichnerowicz, *Existence and invariance of twisted products on a symplectic manifold*, Lett. Math. Phys. **5** (1981), 117–126. [MR82h:58021]

[777] A.Lichnerowicz, *Deformations and quantization*, Dynamical systems and microphysics (Udine 1981), 27–60, Academic Press, New York-London, 1982. [MR85b:58054]

[778] A.Lichnerowicz, *Déformations d'algèbres associées a une variété symplectique (les $*_\nu$-produits)*, Ann. Inst. Fourier **32** (1982), 157–209. [MR83k:58095]

[779] A.Lichnerowicz, *Géométrie différentielle des variétés de contact*, J. Math. Pures Appl. **61** (1982), 345–380. [MR85b:58047]

[780] A.Lichnerowicz, *Deformations of algebras associated with a symplectic manifold*, Differential geometry and mathematical physics (Liege 1980 & Leuven 1981), 69–83, Reidel, Dordrecht-Boston, 1983. [MR84k:58078]

[781] A.Lichnerowicz, *Quantum mechanics and deformations of geometrical dynamics*, Quantum theory, groups, fields and particles, 3–82, Reidel, Dordrecht-Boston, 1983. [MR85e:58056]

[782] A.Lichnerowicz, *Differential geometry and deformations*, Symplectic geometry (Toulouse 1981), 107–121, Pitman, Boston-London, 1983. [MR85h:17007]

[783] A.Lichnerowicz, *Geometry and physics*, Proc. of the international meeting on geometry and physics (Florence 1982), 1–9, Pitagora, Bologna, 1983. [MR87g:01037]

[784] A.Lichnerowicz, *Géométrie des algèbres de Lie locales de Kirillov*, Symplectic geometry and mechanics (Balaruc 1983), 107–124, Travaux en Cours, Hermann, Paris, 1985. [MR87m:58060]

[785] A.Lichnerowicz, *Deformations and quantum statistical mechanics*, Mathematical aspects of classical field theory (Seattle, WA, 1991), Contemp. Math. **132** (1992), 491–501.

[786] A.Lichnerowicz and R.Rubio, *Algèbres de Lie locales de Kirillov transitives et déformations*, C.R.A.S. Paris **299** (1984), 761–765. [MR85k:17015]

[787] W.B.R.Lickorish and K.C.Millett, *The new polynomial invariants of knots and links*, Math. Mag. **61** (1988), 3–23. [MR89d:57006]

[788] S.Lie, *Theorie der Transformationsgruppen*, Vol. 2, Teubner, Leipzig, 1890.

[789] E.H.Lieb, *Exact solution of the problem of the entropy of two-dimensional ice*, Phys. Rev. Lett. **18** (1967), 692–694.

[790] E.H.Lieb, *Some Ginzburg-Landau type vector field equations*, Nonlinear systems of partial differential equations in applied mathematics, Part 2 (Santa Fe, N. M., 1984), 105–107, Amer. Math. Soc., Providence, RI, 1986.

[791] E.H.Lieb and W.Liniger, *Exact analysis of an interacting Bose gas. I. The general solution and the ground state*, Phys. Rev. **130** (1963), 1605–1616. [MR27#6551]

[792] T.D.Schultz, D.C.Mattis and E.H.Lieb, *Two-dimensional Ising model as a soluble problem of many fermions*, Rev. Modern Phys. **36** (1964) 856–871. [MR31#4509]

[793] H.N.V.Temperley and E.H.Lieb, *Relations between the "percolation" and "colouring" problem and other graph-theoretical problems associated with regular planar lattices: some exact results for the "percolation" problem*, Proc. Roy. Soc. London Ser. A **322** (1971), 251–280. [MR58#16425]

[794] E.H.Lieb and F.Y.Wu, *Two-dimensional ferroelectric models*, Phase transitions and critical phenomena, Vol. 1, 331–490, Academic Press, London, 1972.

[795] A.Liulevicius, *Arrows, symmetries and representation rings*, J. Pure Appl. Algebra **19** (1980), 259–273. [MR82j:16020]

[796] A.Liulevicius, *Finite G-sets and Hopf algebras of representation rings*, Symposium on Algebraic Topology in honor of José Adem (Oaxtepec 1981), Contemp. Math. **12** (1982), 183–199. [MR83k:20008]

[797] J.-H.Lu, *Momentum mappings and reduction of Poisson actions*, Symplectic geometry, groupoids, and integrable systems (Berkeley, CA, 1989), M.S.R.I. Publ. **20**, 209–226, Springer, New York, 1991. [MR92j:58036]

[798] J.-H.Lu and T.Ratiu, *On the nonlinear convexity theorem of Kostant*, J. Amer. Math. Soc. **4** (1991), 349–363. [MR92a:58048]

[799] J.-H.Lu and A.Weinstein, *Groupoïdes symplectiques doubles des groupes de Lie-Poisson*, C.R.A.S. Paris **309** (1989), 951–954. [MR91i:58045]

[800] J.-H.Lu and A.Weinstein, *Poisson-Lie groups, dressing transformations, and Bruhat decompositions*, J. Differential Geom. **31** (1990), 501–526. [MR91c:22012]

[801] Z.C.Lu, *Quantum algebras on phase space*, J. Math. Phys. **33** (1992), 446–453. [MR92m:17027]

[802] Z.C.Lu, *Star-products and phase space realizations of quantum groups*, Lett. Math. Phys. **25** (1992), 51–59.

[803] G.Lusztig, *Quantum deformations of certain simple modules over enveloping algebras*, Adv. Math. **70** (1988), 237–249. [MR89k:17029]

[804] G.Lusztig, *Modular representations and quantum groups*, Classical groups and related topics (Beijing 1987), Contemp. Math. **82** (1989), 59–77. [MR90a:16008]

[805] G.Lusztig, *Affine Hecke algebras and their graded version*, J. Amer. Math. Soc. **2** (1989), 599–635. [MR90e:16049]

[806] G.Lusztig, *Finite-dimensional Hopf algebras arising from quantized universal enveloping algebra*, J. Amer. Math. Soc. **3** (1990), 257–296. [MR91e:17009]

[807] G.Lusztig, *On quantum groups*, J. Algebra **131** (1990), 466–475. [MR91j:17019]

[808] G.Lusztig, *Quantum groups at roots of* 1, Geom. Dedicata **35** (1990), 89–113. [MR91j:17018]

[809] G.Lusztig, *Quivers, perverse sheaves, and quantized enveloping algebras*, J. Amer. Math. Soc. **4** (1991), 365–421. [MR91m:17018]

[810] G.Lusztig, *Canonical bases in tensor products*, Proc. Nat. Acad. Sci. U.S.A. **89** (1992), 8177–8179.

[811] G.Lusztig, *Introduction to quantum groups*, Birkhäuser, 1993.

[812] Y.Lykhmus, E.Paal and L.Sorgsepp, *Nonassociativity in mathematics and physics*, Eesti NSV Tead. Akad. Fuus. Inst. Uurim. 1990, no. 66, 8–22. [MR93b:17082]

[813] V.V.Lyubashenko, *Hopf algebras and vector-symmetries*, Uspekhi Mat. Nauk **41** (1986), no. 5, 185–186. [MR88c:58007]

[814] V.V.Lyubashenko, *Tangles and Hopf algebras in braided categories*, Preprint 1991; to appear in J. Pure Appl. Alg.

[815] V.V.Lyubashenko, *Modular transformations for tensor categories*, Preprint 1992; to appear in J. Pure Appl. Alg.

[816] V.V.Lyubashenko and S.Majid, *Fourier transform identities in quantum mechanics and the quantum line*, Phys. Lett. B **284** (1992), 66–70.

[817] V.V.Lyubashenko and S.Majid, *Braided groups and quantum Fourier transform*, Preprint DAMTP/91-26, 1991; to appear in J. Algebra.

[818] A.J.Macfarlane and S.Majid, *Quantum group structure in a fermionic extension of the quantum harmonic oscillator*, Phys. Lett. B **268** (1991), 71–74. [MR92k:81069]

[819] A.J.Macfarlane and S.Majid, *Spectrum generating quantum group of the harmonic oscillator*, Internat. J. Modern Phys. A **7** (1992), 4377–4393.

[820] G.Mack and V.Schomerus, *Quasi-Hopf quantum symmetry in quantum theory*, Nuclear Phys. B **370** (1992), 185–230. [MR93c:81081]

[821] K.Mackenzie, *Lie groupoids and Lie algebroids in differential geometry*, London Math. Soc. Lect. Notes Ser. **124**, Cambridge Univ. Press, 1987.

[822] K.Mackenzie, *Double Lie algebroids and second-order geometry. I*, Adv. Math. **94** (1992), 180–239. [MR93f:58255]

[823] K.Mackenzie, *Generalized Lie theories: Lie algebroids and Lie pseudoalgebras as algebraic invariants in differential geometry*, Preprint 1993.

[824] S.MacLane, *Natural associativity and commutativity*, Rice Univ. Studies **49** (1963), 28–46.

[825] S.MacLane, *Categorical algebra*, Bull. Amer. Math. Soc. **71** (1965), 40–106.

[826] S.MacLane, *Homology*, Springer, Berlin-New York, 1967. [MR50#2285]

[827] S.MacLane, *Coherence and canonical maps*, Symposia Mathematica **IV** (1970), 231–242. [MR46#1879]

[828] S.MacLane, *Categories for the working mathematician*, Springer, New York-Berlin, 1971. [MR50#7275]

[829] S.MacLane, *Origins of the cohomology of groups*, Enseign. Math. **24** (1978), 1–29. [MR81j:01030]

[830] S.MacLane, *Coherence theorems and conformal field theory*, Category theory (Montreal 1991), Canad. Math. Soc. Conf. Proc. **13** (1992), 321–328.

[831] F.Magri and C.Morosi, *A geometrical characterisation of integrable Hamiltonian systems through the theory of Poisson-Nijenhuis manifolds*, Univ. di Milano, 1984.

[832] F.Magri, C.Morosi and O.Ragnisco, *Reduction techniques for infinite-dimensional Hamiltonian systems: some ideas and applications*, Comm. Math. Phys. **99** (1985), 115–140. [MR86k:58051]

[833] S.Majid, *Hopf algebras for physics at the Planck scale*, Classical Quantum Gravity **5** (1988), 1587–1606. [MR90f:81041]

[834] S.Majid, *Liouville theorem for the Yang-Mills self-duality equations*, J. Math. Phys. **29** (1988), 2303–2310. [MR89i:58153]; Erratum: **30** (1989), 1652. [MR90d:58162]

[835] S.Majid, *Matched pairs of Lie groups and Hopf algebra bicrossproducts*, Nuclear Phys. B Proc. Suppl. **6** (1989), 422–424. [MR92f:16048]

[836] S.Majid, *Partial waves for the linearized Yang-Mills equation about a monopole background*, J. Math. Phys. **30** (1989), 1150–1157. [MR90e:81098]

[837] S.Majid, *Fourier transforms on A/G and knot invariants*, J. Math. Phys. **31** (1990), 924–927. [MR91g:58282]

[838] S.Majid, *Matched pairs of Lie groups associated to solutions of the Yang-Baxter equations*, Pacific J. Math. **141** (1990), 311–332. [MR91a:17009]

[839] S.Majid, *More examples of bicrossproduct and double cross product Hopf algebras*, Israel J. Math. **72** (1990), 133–148. [MR92b:16081]

[840] S.Majid, *On q-regularization*, Internat. J. Modern Phys. A **5** (1990), 4689–4696. [MR92e:81159]

[841] S.Majid, *Physics for algebraists: noncommutative and noncocommutative Hopf algebras by a bicrossproduct construction*, J. Algebra **130** (1990), 17–64. [MR91j:16050]

[842] S.Majid, *Quasitriangular Hopf algebras and Yang-Baxter equations*, Internat. J. Modern Phys. A **5** (1990), 1–91. [MR90k:16008]

[843] S.Majid, *Representation-theoretic rank and double Hopf algebras*, Comm. Algebra **18** (1990), 3705–3712. [MR91i:16072]

[844] S.Majid, *Quantum groups and quantum probability*, Quantum probability & related topics, 333–358, World Sci. Publishing, River Edge, NJ, 1991. [MR92m:81108]

[845] S.Majid, *Some physical applications of category theory*, Differential geometric methods in theoretical physics (Rapallo 1990), Lect. Notes Phys. **375** (1991), 131–142. [MR93c:81105]

[846] S.Majid, *Quasi-quantum groups as internal symmetries of topological quantum field theories*, Lett. Math. Phys. **22** (1991), 83–90. [MR92i:81130]

[847] S.Majid, *Braided groups and algebraic quantum field theories*, Lett. Math. Phys. **22** (1991), 167–175. [MR93b:16071]

[848] S.Majid, *Doubles of quasitriangular Hopf algebras*, Comm. Algebra **19** (1991), 3061–3073. [MR92k:16052]

[849] S.Majid, *Examples of braided groups and braided matrices*, J. Math. Phys. **32** (1991), 3246–3253.

[850] S.Majid, *Mass dependence of* U(1) *axial-symmetry breakdown*, Nuovo Cimento A **104** (1991), 617–640. [MR92d:81135]

[851] S.Majid, *Gauge-covariant vacuum expectations and glueballs*, Nuovo Cimento A **104** (1991), 949–960. [MR92g:81210]

[852] S.Majid, *Hopf-von Neumann algebra bicrossproducts, Kac algebra bicrossproducts, and the classical Yang-Baxter equations*, J. Funct. Anal. **95** (1991), 291–319. [MR92b:46088]

[853] S.Majid, *Reconstruction theorems and rational conformal field theories*, Internat. J. Modern Phys. A **6** (1991), 4359–4374. [MR92m:81257]

[854] S.Majid, *Representations, duals and quantum doubles of monoidal categories*, Rend. Circ. Mat. Palermo Suppl. no. 26 (1991), 197–206. [MR93c:18008]

[855] S.Majid, *Rank of quantum groups and braided groups in dual form*, Quantum groups (Leningrad 1990), Lect. Notes Math. **1510** (1992), 79–89.

[856] S.Majid, *An open problem in quantum groups*, Quantum groups (Leningrad 1990), Lect. Notes Math. **1510** (1992), 391–392.

[857] S.Majid, *Tannaka-Krein theorem for quasi-Hopf algebras and other results*, Deformation theory and quantum groups with applications to mathematical physics (Amherst, MA, 1990), Contemp. Math. **134** (1992), 219–232.

[858] S.Majid, *Braided groups and duals of monoidal categories*, Category theory (Montreal 1991), Canad. Math. Soc. Conf. Proc. **13** (1992), 329–343.

[859] S.Majid, **C**-*statistical quantum groups and Weyl algebras*, J. Math. Phys. **33** (1992), 3431–3444.

[860] S.Majid, *Beyond supersymmetry and quantum symmetry (an introduction to braided groups and braided matrices)*, Proc. V Nankai Workshop (June 1992), World Sci. Publ.

[861] S.Majid, *Infinite braided tensor products and 2D quantum gravity*, Proc. XXI DGM (Tianjin 1992), World Sci. Publ.

[862] S.Majid, *Braided groups*, J. Pure Appl. Alg. **86** (1993), 187–221.

[863] S.Majid, *Quantum and braided linear algebra*, J. Math. Phys. **34** (1993), 1776–1196.

[864] S.Majid, *Braided momentum in the q-Poincaré group*, J. Math. Phys. **34** (1993), 2045–2058.

[865] S.Majid, *Transmutation theory and rank for quantum braided groups*, Math. Proc. Cambridge Philos. Soc. **113** (1993), 45–70.

[866] S.Majid, *Cross products by braided groups and bosonization*, Preprint DAMTP/91-11, March 1991; to appear in J. Algebra.

[867] S.Majid, *Anyonic quantum groups*, Preprint DAMTP/91-16, May 1991; to appear in Proc. 2nd Max Born Symp. (Wroclaw 1992), Kluwer Acad. Publ.

[868] S.Majid, *Braided matrix structure of the Sklyanin algebra and of the quantum Lorentz group*, Preprint DAMTP/92-10, January 1992; to appear in Comm. Math. Phys.; available at **ftp.kurims.kyoto-u.ac.jp** as **kyoto-net/92-02-07-majid**.

[869] S.Majid, *Quantum random walks and time-reversal*, Preprint DAMTP/92-20, 1992; available at **ftp.kurims.kyoto-u.ac.jp** as **kyoto-net/92-04-16-majid**.

[870] S.Majid, *The quantum double as quantum mechanics*, Preprint DAMTP/92-48, September 1992; to appear in J. Geom. Phys.

[871] S.Majid, *Free braided differential calculus, braided binomial theorem and the braided exponential map*, Preprint DAMTP/93-3, 1993.

[872] S.Majid, *Quantum and braided Lie algebras*, Preprint DAMTP/93-4, 1993.

[873] S.Majid and M.J.Rodriguez-Plaza, *Quantum and super-quantum group related to the Alexander-Conway polynomial*, to appear in J. Geom. Phys.

[874] S.Majid and M.J.Rodriguez-Plaza, *Universal R-matrix for a non-standard quantum group and superization*, Preprint DAMTP/91-47, November 1991.

[875] S.Majid and Y.Soibelman, *Chern-Simons theory, modular functions and quantum mechanics in an alcove*, Internat. J. Modern Phys. A **6** (1991), 1815–1827. [MR92f:81061]

[876] S.Majid and Y.Soibelman, *Rank of quantized universal enveloping algebras and modular functions*, Comm. Math. Phys. **137** (1991), 249–262. [MR92k:17025]

[877] S.Majid and Y.Soibelman, *Bicrossproduct structure of the quantum Weyl group*, Preprint DAMTP/91-15, May 1991.

[878] G.Maltsiniotis, *Groupes quantiques et structure différentielles*, C.R.A.S. Paris **311** (1990), 831–834. [MR91m:58010]

[879] G.Maltsiniotis, *Groupoïdes quantiqes*, C.R.A.S. Paris **314** (1992), 249–252. [MR93b:16072]

[880] G.Maltsiniotis, *Le language des espaces et des groupes quantiques*, Comm. Math. Phys. **151** (1993), 275–302.

[881] G.Maltsiniotis, *Formes différentielles invariantes à gauche sur le groupe quantique* $GL_{p,q}(2)$, Preprint.

[882] S.Mandelstam, *Determination of the pion-nucleon scattering amplitude from dispersion relations and unitarity. General theory*, Phys. Rev. Lett. **112** (1958), 1344–1360.

[883] Y.I.Manin, *Some remarks on Koszul algebras and quantum groups*, Ann. Inst. Fourier **37** (1987), 191–205. [MR89e:16022]

[884] Y.I.Manin, *Quantum groups and noncommutative geometry*, Université de Montreal, Centre de Recherches Mathématiques, Montreal, PQ, 1988. [MR91e:17001]

[885] Y.I.Manin, *Multiparametric quantum deformation of the general linear supergroup*, Comm. Math. Phys. **123** (1989), 163–175. [MR90h:17018]

[886] Y.I.Manin, *Quantized theta-functions*, Progr. Theoret. Phys. Suppl. no. 102 (1990), 219–228.

[887] Y.I.Manin, *Quantum groups*, Nederl. Akad. Wetensch. Verslag Afd. Natuurk. **100** (1991), no. 5, 55–68.

[888] Y.I.Manin, *Topics in noncommutative geometry*, Princeton University Press, Princeton, NJ, 1991. [MR92k:58024]

[889] Y.I.Manin, *Notes on quantum groups and quantum de Rham complexes*, Preprint.

[890] Y.I.Manin and V.V.Schechtman, *Arrangements of real hyperplanes and Zamolodchikov equations*, Group theoretical methods in physics, Vol. I (Yurmala 1985), 151–165, VNU Sci. Press, Utrecht, 1986. [MR90d:58056]

[891] Y.I.Manin and V.V.Shechtman, *Higher Bruhat orderings connected with the symmetric group*, Funktsional. Anal. Prilozhen. **20** (1986), no. 2, 74–75. [MR88f:06001]

[892] G.H.Weiss and A.A.Maradudin, *The Baker-Hausdorff formula and a problem in crystal physics*, J. Math. Phys. **3** (1962), 771–777. [MR26#4765]

[893] M.Markl, *On the formality of products and wedges*, Rend. Circ. Mat. Palermo Suppl. no. 16 (1987), 199–206. [MR89h:55015]

[894] M.Markl and S.Papadima, *Homotopy Lie algebras and fundamental groups via deformation theory*, Ann. Inst. Fourier **42** (1992), 905–935.

[895] M.Markl and S.Shnider, *Drinfeld algebra deformations, homotopy comodules and associahedra*, Preprint 1993.

[896] M.Markl and J.D.Stasheff, *Deformation theory via deviations*, to appear in Journal of Algebra.

[897] C.M.Marle, *L'oeuvre d'Andre Lichnerowicz en géométrie symplectique*, Physique quantique et géométrie (Paris 1986), 25–42, Hermann, Paris, 1988. [MR89m:01100]

[898] P.P.Martin, *Temperley-Lieb algebra, group theory and the Potts model*, J. Phys. A **21** (1988), 577–591. [MR89c:82061]

[899] P.P.Martin, *Temperley-Lieb algebras and the long distance properties of statistical mechanical models*, J. Phys. A **23** (1990), 7–30. [MR91a:82022]

[900] P.P.Martin, *On Schur-Weyl duality, A_n Hecke algebras and quantum* sl(N) *on* $\bigotimes^{n+1} \mathbf{C}^N$, Infinite analysis (Kyoto 1991), 645–673, World Sci. Publishing, River Edge, NJ, 1992.

[901] V.B.Matveev and A.O.Smirnov, *Star-triangle equations and some properties of algebraic curves that are connected with the integrable chiral Potts model*, Mat. Zametki **46** (1989), no. 3, 31–39.

[902] M.E.Mayer, *Groupoids and Lie bigebras in gauge and string theories*, Differential geometrical methods in theoretical physics (Como 1987), 149–164, Kluwer Acad. Publ., Dordrecht, 1988. [MR90b:17039]

[903] M.E.Mayer, *From Poisson groupoids to quantum groupoids and back*, Differential geometric methods in theoretical physics (Rapallo 1990), Lect. Notes Phys. **375** (1991), 143–154. [MR93e:58207]

[904] J.McCleary, *User's guide to spectral sequences*, Publish or Perish, 1985. [MR87f:55014]

[905] B.M.McCoy, *The chiral Potts model: from physics to mathematics and back*, Special functions (Okayama 1990), 245–259, Springer, Tokyo, 1991. [MR93c:82021]

[906] B.M.McCoy and J.H.H.Perk, *Relation of conformal field theory and deformation theory for the Ising model*, Nucl. Phys. **B285** (1987), 279–294.

[907] B.M.McCoy and T.T.Wu, *Hydrogen-bonded crystals and the anisotropic Heisenberg chain*, Nuovo Cimento **B56** (1968), 311–315.

[908] W.S.McCullough, *Mean-field transfer-matrix analysis of the p-state chiral clock model*, Phys. Rev. B **46** (1992), 5084–5089.

[909] D.McDuff, *On the classifying spaces of discrete monoids*, Topology **18** (1979), 313–320. [MR81d:55020]

[910] J.B.McGuire, *Study of exactly soluble one-dimensional N-body problems*, J. Math. Phys. **5** (1964), 622–636.

[911] H.P.McKean Jr. *Kramers-Wannier duality for the 2-dimensional Ising model as an instance of Poisson's summation formula*, J. Math. Phys. **5** (1964), 775–776. [MR28#5713]

[912] A.Medina and P.Revoy, *Groupes de Lie-Poisson et double extension*, Séminaire Gaston Darboux de Géométrie et Topologie Différentielle 1990–1991, 87–105, Univ. Montpellier II, 1992.

[913] D.Mélotte, *Cohomologie de Chevalley associée aux variétés de Poisson*, Bull. Soc. Roy. Sci. Liège **58** (1989), 319–413.

[914] R.Menikoff and D.H.Sharp, *Representations of a local current algebra: their dynamical determination*, J. Malth. Phys. **16** (1975), 2341–2352. [MR54#2051]

[915] R.Menikoff and D.H.Sharp, *Approximate representations of a local current algebra*, J. Math. Phys. **16** (1975), 2353–2360. [MR54#2052]

[916] J.P.Meyer, *Acyclic models for multicomplexes*, Duke Math. J. **45** (1978), 67–85. [MR80b:55012]

[917] J.P.Meyer, *Bar and cobar constructions. I*, J. Pure Appl. Algebra **33** (1984), 163–207. [MR86g:18010]

[918] W.Michaelis, *Lie coalgebras*, Adv. Math. **38** (1980), 1–54. [MR82b:17016]

[919] W.Michaelis, *The dual Poincaré-Birkhoff-Witt theorem*, Adv. Math. **57** (1985), 93–162. [MR86m:17017]

[920] W.Michaelis, *On the dual Lie coalgebra of the Witt algebra*, XVIIth International Colloquium on Group Theoretical Methods in Physics (Sainte Adele, PQ, 1988), 435–439, World Sci. Publishing, Teaneck, NJ, 1989.

[921] K.Mikami, *Symplectic double groupoids over Poisson* $(ax+b)$*-groups*, Trans. Amer. Math. Soc. **324** (1991), 447–463.

[922] J.W.Milnor and J.C.Moore, *On the structure of Hopf algebras*, Ann. Math. **81** (1965), 211–264. [MR30#4259]

[923] S.Miracle-Sole, *Theorems on phase transitions with a treatment for the Ising model*, Critical phenomena (Sitges 1976), Lecture Notes Phys. **54** (1976), 189–214. [MR58#20101]

[924] B.Mitchell, *Spectral sequences for the layman*, Amer. Math. Mon. **76** (1969), 599–605.

[925] M.J.Stephen and L.Mittag, *A new representation of the solution of the Ising model*, J. Math. Phys. **13** (1972), 1944–1951.

[926] M.J.Stephen and L.Mittag, *Pseudo-Hamiltonians for the Potts models at the critical point*, Phys. Lett. **41A** (1972), 357–358.

[927] I.M.Mladenov, *Reductions and quantization*, Intern. J. Theor. Phys. **28** (1989), 1255–1267.

[928] P.Molin, *Existence de $*$-produits fortement invariants sur les orbites massives de la coadjointe du groupe de Poincaré*, C.R.A.S. Paris **293** (1981), 309–312. [MR83e:58031]

[929] S.Montgomery, *Crossed products of Hopf algebras and enveloping algebras*, Perspectives in ring theory (Antwerp 1987), 253–268, Kluwer Acad. Publ., Dordrecht, 1988. [MR91i:16056]

[930] S.Montgomery, *Simple smash products*, Deformation theory and quantum groups with applications to mathematical physics (Amherst, MA, 1990), Contemp. Math. **134** (1992), 233–234.

[931] S.Montgomery and S.P.Smith, *Skew derivations and* $U_q(\mathrm{sl}(2))$, Israel J. Math. **72** (1990), 158–166. [MR92e:16027]

[932] G.F.Newell and E.W.Montroll, *On the theory of the Ising model of ferromagnetism*, Rev. Modern Physics **25** (1953), 353–389. [MR15,88b]

[933] G.Moore and N.Seiberg, *Polynomial equations in rational conformal field theory*, Phys. Lett. B **212** (1988), 451–460. [MR89m:81155]

[934] G.Moore and N.Seiberg, *Classical and qauntum conformal field theory*, Comm. Math. Phys. **123** (1989), 177–254. [MR90e:81216]

[935] J.C.Moore, *Differential homological algebra*, Actés du Congrès International des Mathématiciens (Nice 1970), Tome 1, 335–339, Gauthier Villars, Paris, 1971. [MR55#9128]

[936] J.Morava, *Some examples of Hopf algebras and Tannakian categories*, Proc. Mahowald Festschrift, Contemp. Math. (1993).

[937] C.Moreno, *Produits star sur certains G/K kählériens. Equation de Yang-Baxter et produits star sur G*, Géométrie symplectique et mécanique (La Grande Motte 1988), Lect. Notes Math. **1416** (1990), 210–234. [MR91g:58094]

[938] J.E.Moyal, *Mean ergodic theorems in quantum mechanics*, J. Math. Phys. **10** (1969), 506–509. [MR39#3785]

[939] S.Sen and P.Murphy, *Constraints on two-dimensional conformal field theories from knot invariants*, Lett. Math. Phys. **18** (1989), 287–297. [MR90k:81246]

[940] A.Nakajima, *Cohomology for comodules*, Math. J. Okayama Univ. **22** (1980), 5–15. [MR81f:18034]

[941] Y.Nambu, in Symmetric and Quark Models (Proc. 1969 Wayne Conf.), 269– , Gordon and Breach, 1970.

[942] J.E.Nelson and T.Regge, *Quantum gravity and quantum groups*, Mechanics, analysis and geometry: 200 years after Lagrange, 503–512, North-Holland, Amsterdam, 1991. [MR92e:81095]

[943] O.M.Neroslavskii and A.T.Vlasov, *Sur les déformations de l'algèbre des fonctions d'une variété symplectique*, C.R.A.S. Paris **292** (1981), 71–73.

[944] K.Newman and D.E.Radford, *The cofree irreducible Hopf algebra on an algebra*, Amer. J. Math. **101** (1979), 1025–1045. [MR80i:16017]

[945] W.Nichols and M.Sweedler, *Hopf algebras and combinatorics*, Umbral calculus and Hopf algebras (Norman, Okla., 1978), Contemp. Math. **6** (1982), 49–84. [MR83g:16019]

[946] J.Niederle and J.Tolar, *Quantization as mapping and as deformation*, Czechoslovak J. Phys. B **29** (1979), 1358–1368. [MR80k:81055]

[947] A.Nijenhuis, *The graded Lie algebras of an algebra*, Indag. Math. **29** (1967), 475–486; Correction: **30** (1968), 253. [MR37#1420a,b]

[948] A.Nijenhuis, *Composition systems and deformations of subalgebras*, Indag. Math. **30** (1968), 119–136. [MR37#1419]

[949] A.Nijenhuis and R.W.Richardson, *Cohomology and deformations of algebraic structures*, Bull. Amer. Math. Soc. **70** (1964), 406–411. [MR31#2299]

[950] A.Nijenhuis and R.W.Richardson, *Cohomology and deformations in graded Lie algebras*, Bull. Amer. Math. Soc. **72** (1966), 1–29. [MR33#4190]

[951] A.Nijenhuis and R.W.Richardson, *Deformations of homomorphisms of Lie groups and Lie algebras*, Bull. Amer. Math. Soc. **73** (1967), 175–179. [MR34#4414]

[952] A.Nijenhuis and R.W.Richardson, *Deformations of Lie algebra structures*, J. Math. Mech. **17** (1967), 89–105. [MR35#5485]

[953] A.Nijenhuis and R.W.Richardson, *Commutative algebra cohomology and deformations of Lie and associative algebras*, J. Algebra **9** (1968), 42–53. [MR37#1421]

[954] M.den Nijs, *The domain wall theory of two-dimensioanl commensuratencommensurate phase transitions*, Phase transitions and critical phenomena, Vol. 12, 219–333, Academic Press, London, 1988.

[955] M.Nomura, *Relations among n-j symbols in forms of the star-triangle relation*, J. Phys. Soc. Japan **57** (1988), 3653–3656. [MR90a:82076]

[956] M.Nomura, *Relations for Clebsch-Gordan and Racah coefficients in* $\mathrm{su}_q(2)$ *and Yang-Baxter equations*, J. Math. Phys. **30** (1989), 2397–2405. [MR90j:17026]

[957] E.Ogievetsky, N.Reshetikhin and P.Wiegmann, *The principal chiral field in two dimensions on classical Lie algebras. The Bethe-ansatz solution and factorized theory of scattering*, Nuclear Phys. B **280** (1987), 45–96. [MR88b:81164]

[958] O.Ogievetsky, *On the homology of Poisson algebra of odd variables*, Duke Math. J. **54** (1987), 27–40. [MR88g:17018]

[959] C.S.O.Yokoi and M.J.de Oliveira, *Chiral Potts model on a Cayley tree with complete and incomplete devil's staircase*, J. Phys. A **18** (1985), L153–L157.

[960] G.I.Olshanskii, *Representations of infinite-dimensional classical groups, limits of enveloping algebras, and Yangians*, Topics in representation theory, 1–66, AMS, Providence, RI, 1991. [MR92g:22039]

[961] G.I.Olshanski, *Quantized universal enveloping superalgebra of type Q and a super-extension of the Hecke algebra*, Lett. Math. Phys. **24** (1992), 93–102.

[962] L.Onsager, *Crystal statistics. I. A two-dimensional model with an order-disorder transition*, Phys. Rev. **65** (1944), 117–149. [MR5,280d]

[963] L.Onsager, *The Ising model in two dimensions*, Critical phenomena in alloys, magnets and superconductors, 3–12, McGraw-Hill, New York, 1971.

[964] S.Ostlund, *Incommensurate and commensurate phases in assymetric clock models*, Phys. Rev. **B24** (1981), 398–405.

[965] H.C.Öttinger, *Mean-field theory of the chiral Potts model*, J. Phys. C **15** (1982), L1257–L1262.

[966] H.C.Öttinger, *Two-dimensional iterated mapping for the mean-field theory of the chiral Potts model*, J. Phys. C **16** (1983), L257–L263.

[967] H.C.Öttinger, *Order-disorder and ferromagnetic-modulated phase transitions of the chiral Potts model*, J. Phys. C **16** (1983), L597–L604.

[968] V.P.Palamodov, *Cohomology of analytic algebras*, Trudy Moskov. Mat. Obshch. **44** (1982), 3–61. [MR83f:13004]

[969] T.G.Pantev and V.V.Tsanov, *The classical Yang-Baxter equation and triple Manin systems*, Differential geometric methods in theoretical physics (Chester 1988), 26–35, World Sci. Publishing, Teaneck, NJ, 1989.

[970] B.Pareigis, *A noncommutative noncocommutative Hopf algebra in "nature"*, J. Algebra **70** (1981), 356–374. [MR83g:16021]

[971] B.Pareigis, *Morita equivalence of module categories with tensor products*, Comm. Algebra **9** (1981), 1455–1477. [MR83d:16044]

[972] G.Patissier, *Déformations différentielles à coefficients constants et produits de Moyal généralisés*, Ann. Inst. H. Poincaré Sect. A **30** (1979), 275–293. [MR81e:58029]

[973] P.A.Pearce, *Temperley-Lieb operators and critical A-D-E models*, Internat. J. Modern Phys. B **4** (1990), 715–734. [MR92b:82049]

[974] H.L.Scott and P.A.Pearce, *Calculation of intermolecular interaction strenghts in the $P_{\beta'}$ phase in lipid bilayers*, Biophys. J. **55** (1989), 339–345.

[975] R.Peierls, *More surprises in theoretical physics*, Princeton University Press, 1991. [MR92i:82002]

[976] J.H.H.Perk, *Bethe-Ansatz en éëndimensionale quantumsystemen*, Doctoraalscriptie, Universiteit van Amsterdam, 1974.

[977] J.H.H.Perk and C.L.Schultz, *Families of commuting transfer matrices in q-state vertex models*, Non-linear integrable systems — classical theory and quantum theory (Kyoto 1981), 135–152, World Sci. Publ., Singapore, 1983.

[978] J.H.H.Perk and F.Y.Wu, *Graphical approach to the nonitersecting string model: star-triangle eqaution, inversion relation, and exact solution*, Physica **138A** (1986), 100-124. [MR87m:82063]

[979] J.H.H.Perk and F.Y.Wu, *Nonintersecting string model and graphical approach: equivalence with a Potts model*, J. Statist. Phys. **42** (1986), 727–742. [MR87e:82065]

[980] W.Pesch and W.Selke, *Depinning and wetting in two-dimensional Potts and chiral clock models*, Z. Phys. B **69** (1987), 295–301

[981] W.S.Piper, *Algebraic deformation theory*, J. Differential Geometry **1** (1967), 133–168. [MR37#2814]

[982] P.Podles and S.L.Woronowicz, *Quantum deformation of Lorentz group*, Comm. Math. Phys. **130** (1990), 381–431. [MR91f:46100]

[983] S.Poisson, *Traité de mécanique*, 2nd ed., Bachelier, Paris, 1833.

[984] G.J.Porter, *The universal H-space of a monoid*, H-spaces (Actés Reunion Neuchatel, 1970), Lect. Notes Math. **196** (1971), 63–67. [MR44#4745]

[985] S.B.Priddy, *Koszul resolutions*, Trans. Amer. Math. Soc. **152** (1970), 39–60. [MR42#346]

[986] A.Proute, *Un contre-exemple à la géométricite du shuffle-coproduit de la cobar-construction*, C.R.A.S. Paris **298** (1984), 31–34. [MR85b:55012]

[987] W.Pusz and S.L.Woronowicz, *Twisted second quantization*, Rep. Math. Phys. **27** (1989), 231–257. [MR93c:81082]

[988] M.Puta, *On the geometric prequantization of Poisson manifolds*, Lett. Math. Phys. **15** (1988), 187–192. [MR89h:58072]; Errata: **17** (1989), 174. [MR90b:58097]

[989] D.E.Radford, *Freeness (projectivity) criteria for Hopf algebras over Hopf subalgebras*, J. Pure Appl. Algebra **11** (1977), 15–28. [MR57#16344]

[990] D.E.Radford, *A natural ring basis for the shuffle algebra and an application to group schemes*, J. Algebra **58** (1979), 432–454. [MR80k:16016]

[991] D.E.Radford, *Embedding commutative irreducible Hopf algebras into cofree irreducible Hopf algebras*, Comm. Algebra **7** (1979), 1211–1243. [MR80m:16008]

[992] D.E.Radford, *On an analog of Lagrange's theorem for commutative Hopf algebras*, Proc. Amer. Math. Soc. **79** (1980), 164–166. [MR81c:16011]

[993] D.E.Radford, *On bialgebras which are simple Hopf modules*, Proc. Amer. Math. Soc. **80** (1980), 563–568. [MR81j:16013]

[994] D.E.Radford, *On the structure of pointed coalgebras*, J. Algebra **77** (1982), 1–14. [MR84h:16009]

[995] D.E.Radford, *The structure of Hopf algebras with a projection*, J. Algebra **92** (1985), 322–347.

[996] D.E.Radford, *Divided power structures on Hopf algebras and embedding Lie algebras into special-derivation algebras*, J. Algebra **98** (1986), 143–170. [MR87d:16010]

[997] D.E.Radford, *On the quasitriangular structures of a semisimple Hopf algebra*, J. Algebra **141** (1991), 354–358. [MR92j:16026]

[998] D.E.Radford, *On the antipode of a quasitriangular Hopf algebra*, J. Algebra **151** (1992), 1–11.

[999] D.E.Radford, *Minimal quasitriangular Hopf algebras*, J. Algebra **157** (1993), 285–315.

[1000] D.E.Radford, *On quantum Yang-Baxter operators derived from finite-dimensional Hopf algebras*, Comm. Algebra **21** (1993), 1141–1152.

[1001] D.E.Radford, *Solutions to the quantum Yang-Baxter equation and the Drinfeld double*, Preprint 1990, to appear in J. Algebra.

[1002] D.E.Radford, *Solutions to the quantum Yang-Baxter equation arising from pointed bialgebras*, to appear in Trans. Amer. Math. Soc.

[1003] D.E.Radford, *Irreducible representations of $U_q(\mathrm{g})$ arising from $Mod^{\bullet}_{C^{1/2}}$*, to appear in Israel Math. Conf. Proc.

[1004] D.E.Radford, *Generalized double crossproducts associated with the quantized enveloping algebras*, Preprint, December 1991.

[1005] D.E.Radford, *On Kauffman's knot invariants arising from finite-dimensional Hopf algebras*, submitted.

[1006] D.E.Radford, E.J.Taft and R.L.Wilson, *Forms of certain Hopf algebras*, Manuscripta Math. **17** (1975), 333–338. [MR53#513]

[1007] D.E.Radford and J.Towber, *Yetter-Drinfeld categories associated to an arbitrary bialgebra*, J. Pure Appl. Alg. **87** (1993), 259–279.

[1008] R.Ree, *Lie elements and an algebra associated with shuffles*, Ann. Math. **68** (1958), 210–220. [MR20#6447]

[1009] R.Ree, *Generalized Lie elements*, Canad. J. Math. **12** (1960), 493–502. [MR22#5698]

[1010] K.-H.Rehren and B.Schroer, *Einstein causality and Artin braids*, Nucl. Phys. **B312** (1989), 715–750. [MR90g:81120]

[1011] N.Yu.Reshetikhin, *The functional eqaution method in the theory of exactly solvable quantum systems*, Soviet Phys. JETP **57** (1983), 691–696. [MR85d:82038]

[1012] N.Yu.Reshetikhin, *Integrable models of quantum one-dimensional magnets with $O(n)$- and $Sp(2k)$-symmetries*, Teoret. Mat. Fiz. **63** (1985), 347–366. [MR97k:82035]

[1013] N.Yu.Reshetikhin, *The spectrum of the transfer matrices connected with Kac-Moody algebras*, Lett. Math. Phys. **14** (1987), 235–246. [MR90b:58114]

[1014] N.Yu.Reshetikhin, *The algebraic Bethe ansatz for* $\mathrm{SO}(N)$*-invariant transfer matrices*, Zap. Nauchn. Sem. LOMI **169** (1988), 122–140. [MR89k:82103]

[1015] N.Yu.Reshetikhin, *Quasitriangular Hopf algebras and invariants of links*, Algebra i Analiz **1** (1989), 169–188. [MR90k:17033]

[1016] N.Reshetikhin, *Quantized universal enveloping algebras, the Yang-Baxter equation and invariants of links*, Comm. Math. Phys. **127** (1990), 1–20.

[1017] N.Reshetikhin, *Multiparameter quantum groups and twisted quasitriangular Hopf algebras*, Lett. Math. Phys. **20** (1990), 331–335. [MR91k:17012]

[1018] N.Yu.Reshetikhin, *Invariants of links and 3-manifolds related to quantum groups*, Proceedings of the International Congress of Mathematicians (Kyoto 1990), 1373–1375, Math. Soc. Japan, Tokyo, 1991.

[1019] N.Reshetikhin, *A new exactly solvable case of an* $\mathrm{O}(n)$*-model on a hexagonal lattice*, J. Phys. A **24** (1991), 2387–2396. [MR93d:82037]

[1020] N.Reshetikhin, *S-matrices in integrable models of isotropic magnetic chains. I*, J. Phys. A **24** (1991), 3299–3309. [MR93b:82025]

[1021] N.Reshetikhin, *Jackson-type integrals, Bethe vectors, and solutions to a difference analog of the Knizhnik-Zamolodchikov system*, Lett. Math. Phys. **26** (1992), 153–165.

[1022] N.Reshetikhin, *The Knizhnik-Zamolodchikov system as a deformation of the isomonodromy problem*, Lett. Math. Phys. **26** (1992), 167–177.

[1023] N.Reshetikhin, *Quantization of Lie bialgebras*, Internat. Math. Res. Notices (inside Duke Math. J.) 1992, no. 7, 143–151.

[1024] N.Yu.Reshetikhin and M.A.Semenov-Tian-Shansky, *Quantum R-matrices and factorization problems*, J. Geom. Phys. **5** (1988), 533–550. [MR92g:17019]

[1025] N.Yu.Reshetikhin and M.A.Semenov-Tian-Shansky, *Central extensions of quantum current groups*, Lett. Math. Phys. **19** (1990), 133–142. [MR91k:17013]

[1026] N.Reshetikhin and F.Smirnov, *Hidden quantum group symmetry and integrable perturbations of conformal field theories*, Comm. Math. Phys. **131** (1990), 157–177. [MR91f:81196]

[1027] N.Reshetikhin and V.Turaev, *Ribbon graphs and their invariants derived from quantum groups*, Comm. Math. Phys. **127** (1990), 1–26. [MR91c:57016]

[1028] N.Reshetikhin and V.Turaev, *Invariants of 3-manifolds via link polynomials and quantum groups*, Invent. Math. **103** (1991), 547–597. [MR92b:57024]

[1029] N.Reshetikhin and P.B.Wiegmann, *Towards the classification of completely integrable quantum field theories (the Bethe-ansatz associated with Dynkin diagrams and their automorphisms)*, Phys. Lett. B **189** (1987), 125–131. [MR88m:82011]

[1030] C.Reutenauer, *Free Lie algebras*, Oxford Univ. Press, 1993.

[1031] A.G.Reyman and M.A.Semenov-Tian-Shansky, *Compatible Poisson structures for Lax equations: an r-matrix approach*, Phys. Lett. A **130** (1988), 456–460. [MR90b:58113]

[1032] R.W.Richardson, *Deformations of subalgebras of Lie algebras*, J. Differential Geometry **3** (1969), 289–308. [MR41#7726]

[1033] M.A.Rieffel, *Strong Morita equivalence of certain transformation group C^*-algebras*, Math. Ann. **222** (1976), 7–22.

[1034] M.A.Rieffel, *Projective modules over higher-dimensional non-commutative tori*, Can. J. Math. **40** (1988), 257–338.

[1035] M.A.Rieffel, *Deformation quantization and operator algebras*, Operator theory: operator algebras and applications, Part 1 (Durham, NH, 1988), Proc. Symposia Pure Math. **51** (1990), 411–423. [MR91h:46120]

[1036] M.A.Rieffel, *Non-commutative tori — a case study of non-commutative differentiable manifolds*, Geometric and topological invariants of elliptic operators, Contemp. Math. **105** (1990), 191–211

[1037] M.A.Rieffel, *Critical points of Yang-Mills for noncommutative two-tori*, J. Diff. Geom. **31** (1990), 535–546.

[1038] M.A.Rieffel, *Lie group convolution algebras as deformation quantizations of linear Poisson structures*, Amer. J. Math. **112** (1990), 657–685. [MR92b:58076]

[1039] M.A.Rieffel, *Some solvable quantum groups*, Operator algebras and topology (Craiova 1989), 146–159, Longman Sci. Tech., Harlow, 1992.

[1040] M.A.Rieffel, *Deformation quantization for actions of* $\mathbf{R}^d$, to appear in Mem. Amer. Math. Soc.

[1041] M.A.Rieffel, *Category theory and quantum field theory*, Preprint.

[1042] D.Riffelmacher, *Sweedler's two-cocycles and Hochschild cohomology*, Pacific J. Math. **89** (1980), 181–188. [MR82g:16025]

[1043] D.S.Rim, *Deformation of transitive Lie algebras*, Ann. Math. **83** (1966), 339–357. [MR33#7463]

[1044] C.M.Ringel, *Hall algebras and quantum groups*, Invent. Math. **101** (1990), 583–591. [MR91i:16024]

[1045] C.M.Ringel, *From representations of quivers via Hall and Loewy algebras to quantum groups*, Proceedings of the International Conference on Algebra, Part 2 (Novosibirsk 1989), Contemp. Math. **131** (1992), 381–401.

[1046] V.Rittenberg and M.Scheunert, *Tensor operators for quantum groups and applications*, J. Math. Phys. **33** (1992), 436–445. [MR92k:16054]

[1047] R.M.Robinson, *Comment to: "A motivated proof of the Rogers-Ramanujan identities" by G.E.Andrews and R.J.Baxter*, Amer. Math. Monthly **97** (1990), 214–215. [MR91j:11086]

[1048] C.Roger, *Cohomology of current Lie algebras*, Deformation theory of algebras and structures and applications (Il Ciocco 1986), 357–374, Kluwer Acad. Publ., Dordrecht, 1988. [MR90b:17025]

[1049] C.Roger, *Déformations universelles des crochets de Poisson*, Géométrie symplectique et mécanique (La Grande Motte 1988), Lect. Motes Math. **1416** (1990), 242–254. [MR91g:58096]

[1050] C.Roger, *Deformations and cohomology of the Lie algebra of currents. Application to the Faddeev anomaly*, Gauge theories of fundamental interactions (Warsaw 1988), 453–468, World Sci. Publishing, River Edge, NJ, 1990. [MR93b:17066]

[1051] C.Roger, *Algèbres de Lie graduées et quantification*, Symplectic geometry and mathematical physics (Aix-en Provence 1990), 374–421, Birkhäuser, Boston, MA, 1991. [MR93f:17045]

[1052] C.Roger, *Déformations algébriques et applications à la physique*, Gazette de Mathématiciens, 1991, no. 49, 75–94.

[1053] A.L.Rosenberg, *The spectrum of the algebra of skew differential operators and the irreducible representations of the quantum Heisenberg algebra*, Comm. Math. Phys. **142** (1991), 567–588. [MR93e:17025]

[1054] J.L.Rosner, *Graphical form of duality*, Phys. Rev. Lett. **22** (1969), 689–692.

[1055] M.Rosso, *Comparison des groupes $SU(2)$ quantiques de Drinfeld et de Woronowicz*, C.R.A.S. Paris **304** (1987), 323–326.

[1056] M.Rosso, *Groupes quantiques et modèles à vertex de V.Jones en théorie des noeuds*, C.R.A.S. Paris **307** (1988), 207–210. [MR90d:57009]

[1057] M.Rosso, *An analogue of P.B.W. theorem and the universal R-matrix for $U_h\mathrm{sl}(N+1)$*, Comm. Math. Phys. **124** (1989), 307–318. [MR90h:17019]

[1058] M.Rosso, *The universal R-matrix for the quantum $\mathrm{SL}(N)$-group*, Knots, topology and quantum field theories (Florence 1989), 497–506, World Sci. Publishing, River Edge, NJ, 1989. [MR92k:17027]

[1059] M.Rosso, *Algèbres enveloppantes quantifiees, groupes quantiques compacts de matrices et calcul différentiel non commutatif*, Duke Math. J. **61** (1990), 11–40. [MR92i:17022]

[1060] M.Rosso, *Koszul resolutions and quantum groups*, Nuclear Phys. B Proc. Suppl. **18B** (1990), 269–276. [MR93b:17049]

[1061] M.Rosso, *Représentations des groupes quantiques*, Astérisque **201-203** (1991), 443–483. [MR93b:17050]

[1062] M.Rosso, *Certaines formes bilinéaires sur les groupes quantiques et une conjecture de Schechtman et Varchenko*, C.R.A.S. Paris **314** (1992), 5–8.

[1063] G.C.Rota, *Hopf algebra methods in combinatorics*, Problèmes combinatoires et théorie des graphes (Orsay 1976), 363–365, CNRS, Paris, 1978. [MR80i:05012]

[1064] M.J.Rothstein, *Deformations of complex supermanifolds*, Proc. Amer. Math. Soc. **95** (1985), 255–260. [MR87d:32041]

[1065] H.Ruegg, *A simple derivation of the quantum Clebsch-Gordan coefficients for* $\mathrm{SU}(2)_q$, J. Math. Phys. **31** (1990), 1085–1087. [MR91f:17012]

[1066] P.Rujän, *A symbolic algorithm for finding exactly soluble statistical mechanical models*, Acta Phys. Hung. **62** (1987), 287–306.

[1067] N.Saavedra Rivano, *Catégories tannakiennes*, Bull. Soc. Math. France **100** (1972), 417–430. [MR47#8556]

[1068] N.Saavedra Rivano, *Catégories Tannakiennes*, Springer, Berlin-New York, 1972. [MR49#2769]

[1069] B.E.Sagan, *The symmetric group*, Wadsworth & Brooks, 1991.

[1070] H.Saleur, *Virasoro and Temperley-Lieb algebras*, Knots, topology and quantum field theories (Florence 1989), 485–496, World Sci. Publishing, River Edge, NJ, 1989. [MR93f:81047]

[1071] A.Salomaa and M.Soittola, *Automata-theoretic aspects of formal power series*, Springer, 1978.

[1072] M.Schaps, *Deformations of algebras and Hochschild cohomology*, Perspectives in ring theory (Antwerp 1987), 41–58, Kluwer Acad. Publ., Dordrecht, 1988. [MR91h:16058]

[1073] M.Schaps, *Deformations of finite-dimensional algebras and their idempotents*, Trans. Amer. Math. Soc. **307** (1988), 843–856. [MR89f:16026]

[1074] V.V.Schechtman and A.N.Varchenko, *Hypergeometric solutions of Knizhnik-Zamolodchikov equations*, Lett. Math. Phys. **20** (1990), 279–283. [MR92g:33027]

[1075] M.Scheunert, *Generalized Lie algebras*, J. Math. Phys. **20** (1979), 712–720.

[1076] M.Scheunert, *The theory of Lie superalgebras*, Lect. Notes Math. **734** (1979). [MR80i:17005]

[1077] M.Schlessinger and J.Stasheff, *The Lie algebra structure of tangent cohomology and deformation theory*, J. Pure Appl. Algebra **38** (1985), 313–322. [MR87e:13019]

[1078] K.Schmudgen, *Eine Einführung in die Theorie der Quantenmatrixgruppen*, Sem. Sophus Lie **1** (1991), 3–21. [MR92g:46087]

[1079] M.P.Schützenberger, *On the definition of a family of automata*, Inform. and Contr. **4** (1961), 245–270.

[1080] A.Schwarz, *The partition function of degenerate quadratic functional and Ray-Singer invariants*, Lett. Math. Phys. **2** (1978), 247–

[1081] G.B.Segal, *The definition of conformal field theory*, Differential geometrical methods in theoretical physics (Como 1987), 165–171, Kluwer Acad. Publ., Dordrecht, 1988. [MR90d:58026]

[1082] G.B.Segal, *Geometric aspects of quantum field theory*, Proceedings of the International Congress of Mathematicians (Kyoto 1990), 1387–1396, Math. Soc. Japan, Tokyo, 1991. [MR93f:81002]

[1083] I.E.Segal, *Quantization of the de Rham complex*, Global Analysis (Berkeley 1968), Proc. Sympos. Pure Math. **XVI** (1970), 205–210. [MR42#1157]

[1084] W.Selke, *The ANNNI model — Theoretical analysis and experimental application*, Phys. Rep. **170** (1988), 213–264.

[1085] W.Selke and J.M.Yeomans, *A Monte-Carlo study of the asymmetric clock model or chiral Potts model in two dimensions*, Z. Phys. B **46** (1982), 311–318.

[1086] M.A.Semenov-Tian-Shansky, *What a classical r-matrix is*, Funktsional. Anal. Prilozhen. **17** (1983), no. 4, 17–33. [MR85i:58061]

[1087] M.A.Semenov-Tian-Shansky, *Classical r-matrices and quantization*, Zap. Nauchn. Sem. LOMI **133** (1984), 228–235. [MR85m:58083]

[1088] M.A.Semenov-Tian-Shansky, *Dressing transformations and Poisson group actions*, Publ. R.I.M.S. **21** (1985), 1237–1260. [MR88b:58057]

[1089] M.A.Semenov-Tian-Shansky, *Classical r-matrices, Lax equations, Poisson-Lie groups and dressing transformations*, Field theory, quantum gravity and strings, II (Meudon/Paris, 1985), Lect. Notes Phys. **280** (1987), 174–214. [MR89g:58098]

[1090] M.A.Semenov-Tian-Shansky, *Monodromy mapping and classical r-matrices*, Zap. Nauchn. Sem. LOMI **200** (1992), 156–166.

[1091] B.S.Shastry, *Decorated star-triangle relations and exact integrability of the one-dimensioanl Hubbard model*, J. Stat. Phys. **50** (1988), 57–79.

[1092] S.Sherman, *Combinatorial aspects of the Ising model for ferromagnetism. I. A conjecture of Feynman on paths and graphs*, J. Math. Phys. **1** (1960), 202–217. [MR22#10273]; Addendum: **4** (1963), 1213–1214. [MR27#6560]

[1093] A.J.L.Sheu, *Quantization of the Poisson* SU(2) *and its Poisson homogeneous space—the* 2*-sphere*, Comm. Math. Phys. **135** (1991), 217–232. [MR91m:58011]

[1094] S.Shnider, *Deformation cohomology for bialgebras and quasi-bialgebras*, Deformation theory and quantum groups with applications to mathematical physics (Amherst, MA, 1990), Contemp. Math. **134** (1992), 259–296.

[1095] S.Shnider and S.Sternberg, *The cobar resolution and a restricted deformation theory for Drinfeld algebras*, to appear in J. Algebra.

[1096] W.M.Singer, *Extension theory for connected Hopf algebras*, Bull. Amer. Math. Soc. **76** (1970), 1095–1099. [MR42#1879]

[1097] W.M.Singer, *A new chain complex for the homology of the Steenrod algebra*, Math. Proc. Cambridge Philos. Soc. **90** (1981), 279–292. [MR82k:55018]

[1098] E.K.Sklyanin, *Some algebraic structures connected with the Yang-Baxter equation*, Funct. Anal. Appl. **16** (1982), 263–270. [MR84c:82004]

[1099] E.K.Sklyanin, *Some algebraic structures connected with the Yang-Baxter equation. Representations of a quantum algebra*, Funct. Anal. Appl. **17** (1983), 273–284. [MR85k:82011]

[1100] E.K.Sklyanin, *On an algebra generated by quadratic relations*, Uspekhi Mat. Nauk **40** (1985), no. 2, 214.

[1101] E.K.Sklyanin, *On the Poisson structure of the periodic classical XYZ-chain*, Zap. Nauchn. Sem. LOMI **150** (1986), 154–180. [MR87k:82036]

[1102] E.K.Sklyanin, *The Poisson structure of the classical XXZ-chain*, Zap. Nauchn. Sem. LOMI **161** (1987), 88-97. [MR89a:82052]

[1103] E.K.Sklyanin and L.A.Takhtajan, *Quantum method of the inverse problem*, Teoret. Mat. Fiz. **40** (1979), 194–220.

[1104] D.-J.Smit, *A quantum group structure in integrable conformal field theories*, Comm. Math. Phys. **128** (1990), 1–37.

[1105] S.P.Smith, *Quantum groups: An introduction and survey for ring theorists*, Noncommutative rings, 131–178, M.S.R.I. Publ. **24**, Springer, 1993.

[1106] S.P.Smith and J.T.Stafford, *Regularity of the four-dimensional Sklyanin algebra*, Compositio Math. **83** (1992), 259–289.

[1107] Y.S.Soibelman, *Gelfand-Naimark-Segal states and Weyl groups for the quantum group* SU(n), Functional Anal. Appl. **24** (1990), 253–255. [MR92f:17019]

[1108] Y.S.Soibelman, *Irreducible representations of the algebra of functions on the quantum group* SU(n) *and Schubert cells*, Soviet Math. Dokl. **40** (1990), 34–38. [MR91d:17019]

[1109] Y.S.Soibelman, *Quantum Weyl group and some of its applications*, Rend. Circ. Mat. Palermo Suppl. no. 26 (1991), 233–235. [MR93d:17022]

[1110] Y.S.Soibelman, *Two problems in quantized algebras of functions*, Quantum groups (Leningrad 1990), 393, Lect. Notes Math. **1510** (1992).

[1111] Y.S.Soibelman, *Selected topics in quantum groups*, Infinite analysis (Kyoto 1991), 859–887, World Sci. Publishing, River Edge, NJ, 1992.

[1112] Y.S.Soibelman, *On a quantum flag manifold*, Functional Anal. Appl. **26** (1992), 225–227.

[1113] Y.Soibelman, *Remark on the representations of braid group arising from quantum Weyl group*, Preprint DAMTP/91-06, 1991.

[1114] L.L.Vaksman and Y.S.Soibelman, *An algebra of functions on the quantum group* SU(2), Functional Anal. Appl. **22** (1988), 170–181. [MR90f:17019]

[1115] L.L.Vaksman and Y.S.Soibelman, *Algebra of functions on the quantum group* $\mathrm{SU}(n+1)$, *and odd-dimensional quantum spheres*, Leningrad Math. J. **2** (1991), 1023–1042. [MR92e:58021]

[1116] Y.S.Soibelman and L.L.Vaksman, *On some problems in the theory of quantum groups*, Representation theory and dynamical systems, 3–55, Amer. Math. Soc., Providence, RI, 1992.

[1117] T.A.Springer and A.V.Zelevinsky, *Characters of* $\mathrm{GL}(n, \mathbf{F}_q)$ *and Hopf algebras*, J. London Math. Soc. **30** (1984), 27–43. [MR86a:16011]

[1118] A.T.Starr, *Electric circuits and wave filters*, Pitman, London, 1934.

[1119] J.D.Stasheff, *On the space-of-loops isomorphism*, Proc. Amer. Math. Soc. **10** (1959), 987–993. [MR22#980]

[1120] J.D.Stasheff, *On homotopy Abelian* H*-spaces*, Proc. Cambridge Philos. Soc. **57** (1961), 734–745. [MR25#4534]

[1121] J.D.Stasheff, *On extensions of* H*-spaces*, Trans. Amer. Math. Soc. **105** (1962), 126–135. [MR31#2726]

[1122] J.D.Stasheff, *A classification theorem for fibre spaces*, Topology **2** (1963), 239–246. [MR27#4235]

[1123] J.D.Stasheff, *Homotopy associativity of H-spaces. I, II*, Trans. Amer. Math. Soc. **108** (1963), 275-312. [MR28#1623]

[1124] J.D.Stasheff, *"Parallel" transport in fibre spaces*, Bol. Soc. Mat. Mexicana **11** (1966), 68–84. [MR38#5219]

[1125] J.D.Stasheff, *Associated fibre spaces*, Michigan Math. J. **15** (1968), 457–470. [MR40#2083]

[1126] J.D.Stasheff, *H-spaces from a homotopy point of view*, Lect. Notes Math. **161** (1970). [MR42#5261]

[1127] J.D.Stasheff, *Families of finite H-complexes—revisited*, H-spaces (Actés Reunion Neuchatel, 1970), Lect. Notes Math. **196** (1971), 1–4. [MR45#6000]

[1128] J.D.Stasheff, *H-spaces and classifying spaces: foundations and recent developments*, Algebraic topology (Madison 1970), Proc. Symposia Pure Math. **XXII** (1971), 247–272. [MR47#9612]

[1129] J.Stasheff, *Homological (ghost) approach to constrained Hamiltonian systems*, Mathematical aspects of classical field theory (Seattle, WA, 1991), Contemp. Math. **132** (1992), 595–609.

[1130] J.Stasheff, *Drinfeld's quasi-Hopf algebras and beyond*, Deformation theory and quantum groups with applications to mathematical physics (Amherst, MA, 1990), Contemp. Math. **134** (1992), 297–307.

[1131] J.Stasheff, *Differential graded Lie algebras, quasi-Hopf algebras and higher homotopy algebras*, Quantum groups (Leningrad 1990), Lect. Notes Math. **1510** (1992), 120–137.

[1132] S.Sternberg, *Group theory in physics*, Cambridge Univ. Press, 1994.

[1133] D.Sternheimer, *Deformation theory applied to quantization and group representations*, Mathematical problems in theoretical physics (Berlin 1981), Lect. Notes Phys. **153** (1982), 314–318. [MR84k:58100]

[1134] D.Sternheimer, *Approche invariante de l'analyse non linéaire et de la quantification*, Systèmes dynamiques non linéaires: intégrabilite et comportement qualitatif, 260–293, Presses Univ. Montreal, 1986. [MR88e:22021]

[1135] A.Stolin, *On rational solutions of Yang-Baxter equation for sl(n)*, Math. Scand. **69** (1991), 57–80.

[1136] A.Stolin, *On rational solutions of Yang-Baxter eqautions. Maximal orders in loop algebra*, Comm. Math. Phys. **141** (1991), 533–548.

[1137] A.Sudbery, *Noncommuting coordinates and differential operators*, Quantum groups (Argonne, IL, 1990), 33–52, World Sci. Publishing, Teaneck, NJ, 1991. [MR92f:17020]

[1138] L.Susskind, *Dual-symmetric theory of hadrons. I*, Nuovo Cimento **69A** (1970), 457–496. [MR42#8813]

[1139] B.Sutherland, *Further results for the many-body problem in one dimension*, Phys. Rev. Lett. **20** (1968), 98-100.

[1140] B.Sutherland, *Two-dimensional hydrogen bonded crystals without the ice rule*, J. Math. Phys. **11** (1970), 3183–3186.

[1141] M.E.Sweedler, *Hopf algebras*, Benjamin, New York, 1969. [MR40#5705]

[1142] W.W.Symes, *Systems of Toda type, inverse spectral problems, and representation theory*, Invent. Math. **59** (1980), 13–51. [MR81g:58019]

[1143] I.Syozi, *Transformation of Ising model*, Phase transitions and critical phenomena, Vol. 1, 269–329, Academic Press, London, 1972.

[1144] E.J.Taft, *Hurwitz invertibility of linearly recursive sequences*, Congr. Numer. **73** (1990), 37–40. [MR91h:16068]

[1145] E.J.Taft, *Quantum deformation of the flag variety*, Quantum groups (Leningrad 1990), Lect. Notes Math. **1510** (1992), 138–141.

[1146] E.J.Taft, *Unsolved problems*, Quantum groups (Leningrad 1990), Lect. Notes Math. **1510** (1992), 394.

[1147] E.J.Taft, *Witt and Virasoro algebras as Lie bialgebras*, J. Pure Appl. Algebra **87** (1993), 301–312.

[1148] E.Taft and J.Towber, *Quantum deformation of flag schemes and Grassmann schemes. I. A q-deformation of the shape-algebra for* GL(n), J. Algebra **142** (1991), 1–36. [MR93e:17028]

[1149] E.J.Taft and R.L.Wilson, *There exist finite-dimensional Hopf algebras with antipodes of arbitrary even order*, J. Algebra **62** (1980), 283–291. [MR81d:16008]

[1150] M.Takeuchi, *The q-bracket product and quantum enveloping algebras of classical types*, J. Math. Soc. Japan **42** (1990), 605–629. [MR91i:17024]

[1151] L.A.Takhtajan, *Quantum inverse scattering method and the algebrized matrix Bethe ansatz*, Zap. Nauchn. Sem. LOMI **101** (1981), 158–183.

[1152] L.A.Takhtajan, *Introduction to algebraic Bethe ansatz*, Exactly solvable problems in condensed matter and relativistic field theory (Panchgani 1985), Lect. Notes Phys. **242** (1985), 175–219.

[1153] L.A.Takhtajan, *Quantum groups and integrable models*, Integrable systems in quantum field theory and statistical mechanics, 435–457, Academic Press, Boston, MA, 1989. [MR92d:58090]

[1154] L.A.Takhtajan, *Lectures on quantum groups*, Introduction to quantum group and integrable massive models of quantum field theory (Nankai 1989), 69–197, World Sci. Publishing, River Edge, NJ, 1990. [MR93c:17031]

[1155] S.Tang, *Commuting transfer matrices for the N state chiral Potts model and algebraic curves of genus higher than one*, PhD Thesis, SUNY at Stony Brook, December 1988.

[1156] H.N.V.Temperley, *The influence of boundary conditions on the Onsager Ising partition function for the plane square lattice*, Proc. Phys. Soc. Sect. B. **70** (1957), 192–197. [MR18,962c]

[1157] H.N.V.Temperley, *An exactly soluble lattice model of the fluid-solid transition*, Proc. Phys. Soc. **86** (1965) 185–192. [MR31#4542]

[1158] J.Y.Thibon, *Hopf algebras of symmetric functions and tensor products of symmetric group representations*, Internat. J. Algebra Comput. **1** (1991), 207–221. [MR92m:05201]

[1159] J.Tolar, *Quantization as deformation: identical particles*, Czechoslovak J. Phys. B **29** (1979), 319–322. [MR80e:81026]

[1160] P.Truini and V.S.Varadarajan, *Universal deformations of simple Lie algebras*, Lett. Math. Phys. **24** (1992), 63–71.

[1161] P.Truini and V.S.Varadarajan, *Universal deformations of reductive Lie algebras*, Lett. Math. Phys. **26** (1992), 53–65.

[1162] V.G.Turaev, *Cocycle for symplectic first Chern class and Maslov indices*, Funkts. Anal. Prilozh. **18** (1984), no. 1, 43–48.

[1163] V.G.Turaev, *The Conway and Kauffman modules of a solid torus*, Zap. Nauchn. Sem. LOMI **167** (1988), 79–89. [MR90f:57012]

[1164] V.Turaev, *The Yang-Baxter equation and invariants of links*, Invent. Math. **92** (1988), 527–553. [MR89e:57003]

[1165] V.G.Turaev, *Algebras of loops on surfaces, algebras of knots, and quantization*, Braid group, knot theory and statistical mechanics, 59–95, World Sci. Publishing, Teaneck, NJ, 1989. [MR90e:82077]

[1166] V.G.Turaev, *The category of oriented tangles and its representations*, Functional Anal. Appl. **23** (1989), 254–255. [MR91b:57009]

[1167] V.G.Turaev, *Operator invariants of tangles, and R-matrices*, Math. USSR Izv. **35** (1990), 411–444. [MR91e:17011]

[1168] V.G.Turaev, *Skein quantization of Poisson algebras of loops on surfaces*, Ann. Sci. École Norm. Sup. **24** (1991), 635–704.

[1169] V.Turaev, *Quantum invariants of 3-manifold and a glimpse of shadow topology*, Quantum groups (Leningrad 1990), Lect. Notes Math. **1510** (1992), 363–366.

[1170] G.M.Tuynman, *Reduction, quantization and nonunimodular groups*, J. Math. Phys. **31** (1990), 83–90.

[1171] K.-H.Ulbrich, *Smash products and comodules of linear maps*, Tsukuba J. Math. **14** (1990), 371–378. [MR91m:16030]

[1172] K.-H.Ulbrich, *On modules induced or coinduced from Hopf subalgebras*, Math. Scand. **67** (1990), 177–182. [MR92e:16030]

[1173] K.-H.Ulbrich, *On Hopf algebras and rigid monoidal categories*, Israel J. Math. **72** (1990), 252–256. [MR92e:16031]

[1174] K.-H.Ulbrich, *Determining bialgebras from k-valued monoidal categories*, Preprint.

[1175] C.Vafa, *Conformal theories and punctured surfaces*, Phys. Lett. B **199** (1987), 195–202.

[1176] I.Vaisman, *On the geometric quantization of Poisson manifolds*, J. Math. Phys. **32** (1991), 3339–3345.

[1177] L.L.Vaksman, *q-analogues of Clebsch-Gordan coefficients, and the algebra of functions on the quantum group* SU(2), Soviet Math. Dokl. **39** (1989), 467–470. [MR90k:33018]

[1178] H.J.de Vega, *Yang-Baxter algebras and integrable models in field theory and statistical mechanics*, Dynamical problems in soliton systems (Kyoto 1984), 81–88, Springer, Berlin-New York, 1985. [MR87i:81109]

[1179] Y.P.Velinov, *Polycategories and strict monoidal categories. I*, Godishnik Vissh. Uchebn. Zaved. Prilozhna Mat. **17** (1981), no. 4, 151–155. [MR84f:18014]

[1180] Y.P.Velinov, *Polycategories and strict monoidal categories. II*, Godishnik Vissh. Uchebn. Zaved. Prilozhna Mat. **20** (1984), no. 2, 37–46. [MR87b:18008]

[1181] G.Veneziano, *Construction of a crossing-symmetric, Regge-behaved amplitude for linearly rising trajectories*, Nuovo Cimento **57A** (1968), 190–197.

[1182] J.-L.Verdier, *Groupes quantiques*, Sém. Bourbaki 1986-1987, Astérisque **152-153** (1987), 305–319. [MR89f:17021]

[1183] E.Verlinde, *Fusion rules and modular transformations in 2D conformal field theory*, Nucl. Phys. **B300** (1988), 360–376. [MR89h:81238].

[1184] E.Verlinde, *A note on braid statistics and the non-abelian Aharonov-Bohm effect*, Proc. Intern. Colloq. Modern Quantum Field Theory (Bombay 1990), 450–461, World Sci. Publ., Singapore, 1991.

[1185] A.M.Vershik, *Algebras with quadratic relations*, Spectral theory of operators and infinite-dimensioanl analysis, 32–57, Inst. Mat. Akad. Nauk Ukr. SSR., Kiev, 1984.

[1186] A.M.Vershik, *The hook formula and related identities*, Zap. Nauchn. Sem. LOMI **172** (1989), 3–20. [MR90j:05022]

[1187] J.Vey, *Déformation du crochet de Poisson sur une variété symplectique*, Comment. Math. Helv. **50** (1975), 421–454. [MR54#8765]

[1188] A.M.Vinogradov, *The union of the Schouten and Nijenhuis brackets, cohomology, and superdifferential operators*, Mat. Zametki **47** (1990), no. 6, 138–140. [MR91h:17019]

[1189] R.M.Vogt, *A remark on A_n-spaces and loop spaces*, J. London Math. Soc. **14** (1976), 321–325. [MR54#11325]

[1190] K.Volpert, *A spectral sequence connecting the Spencer cohomology of a transitive Lie algebra with its deformation cohomology*, Algebras Groups Geom. **7** (1990), 313–344. [MR93b:17068]

[1191] R.Voreadou, *Coherence and non-commutative diagrams in closed categories*, Mem. Amer. Math. Soc. **9** (1977), no. 182. [MR55#8138]

[1192] M.Wambst, *Complexes de Koszul quantiques et homologie cyclique*, C.R.A.S. Paris **314** (1992), 977–982. [MR93d:16011]

[1193] G.H.Wannier, *The statistical problem in cooperative phenomena*, Rev. Modern Phys. **17** (1945), 50–60.

[1194] G.H.Wannier, *Antiferromagnetism. The triangular Ising net*, Physical Rev. **79** (1950), 357–364. [MR12,576e]

[1195] W.C.Waterhouse, *Antipodes and group-likes in Hopf algebras*, J. Algebra **37** (1975), 290–295.

[1196] W.C.Waterhouse, *Introduction to affine group schemes*, Springer, New York-Berlin, 1979. [MR82e:14003]

[1197] E.T.Whittaker and G.N.Watson, *A course of modern analysis. An introduction to the general theory of infinite processes and of analytic functions: with an account of the principal transcendental functions*, Cambridge University Press, New York, 1962. [MR31#2375]

[1198] A.Weil, *Sur les théorèmes de de Rham*, Comment. Math. Helv. **26** (1952), 119–145. [MR14,307b]

[1199] A.Weinstein, *Symplectic manifolds and their Lagrangian submanifolds*, Adv. Math. **6** (1971), 329–346.

[1200] A.Weinstein, *Lagrangian submanifolds and Hamiltonian systems*, Ann. Math. **98** (1973), 377–410.

[1201] A.Weinstein, *On Maslov's quantization conditions*, Lect. Notes Math. **459** (1975), 341–372.

[1202] A.Weinstein, *Symplectic V-manifolds, periodic orbits of Hamiltonian systems, and the volume of certain Riemannian manifolds*, Comm. Pure Appl. Math. **30** (1977), 265–271.

[1203] A.Weinstein, *Lectures on symplectic manifolds*, CBMS Regional Conf. Ser. Math. **29**, AMS, Providence, RI, 1979. [MR82b:58039]

[1204] A.Weinstein, *The local structure of Poisson manifolds*, J. Differential Geom. **18** (1983), 523–557. [MR86i:58059]; Errata and addenda: **22** (1985), 255. [MR87k:58099]

[1205] A.Weinstein, *Poisson structures and Lie algebras*, Astérisque (1985), Numéro Hors Série, 421–434. [MR87k:58100]

[1206] A.Weinstein, *Symplectic groupoids and Poisson manifolds*, Bull. Amer. Math. Soc. **16** (1987), 101–104. [MR88c:58019]

[1207] A.Weinstein, *Some remarks on dressing transformations*, J. Fac. Sci. Univ. Tokyo Sect. IA **35** (1988), 163–167. [MR89j:58036]

[1208] A.Weinstein, *Coisotropic calculus and Poisson groupoids*, J. Math. Soc. Japan **40** (1988), 705–727. [MR90b:58091]

[1209] A.Weinstein, *Affine Poisson structures*, Internat. J. Math. **1** (1990), 343–360. [MR92d:58069]

[1210] A.Weinstein, *Symplectic groupoids, geometric quantization, and irrational rotation algebras*, Symplectic geometry, groupoids, and integrable systems (Berkeley 1989), M.S.R.I. Publ. **20**, 281–290, Springer, New York, 1991. [MR92h:46103]

[1211] A.Weinstein, *Noncommutative geometry and geometric quantization*, Symplectic geometry and mathematical physics (Aix-en Provence 1990), 446–461, Birkhäuser, Boston, MA, 1991. [MR93f:58092]

[1212] A.Weinstein and P.Xu, *Extensions of symplectic groupoids and quantization*, J. Reine Angew. Math. **417** (1991), 159–189. [MR92k:58094]

[1213] A.Weinstein and P.Xu, *Classical solutions of the quantum Yang-Baxter equation*, Comm. Math. Phys. **148** (1992), 309–343.

[1214] H.Wenzl, *Derived link invariants and subfactors*, Operator algebras and applications, Vol. 2, 237–240, Cambridge Univ. Press, 1988. [MR91g:46072]

[1215] H.Wenzl, *Unitarizations of solutions of the quantum Yang-Baxter equation and subfactors*, IXth International Congress on Mathematical Physics (Swansea 1988), 434–437, Hilger, Bristol, 1989. [MR91k:46066]

[1216] H.Wenzl, *Quantum groups and subfactors of type B, C and D*, Comm. Math. Phys. **133** (1990), 383–432. [MR92k:17032]

[1217] H.Wenzl, *Representations of braid groups and the quantum Yang-Baxter equation*, Pacific J. Math. **145** (1990), 153–180. [MR92d:57009]

[1218] H.Weyl, *Gruppentheorie und Quantenmechanik*, Hirzel, Leipzig, 1928.

[1219] J.H.C.Whitehead, *Combinatorial homotopy. II*, Bull. Amer. Math. Soc. **55** (1949), 453–496.

[1220] F.D.Williams, *Quasi-commutativity of H-spaces*, Michigan Math. J. **19** (1972), 209–213. [MR47#5868]

[1221] E.Witten, *Gauge theories and integrable lattice models*, Nucl. Phys. **B322** (1989), 629–697.

[1222] E.Witten, *Quantum field theory and the Jones polynomial*, Comm. Math. Phys. **121** (1989), 351–399.

[1223] E.Witten, *Gauge theories, vertex models, and quantum groups*, Nuclear Phys. B **330** (1990), 285–346. [MR92b:82051]

[1224] S.L.Woronowicz, *Pseudospaces, pseudogroups and Pontriagin duality*, Mathematical problems in theoretical physics (Lausanne 1979), Lect. Notes Phys. **116** (1980), 407–412. [MR82e:46079]

[1225] S.L.Woronowicz, *Duality in the C^*-algebra theory*, Proceedings of the International Congress of Mathematicians, Vol. 2 (Warsaw 1983), 1347–1356, PWN, Warsaw, 1984. [MR88d:46112]

[1226] S.L.Woronowicz, *Group structure on noncommutative spaces*, Fields and geometry (Karpacz 1986), 478–496, World Sci. Publishing, Teaneck, NJ, 1986. [MR90i:46122]

[1227] S.L.Woronowicz, *Compact matrix pseudogroups*, Comm. Math. Phys. **111** (1987), 613–665. [MR88m:46079]

[1228] S.L.Woronowicz, *Twisted* SU(2) *group. An example of a non-commutative differential calculus*, Publ. R.I.M.S. **23** (1987), 117–181. [MR88h:46130]

[1229] S.L.Woronowicz, *Tannaka-Krein duality for compact matrix pseudogroups. Twisted* SU(N) *groups*, Invent. Math. **93** (1988), 35–76. [MR90e:22033]

[1230] S.L.Woronowicz, *Differential calculus on compact matrix pseudogroups (quantum groups)*, Comm. Math. Phys. **122** (1989), 125–170. [MR90g:58010]

[1231] S.L.Woronowicz, *Unbounded elements affiliated with C^*-algebras and noncompact quantum groups*, Comm. Math. Phys. **136** (1991), 399–432. [MR92b:46117]

[1232] S.L.Woronowicz, *A remark on compact matrix quantum groups*, Lett. Math. Phys. **21** (1991), 35–39. [MR92k:46097]

[1233] S.L.Woronowicz, *Solutions of the braid equation related to a Hopf algebra*, Lett. Math. Phys. **23** (1991), 143–145. [MR93b:16075]

[1234] S.L.Woronowicz, *Quantum $E(2)$ group and its Pontryagin dual*, Lett. Math. Phys. **23** (1991), 251–263. [MR93b:17058]

[1235] S.L.Woronowicz, *New quantum deformation of* SL(2, **C**). *Hopf algebra level*, Rep. Math. Phys. **30** (1991), 259–269.

[1236] S.L.Woronowicz and S.Zakrzewski, *Quantum Lorentz group having Gauss decomposition property*, Publ. R.I.M.S. **28** (1992), 809–824.

[1237] F.Y.Wu, *Ising model with four-spin interaction*, Phys. Rev. **B4** (1971), 2312–2314

[1238] F.Y.Wu, *The Potts model*, Rev. Mod. Phys. **54** (1982), 235–268. [MR84d:82033]

[1239] F.Y.Wu, *Potts model and graph theory*, J. Statist. Phys. **52** (1988), 99–112. [MR90b:82075]

[1240] P.Xu, *Morita equivalence of Poisson manifolds*, Comm. Math. Phys. **142** (1991), 493–509. [MR93a:58069]

[1241] P.Xu, *Symplectic groupoids of reduced Poisson spaces*, C.R.A.S. Paris **314** (1992), 457–461.

[1242] P.Xu, *Poisson cohomology of regular Poisson manifolds*, Ann. Inst. Fourier **42** (1992), 967–988.

[1243] H.Yamane, *A Poincaré-Birkhoff-Witt theorem for quantized universal enveloping algebras of type A_N*, Publ. R.I.M.S. **25** (1989), 503–520. [MR91a:17016]

[1244] C.N.Yang, *The spontaneous magnetization of a two-dimensional Ising model*, Phys. Rev. **85** (1952), 808–816.

[1245] C.N.Yang, *Some exact results for the many-body problem in one dimension with repulsive delta-function interaction*, Phys. Rev. Lett. **19** (1967), 1312–1315. [MR41#6480]

[1246] C.N.Yang, *S-matrix for the one-dimensional N-body problem with repulsive or attractive δ-function interaction*, Phys. Rev. **168** (1968), 1920–1923.

[1247] C.N.Yang and C.P.Yang, *One-dimensional chain of anisotropic spin-spin interactions. I. Proof of Bethe's hypothesis for the ground state in a finite system*, Phys. Rev. **150** (1966), 321–327.

[1248] J.M.Yeomans, *Low-temperature analysis of the p-state asymmetric clock model for general p*, J. Phys. C **15** (1982), 7305–7323.

[1249] J.M.Yeomans, *A low-temperature mean-field theory of the three-dimensional, p-state chiral clock model*, J. Phys. C **17** (1984), 3601–3614.

[1250] J.M.Yeomans, *ANNNI and clock models*, Physica **127B** (1984), 187–192.

[1251] D.N.Yetter, *Category theoretic representations of knotted graphs in* $\mathbf{S}^3$, Adv. Math. **77** (1989), 137–155. [MR91c:57018]

[1252] D.N.Yetter, *Quantum groups and representations of monoidal categories*, Math. Proc. Cambridge Philos. Soc. **108** (1990), 261–290. [MR91k:16028]

[1253] D.N.Yetter, *Framed tangles and a theorem of Deligne on braided deformations of Tannakian categories*, Deformation theory and quantum groups with applications to mathematical physics (Amherst, MA, 1990), Contemp. Math. **134** (1992), 325–349.

[1254] D.N.Yetter, *Topological quantum field theories associated to finite groups and crossed G-sets*, J. Knot Theory Ramifications **1** (1992), 1–20.

[1255] A.Zabrodsky, *Hopf spaces*, North-Holland, Amsterdam-New York-Oxford, 1976. [MR55#13416]

[1256] S.Zakrzewski, *Quantum and classical pseudogroups. II. Differential and symplectic pseudogroups*, Comm. Math. Phys. **134** (1990), 371–395. [MR91m:58013]

[1257] S.Zakrzewski, *Poisson-Lie groups and pentagonal transformations*, Lett. Math. Phys. **24** (1992), 13–19. [MR93f:58257]

[1258] A.B.Zamolodchikov, *"Fishing-net" diagrams as a completely integrable system*, Phys. Lett. **97B** (1980), 63–66. [MR92b:81031]

[1259] A.B.Zamolodchikov, *Conformal symmetry in two dimensions: an explicit recurrence formula for the conformal partial wave amplitude*, Comm. Math. Phys. **96** (1984), 419–422. [MR86c:81040]

[1260] A.B.Zamolodchikov and Al.B.Zamolodchikov, *Two-dimensional factorizable S-matrices as exact solutions of certain relativistic quantum field theory models*, Ann. Phys. **120** (1979), 253–291.

[1261] A.V.Zelevinsky, *Induced representations of reductive p-adic groups. II. On irreducible representations of* GL(n), Ann. Sci. École Norm. Sup. **13** (1980), 165–210. [MR83g:22012]

[1262] A.V.Zelevinsky, *A generalization of the Littlewood-Richardson rule and the Robinson-Schensted-Knuth correspondence*, J. Algebra **69** (1981), 82–94. [MR82j:20028]

[1263] A.V.Zelevinsky, *Representations of finite classical groups. A Hopf algebra approach*, Lect. Notes Math. **869** (1981). [MR83k:20017]

[1264] M.Q.Zhang, *How to find the Lax pair from the Yang-Baxter equation*, Comm. Math. Phys. **141** (1991), 523–531. [MR92m:82046]

Index